Cardiac Energetics
Basic Mechanisms and Clinical Implications

Supplement to
Basic Research in Cardiology, Vol. 82 Suppl. 2 (1987)
Editors:
R. Jacob (Tübingen) Th. Kenner (Graz) and W. Schaper
(Bad Nauheim)

R. Jacob, Hj. Just, Ch. Holubarsch (eds.)

Cardiac Energetics

Basic Mechanisms and Clinical Implications

Springer-Verlag Berlin Heidelberg GmbH

ISBN 978-3-662-11291-5   ISBN 978-3-662-11289-2 (eBook)
DOI 10.1007/978-3-662-11289-2

Basic Res. Cardiol. ISSN 0300-8428
Indexed in Current Contents.

Copyright © 1987 by Springer-Verlag Berlin Heidelberg
Originally published by Dr. Dietrich Steinkopff Verlag GmbH & Co. KG, Darmstadt in 1987

Medical editor: Juliane K. Weller – Copy editing: Deborah Marston – Production: Heinz J. Schäfer

Preface

Assessment of cardiac energetics at the level of ATP-synthesis, chemomechanical energy transformation and whole organ dynamics as a function of haemodynamic load, ventricular configuration and oxygen- and substrates supply is basic to understanding cardiac function under physiological and pathophysiological (hypertrophy, hypoxia, ischaemia and heart failure) conditions. Moreover, cardiac energetics should be an important consideration in the choice and application of drugs especially in the case of vasodilators, inotropic agents and in cardioprotective measures. Only by considering energetics at the subcellular, cellular, and whole-heart level we can arrive at a better understanding of cardiac performance and ultimately better clinical judgement and drug therapy. Quantification of myocardial energetics will also help to determine the optimal time for surgical interventions such as valvular replacement or aneurysm resection.

The present volume is the outcome of an international symposium on cardiac energetics held in Gargellen/Montafon (Austria), June 1986. The contributions will certainly help bridge the existing gap between basic research involving isolated structures and that involving the whole organ, on the one hand, and render the results derived from basic research applicable to clinical problems, on the other hand.

Problems of experimental basic research as well as those of pathophysiology and clinical application are discussed in ca. 40 different contributions. In an introductory chapter, the relationship between energetics of cardiac muscle and crossbridge kinetics, metabolism, and the mechanical conditions of muscle contraction are analyzed on the basis of modern physiological and biochemical methods. The second chapter is concerned with the energetic consequences of chronic haemodynamic overload and inotropic interventions. The particularly topical problems of myocardial hypoxia and -ischaemia as well as energetics of the human heart, especially under clinical aspects, are dealt with in the separate consecutive sections.

We are confident that our choice of competent participants has rendered a valuable compendium of the state of the art in the field of cardiac energetics to meet the needs of all those interested: biologists, biochemists, physiologists, pharmacologists, clinical cardiologists, heart surgeons, internists, paediatrists.

It is our wish and obligation to express our thanks to BAYER AG, especially Dr. Günther Albus, for the generous support which made the symposium a successful event.

Finally, we would like to thank Dr. Karl Thomae GmbH for their support in the publication.

R. Jacob
Hj. Just
Ch. Holubarsch

Contents

III. Cardiac energetics in hypoxia and ischaemia

IV. Cardiac energetics in human heart: Clinical implications

I. Cardiac energetics as related to basic mechanisms and mechanical conditions

1. Cardiac energetics as related to basic mechanisms and mechanical conditions

The mechanism of muscle contraction.
Biochemical, mechanical, and structural approaches to elucidate cross-bridge action in muscle

B. Brenner[1] and E. Eisenberg[2]

[1] Institute of Physiology II, University of Tübingen, Tübingen, F.R.G.
[2] National Heart, Lung and Blood Institute, NIH, Bethesda, U.S.A.

Summary

Muscle contraction occurs when the thin actin and thick myosin filaments slide past each other. It is generally assumed that this process is driven by cross-bridges which extend from the myosin filaments and cyclically interact with the actin filaments as ATP is hydrolysed. Current biochemical studies suggest that the myosin cross-bridge exists in two main conformations. In one conformation, which occurs in the absence of MgATP, the cross-bridge binds very tightly to actin and detaches very slowly. When all the cross-bridges are bound in this way, the muscle is in rigor and extremely resistant to stretch. The second conformation is induced by the binding of MgATP. In this conformation the cross-bridge binds weakly to actin and attaches and detaches so rapidly that it can slip from actin site to actin site, offering very little resistance to stretch. During ATP hydrolysis by isolated actin and myosin in solution, the cross-bridge cycles back and forth between the weak-binding and strong-binding conformations. Assuming a close correlation between the behaviour of isolated proteins in solution and the cross-bridge action in muscle, Eisenberg and Greene [12] have developed a model for cross-bridge action where, in the fixed filament lattice in muscle, the transition from the weak-binding to the strong-binding conformation causes the elastic cross-bridge to become deformed and exert a positive force, while the transition back to the weak-binding conformation upon binding of MgATP, causes deformation which, during fibre shortening, leads to rapid detachment of the cross-bridge and its re-attachment to a new actin site. From the results of in vitro experiments, it was furthermore suggested that relaxation occurs when the transition from the weak-binding to the strong-binding conformation is blocked. Results of recent mechanical and X-ray diffraction experiments on skinned fibre preparations are consistent with the assumed close correlation between the behaviour of isolated proteins in solution and the behaviour of cross-bridges in muscle. Furthermore, X-ray diffraction experiments allowed to provide experimental evidence for the postulated structural difference between attached weak-binding and attached strong-binding cross-bridges. Finally, reccent studies have confirmed the prediction of Eisenberg and Greene [12] that the rate limiting step in vitro determines the rate of force generation in muscle.

Structural basis of muscle contraction

It is now generally accepted that muscle contraction occurs when two sets of interdigitating filaments, the thin actin filaments and the thick myosin filaments slide past each other [20, 24]. According to the cross-bridge theory, this sliding process is driven by domains of the myosin molecules, the cross-bridges, which extend from the myosin filaments and cyclically interact with the actin filaments as ATP is hydrolysed [18, 19, 23]. Based on the X-ray diffraction and electron microscope studies of Reedy et al. [32], it was proposed that cross-bridges go through an oar-like cycle of attachment, a change in configuration that leads to filament sliding or force generation, followed by detachment, and then return to the starting point [19, 23].

The challenge at present is to correlate the results of biochemical, mechanical, and structural approaches to form a complete understanding of the cyclic cross-bridge action in muscle, including detailed description of the mechanical, structural and biochemical properties of the states within the cross-bridge cycle.

Kinetic schemes of the actomyosin-ATPase in solution

For most of the biochemical studies, isolated myosin cross-bridges (myosin S-1) have been used. Myosin S-1 is a proteolytic fragment of the myosin molecule which, unlike myosin, is soluble even at low ionic strengths that are required for studying binding of S-1 to actin. Early steady-state experiments showed that each S-1 molecule has one ATP binding site and one actin binding site. These experiments furthermore showed that binding of S-1 to actin increases the myosin ATPase rate more than 200-fold, while at the same time, the binding of ATP to the active site of S-1 greatly weakens the binding of S-1 to actin (Fig. 1 a). The binding constants derived from these plots are $2.2 \times 10^4 \text{ M}^{-1}$, under conditions of 1 mM ATP, 3 mM $MgCl_2$, 1 mM EGTA, 10 mM imidazole, pH 7, 25 °C. Under the same conditions, but in the absence of ATP, the binding constant is greater than $1 \times 10^9 \text{M}^{-1}$. These results suggest that binding of ATP to the nucleotide-free S-1 actin complex leads to cross-bridge detachment [14, 40].

Since binding of S-1 to actin causes an increase in light scattering of the experimental solution, Stein et al. [39] used light scattering in stopped flow experiments to follow the kinetics of S-1 binding to actin in the presence of ATP (Fig. 1 b). Under conditions (1.8 mM MgCl; 1.0 mM ATP; 10 mM imidazole; 33 μm actin; 20 μm S-1; pH 7; 25 °C) during which about 50% of the S-1 is bound to actin during steady state hydrolysis of ATP, the steady state turbidity level is reached as fast as mixing occurs. This indicates that an equilibrium is immediately established between $M \cdot ATP$ and $A \cdot M \cdot ATP$.

The hydrolysis step which follows the binding of ATP to S-1 is accompanied by an increase in fluorescence. This allows us to monitor the time course of the hydrolysis step (Fig. 1 c). Comparison of the fluorescence and light scattering measurements (under the same conditions as in Fig. 1 b) shows that during the transformation of $M \cdot ATP$ and $A \cdot M \cdot ATP$, to $M \cdot ADP \cdot P_i$ and $A \cdot M \cdot ADP \cdot P_i$, respectively, only a very small change in turbidity occurs. This means that in the presence of ATP, or in the presence of the hydrolysis products $(ADP + P_i)$, the binding strength of myosin S-1 to actin is about the same. Measurements of the acto $\cdot$ S-1 ATPase activity at various fractions of S-1 bound to actin indicated that the ATP hydrolysis step occurs when myosin S-1 is both bound to and dissociated from actin [31, 36, 39].

Two kinetic schemes have been proposed which can account for the results of the steady-state and presteady-state biochemical experiments. These models are named according to the number of states with ATP, or the hydrolysis products $(ATP + P_i)$, bound to the myosin S-1. The six-state model [37, 39] shown in Fig. 2 b contains six such states. Rosenfeld and Taylor [33] proposed a four-state model (Fig. 2 a) which is only different from the six-state model in that the rate-limiting step is the hydrolysis step itself. Thus the $M \cdot ADP \cdot P_i^{II}$ and $A \cdot M \cdot ADP \cdot P_i^{II}$ states are omitted.

The basic properties of these two kinetic schemes are identical. In both schemes $M \cdot ATP$ and $M \cdot ADP \cdot P_i$ have a nearly identical affinity to actin, the ATP hydrolysis step occurs both while S-1 is bound or dissociated from actin, and the rate-limiting step pre-

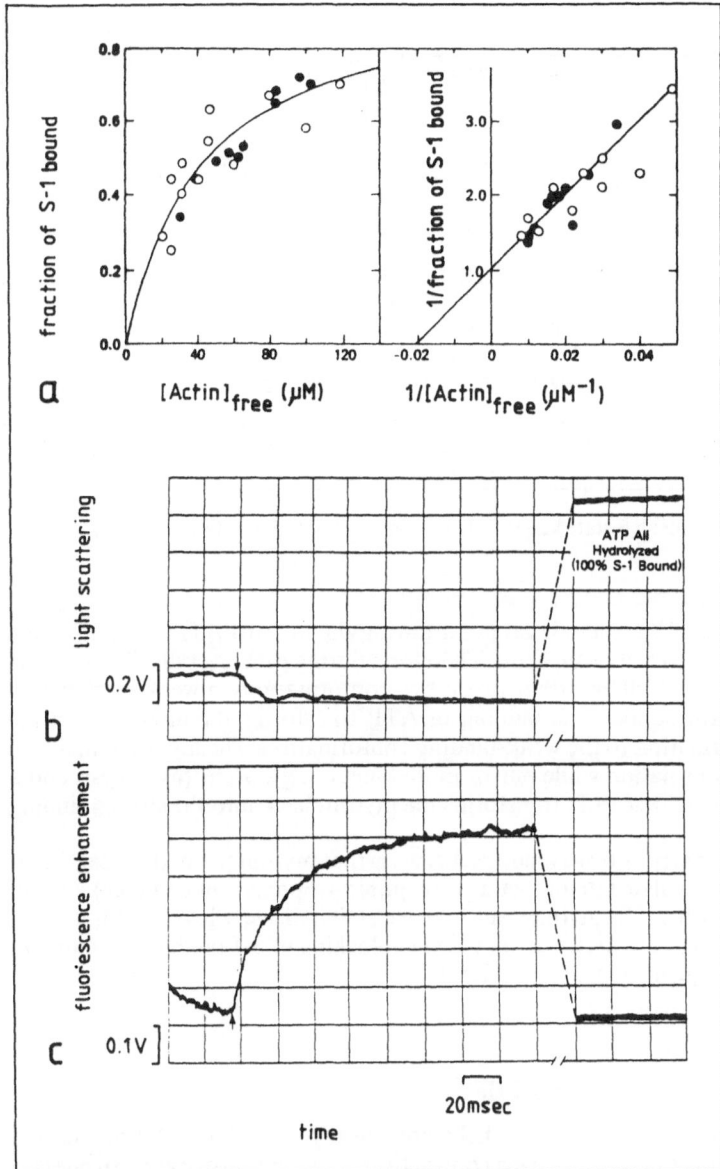

Fig. 1a. Binding of myosin S-1 to actin in the presence of ATP. The direct plot (left) and double reciprocal plot (right) show binding measured using stopped-flow turbidity (o–o) and sedimentation in the airfuge (●–●).

1b. Time course of light scattering immediately after the addition of ATP to acto-myosin S-1. The arrow indicates where the flow stopped.

1c. Time course of fluorescence changes immediately after the addition of ATP to acto-myosin S-1. The arrow indicates where the flow stopped.

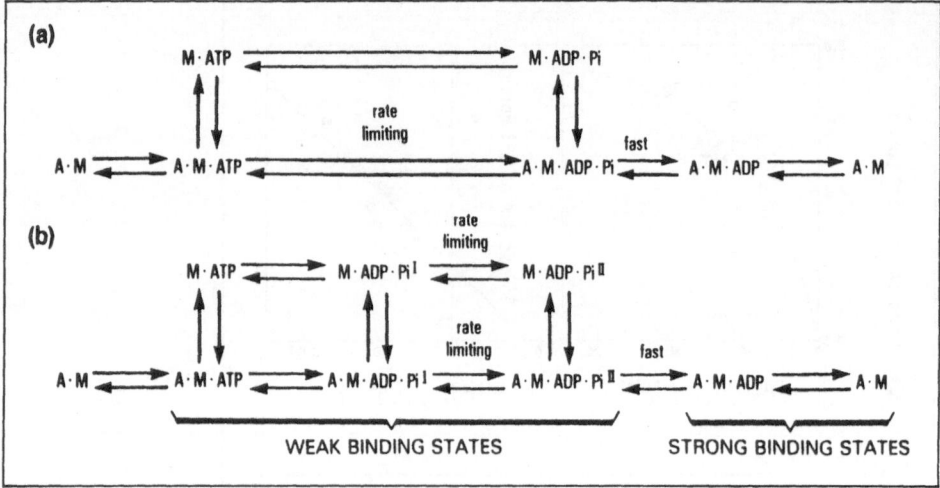

Fig. 2 a. 4-state kinetic scheme; **b** 6-state kinetic schema, M = myosin, A = actin.

cedes the rapid release of P_i. There is no point during the in vitro ATPase cycle where myosin is required to dissociate from actin. Therefore, during each overall ATPase cycle, the myosin molecule simply alternates between two conformations, a weak-binding and a strong-binding conformation. The binding of ATP transforms the myosin from the strong-binding conformation to the weak-binding conformation. The myosin remains in the weak-binding conformation while one or more kinetic steps occur (hydrolysis and a rate-limiting step). The release of P_i transforms the myosin back into the strong-binding conformation.

A further basic property of these kinetic schemes is that myosin S-1 in the weak-binding states can bind to regulated actin (actin + troponin-tropomyosin complex) even in the absence of Ca^{++}, where the actomyosin ATPase activity is very low [10, 11, 41, 42]. This implies that in vitro regulation does not act via blocking of S-1 binding to actin, but rather through a subsequent step, like P_i release.

Implications for cross-bridge action in muscle

Assuming a close correlation between the behaviour of myosin S-1 in solution and the behaviour of cross-bridges in the organized contractile system, Eisenberg and Greene [12] proposed a model for cross-bridge action in muscle, applying the physiological concepts of cross-bridge behaviour developed by Huxley [18] and Huxley and Simmons [21] to the kinetic model of Stein et al. [39]. To do this, Eisenberg and Greene [12] used the theoretical formalism of Hill [17] which provides the framework for relating the biochemical to the physiological properties of the cross-bridge. The model of Eisenberg and Green [12] takes into account the major difference between the situation of S-1 in solution and the situation of cross-bridges in muscle. This is the basic fact that in muscle there is a fixed lattice of actin and myosin filaments that cannot respond to the action of a single cross-bridge, but only to the overall behaviour of an ensemble of cross-bridges.

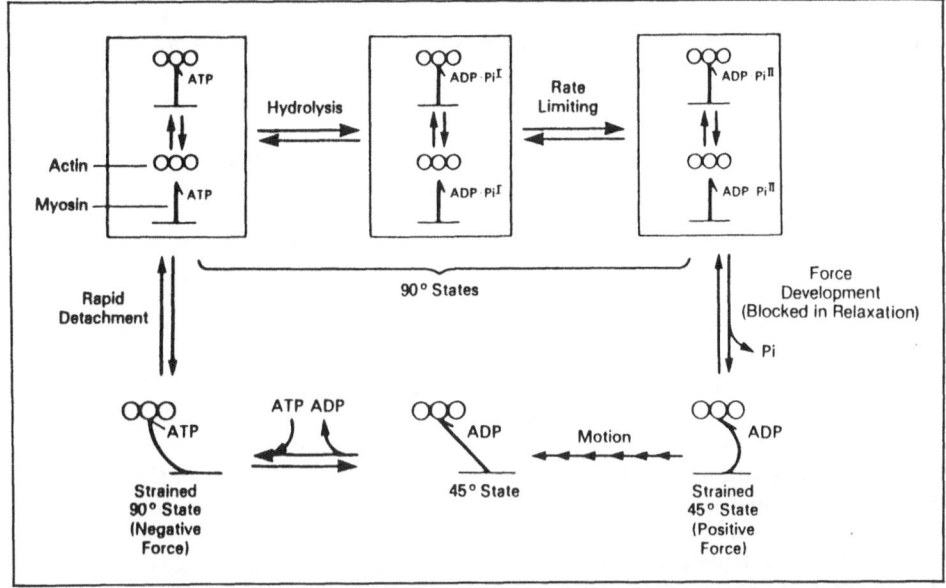

Fig. 3. Cross-bridge model of Eisenberg and Greene. The symbol ←← represents relative motion between the actin and myosin filaments.

On this basis, Eisenberg and Greene [12] proposed that cross-bridges in muscle also exist in two major conformations which are kinetically and structurally different (Fig. 3); a weak-binding conformation ("90 °" states) and a strong-binding conformation ("45 °" states), analogous to the orientation of cross-bridges found in insect flight muscle fibres while relaxed and in rigor [32]. Just as in solution [39], cross-bridges in the weak-binding conformation ($M \cdot ATP$, $M \cdot ADP \cdot P_i^I$ and $M \cdot ADP \cdot P_i^{II}$) are assumed to be in rapid equilibrium between binding to and dissociation from actin ($A \cdot M \cdot ATP$, $A \cdot M \cdot ADP \cdot P_i^I$; and $A \cdot M \cdot ADP \cdot P_i^{II}$ respectively). Since hydrolysis and the rate-limiting step apparently have no great effect on the conformation of S-1 in solution, Eisenberg and Greene [12] assumed that neither of these steps has a great effect on the structure or actin affinity of the cross-bridge in the organized contractile system.

The next step, the P_i release, is accompanied by a large drop in free energy in solution, which is assumed to indicate a large scale change in the structure of the acto-S-1 complex. Unlike the previous steps in the cycle, this step must be different in muscle and in solution. In solution there is no restraint on a change in the structure of S-1 bound to actin. In muscle, however, the actin filament cannot respond to a change in the structure of an attached cross-bridge by moving a short distance; the actin filament can respond only to the average behaviour of all of the attached cross-bridges. Thus, the result of the transition into the strong-binding conformation is a strained cross-bridge that exerts a positive force, as symbolized by its curved shape in Fig. 3. This strain is gradually relieved only as the filaments slide past each other. Hence, the term "strong-binding conformation" no longer refers to a single cross-bridge structure: rather, it refers to a continuum of different structures with a variable amount of strain in the attached cross-bridge.

The following step, the ADP release from the cross-bridge, was assumed to be slow until the work-stroke is completed. This is because, as pointed out by Huxley [18], for any cross-bridge model to work, the cross-bridge must detach from actin slowly until after motion occurs and the strain in the cross-bridge is utilized to do work. Thus, Eisenberg and Greene [12] assumed that ADP release becomes as rapid as it is in solution, only after the filaments move and the cross-bridges reach an unstrained strong-binding conformation.

Finally, after release of ADP, ATP rebinds to the nucleotide-free attached cross-bridge. Eisenberg and Greene [12] assumed that this returns the cross-bridge to the weak-binding conformation. However, again the attached cross-bridge is constrained by the filament lattice and therefore this weak-binding conformation is negatively strained and exerts negative force upon the myofilament lattice. If negatively strained, weak-binding cross-bridges remained attached to actin long enough for significant motion of the filaments to occur, they would reverse the positive work already done in the cross-bridge cycle. Eisenberg and Greene [12] therefore assumed that negatively strained weak-binding cross-bridges rapidly detach from the actin filament before they can do significant negative work. This assumption was based on the biochemical data showing that $31 \cdot ATP$ attaches to, and detaches from, actin very rapidly [25, 39]. The key point is that negatively strained weak-binding cross-bridges detach much faster than positively strained strong-binding cross-bridges. Thus, the return to the weak-binding conformation does not reverse the work performed during the work-stroke even though this transition occurs prior to detachment of cross-bridges from actin. Finally, once the negatively strained cross-bridges detach from actin, they can immediately re-attach to a new actin site and begin a new cycle.

Predictions of the Eisenberg and Greene (1980) model

The proposed cross-bridge model of Eisenberg and Greene [12], schematically outlined in Fig. 3, assumes a close correlation between the behaviour of S-1 in solution and of cross-bridges in the organized contractile system. It therefore makes some specific predictions for cross-bridge action in muscle [13] which can be tested experimentally. In the following section, experimental evidence is presented which is consistent with the presence of two groups of cross-bridge states in muscle which are structurally and kinetically different and exhibit similar properties as described for the weak- and strong-binding states in vitro. Furthermore it is shown that in relaxed fibres cross-bridges can still attach to actin, consistent with the proposal that regulation does not act through control of attachment but rather through control of a subsequent kinetic step, like P_i release. Finally, the proposal that the physiological role of the in vitro rate-limiting step is to control the rate of force generation in muscle, is experimentally confirmed.

Evidence for cross-bridge attachment in relaxed fibres

It has been shown that in vitro, the states $M \cdot ATP$ and $M \cdot ADP \cdot P_i$ can bind to regulated actin in the absence of Ca^{++} [10, 11, 41, 42]. Since the work of Marston [26, 28] demonstated that the chemical cross-bridge states in relaxed muscle are the states with ATP or $ADP + P_i$ bound to the cross-bridges, one would expect to observe cross-bridge attachment in relaxed fibres, if indeed a close correlation exists between the behaviour of S-1 in solution and cross-bridges in muscle.

To probe for cross-bridge attachment, stiffness of relaxed fibres was measured. Since actin affinity of S-1 increases at low ionic strength, the stiffness of relaxed fibres was firstly measured at a low ionic strength of 0.02 M [7]. As shown in Fig. 4a, stiffness of relaxed fibres at 0.02 M ionic strength is about 50% of the stiffness observed in rigor. At the higher ionic strength, stiffness of relaxed fibres is very small, while the rigor stiffness is unchanged.

If stiffness of relaxed fibres at low ionic strength is due to cross-bridge attachment, the stiffness is expected to be proportional to filament overlap, i.e., proportional to the number of potential actin-myosin interaction sites. Since the remaining stiffness and the passive tension at sarcomere lengths around 4.0 μm are relatively insensitive to ionic strength (Fig. 4b), it was assumed that the stiffness at high ionic strength is mainly due to passive parallel elastic components. On this basis, the difference between fibre stiffness at low and high ionic strength would then be a measure of the number of attached cross-bridges, detectable at low ionic strength with stiffness measurements during rapid stretches. Figure 4b demonstrates that stiffness of relaxed fibres at low ionic strength is indeed very closely proportional to the sarcomere length between 2.4 and 3.8 μm, indicating close proportionality between fibre stiffness and filament overlap [7].

Cross-bridge attachment to actin should also lead to a larger amount of mass associated with the actin filaments than could be accounted for by the mass of the actin filaments alone. This was tested by equatorial X-ray diffraction experiments [9]. In these experiments the magnitude of the *11* reflection of relaxed fibres at low ionic strength, is approximately as large as in rigor fibres, where all cross-bridges are attached. This indicates that a much larger amount of mass is associated with the actin filaments than could be accounted for by the actin filaments alone. Quantitative analysis suggests that about the same mass is associated with the actin filaments as in fully Ca^{++} activated fibres at 5 °C [9].

Thus, both stiffness measurements and equatorial X-ray diffraction are consistent with a significant fraction of cross-bridges attached to the actin filaments in relaxed fibres at low ionic strength. Since we have previously shown that under these conditions fibres show no unusual resting tension, active force nor active shortening (unless Ca^{++} is increased above 10^{-6} M), these data indicate that Ca^{++} regulation of muscle apparently does not act through control of cross-bridge attachment, but rather through a subsequent step in the cross-bridge cycle, e.g. P_i release.

Kinetic and structural properties of cross-bridge states in muscle

In vitro weak-binding states are characterized by a rapid equilibrium between binding to, and dissociation from, actin [25, 39]. In contrast, for the strong-binding states detachment from actin is much slower [27, 43]. To probe for the predicted rapid equilibrium between detachment and re-attachment in relaxed fibres at low ionic strength, fibre stiffness was measured at various speeds of stretch. A rapid equilibrium between detachment and re-attachment is expected to result in detachment and re-attachment of cross-bridges during the stiffness measurements, leading to lower force and thus lower stiffness in response to slower stretches [5, 7, 34, 35].

In Fig. 4c, stiffness measurements in relaxed fibres at low ionic strength are shown for different speeds of stretch. It is obvious that during slower stretches the apparent fibre stiffness, that is the slope of the force-displacement plots, is lower than during faster stretches, just as expected when detachment and re-attachment in a less strained position occur during the stiffness measurements. In contrast, in rigor fibres, where presumably all of the cross-bridges are in the strong-binding states, fibre stiffness is almost unaffected by the speed of stretch in the indicated range. More detailed studies show that rigor stiff-

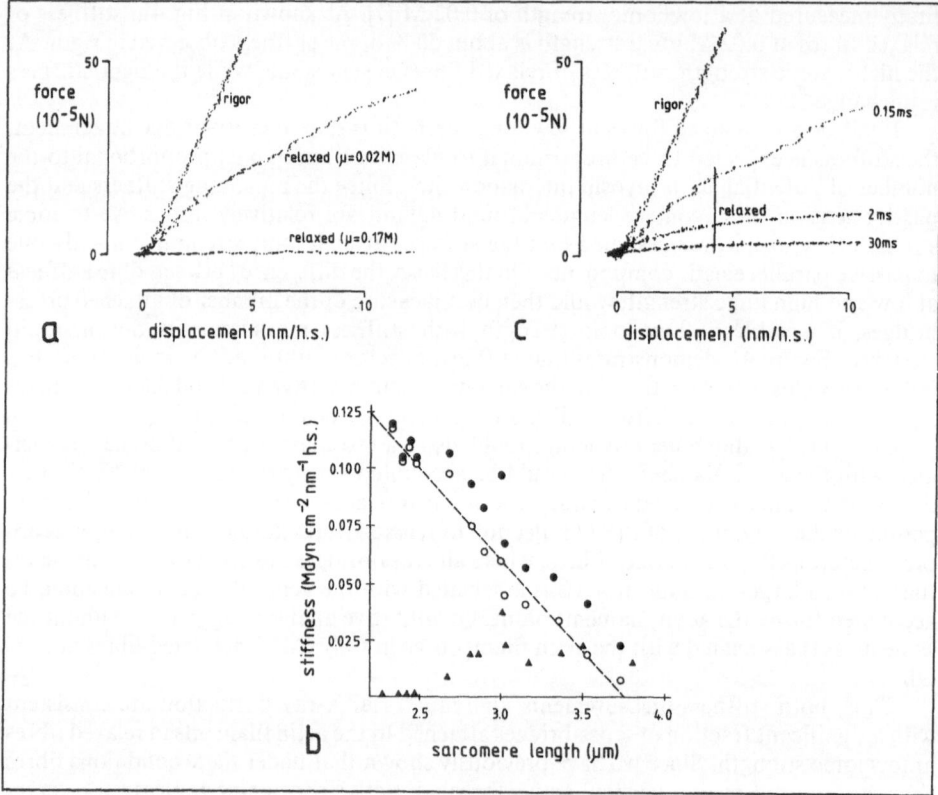

Fig. 4a. Stiffness of single skinned rabbit psoas fibres while relaxed and in rigor. Fibre stiffness is defined as the slope of each trace; speed of stretch is about 3×10 (nm/half-sarcomere)s^{-1}; ionic strength (μ). (Schoenberg et al. [35]).

4b. Stiffness of relaxed single rabbit psoas fibres as a function of sarcomere length. Ionic strengths of 0.17 M (▲–▲) and 0.02 (●–●); difference between stiffness at low and high ionic strength (○–○); and the least squares fit to this (dashed line) are shown. (Brenner et al. [7]).

4c. Effect of speed of stretch on stiffness of relaxed and rigor fibres. Ionic strengths were 0.02 M (relaxing solution) and 0.17 M (rigor solution). Times shown next to traces indicate the time required to stretch the fibre by 5 nm/half-sarcomere. (Brenner et al. [7]).

ness changes at most about 10–15% from the slowest stretches of about 10^{-2} (nm/half-sarcomere)s^{-1} up to the fastest stretches of about 10^{4}(nm/half-sarcomere)s^{-1}, while in relaxed fibres at low ionic strength, stiffness increases up to about 5×10^{4}(nm/half-sarcomere)s^{-1}, the fastest stretches that could be applied [5].

Due to the very different kinetic properties of the weak- and strong-binding states it was suggested that the weak- and strong-binding states differ in their overall structure. To see whether this is the case in the organized contractile system, equatorial X-ray diffraction patterns were recorded of relaxed and rigor fibres [9, 44]. As shown in Fig. 5, the *11* reflections in rigor and relaxed fibres are almost equally strong, while the *10* reflections

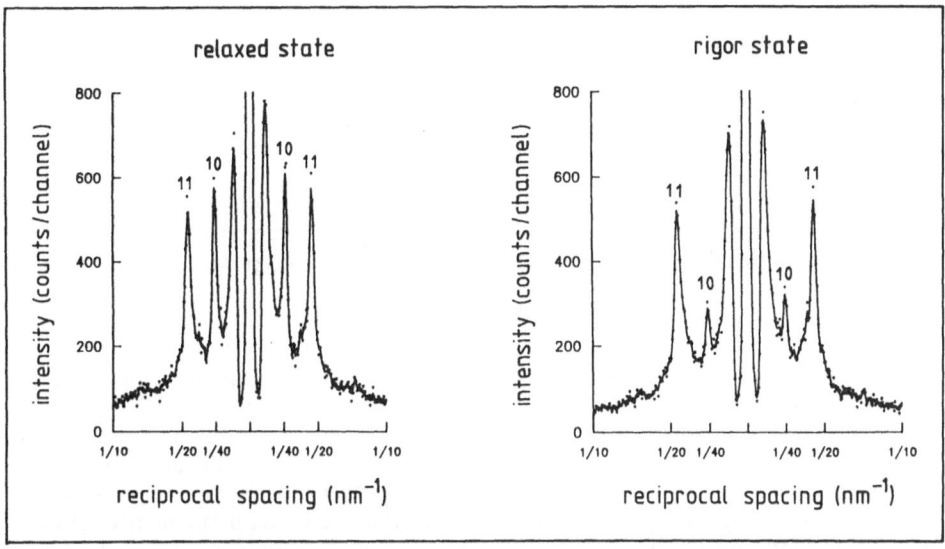

Fig. 5. Equatorial X-ray diffraction patterns recorded from single skinned rabbit psoas fibres while relaxed and during rigor. Original data are shown (dots) and data smoothed by a 3-point weighted average routine (continuous lines): ionic strength = 0.05 M. (B. Brenner and L. C. Yu, unpublished results).

are very different under both conditions. As pointed out above, the strong *11* reflection of relaxed fibres at low ionic strength is consistent with significant cross-bridge attachment. The differences in the *10* reflection cannot be explained by a different number of attached cross-bridges under the two conditions, since changes in the number of attached cross-bridges, at least in rigor, leads to a reciprocal change in the *11* and *10* reflection [8, 16, 22, 45]. In other words, less cross-bridge attachment in rigor, necessary to match the *10* intensities, would lead to a decrease in the magnitude of the *11* in the rigor fibres. This means that for relaxed and rigor fibres the two reflections cannot be matched at the same time, indicating that cross-bridge attachment under these two conditions apparently leads to different mass distributions between the actin and myosin filaments, due to a different structure of the attached cross-bridges. Further evidence for a structural difference between weak- and strong-binding cross-bridges was thought to be provided by recording 2-dimensional X-ray diffraction patterns [29].

The results of mechanical and X-ray diffraction experiments, discussed above, indicate that in muscle, cross-bridge states are present with kinetic properties very similar to those described for the in vitro actomyosin ATPase system. Furthermore, X-ray diffraction experiments suggest that attached weak- and strong-binding cross-bridge have different structures. These findings thus confirm the basic assumption of the model of Eisenberg and Greene [12], that the properties of S-1 in solution closely correspond with the cross-bridge properties in the organized contractile system. The ability of cross-bridges to attach in relaxed fibres, while active force or shortening requires activation by Ca^{++}, indicates that the regulation of muscle does not occur through control of cross-bridge attachment, but rather through regulation of a subsequent kinetic step, e.g. P_i release, as predicted by Eisenberg and Hill [13].

Correlation of the rate of force generation in active muscle with the actomyosin ATPase in solution

Based on the assumption that weak-binding states do not contribute much to isometric force and that the strong-binding states represent the main force generating states, Eisenberg and Hill [13] had postulated that the step which is rate-limiting in solution, and thus controls the transition from the weak- to the strong-binding cross-bridge states, determines the rate of force generation in the cross-bridge cycle in muscle. Since weak- and strong-binding cross-bridge states can be distinguished by mechanical means, such as measurement of force or speed dependence of fiber stiffness, this prediction becomes testable.

To measure the rate of force generation, all or at least most of the cross-bridges have to be accumulated in the weak-binding states. The redistribution of cross-bridges between weak- and strong-binding cross-bridges, can then be followed by monitoring the development of isometric force. Monitoring force development upon step-wise activation (Ca^{++}-jump experiments) cannot be used, since it was shown that in such experiments the rate of force generation is mainly determined by slow steps in the activation of the contractile system [15, 30]. We therefore developed another experimental approach to accumulate cross-bridges in the weak-binding states while fibres are permanently activated [2, 3, 6]. During isotonic shortening near zero load, we had previously found that the number of cross-bridges in the strong-binding states is very much reduced compared to the number in the isometric steady state [1]. The redistribution of cross-bridges between the weak- and strong-binding states can then be followed as the increase in force and fibre stiffness once the shortening has stopped and the fibres were restretched to re-establish the original overlap situation (Fig. 6). The time course of force redevelopment is well fit by a single exponential function [2, 3, 6]. The parallel increase in force and stiffness during force redevelopment indicates that the reincrease in force is due to a reincrease in the number of cross-bridges in the strong-binding states. It was found that any internal fibre shortening during the period of force redevelopment will substantially reduce the rate of force redevelopment [6]. It was therefore necessary to carefully control the sarcomere

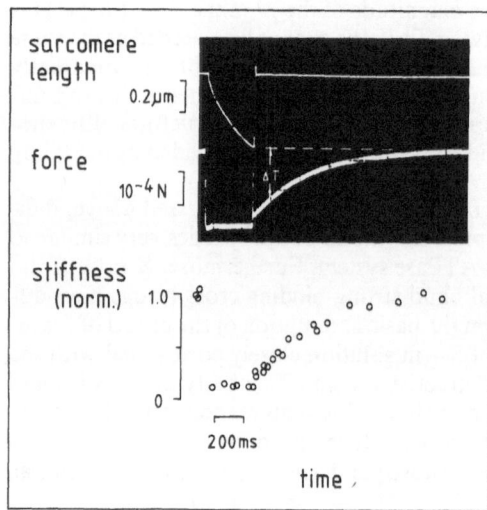

Fig. 6. Protocol for recording development of force and fibre stiffness at 0.17 M ionic strength. (Brenner [3]).

Table 1. Comparison of the rate constants for force redevelopment and cross-linked Acto-S-1 ATPase activity

Conditions	Force redevelopment $(s^{-1(a}$	ATPase $(s^{-1})^b$
5 °C (u = 0.05 M)	3.8 ± 1.3 (mean ± SEM, n = 4)	1.82
5 °C (u = 0.17 M)	4.5 ± 1.9 (mean ± SEM, n = 13)	1.92
15 °C (u = 0.17 M)	14.5 ± 2.4 (mean ± SEM, n = 5)	11.41
25 °C (u = 0.17 M)	26.1 ± 4.0 (mean ± SEM, n = 9)	49.40
35 °C (u = 0.17 M)	50.7 ± 8.2 (mean ± SEM, n = 3)	137

[a] Measurements at 25 °C and 35 °C with overall-length control, all others with sarcomere-length control, during the period of force redevelopment.
[b] All rates are corrected for the 18% reduction in the ATPase activity of S-1 cross-linked to actin, compared to that of uncrosslinked S-1.
Reproduced from Brenner and Eisenberg [6].

length by a servo system using laser light difraction to continuously monitor the sarcomere length.

To test the prediction of Eisenberg and Hill [13], mentioned above, the rate of force redevelopment and the actomyosin ATPase activity in solution were measured under a variety of conditions. To achieve maximum actin activation, even under conditions where binding of S-1 to actin is very weak, the maximum actin activated actomyosin ATPase was measured using myosin S-1 cross-linked to actin [31, 38]. The data are summarized in Table 1. At temperatures ≥ 25 °C, the laser-controlled servo system became unreliable so that internal shortening could not be avoided. For this reason, the rate of force redevelopment is probably underestimated by a factor of more than two [6]. Assuming that the rate constant of force redevelopment is dominated by the transition from the weak- to the strong-binding cross-bridge states, while the subsequent return to the weak-binding states is much slower, Brenner and Eisenberg [6] have suggested that the presented data indicate a close correlation between the two parameters. The assumption that the rate of force redevelopment is mainly determined by the transition of cross-bridges from the weak- to the strong-binding states has been recently tested [4]. It was found to be correct for temperatures above 10–15 °C. At 5 °C, however, the rate constant at which cross-bridges return to the weak-binding states was found to be about 50% of the rate that determines the transition from the weak- to the strong-binding states. Since the rate of force redevelopment is the sum of these two rate constants, this finding would explain a substantially larger rate of force redevelopment at 5 °C compared to the actomyosin ATPase rate at the same temperature, as was found experimentally [6]. Thus, the correlation between actomyosin ATPase and the rate of force redevelopment is consistent with the proposed relation between the rate limiting step in solution, and the rate of force generation in muscle.

Conclusions

The reported studies with skinned rabbit psoas fibres show, at least qualitatively, a close correlation between the behaviour of myosin S-1 in solution and the behaviour of cross-bridges in the organized contractile system in muscle, providing experimental evidence for the fundamental assumption on which the model of Eisenberg and Greene [12] was based.

In muscle, two groups of cross-bridge states are found with kinetic properties very similar to those described for the weak- and strong-binding cross-bridge states of the actomyosin ATPase system in solution.

The ability of cross-bridges to bind to actin in relaxed fibres indicates, independent of the extent of the binding at physiological ionic strength, that regulation of muscle does not occur through control of cross-bridge attachment, but probably through control of a subsequent step, e.g. P_i release.

The comparison of the actomyosin ATPase activity in solution with the rate of force generation in muscle, is consistent with the proposed idea that the step which is rate limiting in solution determines the rate of force generation in the cross-bridge cycle in muscle.

The described mechanical and structural approaches might be useful tools for analysing the kinetics of cycling cross-bridges in active muscle, some preliminary results of which have recently been presented [4].

References

1. Brenner B (1983) Cross-bridge attachment during isotonic shortening in single skinned rabbit psoas fibers. Biophys J 41:33a
2. Brenner B (1984) The rate of force redevelopment in single skinned rabbit psoas fibers. Biophys J 45:155a
3. Brenner B (1985) Correlation between the cross-bridge cycle in muscle and the actomyosin ATPase cycle in solution. J Muscle Res Cell Motil 6:659–664
4. Brenner B (1986) The cross-bridge cycle in muscle. Mechanical, biochemical and structural studies on single skinned rabbit psoas fibers to characterize cross-bridge kinetics in muscle for correlation with the actomyosin ATPase in solution. Basic Res Cardiol 81 (Suppl 1):1–15
5. Brenner B, Chalovich JM, Greene LE, Eisenberg E, Schoenberg M (1986) Stiffness of skinned rabbit psoas fibers in MgATP and MgPP$_i$ solution. Biophys J 50:685–691
6. Brenner B, Eisenberg E (1986) Rate of force generation in muscle: correlation with actomyosin ATPase in solution. Proc Natl Acad Sci USA 83:3542–3546
7. Brenner B, Schoenberg M, Chalovich JM, Greene LE, Eisenberg E (1982) Evidence for cross-bridge attachment in relaxed muscle at low ionic strength. Proc Natl Acad Sci USA 79:7288–7291
8. Brenner B, Yu LC (1985) Equatorial X-ray diffraction from single skinned rabbit psoas fibers at various degrees of activation. Changes in intensities and lattice spacing. Biophys J 48:829–834
9. Brenner B, Yu LC, Podolsky RJ (1984) X-ray diffraction evidence for cross-bridge formation in relaxed muscle fibers at various ionic strengths. Biophys J 46:299–306
10. Chalovich JM, Chock PB, Eisenberg E (1981) Mechanism of action of troponin-tropomyosin. J Biol Chem 256:575–578
11. Chalovich JM, Eisenberg E (1982) Inhibition of actomyosin ATPase activity of troponin-tropomyosin without blocking the binding of myosin to actin. J Biol Chem 257:2431–2437
12. Eisenberg E, Greene LE (1980) The relation of muscle biochemistry to muscle physiology. Ann Rev Physiol 42:293–309
13. Eisenberg E, Hill TL (1985) Muscular contraction and free energy transduction in biological systems. Science 227:999–1006
14. Eisenberg E, Zobel CR, Moos C (1968) Subfragment 1 of myosin: adenosine triphosphatase activation by actin. Biochemistry 7:3186–3194
15. Griffiths PJ, Kuhn HJ, Güth K, Rüegg JC (1979) Rate of isometric tension development in relation to calcium binding of skinned muscle fibres. Pflügers Arch 382:165–170

16. Haselgrove JC, Huxley HE (1973) X-ray evidence for radial cross-bridge movement and for the sliding filament model in actively contracting skeletal muscle. J Mol Biol 77:549–568

17. Hill TL (1974) Theoretical formalism for the sliding filament model of contraction of striated muscle. Part I. Prog Biophys Mol Biol 28:267–340

18. Huxley AF (1957) Muscle structure and theories of contraction. Prog Biophys Chem 7:255–318

19. Huxley AF (1974) Muscular contraction. J Physiol (Lond) 243:1–43

20. Huxley AF, Niedergerke R (1954) Interference microscopy of living muscle fibres. Nature (Lond) 173:971–974

21. Huxley AF, Simmons RM (1971) Proposed mechanism of force generation in striated muscle. Nature (Lond) 233:533–538

22. Huxley HE (1968) Structural differences between resting and rigor muscle. Evidence from intensity changes in the low-angle equatorial X-ray diagram. J Mol Biol 37:507–520

23. Huxley HE (1969) The mechanism of muscular contraction. Science 164:1356–1366

24. Huxley HE, Hanson J (1954) Changes in the cross-striations of muscle during contraction and stretch and their interpretation. Nature (Lond) 173:973–976

25. Lymn RW, Taylor EW (1971) Mechanism of adenosine triphosphate hydrolysis by actomyosin. Biochemistry 10:4617–4624

26. Marston SB (1973) The nucleotide complexes of myosin in glycerol-extracted muscle fibers. Biochim Biophys Acta 305:397–412

27. Marston S (1982) The rates of formation and dissociation of actin-myosin complexes. Biochem J 230:453–460

28. Marston SB, Tregear RT (1972) Evidence for a complex between myosin and ADP in relaxed muscle fibers. Nature New Biol 235:23–24

29. Matsuda T, Podolsky RJ (1984) X-ray evidence for two structural states of the actomyosin cross-bridge in muscle fibers. Proc Natl Acad Sci USA 81:2364–2368

30. Moisescu DG (1976) Kinetics of reaction in calcium-activated skinned muscle fibres. Nature (Lond) 262:610–613

31. Mornet D, Bertrand R, Pantel P, Audemard E, Kassab R (1981) Structure of the actin-myosin interface. Nature (Lond) 292:301–306

32. Reedy MK, Holmes KC, Tregear RT (1965) Induced changes in orientation of the cross-bridges of glycerinated insect flight muscle. Nature (Lond) 207:1276–1280

33. Rosenfeld SS, Taylor EW (1984) The ATPase mechanism of skeletal and smooth muscle acto-subfragment 1. J Biol Chem 259:11908–11919

34. Schoenberg M (1985) Equilibrium muscle cross-bridge behaviour: theoretical considerations. Biophys J 48:467–475

35. Schoenberg M, Brenner B, Chalovich JM, Greene LE, Eisenberg E (1984) Cross-bridge attachment in relaxed muscle. In: Pollack GH, Sugi H (eds) Contractile mechanism in muscle. Plenum, New York, pp 269–279

36. Stein LA, Chock PB, Eisenberg E (1981) Mechanism of the actomyosin ATPase: Effect of actin on the ATP hydrolysis step. Proc Natl Acad Sci USA 78:1346–1350

37. Stein LA, Chock PB, Eisenberg E (1984) The rate-limiting step in the actomyosin adenosinetriphosphatase cycle. Biochemistry 23:1555–1563

38. Stein LA, Greene LE, Chock PB, Eisenberg E (1985) Rate-limiting step in the actomyosin adenosinetriphosphatase cycle. Biochemistry 24:1357–1363

39. Stein LA, Schwarz RP, Chock PB, Eisenberg E (1979) Mechanism of actomyosin adenosine triphosphatase. Evidence that adenosine 5'-triphosphate hydrolysis can occur without dissociation of the actomyosin complex. Biochemistry 18:3895–3909

40. Szentkiralyi EM, Oplatka A (1969) On the formation and stability of the enzymatically active complexes of heavy meromyosin with actin. J Mol Biol 43:551–566

41. Wagner PD (1984) Effect of skeletal muscle myosin light chain 2 on the Ca^{2+}-sensitive interaction of myosin and heavy meromyosin with regulated actin. Biochemistry 23:5950–5956

42. Wagner PD, Giniger E (1981) Calcium-sensitive binding of heavy meromyosin to regulated actin in the presence of ATP. J Biol Chem 256:12647–12650

43. White HD, Taylor EW (1976) Energetics and mechanism of actomyosin adenosine triphospha-
    tase. Biochemistry 15:5818–5826
44. Yu LC, Brenner B (1986) High resolution equatorial X-ray diffraction from single skinned rab-
    bit psoas fibers. Biophys J 49:133–135
45. Yu LC, Hartt JE, Podolsky RJ (1979) Equatorial X-ray intensities and isometric force levels in
    frog sartorius muscle. J Mol Biol 132:53–67

Authors' address:

Dr. B. Brenner, Physiologisches Institut II, Gmelinstr. 5, D-7400 Tübingen

Energetics studies of muscles of different types

M. J. Kushmerick

NMR Division, Department of Radiology, Brigham & Women's Hospital, Boston, Massachusetts U.S.A.

Summary

^{31}P-NMR studies were performed in isolated perfused striated and smooth muscles. Important qualitative and quantitative differences were found in resting muscles. In resting fast-twitch skeletal muscle the chemical potential of ATP obtained from the measured intracellular pH, ATP and inorganic phosphate concentrations and from the ADP concentrations calculated from the position of the creatine kinase equilibrium was -72 kJ/mol ATP. This high value was the result of a very low free ADP and inorganic phosphate content. In resting slow-twitch skeletal muscle, in smooth muscle, and in cardiac muscle at low work rates (literature data), the chemical potential of ATP was lower (approximately -50 to -60 kJ/mol), the difference being primarily due to a much higher inorganic phosphate content (especially in slow-twitch and smooth muscle) and/or a higher ADP concentration (especially in cardiac muscle). Upon stimulation or, for the heart, working at higher work rates, the pattern of chemical changes of phosphocreatine, creatine and inorganic phosphate was the same for all types of muscle. The phosphocreatine levels decreased and the inorganic phosphate concentration increased stoichiometrically without a change in the ATP content so long as the phosphocreatine pool was not totally depleted ($\geq 10\%$). The rate and extent of these chemical changes was dependent on the inherent ATPase and ATP synthesis rates. The exception was in the intracellular pH changes. In fast-twitch and smooth muscle, pH decreased with contractile activity, as expected from the large glycolytic capacity. However, an alkalinization was observed in slow-twitch skeletal muscle and this difference was attributed to the uptake of H^+ during the net hydrolysis of phosphocreatine to creatine plus inorganic phosphate, and to the absence of significant lactate production. The pH of cardiac muscle does not appear to change with work load. The common bioenergetic pattern in all types of muscles is consistent with a graded increase in ADP concentration (from below to well above the apparent K_m for nucleotide translocase ANT) with increasing work as a regulator of mitochondrial respiration. In fast-twitch muscle these changes are also accompanied by large changes in inorganic phosphate concentration (3–30 mM) which may also play a role in metabolic regulation. Based on results of other investigators on the myocardium, it appears that the ADP concentration lies well above the apparent K_m for the adenine nucleotide translocase, so that at high work rates regulation of mitochondrial respiration may be due to other factors (e.g., transport of reducing equivalents or Ca^{2+}-activated dehydrogenases). Finally, the spectra obtained from resting smooth muscle showed the presence of two compounds: phosphorylethanolamine (1 µmol/g) and glycerophosphorylcholine (1.5 µmol/g). The presence of these phospholipid metabolites was not previously known and their function is not yet understood. The other notable difference in smooth muscle compared to striated muscle was a significant upfield chemical shift of the beta peak of ATP in smooth muscle. We interpret this as indicative of a lower plasmic free Mg^{2+} concentration (0.5 mM) in smooth muscle compared to that (1.5 mM) in striated muscle.

I. Introduction

The primary function of muscle is the generation for force, which is essential for locomotion (skeletal muscle), convective flow (cardiac and smooth muscle), and the regulation of this fluid flow (smooth muscle). The muscle machine is a chemomechanical converter that can be examined from two points of view. The first questions the quantitative demands of chemical potential energy placed on the cells by the actomyosin interaction, calcium movements, and other forms of electrical and osmotic work. The second aspect concerns the integrated operation and regulation of metabolic pathways, the net result of which is the generation of chemical potential energy in the form of ATP and other so-called high-energy phosphates, such as phosphorylcreatine (PCr), by oxidative metabolism. In this paper the balance between these two aspects is considered, i.e., the steady-state and dynamic content of high-energy phosphate compounds reflects the balance between generation and utilization.

What distinguishes striated muscle cells from most other cells is that the dynamic range of their metabolic activity is so large. In cardiac and skeletal muscle there is about a 10- and 100-fold dynamic range, respectively, in metabolic rates.

The pattern of energy utilization in mammalian muscles appears to be similar to that observed in amphibians (for reviews, see 9, 15, 18). A detailed study of the energetics of a predominantly fast-twitch muscle, the mouse extensor digitorum longus (EDL), and of a predominantly slow-twitch muscle, the mouse soleus, is available [6], and the data are shown in Fig. 1. The recovery chemical input ($\Delta \sim P_{rec}$) is expressed in µmol $\sim P/g$ and obtained from the following equation: $\Delta \sim P_{rec} = \kappa \zeta O_2 + \lambda \xi lac$, where κ and λ are stoichiometric factors equal to 6.3 and 1.5, respectively. In a majority of the data points for the soleus, the contribution from lactate production is negligible, and for the sake of visual clarity only the points corresponding to the total recovery input were included on the graph in Fig. 1. The lines are linear functions fitted to the data with the following parameters:

soleus: $\Delta \sim P_{rec} = 0.34 \pm 0.71 + 8.73 \pm 0.51 \, L_0 \int Pdt$

EDL: $\Delta \sim P_{rec} = 0.68 \pm 1.3 + 23.4 \pm 3.55 \, L_0 \int Pdt$
for tetani up to 4 s duration, $0 < L_0 \int Pdt < 0.75$

$\Delta \sim P_{rec} = 11.3 \pm 1.5 + 11.1 \pm 1.2 \, L_0 \int Pdt$
for tetani longer than 8 s, $1.0 < L_0 \int Pdt < 2.5$

The energetics of the EDL differ from those of the soleus in two major ways, as the data in Fig. 1 show. The energy cost normalized to the isometric force per cross-sectional area is independent of the tetanic duration in soleus. The longest of these contractions (15 s) is sufficient to deplete most of the high-energy phosphate pool. In the EDL, the energy cost is not constant but depends on the duration of the contractile activity, since there is a decrease in the energy cost for force maintenance after approximately 9 s of tetanus. The mechanism for this decrease in rate of energy utilization is not, as we initially proposed, caused by phosphorylation of the so-called regulatory light chain of myosin (18,000 dalton light chain LC-2f) (compare [3, 7]). We now know that phosphorylation of LC-2f does not alter the maximal velocity of shortening of muscle (under no-load conditions), but appears to increase the sensitivity of phosphorylated myofilaments to calcium ion near the threshold of activation [29].

Because there is definitely a decrease in the rate of cross-bridge turnover as judged from the maximal velocity of shortening [8], it appears that other factor(s) must be involved in the change in rate of energy utilization. Possible mechanisms include regulation by intracellular acidosis, inorganic phosphate (Pi) accumulation, or lowered ATP/ADP

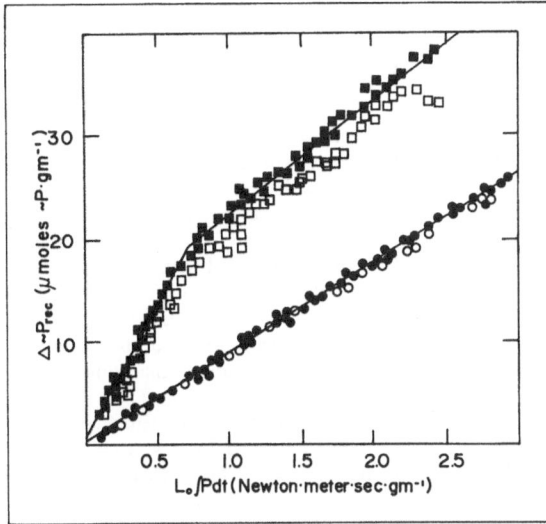

Fig. 1. The relationship between total recovery chemical input and the tension-time integral. Symbols: contribution to the total recovery chemical input from oxygen consumption alone ($\kappa\xi O_2$) in the soleus (○) and EDL (□); total recovery chemical input after the contribution from glycolytic ATP production ($\lambda\xi lac$) was added to the aerobic contribution in soleus (●) and EDL (■). Data from 13 soleus and EDL muscles each. Multiple determinations were performed on each muscle. Muscles were stimulated and allowed to recover for 30 min before stimulating again; figure from (6), which should be consulted for details

ratios [5, 10]. Unfortunately, the mechanism for this important aspect of muscle energetics and mechanics remains unknown in any muscle type.

The second major difference was in the magnitude of the energy cost per unit for force per cross-sectional area. For brief tetani, the energy cost in the fast-twitch EDL was 2.9 times that of the slow-twitch soleus. This was expected from the greater actomyosin-ATPase activity from EDL versus soleus muscle. After some 9 s of tetanus (as indicated above), however, the force-normalized energy cost in the EDL was reduced, so that it was only 1.5 times that of the soleus in prolonged tetani. The mouse soleus contains a significant fraction of fast-twitch fibres and thus the three-fold dynamic range in ATP utilization rates measured in whole muscles probably represents an underestimate of the true range of ATPase rates in mammalian muscle cells from fast- to slow-twitch fibres. In addition to the differences in mechanics, energetics, metabolism and contractile protein isoforms between fast- and slow-twitch fibres, there are a number of known sub-types of fast-twitch fibres, and these have mechanical and metabolic (and probably energetic) properties intermediate between the extremes of "fast" and "slow" [28, 30].

It was clear that in a fully aerobic soleus muscle, a steady state was possible during a tetanus in which the rate of ATP synthesis matched the rate of ATP splitting. That is, in a continuous tetanic stimulation, the oxygenated muscle would develop a steady state wherein there was no further measurable extent of reaction involving Pi, PCr, or ATP. This result is illustrated by the data in Fig. 2 (open symbols). If the soleus were anaerobic, the measured extent of high-energy phosphate splitting would increase continuously during the maintained tetanus, i.e., chemical changes would not be confounded by concurrent aerobic resynthesis.

These observations illustrate a wider dynamic range of aerobic capacity in the slow-twitch oxidative mouse soleus than in the EDL, and indicate that the mitochondrial density and diffusion of oxygen and substrates are sufficient to maintain PCr and ATP levels even during an isometric tetanus. These two muscles of the mouse nicely illustrate the extremes of energetic patterns in striated muscle [18]. The fast-twitch EDL can be described as a "twitch now, pay later" type of cellular oxidative pattern. The slower soleus, which

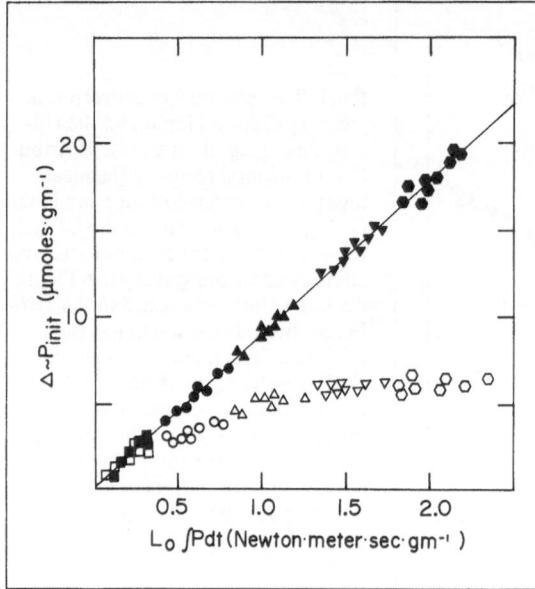

Fig. 2. The relationship between directly-measured initial chemical breakdown and tension-time integral in the soleus. Units as in Fig. 1. Each data point represents the initial chemical breakdown ($\varDelta \sim P_{init}$) in a single muscle compared with its control from observed changes in ATP, PCr, and Pi (see text). Initial chemical change assessed under aerobic conditions (open symbols); and under anaerobic conditions (closed symbols). The solid line is the linear regression function fitting the anaerobic data: $\varDelta \sim P_{init} = 0.03 \pm 0.34 + 8.75 \pm 0.98 \, L_0 \int Pdt$. The symbols refer to muscles stimulated for different tetanus durations: (□, ■), 1 s; (○, ●), 3 s; (△, ▲), 6 s; (▽, ▼), 9 s; (○, ●), 12 and 15 s. Figure from (6).

also has a greater mitochondrial density, follows more of a "pay as you go" strategy of cellular respiration, as do smooth and cardiac muscle. As indicated above, it is possible that a number of intermediate forms exist, perhaps a smooth continuum of mechanical, energetic and metabolic properties.

II. ^{31}P-NMR measurements

The ability of modern nuclear magnetic resonance spectrometers to measure the tissue content of phosphate compounds relevant to muscle energetics has been well established [13]. Sufficiently large magnets exist to monitor these intracellular metabolites in intact human limbs.

The data to be discussed were derived from studies of isolated cat biceps and soleus muscles (striated muscles) and rabbit bladder (smooth muscle) perfused through their arterial tree by a suspension of red cells in Krebs Henseleit buffer [20, 24], of intact rat gastrocnemius muscles [21], and of perfused rat hearts ([12, 17]; examples selected from a large number of possible references). The range of oxygen consumption rates at 27 °–30 °C in both types of in vitro-perfused cat muscles are shown in Fig. 3 A. Contractile activity was induced by applying supramaximal twitches for 5–20 min to obtain the steady-state rate of oxygen consumption, which was measured by arterio-venous differences in the total oxygen content of the perfusate delivered at constant flow. The biceps reached a maximal oxygen consumption rate of about 60 µmol oxygen/min/100 g muscle at 40 twitches/min. In contrast, the maximal rate of oxygen consumption of the soleus did not occur up to stimulation rate of 80 twitches/min. Figure 3 B shows that the levels of PCr and of Pi attained in the steady state are functions of the oxygen consumption; all of these data were obtained under conditions in which we were certain there was no limitation of oxygen supply, as the data in panel A indicate.

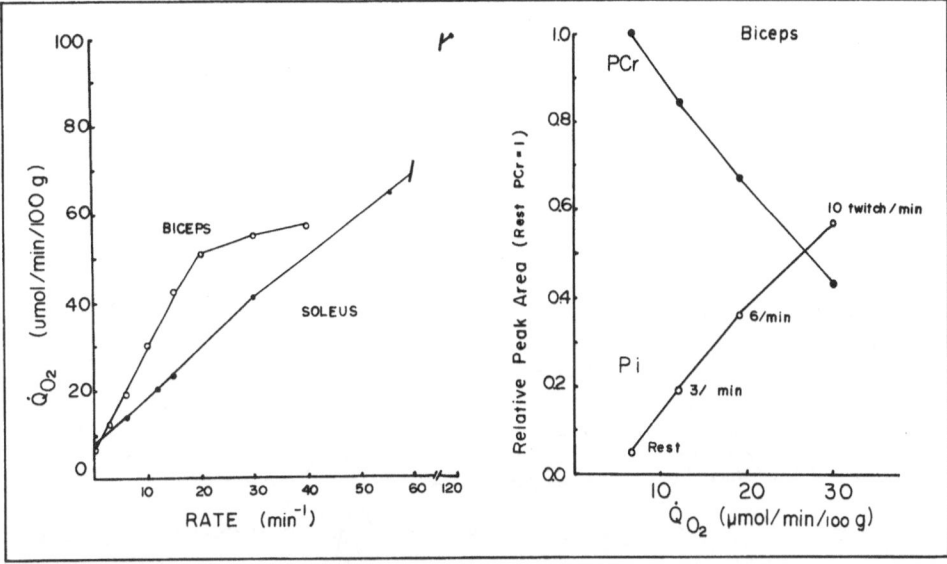

Fig. 3. (A) Shows the steady-state oxygen consumption of cat biceps and soleus muscles isolated and arterially perfused with an erythrocyte suspension at various frequencies of isometric twitch stimulation. (B) Shows the relationship between PCr and Pi levels measured by ^{31}P-NMR (in arbitrary units) during steady state of oxygen consumption given in (A). Data taken from [20].

Table 1. Metabolite content (mM in cell water)

Tissue type	Pi	ATP	PCr	Cr	ADP	ΔG_{obs}
Smooth muscle (rabbit bladder)[a]	2.4	1.5	2.5	2.0	7×10^{-3}	-60
Fast-twitch skeletal (cat biceps)[b]	3.1	8.9	34.9	0.2	0.3×10^{-3}	-72
Slow-twitch skeletal (cat soleus)[b]	10.1	5.0	16.6	7.8	14×10^{-3}	-58
Cardiac[c,d]	0.9	6.2	14.2	11.3	30×10^{-3}	-61
(rat heart)[e]	0.4	10.8	22.6	4.7	14×10^{-3}	-52

[a] Ref. 20; [b] Ref. 24; [c] Control Langendorf-perfused hearts; other muscles under unstimulated "resting" conditions; [d] Ref. 17; [e] Ref. 12; [f] Units kJ/m

The first observation based on NMR results that is noteworthy is that the Pi content of rat muscles in situ and in well-perfused fast-twitch muscles (Table 1; [25]) is substantially lower (i.e., approximately 2 µmol/g) than is usually reported from analyses of extracts of rapidly frozen tissues (i.e., 5–7 µmol/g). It is likely that most if not all of the discrepancy is due to artefactual breakdown of PCr, a much more labile compound than ATP, during the extraction and/or freezing procedures. The Pi content of the slow-twitch soleus is exceptionally high, 8 µmol/g.

The second NMR observation that should be noted is that there are important *qualitative* differences in the intracellular pH change in response to contractile activity in fast versus slow skeletal muscle. It is well established that the cytosolic pH can be measured from the chemical shift or spectral position of the Pi peak [27]. This is due to the fact that

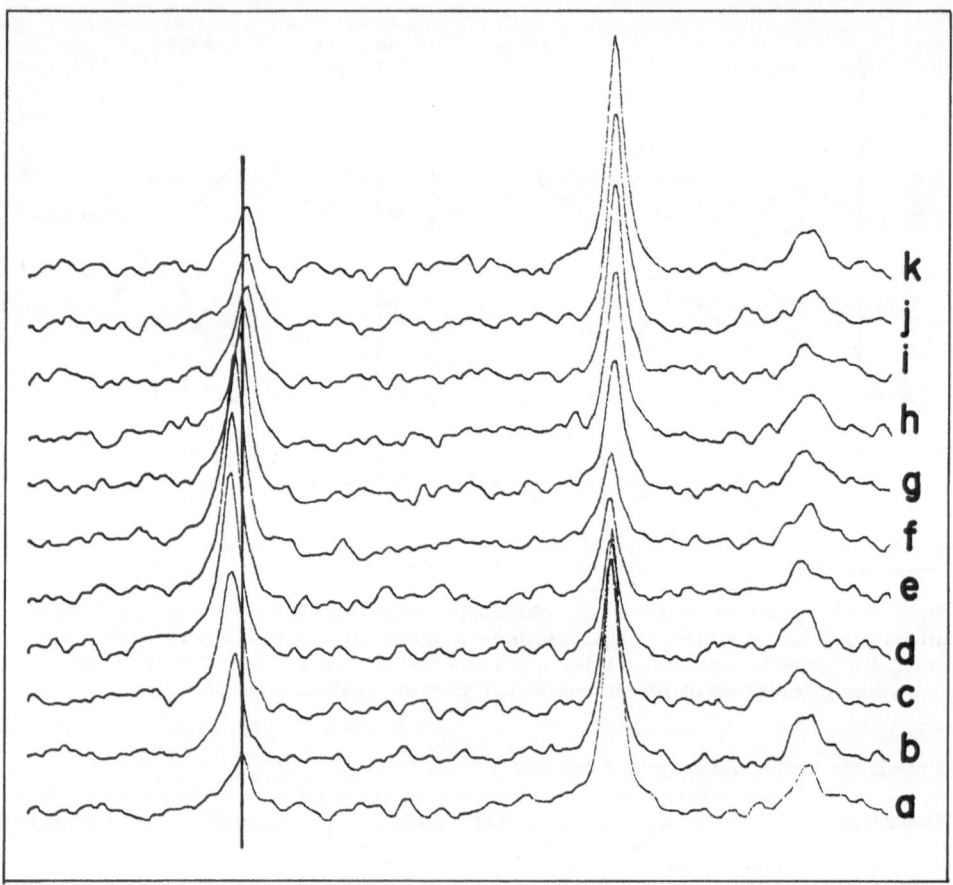

Fig. 4. Portions of ^{31}P-NMR spectra which show the chemical shift of the Pi peak as a measure of intracellular pH during a stimulation-recovery cycle in the cat soleus. The three peaks shown are, left to right, Pi, PCr, and ATP (phosphorus). Each set of spectra is aligned vertically at the frequency of PCr that is independent of pH under the experimental conditions. The vertical line indicates an intracellular pH of 7.1. A shift to the left indicates an alkalinization, and to the right, an intracellular acidification. Symbols: a, unstimulated control; b–f, spectra during 10 min periods of isometric twitches at 60 twitches/min, showing an increase of intracellular pH to 7.4; g–k, spectra during the next 10 min period of recovery. Figure taken from [22].

the doubly protonated species has a chemical shift different from HPO_4^{2-}, that the measured chemical shift is proportional to the average number of each charged species in rapid equilibration with a midpoint at the pKa, and that Pi is predominantly located in the cytosol. During stimulation and subsequent recovery, the intracellular pH changes are characteristically different in the fast-twitch and slow-twitch muscle. For example, during a 15 min period of steady-state twitches, the purely slow-twitch soleus undergoes a marked intracellular alkalinization, from a control value of pH 7.1 to about pH 7.4 (Fig. 4). In the experiments to produce Fig. 4, cat soleus was perfused with an erythrocyte-containing physiological saline [25] at 25 °C and mounted in a Bruker HX-270 spectro-

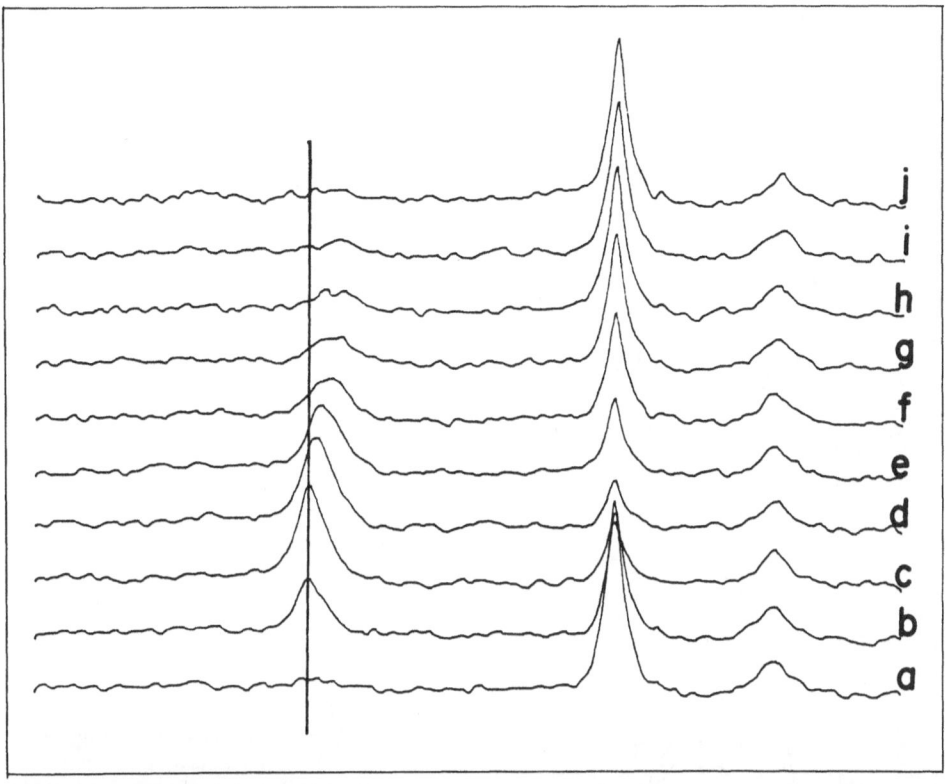

Fig. 5. A similar experiment to Fig. 4 using the perfused fast-twitch cat biceps, showing chemical shift of Pi peak during recovery from 15 min period of stimulation (see text). Each spectrum at 5 min intervals, 20 scans averaged. Symbols: a, unstimulated control; b–d, stimulated at 30 twitches/min; e–j, recovery. Maximal acidification was to pH 6.7. Figure from [22].

meter. The spectra were acquired in the Fourier transform mode using 90 ° pulses at intervals of 15 s at a frequency of 109.3 MHz. Sequential spectra were taken every 2 min, and eight scans averaged.

In these experiments, the extracellular pH was maintained with bicarbonate/CO_2 buffer at pH 7.2. The mechanism for the alkalinization is proton uptake during net PCr breakdown, a reaction to which we will return later. In the recovery period that followed there was a moderate acidification to about pH 6.9. By way of contrast, in the fast-twitch biceps muscle (stimulated in such a way as to deplete PCr to the same extent as in the soleus experiment) little alkalinization was detected, even though the PCr breakdown was more rapid and proceeded to a greater molar extent per unit volume muscle. The PCr content was reduced in both to $^1/_3$ of that at rest, but the content was larger, and therefore the *extent* of PCr decrease greater, in the biceps than in the soleus. The lack of alkalinization is due in part to a greater intracellular buffer capacity in the fast-twitch muscle than in the slow-twitch, and mainly to a rapid onset of glycolysis and concomitant lactate production. During the recovery period, the acidification became quite marked, reaching values as low as pH 6.2 (Fig. 5). These experiments were chosen to emphasize qualitative

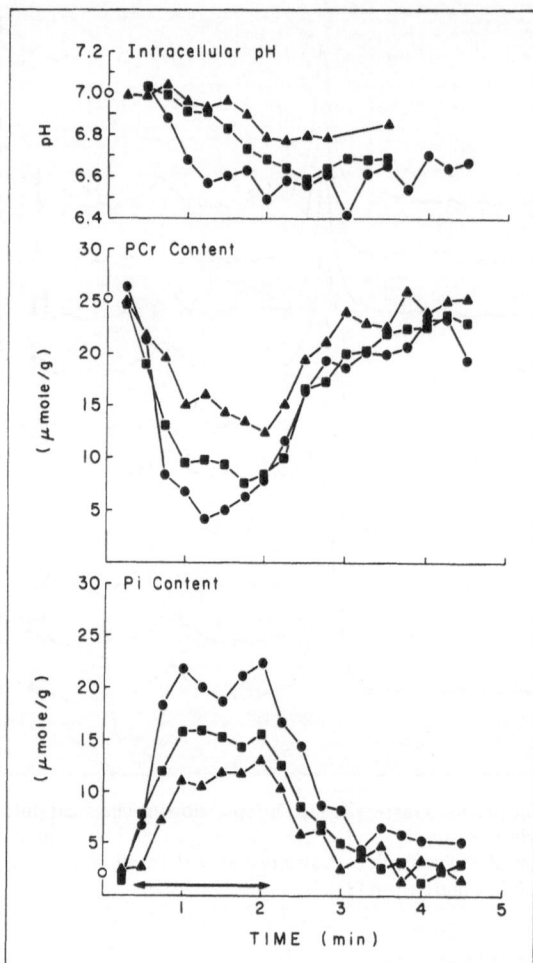

Fig. 6. Chemical changes in a rat lower limb musculature before, during and after 1.8 min stimulation of sciatic nerve at 2 Hz (▲), 4 Hz (■), and 10 Hz (●). Each spectrum is average of 4 FID. Open symbols at origin represent average of control values for this animal obtained from resting muscle before stimulation and at least 15 min after each stimulation period. Figure from [21].

differences between the two muscle types. Under prolonged stimulation of the soleus, at a more rapid twitch rate, intracellular alkalinization yields to acidification, demonstrating a significant glycolytic capacity in this slow-twitch oxidative muscle.

The biceps is a mixed muscle, with approximately equal proportions of FOG and FG fibres. It is noteworthy that the shape of the Pi peak broadened considerably during the recovery period in the biceps, but not in the soleus. This observation clearly indicates a heterogeneity in fibre-to-fibre intracellular pH not present in the homogeneous soleus muscle. Thus there are detectable differences in the kinetics of metabolic responses in FOG and FG fibres, which comprise the heterogeneous cat biceps. There are only slow-twitch fibres in the soleus, and all of them behave homogeneously. In prolonged and intense stimulations of smooth muscle pH decreases only slightly, and in hearts working at various work loads, the pH does not appear to be altered.

The third and most important set of ^{31}P-NMR observations that bear on our topic is that steady-state twitch stimulations at increasing frequencies produce steady-state graded levels of decreased PCr and increased Pi. Provided that the maximal aerobic capacity of the muscle is not exceeded (e.g., to the left of the maxima illustrated in Fig. 3), the graded levels of PCr and Pi are a function of the intensity of the steady-state rate of contractile activity (Fig. 6), and in turn are accompanied by graded levels of cellular respiration, which can be measured as the rate of oxygen consumption. While the smooth muscle was not studied over such a wide range of physiological states, it is clear that the pattern of chemical changes is qualitatively the same, but of much smaller magnitude because of its much smaller content of Pi, ATP and PCr. The same pattern is also observed in well-perfused hearts, where the work load was varied [12, 17].

III. The role of phosphocreatine and creatine kinase

The near-equilibration between PCr and ATP under physiologically achievable steady states is considered to be important in muscle function as a buffer of chemical potential energy, even under conditions of high demand [16, 26]. The PCr content (up to approximately 30 mM in fast-twitch muscle) and creatine kinase (CK) flux (five or more times the maximal ATPase rate) are the bases for this buffer function. What may be the most basic metabolic function of this reaction and the basis for understanding other secondary functions will be discussed, namely, that CK serves to buffer the ADP levels in the μM range and that this aspect of its function is probably more important than its buffering of cellular ATP levels. A subsidiary point is that the combined actions of ATPases and CK also generate quite high concentrations of Pi, which may (in addition to ADP content and ATP/ADP ratios) provide an important metabolic regulator. The cytosolic ADP level is a well known and important metabolic regulator, including that of mitochondrial respiration [23], and Pi may well prove to be important in the regulation of glycolysis and the modification of actomyosin kinetics.

These aspects of CK function are appreciated by considering the relationship between two sets of reactions. The first is the coupled hydrolysis of ATP (the coupled process is not indicated in order to keep the equation simple):

$$ATP \leftrightarrow ADP + Pi + \alpha H^+, \tag{1}$$

where α is a function of pH; near pH 7, $\alpha = 0.7$. This reaction is written reversibly, and this is formally incorrect because, of course, different and spatially separate enzyme systems catalyze the reaction in each direction. Actomyosin ATPase and other coupled ATP hydrolases catalyze the reaction to the right, and mitochondrial respiration catalyzes the reactions to the left. Nonetheless, the intact cell is in a steady state and is well regulated, so that the cell operates as if the reaction were reversible but held far from equilibrium. Substrate level phosphorylation is also a mechanism for ATP generation, but is quantitatively irrelevant for our purposes, since the net extent of reaction is small in the fully aerobic state.

The second reaction to consider is the truly reversible reaction catalyzed by CK:

$$PCr + ADP + H^+ \leftrightarrow ATP + Cr. \tag{2}$$

The evidence for this reaction being reversible and near equilibrium in muscle cytosol, and probably in other cells containing CK, has been well discussed elsewhere [1, 25]. This conclusion is essentially based on the result that the forward and reverse fluxes of the CK reaction [Eq. (2)] are at least several times the fluxes of Eq. (1). The net reaction, which has been demonstrated to occur in a great variety of muscles during contractile activity and discussed so far in this paper, is the following:

$$PCr + (1-\alpha) H^+ \leftrightarrow Cr + Pi. \tag{3}$$

The CK reaction [Eq. (2)] can be considered near equilibrium within the framework of the stationary open metabolic state of muscle cells. By writing the apparent equilibrium constant for Eq. (2) for any particular pH, and rearranging, we obtain:

$$\frac{PCr}{Cr} = \frac{1}{Kck'} \frac{ATP}{ADP}. \tag{4}$$

Notice that the PCr/Cr ratio is directly proportional to the ATP/ADP ratio (at a fixed pH because the constant Kck' includes [H^+]). At pH 7 and 1 mM Mg^{2+}, the value of the apparent equilibrium constant (Kck) is approximately 160. The pH-independent K_{obs} for the CK reaction is 2×10^9 [23]; with that constant, [H^+] must be added to the denominator of Eq. (4). The cellular concentration of PCr, Cr and ATP are in the mM range, typically 35, 0.2, and 9 mM for unstimulated fast-twitch muscles, respectively. These values constrain the ADP concentration to be in the range of 1–9 μM for that muscle at rest [24]. It is generally observed that [ATP] remains constant over a wide range of intensities in all types of muscle during physiological intensity of contractile activity. Thus for a ten-fold decrease in the ratio of PCr/Cr, a ten-fold change in the ratio of ATP/ADP is easily accomplished without observable decrement in ATP. There is obviously an increase in cytosolic ADP, perhaps to several hundred μM, a concentration that is of the order of several times the apparent K_m for the mitochondrial respiration [4], but below the limits of detectability in typical ^{31}P-NMR experiments. Of course, small decreases in ATP must occur when ADP increases, as demanded by conservation of total adenine nucleotide. Note also that with several millimolar concentrations of ATP and several tens of micromolar concentration of ADP, the action of adenylate kinase near equilibrium constrains the [AMP] to the tens of nanomolar level for all types of muscles studied.

There is also evidence that phosphorylation potential is a thermodynamically appropriate quantity to consider in the regulation of mitochondrial respiration in the cell [11]. This is defined as:

$$PP = \frac{ATP}{ADP \ Pi}.$$

It should be emphasized that this quantity ignores the large and differential changes in intracellular pH in specific fibre types, as has been emphasized in the preceding sections. Thus whatever its descriptive usefulness on thermodynamic as well as factual grounds, the phosphorylation potential has its limitations for mechanistic interpretation. Nonetheless, there are expected and observed correlations between the magnitude of the phosphorylation potential and the PCr, Cr and Pi levels and metabolic regulation, i.e., between any metabolite function coupled to the phosphorylation potential and Eq. (3); this is certainly true for those conditions where pH is not changed.

Combining Eq. (1) with Eq. (3) we get:

$$\frac{PCr}{Cr} \cdot \frac{1}{Pi} = \frac{PP}{Kck}. \tag{5}$$

The point of all of these quantitative aspects is to demonstrate that the PCr/Cr ratio is a predictable and quantitative function of the ATP/ADP ratio in the cell and to the thermodynamic potential of ATP. Notice also that if these relationships are valid for all physiological states of the muscle, as the available evidence suggests, then it is clear that functional parameters (force, velocity of shortening, CK flux, cellular rate of O_2 consumption, etc.) will necessary be *interrelated* functions of each other. Thus, debates concerning which of several quantities (PCr/Pi ratios, PP or the true chemical potential) more accurately correlate with cellular function are moot.

IV. Energetics in smooth, skeletal and cardiac muscle

Table 1 contains quantitative analytical data on metabolite content from ^{31}P-NMR work with skeletal [24] and smooth [20] muscle, with some of the available data from rat heart. For this myocardial data the analyses were made of working hearts at low intensity of work [12] or of Langendorf-perfused non-working hearts [17]. The data for the other muscles were for "resting" conditions, i.e., without stimulation. The chemical potential available from ATP and the cytosolic free ADP content was calculated by use of the following formulae:

$$\Delta G'_{obs} = \Delta G^0 - RT \ln \frac{[ATP]}{[ADP][Pi]},$$

where $\Delta G^0 = -32$ kJ/mol at pH 7, and

$$[ADP] = \frac{[ATP][Cr]}{Kck[PCr]},$$

where Kck has the value of 160 at pH 7 [23]. Any differences in Mg^{2+} content between these muscles were ignored. Note the substantial differences between the different muscles in content of ATP (seven-fold), PCr (14-fold), total Cr (eight-fold), Pi (25-fold) and calculated ADP (100-fold).[1] Despite these large variations in contents, the free energy available from ATP for the different muscles is in the range of -50 to -60 kJ/mol; the exception appears to be fast-twitch skeletal muscle in which, because of the very low [ADP], the calculated $\Delta G'_{obs}$ is higher (-72 kJ/mol). This striking uniformly of the ATP chemical potential extends to the case of "working" muscles, i.e., muscles in which a metabolic stress due to contractile activity depletes the PCr content. To illustrate this idea, the results of a simple "back of the envelope" calculation are given in Table 2. Assume that [ATP] is constant during this stress and that the only net reaction in these steady-state stresses is PCr→Cr+Pi, and that the PCr is reduced to 0.5 and 0.2 times the reference values given in Table 1.

This simple calculation is used because there is no appropriate set of directly comparable data for all muscle types for making the desired analysis. However, it has been established in smooth, skeletal and cardiac muscle that steady states of increased metabolic stress are associated with graded decrements in PCr content without significant decrease in ATP content, so that there is a rationale for this calculation. While it must be emphasized that these are simplified calculations, some interesting if speculative features are apparent in Table 2. The increases in [ADP] and [Pi] and decrease in $\Delta G'_{obs}$ for smooth muscle are the smallest observed in all the muscles. Secondly, the high chemical potential

---

[1] The actual numerical value for [ADP] in fast-twitch skeletal muscle works out to be 0.3 µM. However, as discussed in the original paper, this very low value is the result of an extremely low measured free Cr content; analytical errors involved are such that the likely [ADP] is in the range of 0–10 µM, so the range of ADP concentration is more likely to be 6-fold.

Table 2. Calculations of ADP concentration, free energy, and Pi content

Tissue type	Resting	0.5 PCr	0.2 PCr
[ADP] (μM)			
Smooth	7	23	71
Fast-twitch	0.3	54	220
Slot-twitch	14	58	190
Cardiac	30	96	310
	14	92	330
ΔG^0 (kJ/mol)			
Smooth	−60	−56	−52
Fast-twitch	−72	−54	−50
Slow-twitch	−58	−52	−49
Cardiac	−56	−37	−33
	−51	−38	−33
Pi (mM)			
Smooth	2.4	3.6	4.4
Fast-twitch	3	21	31
Slow-twitch	10	18	23
Cardiac	0.9	8	12
	0.4	12	19

in cardiac muscle is due in significant part to the very low [Pi] content. Despite the somewhat higher [Pi] in fast-twitch muscle, the chemical potential is even higher because of the very low [ADP], a consequence of the high PCr/Cr ratio. With substantial decrease in [PCr], the [Pi] increase is 10-fold or greater and is a major in the decline in $\Delta G'_{obs}$ in these muscles. The increase is at least 10-fold over the range considered, and especially in the striated muscles in [ADP] occurs to levels well above the probable K_m for ADP-stimulated mitochondrial respiration [4]. Thus the mechanism for continued stimulation of oxygen consumption at the very large workloads modeled in the calculation (for 0.2 times total PCr content) would require a mechanism other than increase in [ADP]. Indeed, different rate-limiting mechanisms for cellular respiration over the span of low to maximal rates have been discussed for the case of isolated mitochondria [14], for perfused working rat heart [12], and for the in situ canine myocardium [2]. Based on these results and numerical modelling, it would appear that further tests of this concept of multiple mitochondrial regulatory mechanisms will be fruitful in muscles of all types.

Acknowledgements. The work described and the ideas presented are the result of the creative and energetic activities of many fellows and colleagues who have worked in this laboratory. In particular, I would like to single out T. R. Brown, K. E. Conley, M. T. Crow, P. F. Dillon, J. M. Krisanda, E. W. McFarland, R. A. Meyer, T. S. Moerland and H. L. Sweeney. This paper summarizes experimental work and analyses presented in several papers referenced herein, and expands the arguments developed in M. J. Kushmerick, J. Exp. Biol. 115:165 (1985).

References

1. Alger JR, Shulman RG (1984) NMR studies of enzymatic rates in vitro and in vivo by magnetization transfer. Q Rev Biophys 17:83–124
2. Balaban RS, Kantor HL, Katz LA, Briggs RW (1986) Relation between work and phosphate metabolite in the in vivo paced mammalian heart. Science 232:1121–1123
3. Butler TM, Siegman MJ, Mooers SU, Barsotti RJ (1983) Myosin light chain phosphorylation does not modulate cross-bridge cycling rate in mouse skeletal muscle. Science 220:1167–1169
4. Chance B, Williams GR (1956) The respiratory chain in cells and tissues. Adv Enzymol 17:65–134
5. Cooke R, Pate E (1985) The effects of ADP and phosphate on the contraction of muscle fibers. Biophys J 48:789–798
6. Crow MT, Kushmerick MJ (1982a) Chemical energtics of slow- and fast-twitch muscles of the mouse. J Gen Physiol 79:147–166
7. Crow MT, Kushmerick MJ (1982b) Phosphorylation of myosin light chains in mouse fast-twitch muscle associated with reduced actomyosin turnover rate. Science 217:835–837
8. Crow MT, Kushmerick MJ (1983) Correlated reduction of velocity of shortening and the rate of energy utilization in mouse fast-twitch muscle during a continuous tetanus. J Gen Physiol 82:703–720
9. Curtin NA, Woledge RC (1978) Energy changes and muscular contraction. Physiol Rev 58:690–761
10. Edwards RHT, Hill DK, Jones DA (1975) Metabolic changes associated with the slowing of relaxation in fatigued mouse muscle. J Physiol (Lond) 251:287–301
11. Erecinska M, Wilson DF, Nishiki K (1978) Homeostatic regulation of cellular energy metabolism: experimental characterization in vivo and fit to a model. Am J Physiol 234 (Cell Physiol 3):C82–C89
12. From AHL, Petein MA, Michurski SP, Zimmer SD, Ugurbil K (1986) ^{31}P-NMR studies of respiratory regulation in the intact myocardium. FEBS Lett 206:257–261
13. Gadian DG (1982) Nuclear magnetic resonance and its application to living systems. Clarendon Press, Oxford
14. Groen AK, Wanders RJA, Westerhoff HV, van der Meer R, Tager JM (1982) Quantification of the contribution of various steps to the control of mitochondrial respiration. J Biol Chem 257:2754–2757
15. Homsher E, Kean CJ (1978) Skeletal muscle energetics and metabolism. Ann Rev Physiol 40:93–131
16. Jacobus WE, Ingwall JS (1980) (eds) Heart creatine kinase. Williams and Wilkins, Baltimore
17. Kammermeier H, Schmidt P, Jungling E (1982) Free energy change of ATP-hydrolysis: a causal factor of early hypoxic failure of the myocardium? J Mol Cell Cardiol 14:267–277
18. Kushmerick MJ (1983) Energetics of muscle contraction. In: Handbook of physiology. Skeletal muscle. Am Physiol Soc, Bethesda, Maryland, pp 189–236
19. Kushmerick MJ (1986) Lessons for muscle energetics from ^{31}P NMR spectroscopy. In: Brautbar N (ed) Myocardial and skeletal muscle bioenergetics. Plenum Press, New York, pp 647–663
20. Kushmerick MJ, Dillon PF, Meyer RA, Brown TR, Krisanda JM, Sweeney HL (1986) ^{31}P-NMR spectroscopy, chemical analysis and free Mg^{2+} of rabbit bladder and uterine smooth muscle. J Biol Chem 261:14420–14429
21. Kushmerick MJ, Meyer RA (1985) Chemical changes in rat leg muscle by phosphorus nuclear magnetic resonance. Am J Physiol 248 (Cell Physiol 17):C542–C549
22. Kushmerick MJ, Meyer RA, Brown TR (1983) Phosphorus NMR spectroscopy of cat biceps and soleus muscles. In: Bicher HI, Bruley DF (eds) Oxygen transport to tissue, vol 4. Plenum Publishing Corp, New York, pp 303–325
23. Lawson JWR, Veech RL (1979) Effects of pH and free Mg^{2+} on the Keq of the creatine kinase reaction and other phosphate hydrolyses and phosphate transfer reactions. J Biol Chem 254:6528–6537
24. Meyer RA, Brown TR, Kushmerick MJ (1985) Phosphorus magnetic resonance of fast- and slow-twitch muscle. Am J Physiol 248 (Cell Physiol 17):C279–C287

25. Meyer RA, Kushmerick MJ, Brown TR (1982) Application of ^{31}P-NMR spectroscopy to the study of striated muscle metabolism. Am J Physiol 242 (Cell Physiol 11): C1–C11
26. Meyer RA, Sweeney HL, Kushmerick MJ (1984) A simple analysis of the "phosphocreatine shuttle". Am J Physiol 246 (Cell Physiol 5): C365–C377
27. Moon RB, Richards JH (1973) Determination of intracellular pH by ^{31}P magnetic resonance. J Biol Chem 248:7276–7278
28. Reiser PJ, Moss RL, Giulian GG, Greaser ML (1985) Shortening velocity in single fibers from adult rabbit soleus muscles is correlated with myosin heavy chain composition. J Biol Chem 260:9077–9080
29. Sweeney HL, Kushmerick MJ (1985) Myosin phosphorylation in permeabilized rabbit psoas fibers. Am J Physiol 249 (Cell Physiol 18): C542–C549
30. Sweeney HL, Kushmerick MJ, Mabuchi K, Gergely J, Sreter FA (1986) Velocity of shortening and myosin isoenzymes in two types of rabbit fast-twitch skeletal muscle fibers. Am J Physiol 251 (Cell Physiol 20): C431–C434

Author's address:

M. J. Kushmerick, M.D., Ph.D. NMR Division, Department of Radiology, Brigham & Women's Hospital, Boston, Mass. (U.S.A.)

High energy phosphate of the myocardium: Concentration versus free energy change

H. Kammermeier

Abteilung Physiologie, RWTH Aachen, F.R.G.

Summary

About 80% of the energy derived from the oxidation of substrates is stored in the form of ATP in sufficiently oxygenated hearts. This is reflected by a free energy and chemical potential respectively, of ATP of about 60 kJ/mol. This energy level does not need to be correlated with tissue ATP content and can also be reached with markedly lower amount of tissue ATP. With graded hypoxia, this energy level drops to 50 to 40 kJ/mol without a corresponding reduction in tissue ATP, but with a concomitant fall in peak systolic pressure.

Various energy-dependent processes may be responsible for this impairment of cardiac performance. According to experiments with reduced energy demand of the sarcolemmal ion pumping processes and inotropic interventions, the reduced chemical potential of ATP still seems to be sufficiently above that required for the sarcolemmal ion pumping and for the chemo-mechanical energy transformation of the actomyosin system. In contrast, the reduced chemical potential of ATP seems to be no longer sufficient to meet the high level required for normal Ca^{++} accumulation in the sarcoplasmic reticulum.

It has been a well established fact for some decades that chemical energy derived from oxidative phosphorylation and glycolysis is transformed and "stored" in the form of high energy phosphates, i.e. ATP and phosphocreatine. Commonly, the amount of these compounds, i.e. concentration times volume of distribution, is considered the energy reserve of the heart. Thus, for example, 20–30 µmol high energy phosphates per g of heart can be calculated, which should be sufficient for regular performance for 60–90 s. In fact, however, ischaemic failure occurs within 10–20 s when only about 20% of ATP has been consumed.

There are various hypotheses to explain this apparent discrepancy [4], but almost none of them consider that this usual calculation does not respect basic thermodynamic aspects, i.e. the differences between equilibrium and non-equilibrium reactions and the inverse relationship between efficiency and energy flux.

Chemical equilibrium conditions are comparable with the state of connected reservoirs with the same level of fluid on either side. Exchange takes place without energy loss or dissipation, i.e. with 100% efficiency, but energy cannot be drawn from this system. To generate flow or to draw energy from a system a gradient must be created. However, due to energy dissipation, i.e. entropy production, in the form of heat, the efficiency declines accordingly.

The energy metabolism of the cell can be considered to begin with the oxidation of glucose which yields about 2 850 kJ/mol (Fig. 1). Since, mediated by glycolysis and the respiratory chain, 38 mol of ATP are phosphorylated from ADP, 75 kJ are available per mol of ATP synthesized. However, in the cytosol the free energy available from ATP hy-

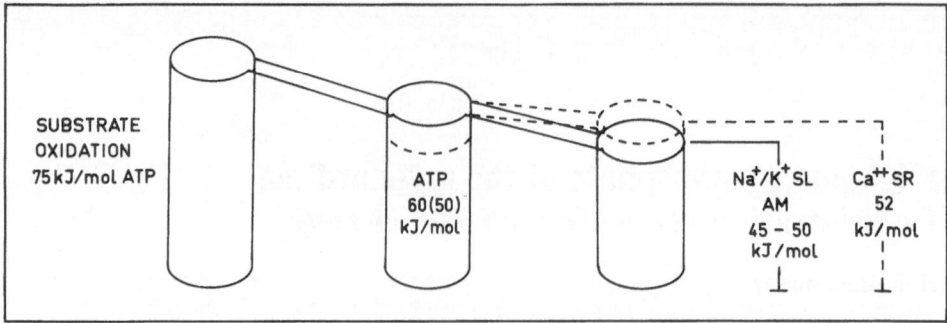

Fig. 1. Energy levels (free energy change, chemical potential) per mol of ATP, by substrate degradation and oxidation (1st column, reservoir); in the cytosol under normoxic and hypoxic (dotted) conditions (2nd reservoir); of ATP-dependent processes [Na/K-transport, actomyosin (AM), Ca^{++}-transport SR (dotted)].

drolysis (at least in heart muscle) was estimated to be about 60 kJ/mol by various investigators. This is about 80% of the afore-mentioned value and somewhat higher than values given in recent textbooks of biochemistry. Thus, apparently 80% of the energy of glucose oxidation is conserved in the form of high energy phosphates. This very high efficiency can be partly explained by the fact that two of the three coupling sites of the respiratory chain are at equilibrium, as far as is currently known.

The energy available from ATP hydrolysis is calculated by the equation in [4], where different terms are used.

$$dG/d\xi = \Delta G^0_{obs} + R \cdot T \cdot \ln \frac{[ADP] \cdot [Pi]}{[ATP]} \, .$$

The formerly used term ΔG – Gibb's energy – is now more correctly termed $dG/d\xi$ nearing free energy change or affinity [1], and the chemical potential of ATP, respectively. The ΔG observable is a constant which amounts to about -30 kJ/mol of ATP for physiological intercellular conditions [4]. The second term also amounts to about -30 kJ, mainly due to the fact that free cytosolic concentration of ADP with 10–50 μM is 2–3 orders of magnitude lower than that of ATP. The free cytosolic concentration of ADP cannot be estimated directly, since it amounts to only about 10% of tissue content and has to be calculated from the CK-equilibrium equation.

$$K_{CK} = \frac{[Cr] \cdot [ATP]}{[PC] \cdot [ADP] \cdot [H^+]} = 10^9 \, [mol^{-1}] \, .$$

Due to the low ADP-concentration, the mass action ratio of ATP-hydrolysis of equation [4] amounts to about 10^{-5}. Such a fraction, far from units, is very sensible to changes in the smallest factor, i.e. ADP. Since an increase in ADP and inorganic P results from ATP hydrolysis, in a closed system a small consumption of ATP, about 10%, would reduce the fraction by one order of magnitude and 20% consumption of ATP would reduce the level of the chemical potential of ATP to a critical extent, as will be shown later on. Nevertheless, this problem is partly compensated in muscle by the phosphocreatine-creatinekinase system.

What can be considered a critical level of chemical potential of ATP, depends on the level demanded by ATP-dependent processes. These are the contractile system and the various ion-pumping processes for sodium, potassium and Ca^{++} transport.

Table 1. Chemical potential of ATP required for ion-transport processes

	$\Delta\Psi$ mV	Ci/Ce	ΔG equilibrium conditions kJ/mol ATP	Minimum efficiency
Na$^+$/K$^+$ATPase	90	40:1 1:10	46	>77%
Ca^{++} ATPase Sarcoplasm. Ret. (3)	–	5:10^{-4}	52	>87%
Ca^{++}/H$^+$-ATPase (4) Sarcolemma	90	1:10^4 2:1	27	>45%

$$\Delta G = R \cdot T \cdot \ln\frac{Ci}{Ce} + Z \cdot F \cdot \Delta\Psi. \quad \Delta\Psi = \text{Membrane potential. Ci/Ce: concentration inside/outside.}$$

The energy level required for ion transport can be estimated from ion concentrations in the different compartments, i.e. extracellular, intracellular and inside the sarcoplasmic reticulum, from published data (Table 1). According to the equation in Table 1, the energy demand depends on the ratios of ion concentrations and on the membrane potential. From the known stoichiometry of the respective transport ATPases, the energy demand for the equilibrium condition and physiological membrane potential amounts to 46 kJ/mol of ATP for sarcolemmal sodium/potassium transport and to 52 kJ/mol of ATP for Ca^{++} transport of the sarcoplasmic reticulum [2]. This level of chemical potential of ATP seems of special interest, respecting that SR contributes up to about 30% to energy turnover of heart muscle [2]. The sarcolemmal Ca^{++}-ATPase due to H$^+$-ion coupling and the stoichiometry of 1:1 [5] exhibits a lower demand of about 27 kJ/mol of ATP. These values are representative of equilibrium conditions, i.e. when no net transport takes place. Consequently, the energy level required for net transfer is still higher.

The 52 kJ/mol of ATP required under equilibrium conditions for the Ca^{++} transport of the sarcoplasmic reticulum, corresponds to 87% of the 60 kJ/mol of ATP available. This is rather close to the 100% characterizing equilibrium conditions and consequently, the Ca^{++} loading capacity of the SR should change with the free energy level of ATP and the SR Ca^{++}-ATPase reaction should also be reversed under certain conditions, as already demonstrated in vitro [2]. The energy level required for the sarcolemmal sodium/potassium ATPase of 46 kJ/mol of ATP (76% of the level available) is also still high, and the transfer can be expected to also depend to a certain extent on the energy level available. In contrast, sarcolemmal Ca^{++}-ATPase (according to the values mentioned above) demands an energy level rather below that available in oxygenated hearts.

The energy level required for the contractile system of the myocardium, i.e. the actomyosin ATPase, cannot be calculated directly in a similar way. However, from the various estimations of efficiencies, the energy required can be expected to be about 45–50 kJ/mol of ATP.

These values, namely the chemical potential of ATP of about 60 kJ/mol and the levels required of 45–55 kJ/mol, necessarily mean that reduction in the chemical potential of ATP (independent of concentration) must be accompanied:
– *either* by reduction in mechanical performance and reduction in concentration gradients for the ions considered
– *or* by increase in efficiencies of the respective processes.

Since essential increase in efficiencies appears not to be possible with very high efficiencies, the former, i.e. dependence on free energy level, has to be expected, at least for the ion transport processes with efficiencies of about 80%.

Table 2. Cardiac ATP-content and performance following 30 min normothermic ischaemia and reoxygenation $\bar{x}, \pm$ SEM

	Control (9)	Postschaemia recovered (10)
ATP, μMol/g$_{ww}$	4.3 ± 0.06	1.9 ± 0.1
Maximum tolerated resistance mmHg	128 ± 6	128 ± 7

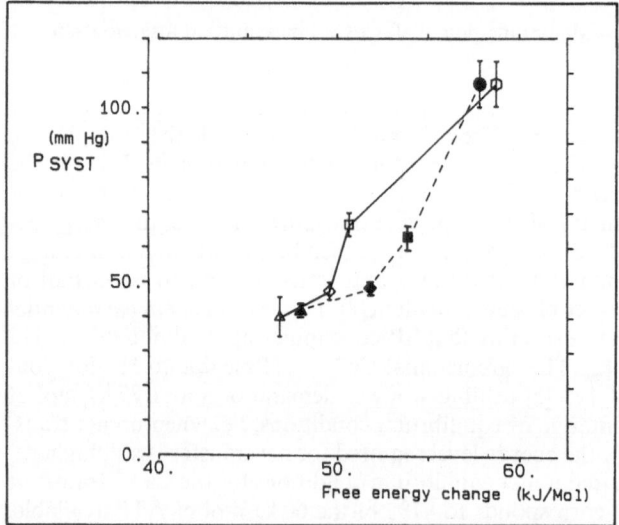

Fig. 2. Free energy change vs peak left ventricular systolic pressure ——— graded PO$_2$ reduction from 736* to 59 mmHg ($\bar{x}, \pm$ SEM, n = 6–8) graded re-oxygenation from 59 to 760* mmHg PO$_2$ (n = 3–7). (– – –curve shifted P$_{syst}$ by +15 mmHg and dG/dξ by +0.9 kJ/mol) to achieve coincidence of controls of two series from 1983 and 1985 * (different slight hyperbaric gassing).

We performed various experiments to find out whether hypoxic failure of the heart depends on the chemical potential of ATP hydrolysis available. With respect to the question of high energy phosphate concentration versus free energy change of ATP, our earlier findings demonstrated that the actual concentration of ATP seems not to be a major determinant of cardiac performance. In guinea pig heart lung preparations, myocardial ATP-content can be reduced by hypoxia and re-oxygenation by more than 50%, without reduction in maximum cardiac performance (Table 2) [3]. The other data from these experiments on restored phosphocreatine levels [3] indicated, however, that in contrast to ATP-content, the chemical potential of ATP was restored.

Two series of experiments were concerned with the relationship between the level of chemical potential of ATP and cardiac performance, in isolated isometrically-working rat hearts. In these experiments, different levels of chemical potential of ATP were induced in perfused hearts by graded hypoxia, ranging from fully oxygenated perfusate with a pO$_2$ of about 700 mmHg, to perfusates equilibrated with 8% oxygen corresponding to

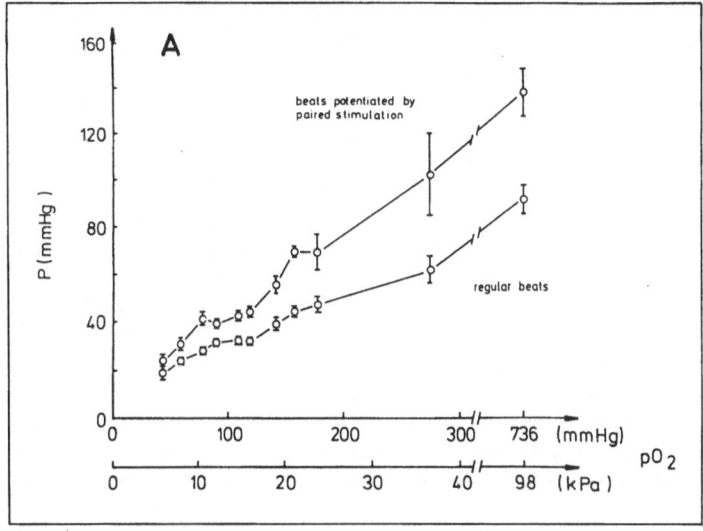

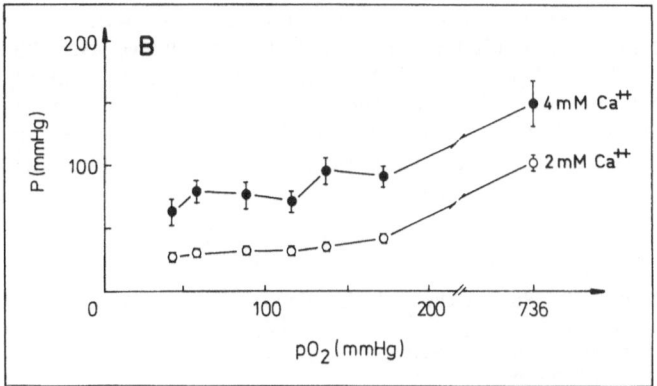

Fig. 3. Peak intraventricular pressure of hearts at different states of hypoxia (3 min perfusion): a) Influence of paired stimulation. Regular beats, pacing with 270 min^{-1}; stimulation interval, 80 ms. b) Influence of short term increase in extracellular Ca^{++} (10 s).

a pO_2 of about 50 mmHg. Figure 2 shows the free energy as chemical potential of ATP-hydrolysis and peak left ventricular systolic pressure (left intraventricular balloon, paced with 270 min). The hearts were analysed by HPLC [4] after freeze-stop following 3 min of hypoxic perfusion.

Cardiac performance, i.e. peak systolic pressure, drops steeply if the free energy change of ATP is reduced by various degrees of induced graded hypoxia. The same relationship is valid if after a 10 min period of anoxic perfusion, stepwise recovery is brought about by graded re-oxygenation (Fig. 2).

This apparent dependence of peak systolic pressure on chemical potential of ATP, however, can be caused by different mechanisms. As previously pointed out, the energy requirement of ion transport processes, as well as that of the actomyosin ATPase, can be the essential factors. To eliminate the influence of sodium/potassium ATPase, experiments with increased extracellular potassium concentration were carried out. Under this procedure, the required energy level of sodium potassium ATPase is reduced to about 39 kJ/mol of ATP, due to partial depolarization.

The relationship between free energy change of ATP and performance, was not changed essentially by 13.5 mM extracellular potassium and was almost identical with that of Fig. 2 and therefore is not demonstrated by another figure. Thus, the Na/K-ATPase seems not to be a primary determinant of the contractile failure.

A second group of experiments was concerned with inotropic interventions during hypoxic failure, using paired stimulation and increased extracellular Ca^{++} concentrations. Paired stimulation enhances peak systolic pressure by approximately the same factor in oxygenated hearts and at each level of hypoxia (Fig. 3a). With short term increase in extracellular Ca^{++} the peak systolic pressure can also be increased to the lowest oxygen tension (Fig. 3b). The increase is even more pronounced than with paired stimulation. Apparently, an inotropic response can be elicited, though the systolic force of contraction is reduced by hypoxia. This indicates that the chemical potential of ATP hydrolysis is still in excess for the respective force of contraction, developed by the contractile system. Accordingly, there seems to be a certain reserve which can be mobilized by improvement in EC-coupling.

Thus, by exclusion of other processes, those of EC-coupling appear to be the primary determinant for hypoxic failure. As pointed out before, the energy level required for Ca^{++} pumping of the sarcoplasmic reticulum is very close to the energy level available. Changes in the level of the chemical potential of ATP should therefore be primarily reflected in a diminished Ca^{++} pumping capacity of the sarcoplasmic reticulum. This, in turn, should reduce the amount of Ca^{++} which can be released and finally affect the activation of the contractile system. All our observations are in accordance with this idea.

References

1. Dawson MJ, Gadian DG, Wilkie DR (1980) Mechanical relaxation rate and metabolism studied in fatiguing muscle by phosphorus nuclear magnetic resonance. J Physiol 299:465–484
2. Hasselbach W, Oetliker H (1983) Energetics and electrogeneity of the sarcoplasmic reticulum calcium pump. Annu Rev Physiol 45:325–339
3. Kammermeier H (1964) Verhalten von Adenin-Nukleotiden and Kreatinphosphat im Herzmuskel bei funktioneller Erholung nach länger dauernder Asphyxie. Verh Dtsch Ges Kreislaufforschung 30:206–211
4. Kammermeier H, Schmidt P, Jüngling E (1982) Free energy change of ATP-hydrolysis: a causal factor of early hypoxic failure of the myocardium? J Mol Cell Cardiol 14:267–277
5. Niggli V, Sigel E, Carafoli E (1982) The purified calcium pump of human erythrocyte membranes catalyses an electroneutral Ca-Proton exchange in reconstituted liposomal systems. J Biol Chem 257:2350–2356

Author's address:

Prof. H. Kammermeier, Abteilung Physiologie der RWTH Aachen, Pauwelstraße, 5100 Aachen, F.R.G.

Cardiac basal and activation metabolism

D. S. Loiselle

Department of Physiology, School of Medicine, University of Auckland,
Private Bag, Auckland, New Zealand

Summary

Cardiac basal metabolism is the rate of energy expenditure of the quiescent myocardium. It is species dependent and increases with pre-load. It has small contributions from membrane-bound cation pumps. The contribution of protein metabolism remains open to question. Calculations show that mitochondrial proton pumping may account for a large fraction of the cardiac basal metabolism. Nevertheless this component remains essentially ill-understood. Cardiac activation metabolism is the supra-basal rate of energy expenditure associated with those processes that activate contraction. In isolated muscle preparations it is typically measured as the rate of heat production or oxygen consumption of a muscle, pre-shortened to a length where active force production is negligible, although it is also estimated by pharmacological intervention. In whole-heart studies it is indexed by the supra-basal rate of oxygen consumption of the empty, beating but non-working heart. Activation metabolism underwrites electrical excitation (the ECG) and excitation-contraction coupling (the cycling of calcium ions). It is increased by agents that increase contractility; it probably increases with pre-load, via the phenomenon of length-dependent activation. The basal and activation components each account for one-quarter to one-third of the total energy expenditure of the heart under normal conditions.

Introduction

With each beat of the heart a complex sequence of electrical and mechanical events occurs. By the end of one cardiac cycle a bewildering number of biochemical reactions underwriting these events have run their course. During the systolic phase the heart has expended a quantity of energy against external mechanical loads and internal mechanical and electrochemical loads. Being a machine of less than perfect efficiency, it has also wasted energy as heat. During the diastolic phase a quantity of energy in excess of that spent during systole has been restored by the (largely oxidative) processes known as recovery metabolism. Viewed energetically then, the cycle is much simplified: each beat of the heart witnesses a measurable quantity of energy expended and restored. It is the task of cardiac energetics to assign, quantitatively, the appropriate fraction of the total cardiac energy expenditure to each of the underlying electrical, mechanical and biochemical events.

This task is simplified by following the conceptual scheme of Gibbs et al. [5, 14, 18, 19, 26]. The total energy expenditure is divided into basal and active components. The basal component is associated with those metabolic processes that oppose the dissipative forces continuously operating within the cell, thereby achieving structural, ionic and electrical homeostasis. The active component is associated with electrical excitation, the

events of excitation-contraction coupling, and mechanical contraction (muscle shortening, the development of pressure and the performance of pressure-volume work). Energy
expenditure associated with the first two of these active processes is known (somewhat
confusingly) as the activation energy. This article concerns itself only with the basal and
activation components of total cardiac energy expenditure – i.e., with those components
not *directly* associated with the performance of external pressure-volume work.

Basal metabolism

The basal metabolism can only be studied by rendering the heart inactive. If a whole heart
is under study, then it is perfused with a cardioplegic solution. If an isolated muscle preparation is under study, then electrical stimulation is withheld. In either case the tissue is
quiescent and its rate of heat production or oxygen consumption defines its basal metabolic rate. We are not yet able to assign fractions of the basal energy expenditure to separate metabolic processes. But any complete description will have to accomodate the following facts.

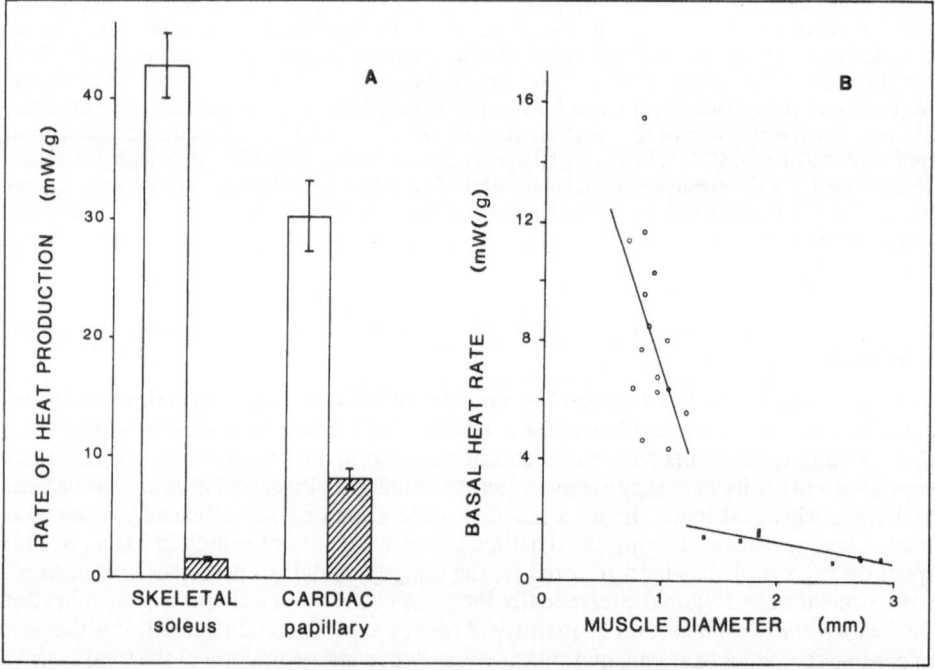

Fig. 1. A) Rate of energy (heat) production of rat soleus (n = 5) and papillary (n = 15) muscles measured myothermally at 27 °C. Open bars: active heat rate of soleus muscles in a 4 s tetanus at a stimulation frequency of 100 Hz; and papillary muscles, pre-incubated in 10 mM Ca^{++} and 10 mM caffeine, measured in a 5 s tetanus at a stimulation frequency of 4 to 10 Hz [24]. Hatched bars: basal
heat rate. B) Rate of basal heat production, as a function of muscle diameter, of papillary (circles)
and soleus (squares) muscle of the rat. Same data as in A except for inclusion of 4 soleus muscles
of large diameter (2.5 to 2.9 mm).

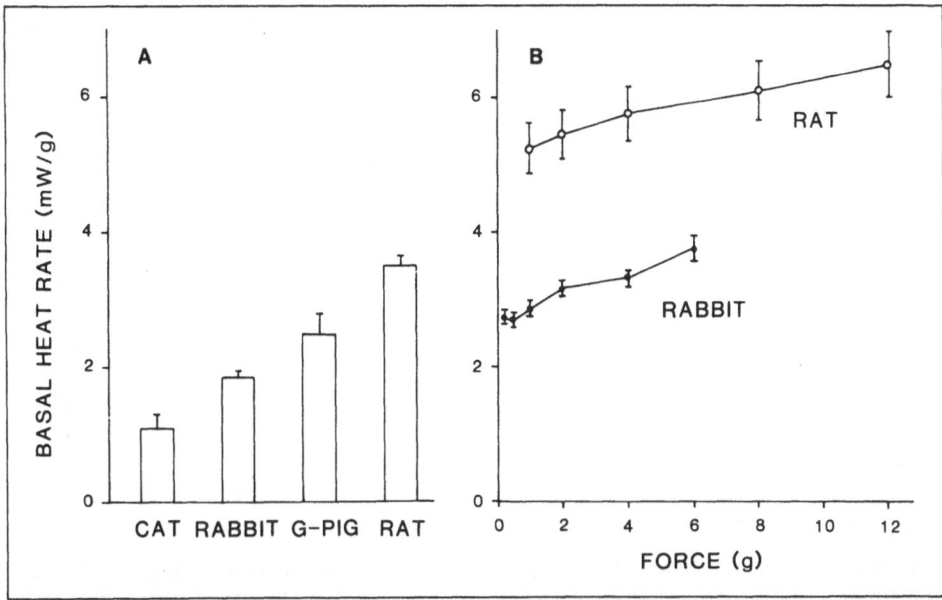

Fig. 2. Rate of resting heat production of papillary muscles. A) Species differences (cat, guinea-pig and rat data from [24]; rabbit data from [6]; 1 g pre-load at 27 °C in 95% O_2 with 10 mM glucose as substrate. B) Effect of pre-load on basal heat production of rat (open circles) and rabbit (closed circles) muscles; at 20 °C to 30 °C in 95% O_2 with 10 mM pyruvate as substrate. Rabbit data from [40]; rat data from [36].

Magnitude

a. *Comparison with Skeletal Muscle.* It is instructive to compare the magnitude of basal metabolism between the two moieties of striated muscle. As shown in Fig. 1 A, the basal metabolic rate is some 5 times higher in cardiac (papillary) than in skeletal (soleus) muscle within the same species (rat). The contrast is even greater if the basal rate is calculated as a fraction of the maximum rate of active heat production during a tetanus. Expressed in this way, the basal metabolism is only 2–3% of the active metabolism in striated muscle but is 25–30% of the active metabolism in cardiac muscle (see Fig. 1). The data presented in Fig. 1 arose from myothermic studies. Hence it is reassuring that Suga et al. [47] have recently reported that in the cross-circulated, blood-perfused dog heart the rate of oxygen consumption of the K^+-arrested heart is 28% of that of the unloaded, beating heart.

b. *Within species.* There is an inverse relation between muscle diameter and the basal heat rate. This has been previously shown for rabbit papillary muscle [40] and is also true for rat papillary as well as rat soleus muscle (Fig. 1 B). This inverse relationship is not an artefact of the myothermic method as it occurs when basal metabolism is indexed as the polarographically-measured rate of oxygen consumption as well [27]. The cause of this relationship is not understood but seems to be unrelated to the degree of oxygenation of the muscle [38].

c. *Across species.* Cardiac basal metabolism varies inversely with body weight across species. The data of Fig. 2A show a 2 to 3-fold increase in rate of cardiac basal heat production over the range from cat and rabbit through guinea-pig to the rat. The data of Fig. 2A show that the species difference in cardiac basal metabolism persists across different pre-loads. The difference in absolute magnitude between the values in Fig. 2 A and 2 B may be attributed to the difference in metabolic substrates employed (glucose and pyruvate, respectively).

Pre-load Dependency

When the pre-load or passive stress applied to a quiescent papillary muscle is increased, its rate of resting heat production also increases. This was first shown by Gibbs et al. [26] who also showed the magnitude of the effect to be quite variable from muscle to muscle. The effect occurs in all species examined (see Fig. 2 B) and appears to be the equivalent of the stretch effect in skeletal muscle studied extensively by Feng over fifty years ago [13]. The stretch-effect has not yet been demonstrated in the whole heart – in part because even small increases in diastolic volume can greatly disturb the pattern and extent of coronary flow (Loiselle, unpublished observations).

Temperature Insensitivity

The cardiac basal metabolism has a Q_{10} value of only about 1.3 to 1.4 which is unusually low for most biological processes. This relative insensitivity to temperature has been demonstrated for the rate of resting heat production of rabbit [40] and rat [36] papillary muscles resting on a thermopile, for the rate of oxygen consumption (measured polarographically) of thin slices of rat ventricular tissue [37], and for the rate of oxygen consumption of isolated, Langendorff-perfused, K^+-arrested dog [3], rabbit [37] and guinea-pig [32, 35] whole hearts.

Substrate Dependency

The metabolic rate of the quiescent myocardium, even for a given species under conditions of fixed temperature and fixed pre-load, is not a constant. It varies with metabolic substrate as already hinted in Fig. 2 B. Chapman and Gibbs [6] showed that pyruvate increased the basal heat rate of rabbit papillary muscle by nearly 50% with respect to glucose. Lactate and acetate had intermediate effects (Fig. 3 A). Gibbs & Kotsanas [23] have confirmed this result in the rabbit heart arrested by perfusion with low Ca^{++}; a mixture of glutamate, pyruvate and fumarate caused a 30% increase in basal oxygen consumption despite a 40% decrease in coronary flow.

Burns and Reddy [4] have shown that a mixture of amino acids causes a 2.4-fold increase in the rate of oxygen consumption of isolated cardiac myocytes. Addition of amino acids, in the form of Aminosol [a commercially available mixture of 19 amino acids (Vitrum, Sweden)], had no effect on the rate of heat production of rat papillary muscles even when they were subjected to increased passive stress [36]. However, presentation of Aminosol to the saline-perfused guinea-pig heart causes a modest increase in rate of oxygen consumption (Fig. 3 B). Data came from 8 guinea pig hearts, Langendorff perfused with saline at constant pressure (50–55 torr) at 37 °C with 10 mM glucose as substrate. (Cardiac arrest was achieved by elevating $[K^+]$ to 20 mM.) The increase in oxygen consumption, which averaged 20% but was somewhat lower in the absence and higher in the presence of insulin, occurred despite an increase in coronary vascular resistance. It is not known whether this stimulus to oxygen consumption reflects an increased rate of protein

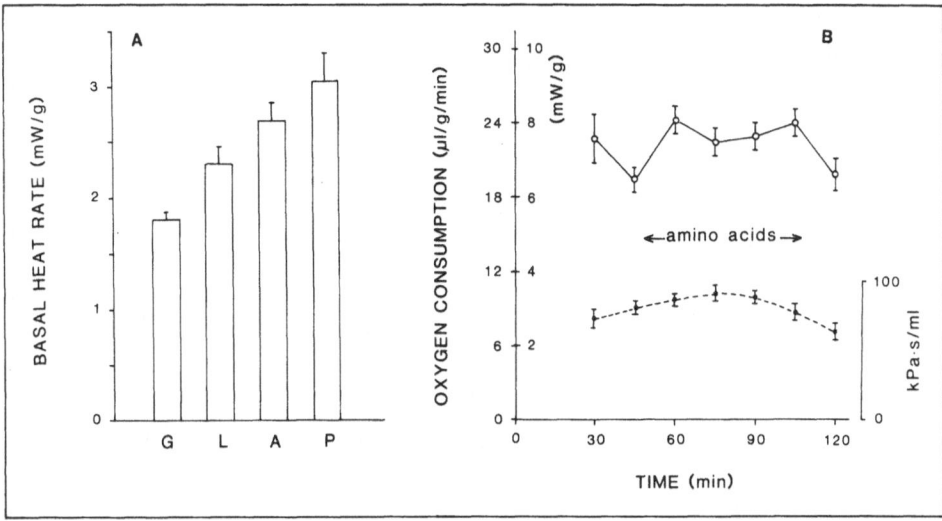

Fig. 3. Effect of substrate on cardiac basal metabolism. A) Data [6] for 8 rabbit papillary muscles incubated at 21 °C in glucose (G), lactate (L), acetate (A) and pyruvate (P). B) Effect of Aminosol (amino acids) on rate of oxygen consumption (open circles with solid lines; left-hand scale shows units of both oxygen consumption and energy flux assuming 20 kJ per l of O_2) and coronary vascular resistance (ratio of perfusion pressure to coronary flow – filled circles and broken lines and right-hand scale) as a function of time following cardiectomy. Note difference in vertical scale (mW/g) between A and B.

synthesis or an increased rate of oxidative metabolism via the tricarboxylic acid cycle akin to the Chapman & Gibbs [6] results presented above (see Fig. 3 A).

Interpretation

What interpretation can be placed on the distinguishing features of cardiac basal metabolism listed above? It may be helpful to consider the various processes that are imagined to continue, perhaps at a reduced rate, in an inactive myocardial cell.

Role of Membrane-Bound Ion Pumps. In the quiescent state the cardiac muscle cell must continue to experience passive leakage of Na^+ and K^+ across the sarcolemma. It may well experience leakage of Ca^{++} into the cytoplasm from both extracellular and sarcoplasmic reticular sources. Thus some contribution to the basal energy expenditure may arise from both the Na^+-K^+-ATPase and the various Ca^{++}-ATPases.

Gibbs & Chapman [19], using published values of sodium flux, estimated that Na^+-K^+-ATPase activity would account for only about 10% of the observed basal metabolism. Using more recent data for sodium flux in cultured rat heart cells, Gibbs [15] has suggested that the sodium pump might account for as much as 25% of the resting cardiac metabolism. Experimental support comes from the work of Lochner & Dudziak [33] who showed that sodium depletion dramatically reduced (by about 2/3) the rate of oxygen consumption of the K^+-arrested rat heart. Likewise elevation of the potassium ion concen-

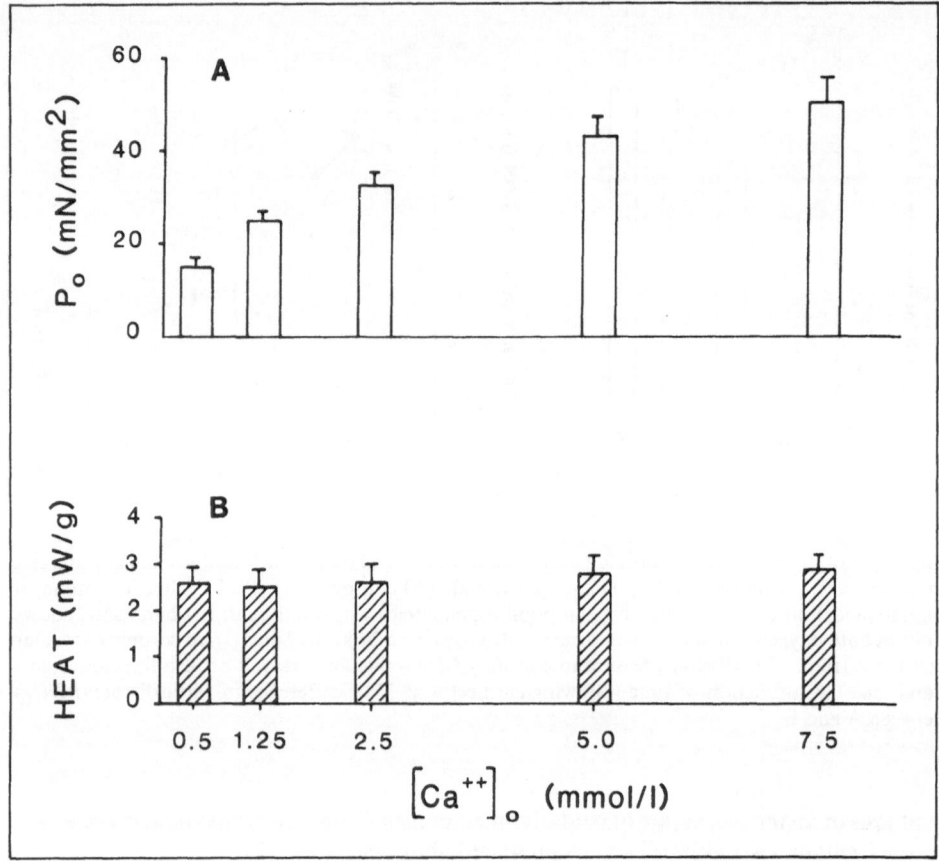

Fig. 4. Peak isometric twitch force (P_o) (A) and resting heat rate (B) as a function of external Ca^{++} concentration for rabbit papillary muscles (n = 10). Data of [51].

tration from 5 to 20 mM in the 17 °C arrested guinea-pig heart causes a small (about 20%) but statistically significant elevation of oxygen consumption (Loiselle, unpublished data). On the other hand, Coleman [9] could measure no effect upon the rate of resting heat production of cat papillary muscles when the sodium pump was poisoned by acetyl-strophanthidin. Thus the contribution of the resting rate of Na^+-K^+-ATPase activity remains imperfectly characterized.

No such doubt attends the resting contribution of calcium pumps. In a study performed in collaboration with Dr. Igor Wendt, rabbit papillary muscles were incubated in 10 mM glucose at 27 °C, in various saline solutions whose calcium concentrations ranged from 0.5 to 7.5 mM. The results are shown in Fig. 4. The rate of resting heat production remained constant, independent of Ca^{++}, despite a 2.5-fold increase in peak isometric force production [51]. Suga et al. [47] have reported a similar result in the cross-circulated, blood-perfused, K^+ arrested dog heart. The null effect of external calcium concentration has been confirmed in studies in which papillary muscles have been tetanized in the presence of 10 mM Ca^{++} and caffeine. Despite profound effects on mechanical performance, the tetanizing medium does not consistently increase the rate of basal heat production [8, 24, 41]. Even the catecholamines, which achieve their inotropic effect at least in part by an increase in intracellular calcium, have no effect on either the basal rate of heat production of isolated papillary muscles [8] or the rate of oxygen consumption of the arrested dog heart [47, but see 30].

Role of Protein Metabolism. Variation of the rate of protein synthesis both within and across species [12] has provided circumstantial evidence that protein metabolism (synthesis and degradation) may account for the intra- and inter-species variations shown in Figs. 1 A and 2 A. Gibbs [15] has calculated, on the basis of these data, that protein synthesis could account for about 1 mW/g of the basal cardiac heat rate observed in the rat. Kira et al. [29] have since shown, however, that during cardiac arrest an increased rate of protein synthesis can occur with no effect on basal oxygen consumption. Schreiber et al. [44] report data suggesting that protein synthesis makes, at best, but a small contribution to cardiac basal metabolism.

Mitochondrial Proton Pumping. Given the continuing inability to explain the cardiac basal metabolism (and, in particular, its inherently high rate) perhaps there is room in the literature for yet another suggestion. A significant component of the basal cardiac metabolism may reflect the mitochondrial response to the continuous passive influx of hydrogen ions. At present, the evidence is entirely circumstantial. (i) The five to eight-fold higher rate of basal metabolism in cardiac than in skeletal muscle (Fig. 1) reflects a similar difference in mitochondrial content. (ii) The inverse relation between body size and myocardial mitochondrial volume [28] is reflected in the inverse relation between heart (or body) size and basal metabolic rate (Fig. 2). (iii) The low Q_{10} of basal cardiac metabolism relfects the relative temperature insensitivity of diffusive processes (passive leakage of protons).

The calculation, which is based on Mitchell's chemiosmotic hypothesis [42], involves the following assumptions[1]:

(1) Hydrogen ions continuously leak across the inner mitochondrial membrane into the matrix. Leakage is through an electrical gradient of 150 mV [31] through a membrane of resistance 2 Mohm $\cdot$ cm^2. The basal current is thus 7.5×10^{-8} amp/cm^2 or 7.8×10^{-13} mol H^+/cm^2/s.

(2) In rat left ventricular cells, the ratio of mitochondrial (inner membrane plus cristae) surface area to mitochondrial volume is 57/μm and the ratio of mitochondrial volume to cell volume is 0.351 [45]. Myocardial specific density is assumed to be 1 g/cm^3.

(3) The P:O ratio of oxidative phosphorylation is 3; the P:H^+ ratio of electron transport is 2. Hence 1 mol of O_2 is consumed per 12 mol H^+ translocated.

(4) The energetic equivalent of oxygen is 20 kJ/l.

The product of these numeric values (and the standard molar gas volume of 22.4 l) yields an estimated cost of basal mitochondrial proton pumping of 5.8 mW/g. Clearly this estimated value could account for a large fraction of the observed basal metabolic rate, even in the rat. The hypothesis awaits experimental testing. In the meantime, it is probably honest to conclude that the cardiac basal metabolism remains essentially unexplained.

---

[1] I am grateful to Dr. Brian Chapman for assistance in formulating this model.

Activation metabolism

The activation component of active metabolism can be defined in two somewhat different ways, each of which arises quite naturally from its own experimental model. Myothermic studies measure the energy expenditure of isolated cardiac muscle preparations – either papillary muscles or trabecula carnae. The basal component is nulled electronically. The activation heat is then the heat liberated, following stimulation, under the special circumstance when no measurable force is generated. This is done either by pre-shortening the muscle to a length where active force production is zero (for reviews see 14, 18, 19) or by the use of some agent that inhibits active force production – typically mannitol [1]. In either case the suprabasal heat production, in response to excitation but in the absence of contraction, is attributed primarily to the events associated with the cycling of calcium ion. (The metabolic cost of the electrocardiogram is usually considered to be negligible.) To be more specific, the net cost is attributed to the hydrolysis of ATP by the sarcolemmal and sarcoplasmic retricular Ca^{++}-pumps. (The thermal consequences of the release, diffusion and binding of calcium are energetically neutral over a complete cycle.) In the Gibbs scheme, the activation heat is given by the intercept of the enthalpy-stress relation (Fig. 5 A).

In whole heart studies it is usual to measure the oxygen consumption, by the Fick method, although the technically more difficult method of whole-heart calorimetry has also been applied [10, 11] (Fig. 5 D). In either case it is natural to define the activation metabolism as the supra-basal oxygen consumption (or heat production) of the heart in the absence of external pressure-volume work. In this case the heart is still visibly con-

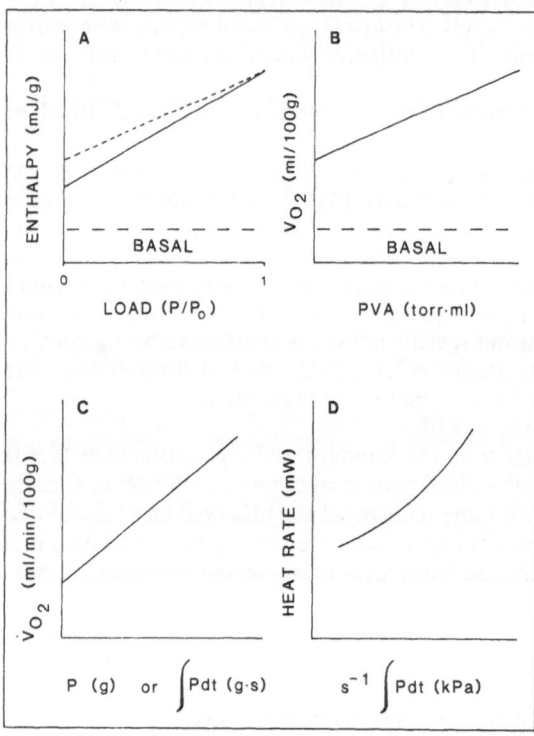

Fig. 5. Stylized diagrams of energy expenditure as a function of mechanical performance. A) Per beat heat production of isometric twitch contractions of isolated papillary muscle as a function of relative load (P/P₀) varied by pre-shortening the muscle (solid line) or by a quick-release technique. From Gibbs et al. [14, 18, 19] (see also Gibbs, this supplement).
B) Per beat oxygen consumption of the cross-circulated, blood-perfused dog heart as a function of pressure-volume area (PVA). From Suga et al. [46–49]. C) Rate of oxygen consumption as a function of wall force (P) or force-time integral, in the isolated, cross-perfused dog heart [50].
D) Data of Coulson [10, 11]. Heat production of isovolumetric contractions in the isolated, Langendorff-perfused rabbit heart as a function of the product of heart rate and the pressure-time integral per beat.

tracting but the left ventricle is empty. In the scheme of Suga et al. ([46, 47, 48], for a brief review see [49]) this is given by the supra-basal intercept of the oxygen consumption-pressure volume area (PVA) relation (Fig. 5B), where PVA is the area between the end-systolic and end-diastolic pressure-volume curves and the systolic segments of the instantaneous pressure-volume loop. It has also been presented as the intercept of either the oxygen consumption vs force, or oxygen consumption vs force-time integral relation (calculated from a simple spherical model of the left ventricle [50]), where the wall force has been deduced from the developed pressure (Fig. 5C). These various formulations are presented schematically in Fig. 5, in all of which the supra-basal intercept of the energy-mechanical performance relation yields the energy cost of unloaded contractions, i.e. the activation energy.

It could be objected that the first of these indices of activation metabolism (the intercept of the enthalpy-stress relation of isometrically contracting papillary muscles, Fig. 5 A) and the second of these indices (the intercept of the VO_2-PVA relation, Fig. 5 B) are not equivalent since in the latter case the heart is visibly beating. Indeed, the isolated Langendorff-perfused heart often twists quite vigorously about its aortic suspension – presumably due to the spiral arrangement of the ventricular muscle fibres. Nevertheless, like its pre-shortened or quick-released papillary muscle counterpart resting on a thermopile, the heart does no *external* work in this situation. It is instructive then to compare results obtained by the two methods.

Qualitative Comparison of the Indices of Activation Metabolism

Results from the two models (Gibbs vs Suga) are in good qualitative agreement. Agents that enhance contractility achieve a parallel elevation of both the heat-stress (Gibbs) and the VO_2-PVA relation (Suga). The intercepts of these relations, which index the activation metabolism, are both increased in the presence of catecholamines (Figs. 6A and 6B) and in the presence of raised external calcium concentration (Figs. 6C and 6D) [7, 11, 25].

In both models an increase in heart rate leads to an increase in mechanical performance (isometric stress and pressure-volume area, respectively) with no change in the intercept [2, 21, 48]. Similarly, lowering the temperature, which leads to an elevation of the stress-independent heat in isolated papillary muscles of the rabbit [20], cat [2] and rat [34] also raises the oxygen consumption per beat of the empty, spontaneously beating, Langendorff-perfused rabbit heart [37].

Quantitative Comparison of the Indices of Activation Metabolism

At first glance the two models seem to yield quantitatively different estimates of activation metabolism. Comparison of Fig. 6A (Gibbs model) with Figs. 6B or 6C (Suga model) yields intercept values for control (C) conditions of about 2 mJ/g and 5 or 6 mJ/g respectively. But this comparison is misleading because the intercept value in the Suga model includes the basal metabolism whereas in the Gibbs model it does not. Suga et al. [47] report that the basal component accounts for 28% of the oxygen consumption of the beating but unloaded heart. Thus the activation component becomes 72% of the intercept value of the VO_2-PVA relation. It thus reduces to 3.6 to 4.3 mJ/g. This value is still about twice as high as the estimate arising from the Gibbs model. Can the two estimates be reconciled?

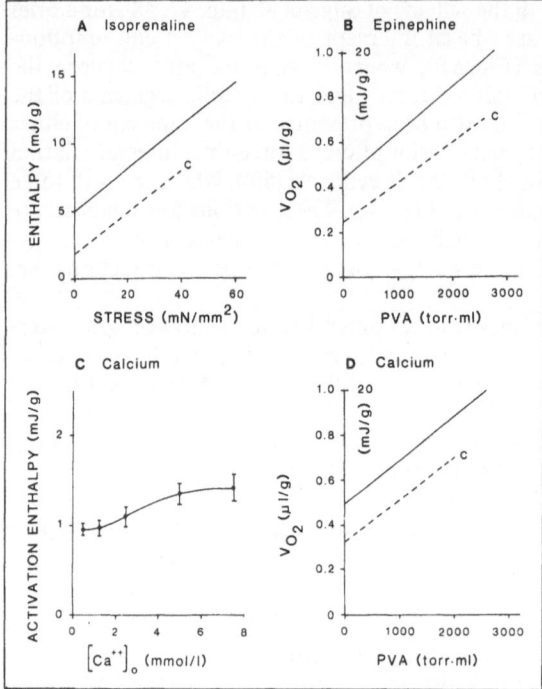

Fig. 6. Effect of agents that alter contractility upon active metabolism.
A) Isoprenaline upon the enthalpy-stress relation of isolated rabbit papillary muscle [22]. B and D) Epinephrine and elevated calcium concentration, respectively, upon per beat oxygen consumption-pressure volume area (PVA) relations [47]; scales show both oxygen consumption (μl/g) and equivalent energy (mJ/g) units. A, B and D show control response ("C" line); effect of inotropic agent (solid line) – in each case an increase in the suprabasal intercept or activation component.
C) External calcium ion concentration on the activation heat (mJ/g/beat) given by the intercept of enthalpy-stress relations separately determined at each calcium level. Data of [51].

Firstly, there exist species differences in the activation heat component. The value for cat and guinea-pig (2.6 mJ/g) is over 50% greater than for the rat (1.6 mJ/g) [39]. Corresponding data for dog papillary muscles do not exist but a species size-dependent (higher) value would reduce the discrepancy.

Secondly, and more importantly, the activation component of cardiac metabolism has probably been underestimated in previous myothermic studies. As the intercept of the heat-stress relation, it has always been measured in muscles pre-shortened to a length where force is zero. Now in skeletal muscle preparations, a pre-shortened muscle yields a considerably higher estimate of activation heat than a pre-stretched muscle – presumably due to residual cross-bridge activity [43]. But it is not possible to stretch papillary muscle enough to reliably estimate the intercept of the heat-stress relation because stretching much beyond l_{max} damages the muscle. Hence it has often been suggested that the myothermic technique overestimates the activation component of active metabolism. In fact, it seems to underestimate it.

Recent experiments by Gibbs [16, 17] make use of an ergometer to achieve a quick release (within the latency period) of papillary muscles to some pre-determined muscle length but always starting from l_{max}. It is thus possible to plot a new enthalpy-stress relation using the latency quick release technique [17]. The resulting estimate of activation heat is some 70% greater than that arising from the technique of pre-shortening the muscle. It thus appears that shortening-induced deactivation more than compensates for any residual cross-bridge activity in pre-shortened muscle.

An increase of 70% in the intercept of the enthalpy-stress relation (compare broken and solid lines, Fig. 5 A) yields an estimate of activation heat of about 3.4 mJ/g, a value which is now in accord with the quite independent estimates achieved by Suga et al. [47]. The Gibbs and Suga models of cardiac energy expenditure are thus in good quantitative agreement.

Implications

Gibbs [14] has estimated that the heart typically operates in a region corresponding to about 0.5 P_0 on the enthalpy-load relation (dotted line Fig. 5 A). At this point it can be seen that each of the basal and activation components account for one-quarter to one-third of the total enthalpy. Likewise a dog heart normally operates with a PVA value in the vicinity of 1 500 torr · ml (Figs. 6 B, 6 D). Under control conditions, at least one-half of the total cardiac oxygen consumption is again seen to consist of the basal and activation components. The Gibbs and Suga models are hence in agreement in attributing something in excess of 50% of the total cardiac energy expenditure to processes not directly related to the generation of external force or pressure or to the performance of pressure-volume work. Since the basal metabolic rate varies inversely with animal (species) size [24], this fraction is expected to be larger in hearts of smaller animals (which have higher basal rates [24] but lower total energy expenditure per beat [39]). Indeed, Wilkman-Coffelt et al. [52] report data showing that, in the rat heart, Langendorff perfusion lends to a rate of oxygen consumption over 70% of that of the working heart preparation. The

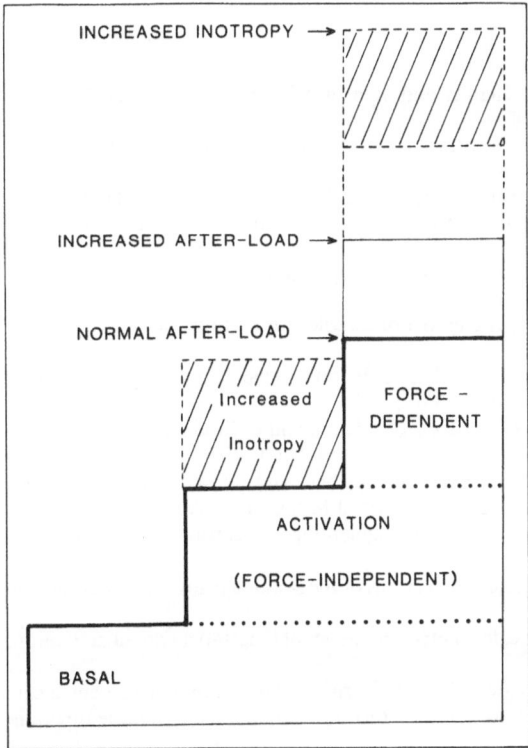

Fig. 7. Schematic diagram of relative magnitude (vertical height) of the 3 components of cardiac energy expenditure. Heavy solid line = normal conditions; light solid line = increased after-load; broken line = increased inotropy. For details see text.

relative contribution of the basal and activation components must decline (provided that each remains constant) as after-load increases (see Control curves, Figs. 6A, 6B and 6D).

Figure 7 shows a schematic diagram based on the data from Figs. 5 and 6, representing the relative magnitude of the components, basal, activation or force-independent, and force-dependent metabolism, of cardiac energy expenditure per beat. Under normal conditions (P = 0.4 to 0.5, as in Fig. 5A; PVA = 1 500 torr ml, as in Fig. 6B) of cardiac energy expenditure, the basal component comprises about one third of the metabolism of the beating but non-working heart: basal plus activation metabolism about one half to two thirds of the total. Under control conditions, the force-dependent component can be increased by increasing the afterload. An increase in inotropic state (regions bounded by broken lines) can approximately double the activation component (shaded region) as well as the force-dependent component (unshaded region). Basal metabolism now accounts for about one-fifth of the total; activation metabolism still accounts for about one-third. Nevertheless, as the schematic summary in Fig. 7 shows, the non-force related components of cardiac metabolism are clearly substantial and should not be underrated in any description of cardiac energetics.

Acknowledgements. I am grateful to Dr. Brian Chapman for assistance in calculating the hypothesized contribution of passive proton leakage currents to basal mitochondrial oxidation, to Dr. Igor Wendt for permission to publish the data presented in Figs. 4 and 6C and to Dr. Colin Gibbs for permission to refer to the paper of Gibbs and Kotsanas which is currently in press. It is a pleasure to thank, also, Miss Sue Mills for her excellent secretarial work.

References

1. Alpert NR, Mulieri LA (1982) Heat, mechanics and myosin ATPase in normal and hypertrophied heart muscle. Fed Proc 41:192–198
2. Barclay JK, Gibbs CL, Loiselle DS (1979) Stress as an index of metabolic cost in papillary muscle of the cat. Basic Res Cardiol 74:594–603
3. Bretschneider HJ, Hubner G, Knoll D, Lohr B, Nordbeck H, Spieckermann PG (1975) Myocardial resistance and tolerance to ischemia: physiological and biochemical basis. J Cardiovasc Surg 16:241–260
4. Burns AH, Reddy WJ (1978) Amino acid stimulation of oxygen and substrate utilization by cardiac myocytes. Am J Physiol 235:E461–E466
5. Chapman JB, Gibbs CL (1972) An energetic model of muscle contraction. Biophys J 12:227–236
6. Chapman JB, Gibbs CL (1974) The effect of metabolic substrate on mechanical activity and heat production in papillary muscle. Cardiovasc Res 8:656–667
7. Chapman JB, Gibbs CL, Gibson WR (1970) Effects of calcium and sodium on cardiac contractility and heat production in rabbit papillary muscle. Circ Res 27:601–610
8. Chapman JB, Gibbs CL, Loiselle DS (1977) Simultaneous heat and fluorescence at high rates of energy expenditure: effects of caffeine and isoprenaline. J Mol Cell Cardiol 9:715–732
9. Coleman HN (1967) Role of acetylstrophanthidin in augmenting myocardial oxygen consumption. Circ Res 21:487–495
10. Coulson RL (1976) Energetics of isovolumic contractions of the isolated heart. J Physiol 260:45–53
11. Coulson RL (1982) Isolated whole heart calorimetry: energetics of length-dependent activation. Fed Proc 41:199–203
12. Earl CA, Laurent GJ, Bonnin CM, Sparrow MP (1978) Turnover rates of muscle protein in cardiac and skeletal muscles of dog, fowl, rat and mouse: turnover rate related to muscle function. Aust J Exp Biol Med 56:265–277

13. Feng TP (1932) The effect of length on the resting metabolism of muscle. J Physiol 74:441–454
14. Gibbs CL (1978) Cardiac energetics. Physiol Rev 58:174–254
15. Gibbs CL (1983) Thermodynamics and cardiac energetics. In: Dintenfass L, Julian DG, Seaman GVF (eds) Heart perfusion, energetics and ischemia. NATO ASI Series: Life Sciences, vol 62. Plenum Press, New York, pp 549–576
16. Gibbs CL (1986a) The dependence of activation heat on extracellular calcium. J Mol Cell Cardiol 18 [Suppl 1]:298P
17. Gibbs CL (1986b) Cardiac energetics and the Fenn effect. Basic Res Cardiol (this supplement)
18. Gibbs CL, Chapman JB (1979a) Cardiac heat production. Ann Rev Physiol 41:507–519
19. Gibbs CL, Chapman JB (1979b) Cardiac energetics. In: Berne RM, Sperelakis N (eds) The cardiovascular system. Handbook of Physiology, Ch 22, Am Physiol Soc, Bethesda, MD, pp 775–804
20. Gibbs CL, Gibson WR (1969) Effect of ouabain on the energy output of rabbit cardiac muscle. Circ Res 24:951–967
21. Gibbs CL, Gibson WR (1970) Effect of alterations in the stimulus rate upon energy output, tension development and tension-time integral of cardiac muscle in rabbits. Circ Res 27:611–618
22. Gibbs CL, Gibson WR (1972) Isoprenaline, propranolol, and the energy output of rabbit cardiac muscle. Cardiovasc Res 6:508–515
23. Gibbs CL, Kotsanas G (1986) Factors regulating basal metabolism of the isolated perfused rabbit heart. Am J Physiol (in press)
24. Gibbs CL, Loiselle DS (1978) The energy output of tetanized cardiac muscle: species differences. Pflug Arch 373:31–38
25. Gibbs CL, Vaughan P (1968) The effects of calcium depletion upon the tension-independent component of cardiac heart production. J Gen Physiol 52:532–549
26. Gibbs CL, Mommaerts WFMH, Ricchiuti NV (1967) Energetics of cardiac contractions. J Physiol 191:25–46
27. Gibbs CL, Woolley G, Kotsanas G, Gibson WR (1984) Cardiac energetics in daunorubicin-induced cardiomyopathy. J Mol Cell Cardiol 16:953–962
28. Hoppeler H, Linstedt SL, Claassen H, Taylor CR, Mathieu O, Wiebel ER (1984) Scaling mitochondrial volume in heart to body mass. Resp Physiol 55:131–137
29. Kira Y, Kochel PJ, Gordon EE, Morgan HE (1984) Aortic perfusion pressure as a determinant of cardiac protein synthesis. Am J Physiol 246:C247–C258
30. Klocke FJ, Kaiser GA, Ross J Jr, Braunwald E (1965) Mechanism of increase of myocardial oxygen uptake produced by catecholamines. Am J Physiol 209:913–918
31. Lehninger AL (1979) Biochemistry (2nd edn). Worth Publishers Inc, NY
32. Lochner W, Arnold G, Muller-Ruchholtz ER (1968) Metabolism of the artificially arrested heart and of the gas-perfused heart. Am J Cardiol 22:299–311
33. Lochner W, Dudziak R (1965) Stillstandumsatz und Ruheumsatz des Herzens. Pflug Arch 285:169–177
34. Loiselle DS (1979) The effects of temperature on the energetics of rat papillary muscle. Pflug Arch 379:173–180
35. Loiselle DS (1983) Some factors modifying the metabolism of the K$^+$-arrested guinea-pig heart. J Mol Cell Cardiol 15 [Suppl 1]:286P
36. Loiselle DS (1985a) The rate of resting heat production of rat papillary muscle. Pflug Arch 405:155–162
37. Loiselle DS (1985b) The effect of temperature on the basal metabolism of cardiac muscle. Pflug Arch 405:163–169
38. Loiselle DS (1985c) Simulation of simple and myoglobin-facilitated oxygen diffusion in resting papillary muscle. J Mol Cell Cardiol 17(5):24P
39. Loiselle DS, Gibbs CL (1979) Species differences in cardiac energetics. Am J Physiol 237:H90–H98
40. Loiselle DS, Gibbs CL (1983) Factors affecting the metabolism of resting rabbit papillary muscle. Pflug Arch 396:285–291
41. Loiselle DS, Wendt IR, Hoh JFY (1982) Energetic consequences of thyroid-modulated shifts in ventricular isomyosin distribution in the rat. J Mus Res Cell Motil 3:5–23

42. Mitchell P (1961) Coupling of phosphorylation to electron and hydrogen transfer by a chemiosmotic type of mechanism. Nature 191:144–148
43. Rall JA (1982) Energetics of Ca^{2+} cycling during skeletal muscle contraction. Fed Proc 41:155–160
44. Schreiber SS, Evans C, Oratz M, Rothchild M (1986) The basal level of cardiac protein synthesis. J Mol Cell Cardiol 18 [Suppl 1]:26P
45. Smith HE, Page E (1976) Morphometry of rat heart mitochondrial subcompartments and membranes: application to myocardial cell atrophy after hypophysectomy. J Ultrastruct Res 56:31–41
46. Suga H, Hayashi T, Shirahata M (1981) Ventricular systolic pressure-volume area as a predictor of cardiac oxygen consumption. Am J Physiol 240:H39–H44
47. Suga H, Hisano R, Goto Y, Yamada O, Igarashi Y (1983a) Effect of positive inotropic agents on the relation between oxygen consumption and systolic pressure volume area in canine left ventricle. Circ Res 53:306–318
48. Suga H, Hisano R, Hirata S, Hayashi T, Yamada O, Ninomiya I (1983b) Heart-rate independent energetics and systolic pressure-volume area in dog heart. Am J Physiol 244:H206–H214
49. Suga H, Yamada O, Goto Y (1984) Energetics of ventricular contraction as traced in the pressure-volume diagram. Fed Proc 43:61–63
50. Weber KT, Janicki JS (1977) Myocardial oxygen consumption: the role of wall force and shortening. Am J Physiol 233:H421–H430
51. Wendt IR, Loiselle DS (1981) The effect of external calcium concentration on activation heat in cardiac muscle. J Mol Cell Cardiol [Suppl 3] 13:8P
52. Wilkman-Coffelt J, Sievers R, Coffelt RJ, Parmley WW (1983) The cardiac cycle: regulation and energy oscillations. Am J Physiol H354–H362

Author's address:

D. S. Loiselle, Department of Physiology, School of Medicine, University of Auckland, Private Bag, Auckland, New Zealand

Mechanical determinants of myocardial energy turnover

G. Elzinga, C. M. B. Duwel, F. Mast, and N. Westerhof

Laboratory for Physiology, Free University Amsterdam, The Netherlands

Summary

Energy turnover of the left ventricle does not differ in isovolumic contractions and contractions where pressure is released from peak to zero. This experimental result corresponds to predictions from a time varying elastance model of the mechanical and energetic properties of the left ventricle. To assess the validity of this model for cardiac muscle in general, experiments were designed to investigate whether mechanical and energetic behaviour of isolated cardiac muscle preparations could also be predicted from the time varying elastance model. The results obtained so far indicate, however, that not all experimental results can be accommodated by the model. This suggests that the value of the model may be limited.

Introduction

In 1964, Monroe [7] showed that oxygen consumption in isolated dog hearts in a steady state of isovolumic contractions, was only slightly higher than for steady state contractions during which ventricular volume was suddenly reduced at peak pressure so that pressure fell to zero (Fig. 1). This observation was in sharp contrast to the notion generally accepted then, that the pressure integral was a reliable measure of the amount of oxygen consumed; the pressure integrals of the types of contractions studied by Monroe differed by a factor of about two. In Monroe's experiments, use was made of a balloon in the left ventricle. The balloon was filled with air – an experimental detail which led to criticism in the following years because of the volume changes during contraction caused by compression of air. Moreover, Monroe did not evaluate the energetic effects of the shortening required to reduce pressure to zero.

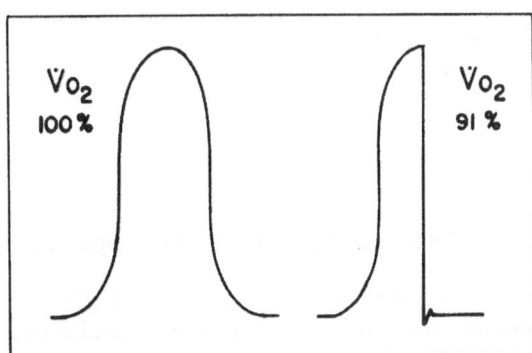

Fig. 1. Oxygen consumption during steady state isovolumic contractions (left) and contractions in which pressure was released at its peak to zero (right); air filled balloon in the dog. Modified from Monroe [7].

Now, more than 20 years later, the idea that the pressure integral is a reliable measure of cardiac energy turnover has been shown to be too simple a hypothesis. Several new postulates have been brought forward to predict the energy turnover from pressure and output [11, 13]. One of the most attractive hypotheses has been proposed by Suga [13]. He based his predictions on the time varying elastance concept, proposed by him and Sagawa, as a description of the haemodynamic behaviour of the left ventricle [12]. Making use of the pressure volume area (PVA) hypothesis (see below for details) it appears to be possible to predict reliably cardiac oxygen consumption under many different haemodynamic conditions. The predictive power of the PVA hypothesis, in combination with the fact that the time varying elastance concept can describe the mechanical behaviour of the ventricle, may suggest that the time varying elastance behaviour is a fundamental property of cardiac muscle.

In this study, we set out, in view of the above criticism, to confirm Monroe's study and to address the question of whether a basic muscle property can be held responsible for the time varying elastance behaviour. The results are important as to what extent the concept can be extrapolated to other species (including man) and to situations other than those under which the results were obtained (such as disease).

Methods

A. Isolated supported cat heart

Arterial blood of a donor cat was used to perfuse an isolated cat heart in a Langendorff perfusion set up. A fluid filled balloon was introduced into the left ventricular cavity of the isolated heart and connected to a piston mounted on an electromagnetic shaker. The position of the piston was controlled by a feedback system allowing programmed changes of ventricular volume during the cardiac cycle.

Left ventricular oxygen consumption was determined by measuring continuously the arterio-venous oxygen content difference (A-VOX-systems, San Antonio, USA), and multiplying that difference by coronary blood flow. Left ventricular pressure was measured inside the balloon with a micromanometer (Millar, Houston, USA). Ventricular volume was derived from the position of the piston.

Two experimental protocols were followed: (a) Oxygen consumption during isovolumic contractions was compared with that found for contractions where peak pressure was suddenly reduced to zero. Both types of contractions were maintained for at least 2 min to allow the measurement of steady state oxygen consumption. (b) Oxygen consumption was measured during isovolumic contractions at different end-diastolic volumes and compared with that of contractions, starting at different end-diastolic volumes, where no pressure development was allowed by timely withdrawal of fluid.

B. Isolated cardiac muscle

1. Isolated cat trabeculae

Since this preparation has been described in detail elsewhere [1] a short description will suffice here.

Cardiac trabeculae (diameter < 0.35 mm) isolated from the feline right ventricle, were mounted between a force transducer and a movable arm, connected to a loudspeaker coil.

The position of the coil, and thus the length of the muscle, was controlled by a feedback system enabling length or force control. The feedback system was designed to enable the contraction pattern of the preparation to mimick the physiological contraction and relaxation sequence in the ventricular wall: isometric contraction, isotonic shortening (ejection), isometric relaxation, and isotonic lengthening (filling). The experiments were done at 37 °C and in a Tyrode solution with a calcium concentration of 1.35 mM.

During steady state contractions at a frequency of 2 Hz, the load during the shortening period resembled the arterial input impedance. At every tenth contraction the preparation underwent a steady load during shortening, the level of which was different each time. In these contractions the end of systole could be easily seen in the tracings, since it is the moment at which isometric relaxation begins.

2. Isolated rabbit papillary muscle

Preliminary experiments were done on papillary muscles isolated from rabbit right ventricles and mounted on a thermopile as described by Mulieri et al. [8]. Isometric force and temperature change were measured. Temperature change during the twitch was converted into heat by correcting for heat loss and accounting for the thermal mass involved in the temperature change. The experiments were done at 20 °C. The Tyrode solution had a calcium concentration of 1.35 mM.

Two experimental protocols were followed: (a) Heat per twitch and peak force were measured during steady state contractions (0.2 Hz) at l_{max} and at three shorter lengths. (b) Heat per twitch and peak force were measured at l_{max} during short irregular twitch trains (10–15 contractions). The potentiation caused by this irregularity was used as a means to vary the inotropic state.

In the analysis, the force length area (FLA, see Fig. 3) was obtained from the force length diagram constructed on the basis of the results obtained with the first protocol. Moreover, the heat produced during the twitch was divided into two parts: heat produced during contraction and heat produced during relaxation (H_{ir}, see inset Fig. 5).

Results and discussion

A. Isolated supported cat heart

Cardiac oxygen consumption during a steady train of isovolumic contractions and that for contractions during which pressure was suddenly released at its peak to zero, were not significantly different (Fig. 2). The average (n = 8) value of released beats was 98% of that of isovolumic contractions, during which oxygen consumption was on average $8.51 \, ml \cdot 100 \, g^{-1} \cdot min^{-1}$.

This result (cf Fig. 1) confirms what Monroe found using air-filled balloons in isolated supported dog left ventricles [7]. It implies that in his experiments, the work on the air in the balloon during contraction did not affect the conclusion he drew from his findings, i.e. that oxygen consumption is almost completely determined by processes occurring before the moment of peak isovolumic pressure.

However, this conclusion is based on the assumption that the shortening required to drop pressure to zero has no effect on oxygen consumption. If shortening caused an appreciable breakdown of ATP, it could be that the lack of change in oxygen consumption

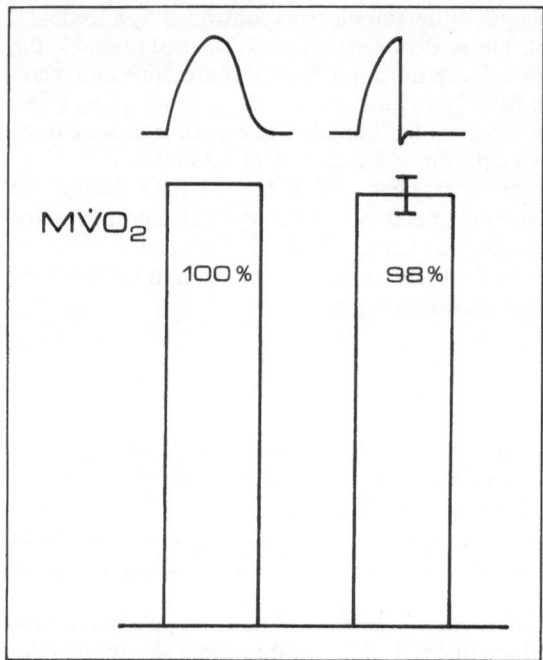

Fig. 2. Oxygen consumption, as in Fig. 1, but in isolated supported cat left ventricles with fluid filled balloons.

with pressure release is a result of two processes (e.g. shortening and pressure reduction) with opposite effects on the rate of ATP hydrolysis, leading to no change in the overall rate of energy turnover.

Although Monroe measured oxygen consumption from beats during which no pressure was developed in the ventricular balloon, this does not allow the estimation of energy turnover per unit shortening. For that purpose, the effect of different amounts of shortening on oxygen consumption would have to be studied. Only in this way does it become possible to determine a value for the energy turnover per unit of shortening. In our experiments (n = 11) we varied left ventricular end-diastolic volume and made the ventricle eject in such a way that pressure remained low during the whole cycle. For these isotonic contractions we found that oxygen consumption ($M\dot{V}O_2$ in ml $\cdot$ 100 g^{-1} $\cdot$ min^{-1}) changed little with changes in end-diastolic volume (EDV in ml; balloon volumes between 0–5 ml). When the results were fitted with a straight line it was found that on average $M\dot{V}O_2 = 3.39 + 0.26$ EDV. The dependency on end-diastolic volume was about 7 times less than that for isovolumic contractions.

The results of these experiments suggest that Monroe was right to conclude that the mechanical events during relaxation hardly affected oxygen consumption, and thus the energy turnover of the ventricle. This conclusion is in accordance with predictions made on the basis of the time varying elastance concept [12], and with experiments designed to test the energetic aspects of that model [13].

To understand the energetic implications of the time varying elastance, E(t), somewhat better, its behaviour is presented schematically in Fig. 3. The schematic diagram has been made up to discuss the concept in terms of isolated cardiac muscle properties, but it has been extrapolated from results from isolated dog whole left ventricles [12]. There-

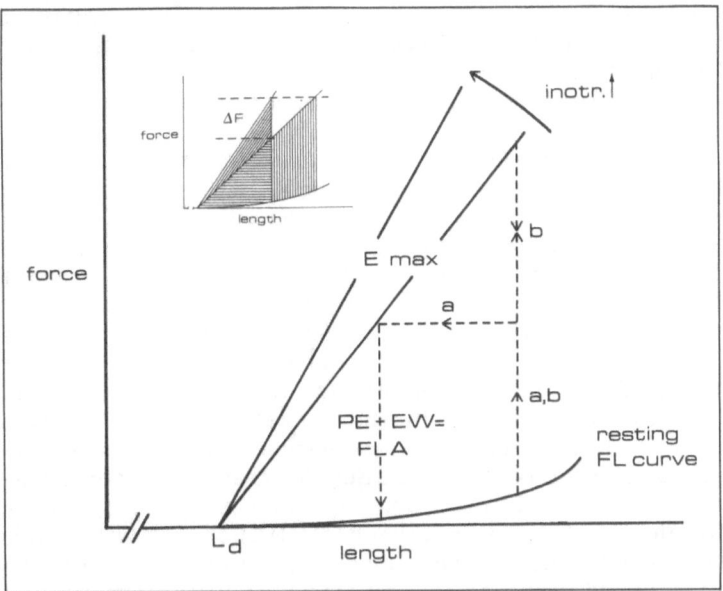

Fig. 3. A schematic representation of Equation (1), in terms of force and length, based on results from isolated dog left ventricles [12] and extrapolated to cardiac muscle. Inset: force length area under isometric conditions (horizontal lines); increase in FLA due to increased resting length (vertical lines), and due to increased inotropic state (diagonal lines).

fore in the case of Monroe's experiment, we have to read pressure (P) for force (F) and volume (V) for length (L).

In the time varying elastance concept it is proposed [12] that left ventricular elastance behaviour is rather simple and can be described by the equation

$$E(t) = P(t)/[V(t) - V_d] \qquad (1)$$

The equation implies that the ventricular pressure-volume relationship at the end of systole can be represented by a single straight line (E_{max}) with an intercept on the volume axis (V_d). The intercept and slope of this line do not depend on the history of contraction, although the location of the ventricle on the line at the end of systole will, of course, vary with arterial load.

During contraction cardiac muscle follows a series of elastance isochrones of which the diastolic one (resting FL curve) and the end-systolic one (E_{max}) are shown here.

The (physiologically) afterloaded contraction a forms a loop (EW) enclosing an area the magnitude of which is equal to the external work done. With an increase in inotropic state the E(t) lines, as is shown for E_{max}, rotate around an intercept on the length axis (L_d).

According to the concept, the energy required by the ventricle for its mechanical activity is proportional to the area enclosed by the systolic pressure-volume trajectory, the E_{max} line and the end-diastolic pressure-volume curve. This pressure volume area (PVA, or in terms of isolated muscle, FLA, see Fig. 3) is equal to the sum of the area representing the external work (EW) done by the ventricle and an area (enclosed by the E_{max} line and the left side of the EW loop) representing the potential energy (PE) stored at the end of systole in the ventricular wall, and is assumed to be proportional to energy turnover related to shortening and force production. In the case of isovolumic contractions (contraction *b* in Fig. 3) PVA is equal to the area PE.

Tests, making use of isolated supported dog ventricles, have verified the energetic predictions based on this time varying elastance concept [12, 13]. Changes in load do not affect the elastance-time [E(t)] function. Therefore the afterloaded contraction *a* and the isometric contraction *b* reach at the end of systole the same E_{max} line.

The proportional relationship between the oxygen consumed for the contractile processes proper and the pressure volume area implies that the amount of energy required for a given contraction is determined at the moment the E_{max} line is reached, i.e. at the end of systole. Whatever happens in terms of mechanical phenomena thereafter does not affect the pressure volume area and therefore does not affect the contraction-related energy turnover. Our findings, described above, and those of Monroe [7], reported earlier, confirm the prediction made on the basis of the time varying elastance concept. However, can these results be regarded as a sufficiently critical test of the time varying elastance of Equation (1) as the basic mechanism determining myocardial energetics?

The time varying elastance model of Sagawa and Suga has characteristics resembling those of the viscoelastic model of muscle contraction which was popular at the beginning of this century [9]. More than 60 years ago, Fenn showed [3, 4] that the viscoelastic model of muscle contraction could not explain his finding that heat plus work during shortening was more than the heat produced during an isometric contraction. It is true that this finding has not been confirmed for heart muscle and that it is unlikely on the basis of the prevailing evidence that it will be [10]. However, it is a rather weak argument to regard the time varying elastance as a true basis for muscle contraction, even in the case of the myocardium.

We therefore tried to design experiments to critically test the concept of the time varying elastance as a fundamental property of cardiac muscle. Two tests were performed, one of the mechanical implications of Equation (1), the other based on energy considerations. In view of the rather fundamental nature of the question posed, the experiments were done on isolated cardiac muscle, thereby reducing complexities resulting from ventricular geometry.

B.1. Isolated cat trabeculae

The mechanical prediction made was that according to Equation (1) the moment at which E_{max} (Fig. 3) is reached during the cycle, does not depend on the course the muscle takes in the force-length plane or on the position of the end-systolic force-length point. These experiments were executed such that the moment of E_{max} could be easily identified in the force tracing [1]. In intact hearts this moment is usually not easily defined because the ratio of pressure to volume determines the moment at which E_{max} occurs. Since both vary with time, the moment of E_{max} cannot be assessed from one of the variables only.

It was found for these isolated trabeculae (n = 5) that the moment of E_{max} was strongly load dependent (Fig. 4). This observation contrasts with the prediction of Equation (1), and therefore provides evidence that time varying elastance behaviour, as described by this formula, is not a fundamental muscle property.

B.2. Isolated rabbit papillary muscle

To fully understand the second test, based on an energetic prediction it is useful to reconsider the graphical presentation of Equation (1) given in Fig. 3. In the inset the force

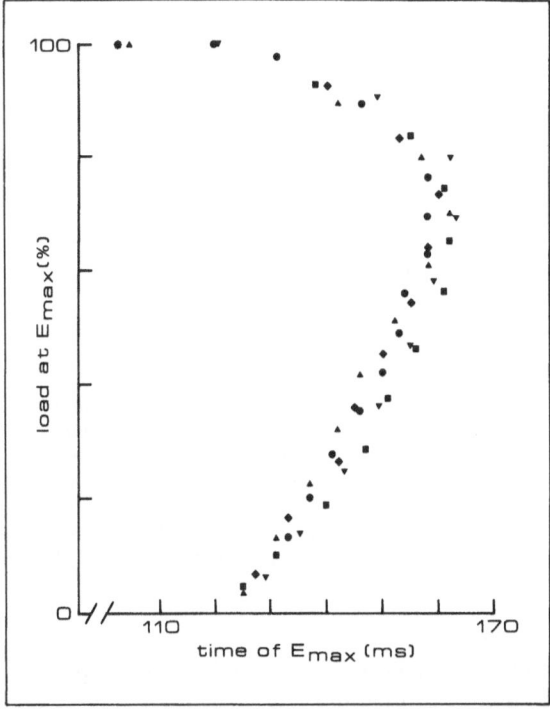

Fig. 4. Modified from Elzinga and Westerhof, 1984 [2].

length area under control isometric conditions is indicated by horizontal lines. The increase in FLA due to an increase in resting length is indicated by vertical lines; the increase in FLA due to an increase in inotropic state by lines parallel to the control E_{max} line. It can be seen that for the same increase in force, the increase in FLA depends on the mechanism from which it results: an increase in inotropic state causes FLA to increase less than an increase in initial muscle length does.

At the end of systole, i.e. at the moment of E_{max}, during an isometric contraction the amount of energy stored as potential energy is (in joules) equal to the force length area. Two predictions can be derived from this statement:

1. Since in an isometric contraction, no energy leaves the muscle as work, the amount of heat liberated after peak force should at least be equal to the force length area.

2. When force is increased by increasing muscle length, the force length area increases more than when force increases through an increase in contractility. Consequently, the heat liberated during relaxation increases more when a given increase in force is brought about by an increase in length, than when this occurs through an increase in inotropic state.

These two predictions were tested in experiments on isolated rabbit papillary muscles (n = 5). The preliminary results obtained so far show that under the chosen experimental conditions (see Methods) the force length area is larger than the amount of heat liberated during relaxation, 14.7 ± 2.4 and 9.8 ± 2.4 μJ, respectively. Moreover, the relation between force and the amount of heat liberated during relaxation was the same for changes in initial resting length (shown as circles in Fig. 5) as for changes in inotropic state (filled circles) obtained by varying the contraction interval (Fig. 5).

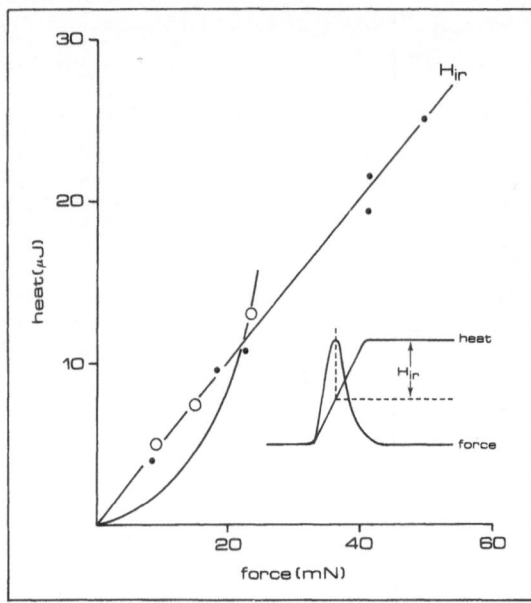

Fig. 5. Inset: Heat production during mechanical activity (initial heat) subdivided into heat produced during contraction and during relaxation (H_{ir}). H_{ir} is porportional to the force produced, whether by increase in inotropic state (●) or in muscle length (○).

These findings do not support the time varying elastance concept as a description of the fundamental mechanism of cardiac muscle contraction. This conclusion implies that the behaviour described by Equation (1), represents a reliable analog of left ventricular properties under certain conditions only. The analogy then holds true because of a specific combination of geometrical and contractile properties, occurring in the canine left ventricle. If one of the determining factors changes, the analogy may no longer hold. This is the case for instance, for the canine right ventricle [6] which shows a much more complex behaviour than that given by Equation (1). It may be expected, however, that not only geometry but also other factors, such as age and disease, affect the validity of the concept and cause ventricular behaviour to be much more complex than the time varying elastance of Equation (1) predicts [5].

References

1. Elzinga G, Westerhof N (1982) Isolated cat trabeculae in a simulated feline heart and arterial system: contractile basis of cardiac pump function. Circ Res 51:430–438
2. Elzinga G, Westerhof N (1984) Does the history of contraction affect the pressure-volume relationship? Fed Proc 43:2402–2407
3. Fenn WO (1923) A quantitative comparison between the energy liberated and the work performed by the isolated sartorius muscle of the frog. J Physiol (Lond) 58:175–203
4. Fenn WO (1924) The relation between the work performed and the energy liberated in muscular contraction. J Physiol (Lond) 58:175–203
5. Kissling G, Takeda N, Vogt M (1985) Left ventricular end-systolic pressure-volume relationships as a measure of ventricular performance. Basic Res Cardiol 80:594–607
6. Maughan WL, Shoukas AA, Sagawa K, Weisfeldt ML (1979) Instantaneous pressure-volume relationship of the canine right ventricle. Circ Res 44:309–315
7. Monroe RG (1964) Myocardial oxygen consumption during ventricular contraction and relaxation. Circ Res 14:294–300

8. Mulieri LA, Luhr G, Trefry J, Alpert NR (1977) Metal-film thermopiles for use with rabbit right ventricular papillary muscles. Am J Physiol 233:C146–C156
9. Needham DM (1971) Machina carnis. Cambridge University Press, Cambridge
10. Rall JA (1982) Sense and nonsense about the Fenn effect. Am J Physiol 24:H1–H6
11. Rooke GA, Feigl EO (1982) Work as a correlate of canine left ventricular oxygen consumption, and the problem of catecholamine oxygen wasting. Circ Res 50:273–286
12. Suga H, Sagawa K, Shoukas AA (1973) Load independence of the instantaneous pressure-volume ratio of the canine left ventricle and effects of epinephrine and heart rate on the ratio. Circ Res 32:314–322
13. Suga H (1979) Total mechanical energy of a ventricle model and cardiac oxygen consumption. Am J Physiol 236:H498–H505

Authors' address:

G. Elzinga, M.D., Ph.D., Laboratory for Physiology, Free University Amsterdam, Van der Boechorststraat 7, 1081 BT Amsterdam, The Netherlands

Cardiac energetics and the Fenn effect

C. L. Gibbs

Department of Physiology, Monash University, Clayton, Victoria, Australia

Summary

The energy output of a cardiac contraction can be divided into several phenomenologically measured components, although it must be emphasized that such subdivisions are often thermodynamically misleading. There is an activation term that relates to Ca^{++} release and retrieval, a work term and a stress or load-dependent heat term. The work and load-dependent energy terms presumably have their origin in the actin-activated myosin ATPase. It can be shown that the enthalpy: load relationship has a similar format across both mammalian and amphibian hearts: the scaling of both the energy and load axes is however altered by changes in contractility. The fact that enthalpy production is so clearly load-dependent indicates that there is a Fenn effect in cardiac muscle, although the discovery that energy output is greatest in an isometric contraction clearly contradicts one of the two central findings of Fenn's skeletal muscle investigations. Cardiac oxygen consumption per beat can be linearly correlated with ventricular systolic pressure – volume area (PVA) which is defined in terms of stroke work and potential energy components. If the basal and activation components are subtracted out cardiac muscle can be shown to operate at a constant PVA efficiency. The existing myothermic and polarographic data can be reconciled with the PVA concept.

Cardiac energetics

Fenn's paper [8] is widely regarded as a classic in muscle physiology. He examined the energy liberated and the work performed by frog sartorius muscles in either twitches or brief (0.2 s) tetani. There were ten summary statements but we need concentrate on only two, namely: (A) Whenever a muscle shortens upon stimulation and does work in lifting a weight, an extra mount of energy is mobilised which does not appear in an isometric contraction. Hence less energy is liberated in an isometric contraction than in any contraction in which the muscle is allowed to shorten. (B) The energy liberated by the contraction of a single muscle fibre for a given stimulus is not dependent solely upon the initial mechanical and physiological condition of the muscle but can be modified by the nature of the load which the muscle "discovers" it must lift after the stimulus is over.

In terms of cardiac muscle statement B can be accepted but the second part of statement A clearly cannot and even the first part needs to be modified by adding "where a force equal to the load is developed" after...isometric contraction.

Figure 1 shows the enthalpy (heat + work) versus load relationship that is obtained in all mammalian hearts so far examined (Fig. 1, left) and that obtained in an amphibian heart (Fig. 1, right). It should be noted that in these curves, unlike those plotted by Fenn [8], the enthalpy values include both the initial and recovery metabolisms. Since the recovery heat is stoichiometrically related to the initial enthalpy the argument is not affected. We believe the recovery energy is approximately 70% of the initial energy; see Chapman & Gibbs [5] for the theoretical and experimental basis of this statement.

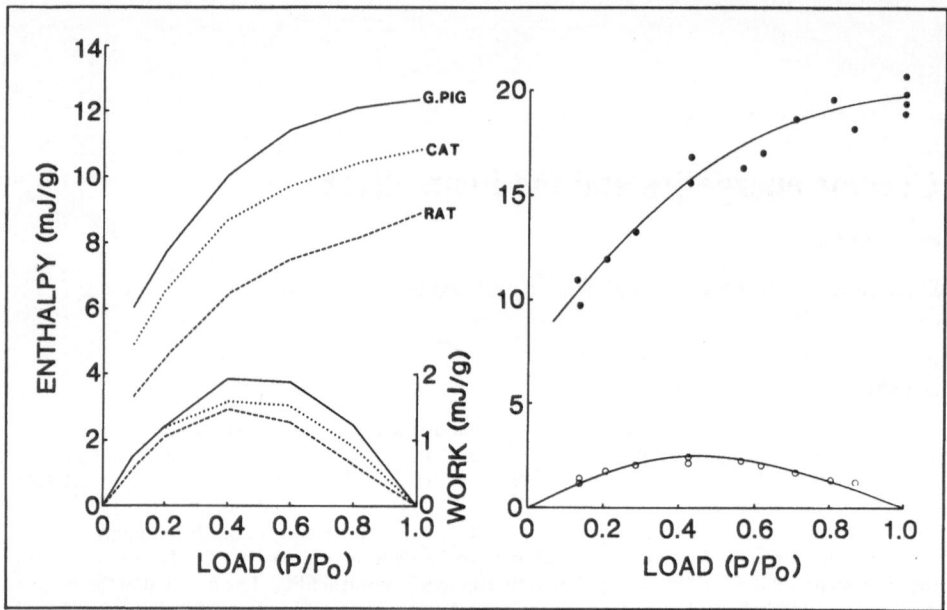

Fig. 1. Total isotonic enthalpy (heat + work, mJ·g^{-1}) against relative load. Left: averaged data for guinea pig, cat and rat, taken from [23]. Right: unpublished data collected from a toad ventricular strip by Dr. I. Wendt. Both sets of data obtained at 27 °C.

The major difference between Fenn's skeletal data and that found in the cardiac literature, is that energy output is maximal in isometric cardiac muscle. It is noteworthy that there is a steep fall in enthalpy output per beat moving from heavy to light afterloads. This fall off in energy flux with load has significant beneficial effects for cardiac muscle in terms of mechanical efficiency w/(w + heat) which is maximal for loads in the 0.2 to 0.4P_o range. For a more detailed discussion of this point see references [4, 9, 10, 14, 16].

Over the years my colleagues and I have made it clear that we believe the only sensible way to discuss energy metabolism it to relate it to the underlying biochemistry [4, 10, 14]. In particular we have, in terms of initial metabolism, concentrated on the three major ATPases found in muscle, namely:

i.   the Na$^+$-K$^+$-ATPase of the sarcolemmal Na$^+$ pump
ii. the sarcoreticular and sarcolemmal Ca^{++}-ATPases and
iii. the myofibrillar (actomyosin) ATPase.

This listing does not mean that there are no other chemical reactions and in particular other ATP-using processes occurring during a contraction, but the present evidence suggests that these three enzyme systems account for nearly all the initial metabolism.

The initial energy, associated with the ATPases listed above, is followed by the recovery heat which can be attributed to the oxidation of carbohydrates, fatty acids and lactate, supplemented to a small extent by substrate level phosphorylations occurring in the cytoplasmic (glycoytic) and mitochondrial (oxidative) pathways. In amphibian skeletal muscle there is temporal separation of the initial and recovery metabolisms but the situation is more complex in mammalian skeletal muscle [7, 30]. In cardiac muscle we believe [5] that there is no clear separation of the two components (even at temperatures as low as 20 °C) but this point is not universally accepted [2].

On biochemical and thermodynamic grounds active energy output can be subdivided into heat associated with muscle activation and heat and external work associated with cross-bridge turnover.

Activation metabolism

The activation heat we ascribe to the cost of sarcoreticular and sarcolemmal Ca^{++} pumping. The magnitude of this component is clearly dependent upon the Ca^{++} load released into the cytoplasm [18] but we also have evidence that there is some contribution from the Na^+ pump. If the ratio of $[Ca^{++}]o$ to $[Na^+]o$ is altered in the physiological saline then it is possible to produce marked changes in the magnitude of the activation heat component [14].

In my laboratory the activation heat has customarily been measured by shortening papillary muscles to points on their length: tension relation at which they cannot develop active force. This method, which gives reproducible results without any damage to the muscle preparations and which satisfactorily detects changes produced by pharmacological agents [11], is open to criticism. On the one hand it can be argued that the method will overestimate the real magnitude of the activation heat because there may still be some crossbridge activity occurring (sarcomere shortening and internal work) whilst on the other hand it can be argued that the shortening will deactivate Ca^{++} release [1, 22].

Recently I have used a rapid release technique in an attempt to get another estimate of the activation heat. A preparation is held at l_{max} for 15 ms after a stimulus and is then rapidly shortened by an ergometer to a length where no active force development can occur before subsequently being returned to l_{max}. This technique produces activation heat values which are about 70% higher than those obtained in pre-shortened preparations.

Cross-bridge metabolism

By varying muscle length or by varying the release distance it is possible to plot out isometric enthalpy against peak developed stress (force/CSA). Note that if the magnitude of the activation heat varies it alters the apparent dependency of enthalpy production upon stress (see Fig. 2). For many years I have considered the heat above the activation heat baseline to be caused by cross-bridge activity and called this heat the stress-dependent component: in our protocols it includes both the initial and recovery fractions. In Hill's analysis it would include degraded internal work, shortening heat and heat associated with the persistance of tension [20]. In a recent alternative analysis it would be considered to be degraded potential energy [28].

From a thermodynamic viewpoint the measured enthalpy (stress-dependent heat) must represent i) degraded internal work i.e. some fraction of the free energy made available when ATP is hydrolysed; ii) degraded free energy i.e. the balance of the free energy not converted into work and iii) the heat of the accompanying entropy change.

Under physiological conditions it should be realised that if ΔG varies considerably in vivo, then the entropic contribution will also change [14].

In afterloaded isotonic contractions additional energy is liberated, shown schematically by the dotted line in Fig. 2, and the question arises as to whether this extra energy can be related to the external work. In [17] I subtracted the external work from the total enthalpy and this left an energy surplus which at the time was interpreted in terms of a shortening heat component [19, 21] but by 1970, I realised I had probably made a mistake and was suggesting that there had to be an allowance made for a recovery heat counterpart of the work term [16]. When such a correction was made there was scarcely any en-

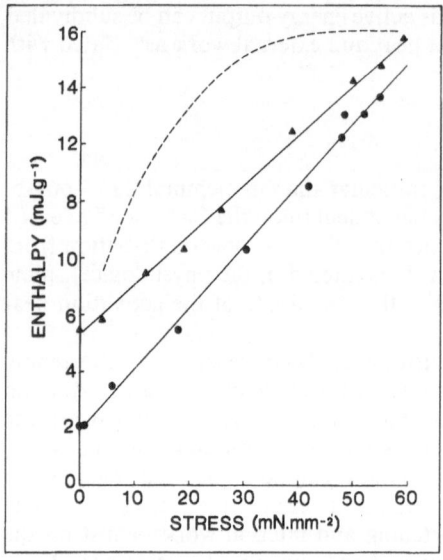

Fig. 2. Isometric heat versus peak stress data from rabbit papillary muscle, obtained by gradually shortening the muscle down its length : tension curve (●—●) and using the latency release technique (▲—▲). The position of the isotonic enthalpy versus stress (load) is also shown (dotted line).

ergy surplus except at the light load end of the enthalpy: load relationship. In 1974 Chapman and I suggested, on theoretical and experimental grounds, that the recovery heat component should be about 71% of the initial energy magnitude [5]. Recently I have collected some more afterload enthalpy data and corrected the total enthalpy for both the work term and its recovery heat counterpart. As suspected, there is essentially no surplus heat over the medium to heavy load range: the assumption of a higher recovery heat value would make significant inroads into the stress-dependent heat. There still tends to be some heat unaccounted for in the low (0.05 to 0.3 P_o) load range but this surplus almost completely disappears if the latency release activation heat value is used in the analysis (this, as mentioned above, necessitates a somewhat less steep *isometric* enthalpy: load relationship). There can of course be arguments about whether the isometric heat: load (stress) relationship is the proper baseline to use in the above analysis. Since no load can be lifted until a muscle develops a force equal to the load it seems intuitively obvious that more crossbridges and hence more ATP will turnover in a contraction where shortening occurs.

The pressure volume area (PVA) concept

I would now like to show that the papillary muscle energy data collected by myself and others [6, 25] is consistent with data obtained on the cross-perfused working dog heart. Since the early 1970's Sagawa et al. have been looking at cardiac mechanics in terms of work diagrams and pressure-volume relationships. They believe (see [27] for a review) that the behaviour of a cardiac chamber can be described by considering it to be a time varying compliance in which the end-systolic pressure volume points fall on a straight line, regardless of pre- and afterload magnitude (provided contractility is constant). Recently Suga et al. [27, 28] have shown that the per beat cardiac oxygen consumption is linearly related to the left ventricular end-systolic pressure volume area, PVA. PVA in an ejecting heart is the sum of stroke work (W) and a potential energy (PE) term: the details are shown in Fig. 3 (left plot).

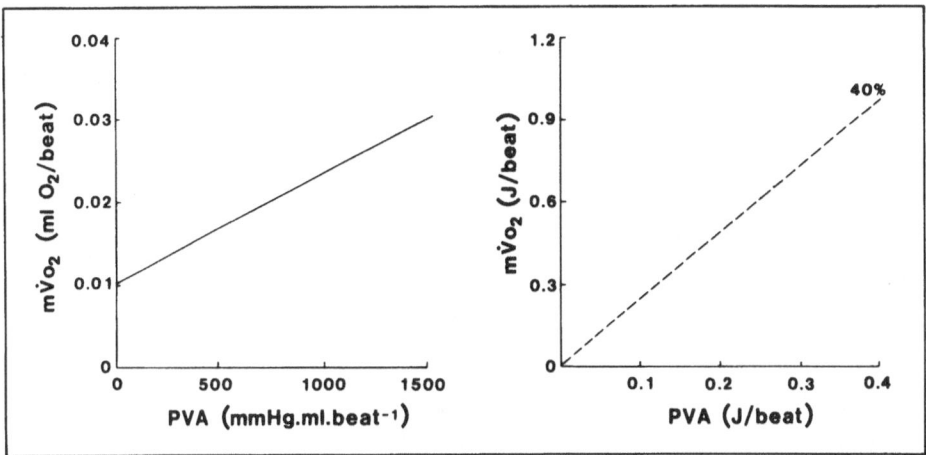

Fig. 3. Left: schematic diagram of the linear relationship between myocardial oxygen consumption per beat (mVo$_2$) and pressure-volume area, PVA. Right: data converted into standard energy units but with basal and activation components subtracted from mVo$_2$. Based on the data of [27, 28].

Suga has shown that if the basal and activation oxygen consumption terms are separated from the total oxygen cost per beat, then the ratio of PVA to the residual or corrected oxygen cost is *constant* regardless of the afterload size; see Fig. 3 (right plot). The slope of the relationship suggests an in vivo PVA efficiency of approximately 40%.

Chapman and I have shown, on theoretical grounds, that this result is consistent with existing myothermic and polarographic papillary muscle data [12, 15], but it seems to present problems for current muscle models. In reality Suga's dog heart data are compatible with the papillary muscle data for two reasons:
i) the relationship between PVA and load is curvilinear (convex) provided the tension:length, or pressure:volume relationship is approximately triangular and
ii) PVA efficiency defined as above is constant.

Suga et al. are actually detecting an initial constant crossbridge efficiency of about 68% [15]. It should be noted that by mistake in our paper, I constructed an enthalpy:load relationship using the thermodynamic i.e. free energy efficiency (56%) rather than the enthalpy efficiency. It is quite clear from Suga's and our data that there is a pronounced Fenn effect, at least in terms of statement ii, i.e. the nature of the load is determining the energy output.

Over the last 18 months I have been able to make a more searching test of the apparent agreement between the two sets of data. Up to the present, I have not varied either pre-load or contractility, only afterload. In terms of experimental analysis it is necessary to make two points. Firstly I have obtained stress-dependent heat:load relationships by latency releases of different magnitudes and fitted the data by linear regression. Secondly I have calculated PVA as the sum of the measured external work plus the potential energy. The latter value has been calculated from the length:tension relationship measured directly in various afterloaded isotonic contractions. A typical experimental result is shown in Fig. 4[1].

---

[1] The latency release heat:load relationship is shown (solid line) and data obtained when energy equivalent to 1.71. (We) is subtracted from each isotonic enthalpy point shown (squares) [5].

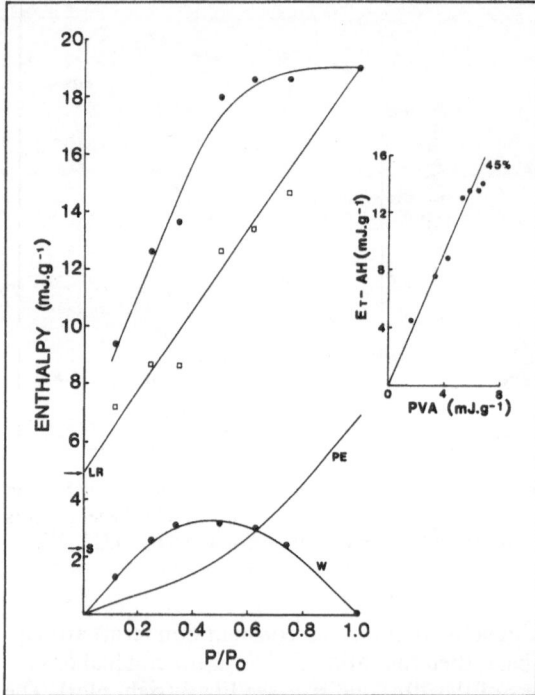

Fig. 4. Total isotonic enthalpy (E_T), external work (We), potential energy (PE), against relative load, from a rabbit papillary muscle, $l_{max} = 4.2$ mm, mass $= 2.9$ mg. The activation heat measured with the shortening protocol is also shown (S). Inset: Isotonic enthalpy minus activation heat, E_T-AH, against the sum of We and PE for different loads.

There seems to be reasonable agreement with Suga's data. The PVA efficiency calculated here as We + potential energy/total active heat-activation heat is fairly constant, (inset to Fig. 4), particularly over the 0.5 to 1.0 Po range [9]. There is a tendency in the preparations examined so far for the PVA efficiency to fall as the afterload tends to zero. The goodness of fit in the light load region will be critically dependent upon the real in vivo magnitude of the activation heat. High values will make the agreement good, low estimates will make it poor, at least in the low load range.

Since at the sarcomere level there is not much difference between cardiac and skeletal muscle the question arises as to whether this type of analysis can be extended to skeletal muscle. I am convinced that i) it can only be applied to twitch contractions and ii) that it will only be useful in those muscles that exhibit a steep fall off in their enthalpy output over the medium to light load range.

I have re-analysed some twitch skeletal data obtained from frog and toad sartorii at 10 °C [13]. The potential energy (PE) term was calculated, as above, from the shortening measured under different afterloads. The activation heat was taken to be 25% of the isometric initial heat at l_{max} and then initial isotonic enthalpy minus the activation heat was plotted against the "PVA" equivalent (W + PE) (Fig. 5). I have drawn in the 68% isoefficiency line to allow comparison with cardiac data. The data points are for 0.05, 0.2, 0.4, 0.6, 0.8 and 1.0% loads, where P_o is the *twitch* peak force. Clearly the data don't fall on an isoefficiency line but as the afterload increases the "PVA" efficiency does climb towards the value recorded in cardiac muscle. It is also apparent that the divergence from isoefficiency is much more evident in the faster (frog) muscle.

I have long been attracted to the hypothesis that transduction efficiency is practically the same in all muscle types, see also [24]. In the context of this paper, I find it interesting

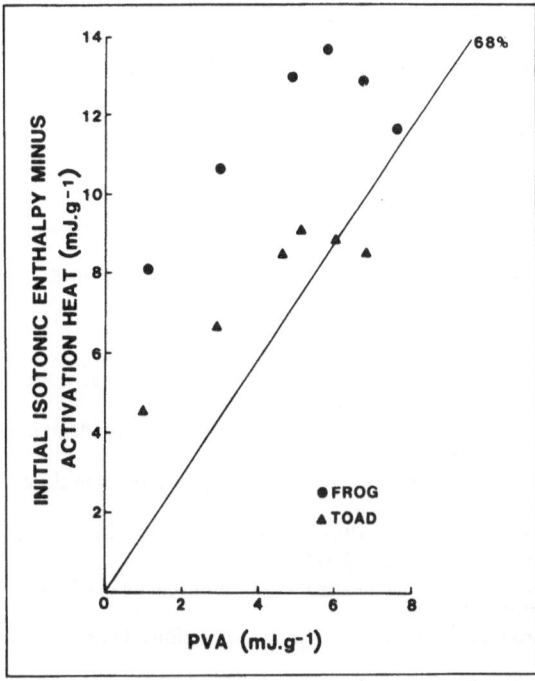

Fig. 5. Initial isotonic enthalpy minus activation heat against the sum of We and PE for frog (●) and toad (▲) sartorii. Data from Tables 1 and 2 of [13].

that when Brandt and Orentlicher [3] re-analysed Hill's frog sartorius data [19] and Woledge's tortoise data [31] they found essentially constant (0.69–0.76) transduction efficiency (W/W + shortening heat): it only decreased over the 0 to 0.2 P_o load range. I have some reservations about this comparison, as Hill and Woledge's data were essentially obtained in tetanized muscles released to different afterloads but there are certain similarities, e.g. a maximum efficiency value of about 70% was obtained and the measured efficiency was practically independent of load.

In an interesting article on the Fenn effect, Rall [26] concluded that the "cardiac results are strikingly different and cannot lead to the same (i.e. Fenn's) conclusion" but I hope that I have been able to argue that it is certainly possible to accept the second (B) of Fenn's two main conclusions. Perhaps we should have reservations about using amphibian skeletal muscle data as the "gold standard" in muscle energetics?

Acknowledgements. It is a pleasure to acknowledge the help I have received over the years from my colleagues Dr. B. Chapman, Dr. D. Loiselle and Dr. I. R. Wendt. I am indebted to Dr. Wendt for the toad ventricular energetic data shown in Fig. 1. This article was written whilst on sabbatical leave and I would like than Professor John Gavin of the Pathology Department, Auckland University for the facilities made available to me.

References

1. Allen DG, Kentish JC (1985) The cellular basis of the length-tension relations in cardiac muscle. J Mol Cell Cardiol 17:821–840
2. Alpert NR, Mulieri LA (1982) Increased myothermal economy of isometric force generation in compensated cardiac hypertrophy induced by pulmonary artery constriction in rabbit. A characterization of heat liberation in normal and hypertrophied right ventricular papillary muscles. Circ Res 50:491–500

3. Brandt PW, Orentlicher M (1972) Muscle energetics and the Fenn effect. Biophys J 12:512–527
4. Chapman JB (1982) Heat production. In: Drake-Holland AJ, Noble MM (eds) Cardiac metabolism. Wiley, Chichester, pp 239–256
5. Chapman JB, Gibbs CL (1974) The effect of metabolic substrate on mechanical activity and heat production in rabbit papillary muscle. Cardiovasc Res 8:656–667
6. Coleman HN, Sonnenblick EH, Braunwald E (1969) Myocardial oxygen consumption associated with external work: the Fenn effect. Am J Physiol 217:291–296
7. Crow MT, Kushmerick MJ (1982) Chemical energetics of slow- and fast-twitch muscles of the mouse. J Gen Physiol 79:147–166
8. Fenn WO (1923) A quantitative comparison between the energy liberated and the work performed by the isolated sartorius of the frog. J Physiol (Lond) 58:175–203
9. Ford LE (1980) Effect of afterload reduction on myocardial energetics. Circ Res 46:161–166
10. Gibbs CL (1978) Cardiac energetics. Physiol Rev 58:174–254
11. Gibbs CL (1982) Modification of the physiological determinants of cardiac energy expernditure by pharmacological agents. Pharmacol Ther 18:133–152
12. Gibbs CL (1985) Physiological factors determining cardiac energy expenditure. In: Sideman S, Beyer R (eds) Simulation and imaging of the cardiac system. Karger, pp 358–377
13. Gibbs CL, Chapman JB (1974) The effect of stimulus conditions and temperature upon the energy output of frog and toad sartorii. Am J Physiol 227:964–971
14. Gibbs CL, Chapman JB (1979) Cardiac energetics. Handbook of Physiology, Cardiovascular System. Bethesda MD: Am Physiol Soc, vol 2, chap. 22, pp 775–804
15. Gibbs CL, Chapman JB (1985) Cardiac mechanics and energetics: chemomechanical transduction in cardiac muscle. Am J Physiol 249:H199–H206
16. Gibbs CL, Gibson WR (1970) Energy production in cardiac isotonic contractions. J Gen Physiol 56:732–750
17. Gibbs CL, Mommants WFHM, Ricchivti NV (1967) Energetics of cardiac contractions. J Physiol (Lond) 191:25–46
18. Gibbs CL, Vaughan P (1968) The effect of calcium depletion upon the tension-independent component of cardiac heat production. J Gen Physiol 52:533–549
19. Hill AV (1964) The effect of load on the heat of shortening of muscle. Proc R Soc (Lond) Ser B 159:297–318
20. Hill AV (1964) The effect of tension in prolonging the active state in a twitch. Proc R Soc (Lond) Ser B 159:589–595
21. Hill AV (1964) The variation of total heat production in a twitch with velocity of shortening. Proc R Soc (Lond) Ser B 159:596–605
22. Jewell BR (1977) A re-examination of the influence of muscle length on myocardial performance. Circulation 40:221–230
23. Loiselle DS, Gibbs CL (1979) Species differences in cardiac energetics. Am J Physiol 237:H90–H98
24. Oplatka A (1972) On the mechanochemistry of muscular contraction. J Theor Biol 34:379–403
25. Pool PE, Chandler BM, Seagren SC, Sonnenblick EH (1968) Mechanochemistry of cardiac muscle II. The isotonic contraction. Circ Res 22:465–472
26. Rall JA (1982) Sense and nonsence about the Fenn effect. Am J Physiol 242:H1–H6
27. Sagawa K (1978) The ventricular pressure-volume diagram revisited. Circ Res 43:667–687
28. Suga H, Hayashi T, Shirahata M (1981) Ventricular systolic pressure-volume area as predictor of cardiac oxygen consumption. Am J Physiol 240:H39–H44
29. Suga H, Hisano R, Goto Y, Yamada O, Igashari Y (1983) Effect of positive inotropic agents on the relation between oxygen consumption and systolic pressure-volume area in canine left ventricle. Circ Res 53:306–318
30. Wendt IR, Gibbs CL (1976) Recovery heat production of mammalian fast- and slow-twitch muscles. Am J Physiol 230:1637–1643
31. Woledge RC (1968) The energetics of tortoise muscle. J Physiol (Lond) 197:685–707

Author's address:
Dr. C. Gibbs, Department of Physiology, Monash University, Clayton, Victoria 3168, Australia

Cardiac energetics: significance of mitochondria

M. Siess

Department of Pharmacology, Faculty of Theoretical Medicine, University of Tübingen, F.R.G.

Summary

The mitochondrial activity as the energy producing step during biological oxidation was observed at rest and its regulation by the energy consuming auxotonic contractile work, depending on the preload, afterload and beat rate in isolated superfused left guinea pig atria. The mitochondrial activity was measured by (1) continuous determination of the O_2 uptake rate, (2) the rate of $^{14}CO_2$ production from labelled glucose or FFA and (3) separate measurements of the atrial ATP-, ADP-, AMP-, CP- and NAD-concentrations, for determination of the energy state. Some results, with points of general interest, are reported and discussed, including this model, former studies about cardiac energetics and the efficiency of cardiac work, reviewed recently [21].

A. Influences on resting O_2 uptake

1. Addition of the FFA increases the resting O_2 uptake by a direct activating effect in the respiratory chain, which cannot be inhibited by nifedipine.

   2. K^+ depolarization enhances the O_2 uptake after addition of 10 mmol/l to 60 mmol/l KCl between 10 and 100% by activation of the slow Ca^{++} channels with influx of Ca^{++} completely inhibited by nifedipine. The energy quotient is not improved by a 100% increased resting O_2 uptake and the new synthesis of adenine nucleotides and NAD from precursors is inhibited, indicating a lowered energy coupling in the respiratory chain.

   2.1. Between 90 mmol/l and 250 mmol/l KCl the O_2 uptake increases to 100 and 250% O_2 not inhibited by nifedipine and presumably due to an increased Na^+/Ca^{++} exchange: high external Ca^{++} or low Na^+ enhances the O_2 uptake together with an activation of the actomyosin system (contracture).

   3. External Ca^{++}: Concentrations between 0–7.5 mmol/l Ca^{++} do not influence the resting O_2 uptake during a normal resting potential. After K^+ depolarization (30 mmol/l KCl) addition of 1.5 mmol/l Ca^{++} or 10^{-6} mol/l adrenaline enhances the increased O_2 uptake.

   3.1. Activation of the Na^+/Ca^{++} exchange by cardiac glycosides (10^{-6} mol/l g-strophanthin = ouabain) increases the resting O_2 uptake to $\sim 10\%$ during 30 min: Ca^{++} enhances this effect not reversed by nifedipine.

   4. Resting tension and O_2-consumption ("Feng" or "Stretch" effect): Stretch with 15 mN of left guinea pig atria or rat papillary muscles causes no enhancement of resting O_2 uptake. The Feng effect can be observed only after activation of the slow Ca^{++} channels by K^+ depolarization (30 mmol/l) or perhaps due to ischaemia and anoxia.

B. Regulation of mitochondrial activity by auxotonic contractile work

1. The auxotonic contractile work was measured by displacement of a calibrated spring blade. The external work can be calculated in mm × mN, counteracting by force and heat, the energy of the elastic elements in the muscle preparation, stretched with the preload tension and also the external con-

tractile work by shortening and movement of the spring blade (mm) by force (mN). The total energy of the activated actomyosin system (total work) is stored in the displaced spring blade, due to the constant of the spring blade, the preload tension and the afterload force.

2. Frank-Starling effect: Increased preload tension enhances by positive inotropy the external auxotonic contractile work/beat at 2.5–10 mN, with a maximum inotropy at a frequency of 2 Hz between 7.5–10 mN preload and at 3 Hz between 5–7.5 mN, respectively. The O_2 uptake increases related to the work with a lower percentage providing at the maximum of the inotropy the best efficiency of external contractile work which decreases again with the fading inotropy between 10 and 15 mN preload.

3. Auxotonic contractile work and O_2 uptake at the same preload by a changed afterload resistance. A considerably increased total auxotonic contractile work with a shift in the ratio of internal/external work to higher values can be observed by switching the afterload resistance from a hard spring blade (isometric type of auxotonic contraction) to a soft one (isotonic type) without any change of the O_2 uptake at a constant beat rate and the same preload (5 mN) providing a constant lengthening of the muscle fibres. The same amount of ATP dependent energy is therefore transferred with the hard spring blade in very low internal work (heat), low external contractile work but high force compared with the movement of the soft one: here a 6-fold higher internal work (heat) as an explanation for the Fenn effect and a higher external contractile work with a lower force can be observed with an improvement in the efficiency of the external contractile work. With a beat rate of 1 and 2 Hz, the same can be principally observed. A general view of the cardiac energetics with the different sources of heat delivery during energy production and energy consumption is demonstrated.

The findings observed with this model are compared with the cardiac energetics of heart failure and hypertension.

I. Introduction and methods

Mitochondria are subcellular organelles and provide, as a fundamental property, the regulation of intracellular aerobic energy production by the biological oxidation of FFA, carbohydrates and various metabolites to CO_2 and H_2O, resulting in production of high energy phosphates and heat (I. energy coupling step, Fig. 1). The activity of mitochondrial function in cardiac muscle preparations can be observed by measuring: 1. the O_2 uptake, 2. the $^{14}CO_2$ production from ^{14}C-labelled substrates and 3. the cytosolic phosphorylation potential. It would be necessary to determine additionally 4. the cardiac work and 5. the developed heat [6] to have the total balance account of produced and consumed cardiac energy. However, it is difficult for technical reasons to observe all these parameters simultaneously and continuously. We have measured the O_2 uptake as well as the $^{14}CO_2$ production rate at rest and together with the auxotonic contractile work in superfused isolated left guinea pig atria. We observed the mitochondrial activity at rest as the energy producing step for the basal energy need and the regulation by the energy consuming contractile work (II. excitation-contraction coupling step, Fig. 1) in dependance of the preload, the afterload and the beat rate. High energy phosphate concentrations (ATP, ADP, AMP, CP and NAD) have been determined in separate studies to observe changes in the atrial energy state by calculation of the energy quotient [ATP]/([ADP]+[AMP]) as one parameter of the phosphorylation potential [7, 8, 9]. The methods in detail can be found in [3, 4, 18, 20, 22] and most of our results previously published [15, 16, 17, 19]. Since one review about our findings [21] is close to the topic of this book and has been published here only recently, I would like to mention some important points of general interest.

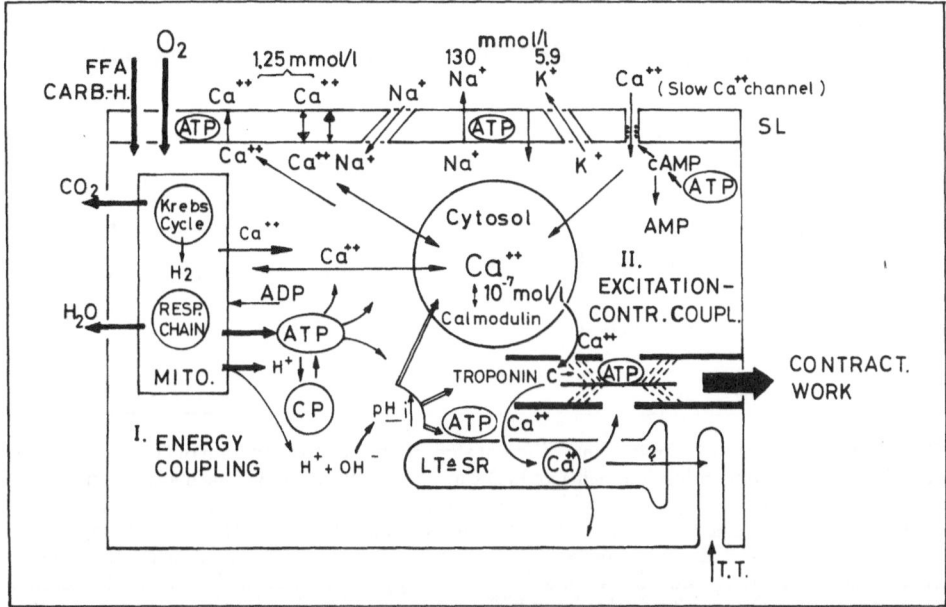

Fig. 1. Simplified scheme of the cardiac muscle cell as an energy-producing and -consuming system with transfer in contractile work and heat. Energy production in the I. energy coupling step in mitochondria during biological oxidation of FFA and carbohydrates, producing high energy phosphates (ATP stored in CP). The most energy consuming step is the II. excitation-contraction coupling with the energy need for ion- and electrolyte-fluxes and gradients in the sarcolemma (SL). From left to right 1. Ca^{++} pump, 2. Na^+/Ca^{++} exchange, 3. fast Na^{++} channel, 4. Na^+/Ka^+ pump, K^+ channel and the slow Ca^{++} inward channel. The actomyosin system producing contractile work is activated by increased cytosolic free Ca^{++} ($10^{-7} < 10^{-5}$ mol/l) connected with the calmodulin system and the pH_i. Relaxation occurs by uptake of Ca^{++} into the longitudinal (LT) and transversal (TT) tubuli of the sarcoplasmatic reticulum (SR) and by Ca^{++} efflux mechanisms through and into the SL (glycocalyx).

II. Results and discussion

A. Resting O_2 consumption

The resting O_2 consumption is dependent on temperature, pH, external K^+ concentration, intracellular Ca^{++} transport mechanisms, as well as the kind of substrate oxidized in the mitochondria [5, 10, 21].

1. Influence of FFA

Addition of FFA to the Krebs-Henseleit-solution very quickly increases the resting O_2 consumption of left atria to an ~13% higher level (Fig. 2). In this way, it could be shown that the efficiency of cardiac work is lowered [15, 16, 18, 19, 20, 21]: Since the competitive inhibition of the exogenous ^{14}C glucose oxidation rate (-90%) by FFA is a long lasting process reaching a new steady state after ~1.5 h, we assume a direct regulatory effect of FFA in the respiratory chain. This could be explained as a compensation for the lower

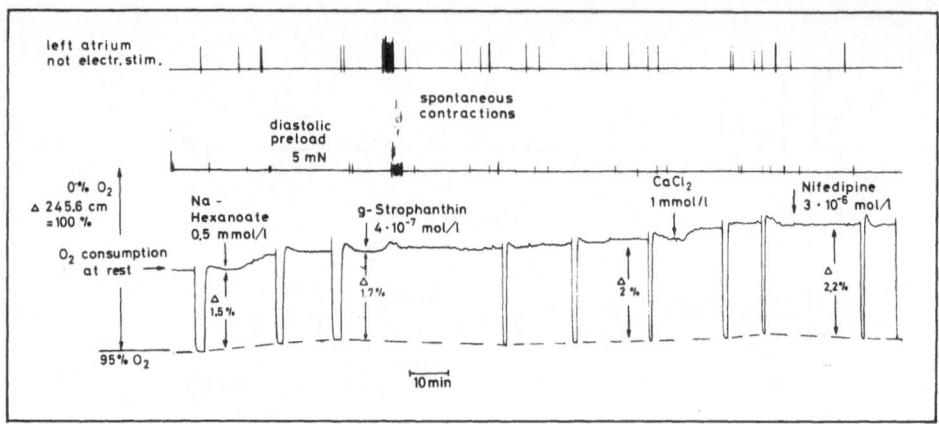

Fig. 2. Increase of resting O_2 uptake in 3 different steps by cumulative addition of (1) Na-hexanoate (C_6-FFA), (2) g-strophanthin (ouabain) and (3) $CaCl_2$. This effect of all 3 agents is not influenced by nifedipine. Original curve taken with a superfused resting left guinea pig atrium (K.-H.-solution + 15 mmol/l glucose, 30°C, pH 7,4. From top to bottom: 1. beat rate: 0 min^{-1} with single spontaneous contractions 2. diastolic preload tension 5 mN: some single spontaneous contractions recorded by a separate channel. 3. and 4. difference of O_2 pressure measured in the K.-H.-solution (gassed with 95% O_2/5% CO_2) behind and in front of the muscle chamber at 1 ml/min pump flow (see text).

combustion value of FFA and the decreased yield of ATP during oxidation of FFA calculated for 1 mol O_2 and compared with glucose [21]. This sudden increase of O_2 uptake due to FFA cannot be inhibited by nifedipine (Fig. 2).

2. Influence of external K^+ concentration (K^+ depolarization)

It could be shown that 10 mmol/l KCl added to normal KH solution in the left atria, stimulated with 0.5 Hz causes a short positive inotropic peak before standstill occurs. The oxygen consumption at rest increases here to $\sim 10\%$ compared with the initial resting O_2 uptake, due to K^+ depolarization. Nifedipine added in previous studies completely inopening of the slow Ca^{++} channels). In Ca^{++} free solution there is no increase of O_2 uptake, due to K^+ depolarization. Nifedipine, added in previous studies completely, inhibits the K^+-induced enhancement of O_2 uptake in the range between 10 and 90 mmol/l KCl [16, 20, 21]. However, when nifedipine is added at the peak of the increased O_2 uptake, reached 5–15 min after K^+ depolarization, the increased O_2 uptake rate decreases very slowly to the initial value during ~ 90 to 120 min (Fig. 3). The enhanced O_2 uptake, due to K^+ depolarization, cannot be further reversed to the initial value by blocking the slow Ca^{++} channels when nifedipine is added ~ 2 h after K^+ depolarization (Fig. 3). The increased O_2 uptake between 15.9 and 65.9 mmol/l KCl is not connected with an improved energy quotient ([ATP]/[ADP]+[AMP]) and KCl inhibits also significantly and concentration dependent NAD- and adenine nucleotide – (ATP, ADP, AMP) – synthesis from precursors (adenine, ribose, nicotinic acid) during 5 h incubation time [17, 22] indicating uncoupling effects in the respiratory chain.

Higher concentrations of KCl (100 to 200 mmol/l) enhance, independent of nifedipine, the resting level of O_2 uptake from 220% to 350% above the initial value

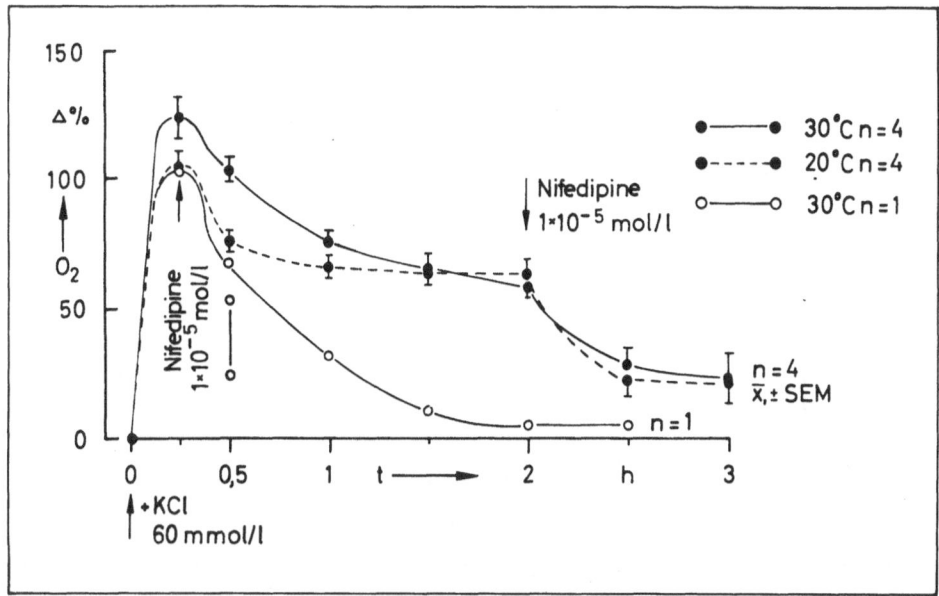

Fig. 3. Time course of the increased resting O_2 uptake of left guinea pig atria as a percentage of the initial value at 30 °C or 20 °C due to K^+ depolarization and inhibition by nifedipine. Addition of 60 mmol/l KCl increases the resting O_2 uptake at 30 °C and 20 °C by the same degree to 100–125%, reaching this peak after 5–15 min, declining to a 60% plateau level above the initial value during 1–2 h. Nifedipine, given after 2 h lowers this 60% level to only 25% above the initial values. Nifedipine, however, added at the peak after 5–15 min, decreases slowly the O_2 uptake rate and completely to initial values after 2 h. Nifedipine given before K^+ depolarization completely inhibits (up to 60–90 mmol/l KCl) the increase of O_2 uptake: the initial value remains unchanged (here not shown). The resting O_2 uptake at 20 °C is $\sim 20\%$ lower compared with the value at 30 °C (Siess et al. [17] modified).

(Fig. 4). Here obviously an increased Ca^{++} influx by an activated Na^+/Ca^{++} exchange takes place, whereas the influx of Ca^{++} through the nifedipine-dependent slow Ca^{++} channels at higher KCl concentrations is no more effective. This is in agreement with direct measurements of changes of the intracellular Ca^{++} concentrations under similar conditions in rat myocytes [14]. Perhaps, even at lower KCl concentrations, the Na^+/Ca^{++} exchange is activated time dependent, for the reason that nifedipine does not reduce the increased O_2 uptake to the initial values at a later stage (Fig. 3). The activation of the actomyosin system (contracture) occurs in concentrations above 100 mmol/l KCl if the external Ca^{++} concentration is enhanced (Fig. 4) or if that of Na^+ is lowered. We must therefore assume that activation of the mitochondria by Ca^{++} influx is separated from the activation of the actomyosin system (Fig. 1). This could be due to a different sensitivity of each system against free Ca^{++} or different Ca^{++} concentrations locally released at actomyosin and mitochondria, perhaps regulated by calmodulin. At high concentrations of KCl irreversible damage of mitochondria occurs and the O_2 uptake, stimulated to a maximum, decreases without influence on the degree of contracture (Fig. 4).

It is of special interest that even a low concentration, of 30 mmol/l KCl, sensitizes the resting atria for an additional increase of the resting O_2 uptake rate by raising the external

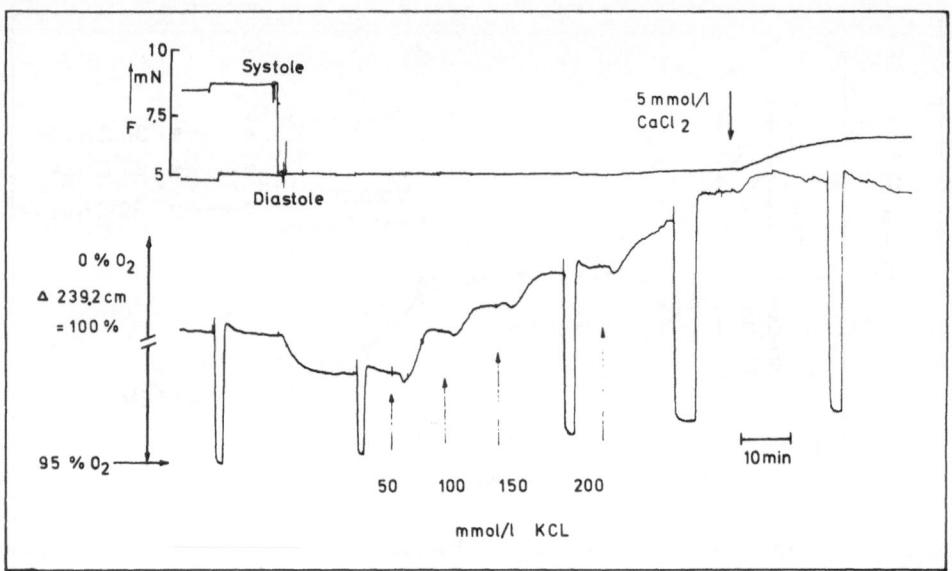

Fig. 4. Increased resting O_2 uptake with high K^+ concentrations and increased diastolic tension (contracture) by Ca^{++}: the electric stimulation of left guinea pig atria (1 Hz) was stopped and the steady state of resting O_2 uptake as initial value (100%) determined (resting potential: RP calculated to -83 mV); addition of 50 mmol/l KCl (RP -26 mV), 100 mmol/l KCl (-7 mV), 150 mmol/l ($+3$ mV), 200 mmol/l ($+10$ mV) increases the resting O_2 uptake from 100–350% without change of the preload tension of 5 mN. Only addition of 5 mmol/l $CaCl_2$ activates the actomyosin system with contracture and additional enhancement of O_2 uptake. This can be observed at depolarization above 90 mmol/l KCl by high external Ca^{++} or by lowering of the Na^+ concentration (here not shown). As a sign of mitochondrial damage by these high concentrations the O_2 uptake decreases slowly whereas the contracture remains on the same level (see text). O_2 uptake and contracture are not inhibited by nifedipine above 90 mmol/l KCl (here not shown).

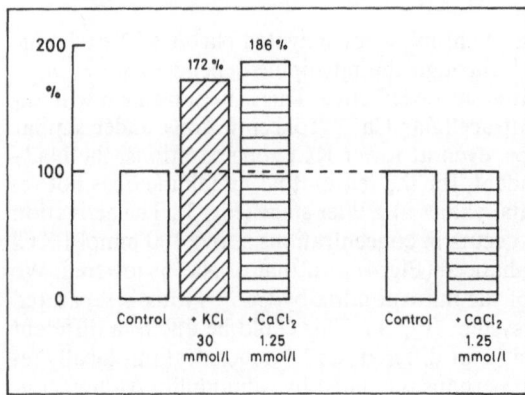

Fig. 5. The resting O_2 uptake (left atria) increased by KCl 30 mmol/l can be slightly enhanced by a higher external Ca^{++} concentration as a sign that the slow Ca^{++} channels are opened by K^+ depolarization. Without K^+ depolarization an increase of external Ca^{++} has no effect on the O_2 uptake.

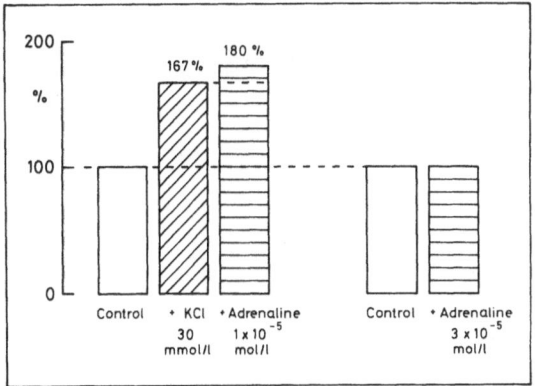

Fig. 6. Whereas addition of adrenaline without K^+ depolarization has no effect of the resting O_2 uptake during 30–60 min, the increased O_2 uptake after K^+ depolarization is enhanced by adrenaline slightly but significantly.

Ca^{++} concentration, or by addition of epinephrine (Figs. 5 and 6); neither of them is effective here without K^+ depolarization. Obviously, epinephrine enhances the Ca^{++} influx only after activation of the slow Ca^{++} channels by K^+ depolarization (Fig. 6).

3. Influence of external Ca^{++}

Enhancement of external Ca^{++} from 2.5 to 3.75 mmol/l Ca^{++} in the modified Krebs-Henseleit-solution does not influence the resting O_2 uptake during 30 min (Fig. 5). In the same way, an increase in external Ca^{++} from 0 to 7.5 mmol/l in resting atria is ineffective on O_2 uptake. During rest, the sarcolemma seems to act as a barrier to the influx of external Ca^{++}. Even addition of epinephrine to resting atria does not influence the initial O_2 uptake rate of atria incubated with 2.5 mmol/l Ca^{++} containing solutions during ~ 30 min (Fig. 6). However, an increase of resting O_2 to a 17% higher level during 2 h can be observed after addition of cardiac glycosides (for example, ouabain = g-strophanthin 10^{-6} mol/l) without contracture (Fig. 2). Since cardiacglycosides act by inhibition of the sarcolemmal Na^+/K^+ pump, this finding could be explained by an increase of the cytosolic Ca^{++} due to an activated Na^+/Ca^{++} exchange and therefore the O_2 uptake can be additionally enhanced by elevation of external Ca^{++} (Fig. 2). This effect of g-strophanthin on the resting O_2 uptake cannot be inhibited by nifedipine (Fig. 2) and can also be observed if glucose is the preferentially oxidized substrate.

4. Resting tension and O_2 consumption

("Stretch" or "Feng" effect)

We have reported [15, 16, 19, 21] that we have never seen an increased O_2 consumption caused by "stretching" of the resting left atria of guinea pigs. We have repeated these experiments with papillary muscles of rats, and also in this preparation we could not observe by "stretch" an enhanced O_2 consumption. It has been reported very recently [1] that after standstill of beating rat hearts with 30 mmol/l KCl and stretch by increased volume pressure, the O_2 uptake increased together with an activation of the creatine kinase. We have therefore repeated our experiments with the left atria of guinea pigs after K^+ depolarization. Figure 7 shows clearly that without K^+ depolarization an increased tension of 15 mN as well as a lowering to 1 mN has no effect on the resting O_2 uptake. Addition of 30 mmol/l KCl at 1 mN preload tension enhances, in this experiment, the rest-

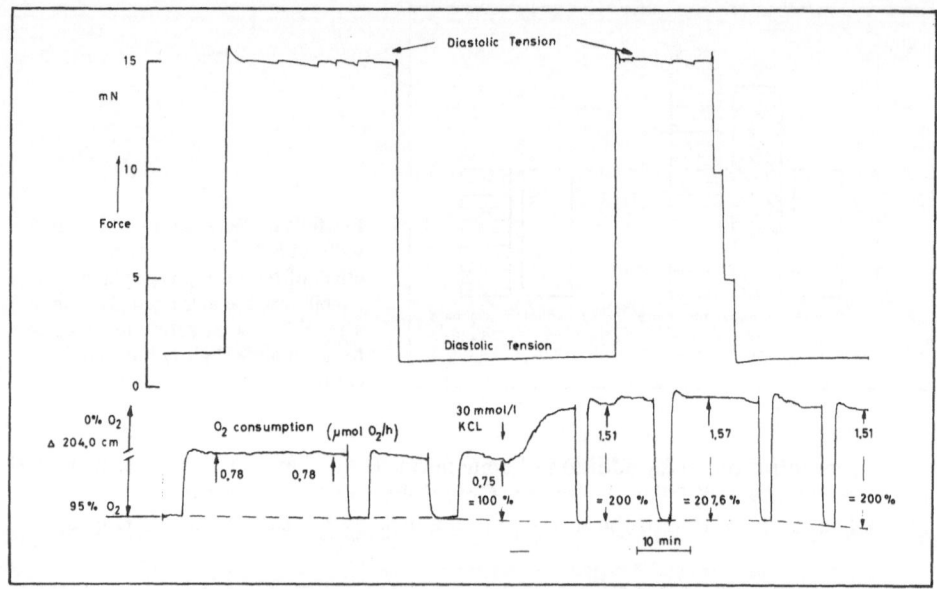

Fig. 7. Resting O_2 uptake in dependence on the diastolic tension: "Stretch" or "Feng"-effect (see text).

ing O_2 uptake from 100 to 200%. After repetition of the stretch with 15 mN preload we can now observe a small but significant additional increase in the resting O_2 uptake from 200 to 207.6% reversed by a lowered preload tension to 1 mN. This effect cannot be inhibited by β-adrenergic blockade, but by nifedipine. We assume, therefore, that the "stretch"- or "Feng"-effect ([21], reviewed) can be observed only in cardiac muscle preparations, arrested by K^+ depolarization, or perhaps also at a reduced resting potential due to hypoxic, anoxic or ischaemic conditions. The stretch, then, causes an additional influx of external Ca^{++} through the opened slow Ca^{++} channels activated by the conditions mentioned above. However, stretching of resting cardiac muscle preparations at normoxic energy conditions and at a normal resting potential, does not activate the mitochondrial activity measured with our very sensitive method of O_2 uptake: 1 nmol of O_2/ml difference in a 1113 nmol O_2/ml containing solution (at an airpressure of 730 mm Hg and 95% O_2), can be determined significantly see [16] and [20].

B. Regulation of the mitochondrial activity by the auxotonic contractile work

1. Energetics of the spring blade during auxotonic contraction

The auxotonic contraction moving a calibrated spring blade allows the measurement of the internal work, stored as mechanical energy in the spring blade (mm × mN) by the preload tension (mN) and the correlated displacement (mm) of the spring blade, together with the external contractile work (mm × mN) during contraction. In Fig. 8 the calibrated spring blade (10 mN/mm) is moved by stretching the resting left atrium with 5 mN preload in position II and the displacement of the spring blade in position I (0.5 mm = S1).

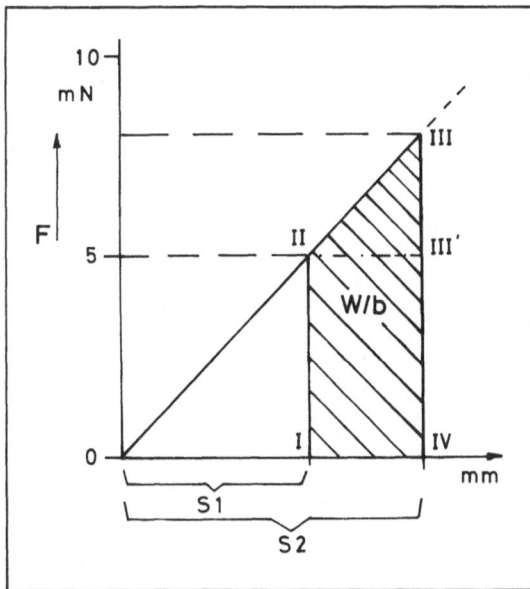

Fig. 8. Scheme of the mechanical energy stored in a calibrated spring blade (10 mN/1 mm) during auxotonic contraction of an atrium at 5 mN preload tension correlated with 0.5 mm displacement of the spring blade (see text).

At electric stimulation (0.1 ms, 0.3–2.0 mA, 15% above threshold) the sliding filaments in the actomyosin system are activated and force will be developed in direction I–II by internal shortening, without change of the length of the stretched muscle fibre, counteracting the preload tension of the stretched elastic elements in the muscle fibres. The activation energy of the atrial muscle required to overcome the 5 mN preload tension by 5 mN force is equal to the mechanical energy loaded in the spring blade by the preload and expressed in the white triangle 0-I-II as mm × mN. External mechanical contraction will start when the developed force exceeds the preload tension and external shortening moves the spring blade (I-IV) by the force increasing from II to III (mN). The hatched area I-II-III-IV represents in mm × mN the external contractile work produced by the activated actomyosin system, splitting ATP to ADP in the oscillating cross-bridge reactions. Visible on the chart is the "contractile amplitude" III'-III, shown as force with the energy of the triangle II-III-III' (Fig. 8). The activation energy as "internal work" to overcome the preload tension by force (white triangle) and the "external mechanical contractile work" (hatched area) together represent the total energy (0-III-IV) in mm × mN transferred from the liberated energy of the contracting muscle preparation into the displaced spring blade during one contraction cycle. This amount of mechanical energy stored in the spring blade stems from ATP split to ADP in the muscle fibres, and this energy will be transferred: 1. *during activation* as "internal" work only in heat and 5 mN force (preload tension) and can be measured by the constant of the spring blade (10 mN/1 mm) as 1.25 mm × mN mechanical energy representing the white triangle area and can be calculated by the mechanical heat equivalent to 1.25×10^{-6} kJ or 0.299×10^{-6} kcal; 2. *during contraction* as external contractile work (2.0 mm × mN as an example, Fig. 8) due to the developed force and shortening of the muscle fibres with an unknown amount of heat developed during contraction. At *inactivation* the force decreases to 0, the mechanical energy of the spring blade is released (hatched area) and the spring blade moves back into the position II of the 5 mm/mN preload tension due to the elastic elements in the muscle fibre and the correlated displacement of 0.5 mm (I), stretching in this way the shortened actomyosin system to the initial resting length of the atria. The energy of the white triangle stored in the spring blade represents the energy of the stretched elastic elements in the muscle fibres which has to be overcome during the next contraction cycle by the internal work. The degree of displacement of the spring blade by the preload tension at the same spring constant is correlated, but not equal, to the unknown lengthening of the muscle fibre, which depends on the (here unmeasured) resting length-tension relationship. The observed displacement of the spring blade at 5 mN preload tension is low with a hard spring blade, and increases considerably with a soft one, and also the energy of the loaded spring blade will be here considerably higher to antagonize the 5 mN preload tension by 5 mN force to start auxotonic shortening. However, with the same preload tension of 5 mN in the hard and in the soft spring blade, the lengthening of the atrium will be the same according to the resting length-tension relationship, in contrast to the different displacement of the spring blade. The release of different amounts of activation energy in the atrial muscle cell as internal work to produce 5 mN force and heat will be regulated by the *external* amount of energy loaded in the spring blade at rest, dependent on the spring blade constant and the preload. The directly correlating molecular equivalents of this "elastic" energy, loaded in the muscle fibres, and the connective tissue at rest as phenomena of *living* cardiac tissue, are unknown.

2. Frank-Starling mechanism and oxygen consumption
(Auxotonic contractile work and O_2 uptake in dependance of the preload)

We have seen above that an increased diastolic preload tension at the same constant as the spring blade has no effect on the resting O_2 uptake, in spite of the enhancing energy stored in the displaced spring blade and the increased stretch of the muscle fibres. However, an increased preload tension in stimulated or spontaneously beating atria enhances the internal and external contractile work/beat, in accordance with the Frank-Starling mechanism, together with the oxygen consumption in the range between 0, 2.5, 7.5, and 10 mN preload. Further enhancement of the preload to 12 and 15 mN decreases the contractile work/beat, however, without further increase of the O_2 uptake, which remains at the level reached with 10 mN and a tendency to a slow decrease after 1 h. At a beat rate of 2 Hz the maximum effect of inotropy is reached with 7.5–10 mN, at 3.5 Hz this maximum is shifted to a lower preload of about 5–7.5 mN [15, 16, 19, 21]. The oxygen consumption increases to a lower percentage than the auxotonic work/beat during the different steps raising the preload. Therefore the efficiency of contraction increases with preload and shows, at the maximum of Frank-Starling at 7.5 mN to 10 mN preload, the best efficiency, and at a higher frequency corresponding to between $\sim$5–7.5 mN. Also, the Frank-Starling-effect observed with a higher beat rate is connected with a lower efficiency, due to an increased need for pico-mol O_2 for the same contraction work/beat of 1 mm × mN as a multiplicating effect of the increased amount of contraction cycles [16, 19, 21].

The increased work/beat due to increased lengthening of the muscle between a distinct range of preload tension is well known as "Frank-Starling mechanism" and the maximum of inotropy is explained as the maximum activation of the cross-bridge reactions, at a distinct distance between the myosin and the actin filaments, by lengthening the sarcomeres, and related to Ca^{++} and ATPase activity [13]. The contraction work is the most ATP consuming step compared with resting conditions. The increased O_2 uptake in mitochondria will be regulated here especially by a lowered ATP/ADP quotient [2], perhaps also by an increasing cytosolic free Ca^{++} [11]. It is of interest that the O_2

uptake with increasing inotropy and work/beat is lower in percentage than the O_2 uptake rate. Therefore the efficiency of cardiac work related to O_2 uptake increases more with inotropy than with chronotropy. This shows the higher oxygen need for the same 1 mm/mN work, which increases considerably with the beat rate [19, 21]. Other regulatory factors of mitochondrial activity might also be important here, such as the $NAD^+/NADH$ quotient, the rate of oxidation of cytochrome c by the activity of the cytochrome c oxidase, the pH, and the total cytosolic phosphorylation potential, as we have reviewed previously [21].

3. Auxotonic contractile work and O_2 uptake at the same preload and a different afterload due to a change of the spring blade constant

We have recently reported [21] experiments on spontaneously beating atria (115 bpm), stretched with the same preload of 5 mN to the same length of the sarcomeres in the muscle fibres, measuring the auxotonic contractile work and the oxygen consumption by a change in the afterload resistance, using hard and soft spring blades with different constants. Here it could be shown that the O_2 consumption of these atria at the same 5 mN preload tension remains constantly at the hard and the soft spring blade in spite of the different produced force and work of contraction: The force (mN) decreased by change from the hard spring blade to the soft one from 10.3 to 7.7 mN whereas the "external" work (+269%) and the "internal" work (+704%) and the total work (+365%) increased considerably and here also the efficiency [21]. In conclusion, we can assume that the afterload resistance against contraction is decisive for the efficiency of the total work and especially for the most important external contractile work whereas the released energy in the muscle is the same at the same lengthening of the sarcomeres by the same preload.

We have now repeated these experiments with left guinea pig atria stimulated with a low beat rate of 1 Hz at 30 °C, moving with the same atria a hard and a soft spring blade to compare the auxotonic work with the O_2 uptake. The atria had been stretched with 5 mN preload to the same length. Figure 9 shows the results of one experiment: according to the high resistance of the hard spring blade against contraction of the sliding filament system, the developed force with 15 mN is considerably higher compared with the force of the soft one (8 mN). The external work increased, by the switch from the hard to the soft spring blade, by 30%, the internal work by 550%, the energy of the total work connected with the contraction cycle increased by 90%. The amount of energy needed to develop 5 mN force by activation of the crossbridge reactions counteracting the 5 mN preload tension of the elastic elements, is considerably lower in the hard spring blade compared with the soft one (Fig. 9, white triangle areas). We can observe with the soft spring blade that 5 mN preload displaces this more isotonic type of spring blade much more than 5 mN preload moving the hard isometric type. Therefore, the amount of energy loaded with 5 mN preload in the soft spring blade is 5.5 fold higher to produce 5 mN force compared with the hard isometric type. It is of interest that the "visible" part of the energy of the contractile amplitude is, however, higher at the hard isometric type of auxotonic contraction. In Fig. 10, it is shown that the resting O_2 uptake is increased by the auxotonic work with 60 bpm (1 Hz) from 100% to 214%. In our previous study [21] we could show, that the resting O_2 uptake increased from 100% to 400% due to a higher beat rate of 115 bpm of the spontaneously beating atrium. We have seen in both experiments that the O_2 consumption in the same atrium does not change if we switch, at the same preload and the same beat rate, the auxotonic type of contraction from the hard isometric to the soft more isotonic type. Therefore we can confirm, even at the low beat

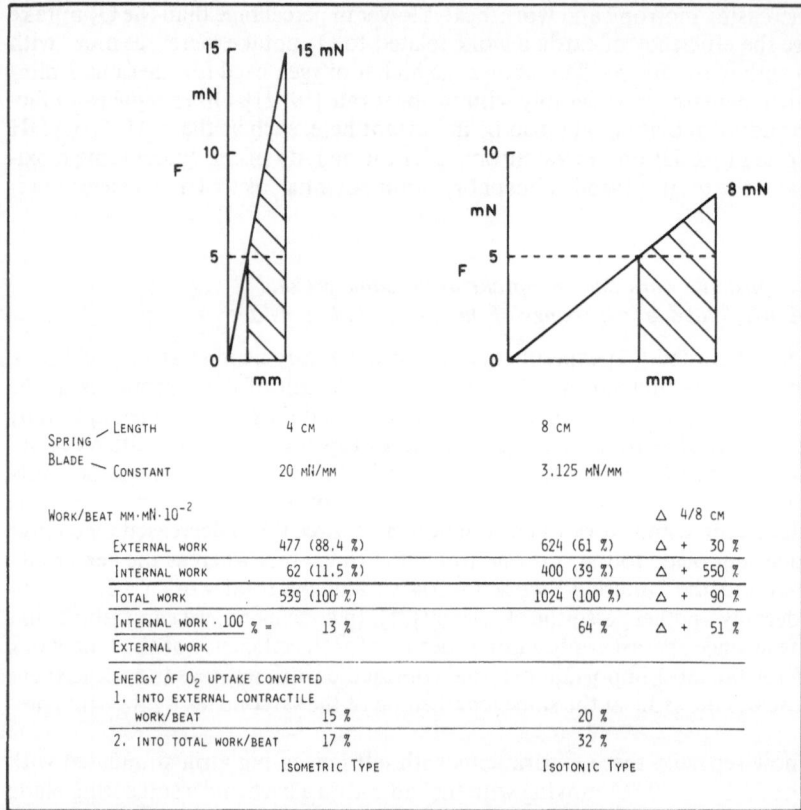

Fig. 9. Energy converted into contractile work at a low beat rate (60 bpm, 1 Hz). Change of the auxotonic contractile work of the left atrium stimulated with 1 Hz at the same preload tension (5 mN) depending on the afterload resistance moving a hard or a soft spring blade. In spite of the different force and work, the oxygen consumption does not change at the same beat rate and at the same preload tension with the same lengthening of the sarcomeres in the muscle fibres. In this way the efficiency of the external contractile work is lowered at the high afterload. However, the ratio of internal work to external contractile work increases with the lower afterload, also impairing the efficiency of the external cardiac work in a different way (see text).

rate of 60 bpm, our previous findings, that at the same lengthening of the sarcomeres due to the same preload and at the same beat rate the energy supply of ATP released during one contraction cycle is the same, in spite of the produced different force and work by change of the afterload.

If we suppose a 50% efficiency of energy coupling in the mitochondria [21] there will be, transferred from a 200% increased O_2 uptake, 100% increase in ATP for basal metabolism and the auxotonic work (Fig. 10); 100% is lost as "metabolic" heat [21]. According to the different total work at the "isometric" and the "isotonic" type of auxotonic contraction, we can calculate that 35% out of 100% ATP is consumed at the isometric type for total work and 65% for ATP dependent heat and basal metabolism, whereas at the isotonic type, 65% energy of ATP is transferred in total work and only 35% in heat

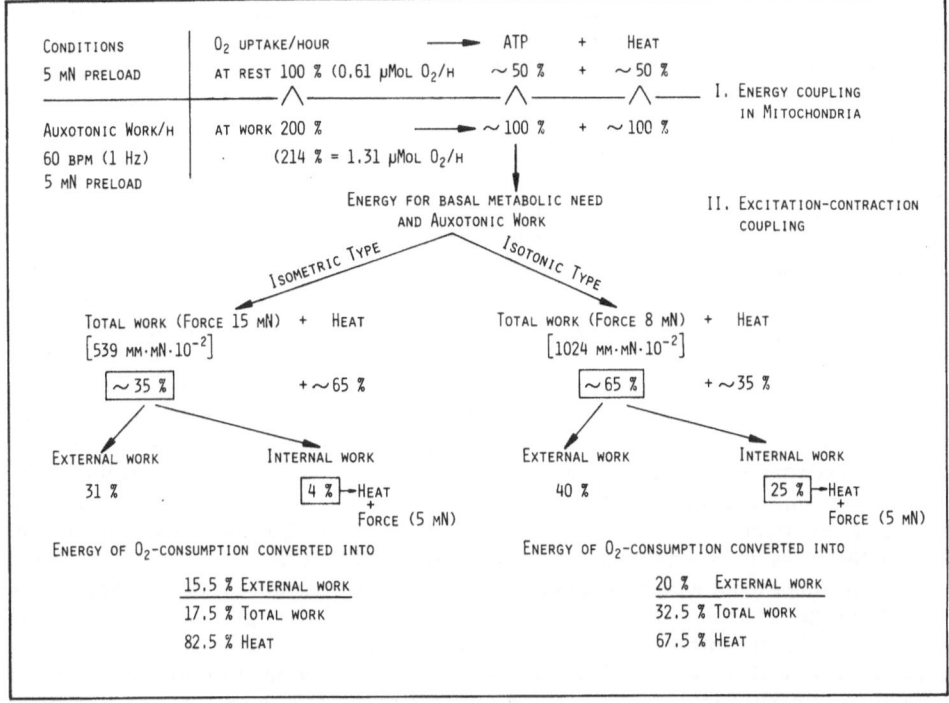

Fig. 10. Calculated energy balance and distribution of the produced energy by O_2 uptake transfered into work, and heat of the same atrium moving at the same preload tension a hard (isometric type) or soft (isotonic type) spring blade.

and basal energy demand. It is understandable that the isometric type of auxotonic contraction with a low total work and high force will produce more heat than the more isotonic type with the low resistance afterload against shortening by contraction. This seems, at first sight, not to be in agreement with the Fenn effect (reviewed in ref. 21), but if we now differentiate the energy need for external and internal work, we see that from the same released energy of 100% ATP at the isometric type, only 4% of the 35% produced total work is needed to develop 5 mN force and heat, whereas at the isotonic type a 6-fold higher amount (25%) of energy is converted into 5 mN force and heat: The external load in the spring blade governs the internal work to produce 5 mN force. This is in agreement with our previous findings at a higher beat rate and could explain the Fenn effect that "a muscle by change from the isometric contraction to isotonic shortening develops additional heat". Also it confirms the definition of Mommaerts [12] "What is the Fenn effect? Muscle is a regulatory engine, the energy output of which is governed by the load" [12], and we would differentiate "governed by the preload *and* by the afterload".

If we relate the consumed energy to 100% O_2 uptake, we find that this energy is converted by the isometric type of auxotonic contraction to 15.5% in external work, and 17.5% into the total work, whereas by the isotonic type, 20% is converted into external work and 32.5% into total work during activation and contraction of the actomyosin system. The rest is partially lost as metabolic heat during energy coupling or needed for ATP dependent basal metabolic reactions or directly lost as ATP dependent heat (internal work) or indirectly connected with auxotonic contraction and relaxation.

The percentage of the energy of O_2 consumption converted into external work is not identical to the efficiency of the external cardiac contractile work. The real efficiency of our preparations is considerably lowered by the hydrostatic pressure on the atria by the solution and the friction of the system, and the disarray of muscle fibres which are not all parallel with the direction of the tension and the contraction.

If we transfer these findings of our model to the whole heart in the organism, this would mean that with an increasing preload in the failing heart, the uneconomical internal work will increase. In this way, the energy of activation of force to overcome the preload tension will be higher and therefore worsen the efficiency of the external contractile work and the ejection volume of blood. There is only a small range for the preload and the maximum inotropy of the Frank-Starling mechanism, which is shifted at a high beat rate to a lower preload. We observed that by increase of the preload above 10 mN, the force and the external contractile work decreases very quickly if the maximum of Frank-Starling inotropy is exceeded, presumably by an increased internal work. The ratio of internal to external work will increase during heart failure and the efficiency will be worsened in this way if the Frank-Starling maximum is trespassed. This seems to be especially important as *circulus vitiosus* for the increased energy needed during high end-diastolic ventricular pressure of the dilated ventricle during heart failure.

On the other hand we can observe that the afterload resistance is decisive for the efficiency of the external work during the same preload and the same energy supply: this means that a heart contracting against a high blood pressure resistance at a compensated preload works with a lower efficiency compared with contraction at a normal blood pressure, the end-diastolic ventricular pressure and the beat rate remain in the same normal range. In both cases the energy need shown by O_2 uptake, remains unchanged, governed by the same lengthening of the sarcomeres due to preload tension.

Acknowledgements. The investigations mentioned in this paper have been partly supported by the DFG (Si 103) or the Dr. Karl-Kuhn-Foundation. I thank all my coworkers mentioned in the references for their excellent experimental work and the good discussions. I appreciate gratefully the skillful technical assistance of Mrs. A. Weible, performing and analysing experiments and completing figures, graphs and photographs as well as the excellent help of Mrs. Weber for typing the manuscript and Dr. U. Delabar and Dr. I Teutsch for reading and improving the text.

References

1. Bittl JA, Ingwall JS (1986) The energetics of myocardial stretch, creatine kinase flux and oxygen consumption in the noncontracting rat heart. Circ Res 58:378–383
2. Chance B, Williams GR (1956) Respiratory enzymes in oxidation phosphorylation. J Biol Chem 217:409–427
3. Delabar U, Siess M (1979) Synthesis and degradation of NAD in guinea pig cardiac muscle: I. Dependence upon the extracellular concentration of nicotinamide and nicotinic acid. Basic Res Cardiol 74:528–544
4. Delabar U, Siess M (1979) Synthesis and degradation of NAD in guinea pig cardiac muscle: II. Studies about the different biosynthetic pathways and the corresponding intermediates. Basic Res Cardiol 74:571–593
5. Fiskum G, Lehninger AL (1982) Mitochondrial regulation of intracellular calcium. In: Wai Yiu Cheung (ed) Calcium and cell function, vol II. Academic Press, New York, pp 39–80

6. Gibbs CL (1983) Thermodynamics and cardiac energetics. In: Dintenfass L, Julian DG, Seaman GVF (eds) Heart perfusion, energetics and ischemia. Plenum Press, New York, NATO Scientific Affairs Division, Series A, vol 62, pp 549–576
7. Gibbs CL (1985) The cytoplasmatic phosphorylation potential. Its possible role in the control of myocardial respiration and cardiac contractility. J Mol Cell Cardiol 17:727–731
8. Giesen J, Kammermeier H (1980) Relationship of phosphorylation potential and oxygen consumption in isolated perfused rat hearts. J Mol Cell Cardiol 12:891–907
9. Kammermeier H, Schmid P, Jüngling E (1983) Free energy change of ATP hydrolysis: a causal factor of early hypoxic failure of the myocardium? J Mol Cell Cardiol 14:267–277
10. Loiselle DS (1985) The effect of temperature on the basal metabolism of cardiac muscle. Pflügers Arch 405:163–169 (Europ J Physiol)
11. McCormack JG, Denton RM (1986) Ca^{2+} ions as a link between functional demands and mitochondrial metabolism in the heart. In: Rupp H (ed) The regulations of heart function, basic concepts and clinical applications. Thieme Inc, New York, pp 186–200
12. Mommaerts WFHM (1970) What is the "Fenn"-effect? Muscle is a regulatory engine, the energy output of which is governed by the load. Naturwissenschaften 57:326–330
13. Rupp H (1986) The Ca^{++} responsiveness of myofilaments in terms of ATPase activity, shortening velocity, and tension generation. In: Rupp H (ed) The regulation of heart function. Thieme Inc, New York, pp 234–248
14. Sheu S-S, Sharma VK, Uglesity A (1986) Na^{+}–Ca^{++} exchange contributes to increase of cytosolic Ca^{++} concentration during depolarization in heart muscle. Ann J Physiol (Cell Physiol 19) 20:C651–C656
15. Siess M (1977) Influences on the efficiency of cardiac work. Basic Res Cardiol 72:299–305
16. Siess M (1983) Influences on the mitochondrial function of cardiac tissue. In: Sono KH and Nagano M (eds) Cardiac structure and metabolism. Tokyo, pp 1–42
17. Siess M, Delabar U, Stieler K, Leuchtner J, Teutsch I, Khattab A, El Hawary MB (1987) Protective and nonprotective effects of drugs on cardiac contractile activity and high energy phosphates during anoxia and after reoxygenation. In: Dhalla NS, Innes IR, Beamish RE (eds) Myocardial ischemia. Martinus Nijhoff Publ, Boston Mass, USA, pp 20 (in press)
18. Siess M, Keller HJ, Scharre E, Geisler J, Müller G (1970) The continuous and simultaneous measurement of O_2 consumption, rate of decarboxylation of ^{14}C substrates and the performance of spontaneous beating isolated heart atria of guinea pigs. J Mol Cell Cardiol 1:261–289
19. Siess M, Mensing HJ, Stieler K (1976) Investigations about the determinants of the myocardial oxygen consumption. In: Knoll J, Szekeres L, Papp JGy (eds) Symposium on pharmacology of the heart. Akadémiai Kiado, Budapest, pp 65–73
20. Siess M, Stieler K (1984) Methods for studying mitochondrial function in superfused cardiac muscle preparations. In: Dhalla NS (ed) Methods in studying cardiac membranes, vol I, CRC Press Inc, Boca Raton, FL/USA, pp 87–109
21. Siess M, Stieler K, Leuchtner J, Delabar U (1986) Some problems of cardiac energetics. In: Jacob R (ed) (1986) Controversial issues in cardiac pathophysiology. Basic Res Cardiol [Suppl 1] 81:79–94
22. Zeitler N (1986) Untersuchungen zum Gehalt energiereicher Phosphate und NAD im Herzvorhof bei Kalium-Depolarisation. Inaug Dissertation Medizinische Fakultät (Theoret Medizin), Tübingen, pp 1–76

Author's address:

Dr. M. Siess, Department of Pharmacology, Faculty of Theoretical Medicine, University of Tübingen, Wilhelmstraße 56, 7400 Tübingen, F.R.G.

Heat production and oxygen consumption following contraction of isolated rabbit papillary muscle at 20 °C

F. Mast and G. Elzinga

Laboratory for Physiology, Free University, Amsterdam, The Netherlands

Summary

The time course of oxygen uptake following isometric twitch contractions of isolated rabbit papillary muscles was measured using a polarographic oxygen electrode.

Using a diffusion model we eliminated the effect of oxygen storage on the measured time course of oxygen uptake to determine the time constant of mitochondrial "off" kinetics. Two different approaches were followed. In Method 1, two steady-state levels were compared, whereas in Method 2, the time course of mitochondrial "off" kinetics was studied. Using Method 1 we found $\tau = 20 \pm 8$ seconds (n = 7), whereas Method 2 yielded $\tau = 26 \pm 9$ seconds (n = 11).

These findings were compared with preliminary measurements of recovery heat production of the same preparation and at the same temperature. Heat produced after a train of 10 twitch contractions appeared to follow a monoexponential time course with a time constant of 24.9 ± 9.5 seconds (n = 9).

These results suggest that aerobic metabolism in isolated rabbit papillary muscle constitutes the only recovery process.

Introduction

For amphibian skeletal muscles, contracting at low temperatures, a temporal dissociation is found between the heat production accompanying initial and recovery metabolism [10]. The time course of recovery processes following contraction of amphibian and mammalian skeletal muscle have been extensively studied using variables, such as oxygen [5, 7, 12, 16, 18], heat [8, 13], fluorescence [8] and phosphocreatine [16].

At higher temperatures and for mammalian muscles the temporal separation between initial and recovery metabolism becomes less distinct. In highly oxidative muscles, such as the heart, the aerobic resynthesis of ATP is relatively fast and may already start during contraction. So far only a few attempts have been made to describe the time course of oxidative metabolism in isolated papillary muscle of the rabbit, monitoring either NADH-fluorescence [3] or heat production [4]. In particular the time course of recovery oxygen consumption of isolated heart muscle has not yet been investigated in any detail. This may well be due to the fact that the accuracy of determination of the time course is limited by the large response time of the measuring system as compared to the time constant(s) to be measured. Moreover, with thicker muscles, changes in the oxygen gradient in the muscle might distort the true recovery signal.

This study has been designed to determine the time course of recovery under aerobic conditions following isometric twitch contractions of isolated rabbit papillary muscle.

For that purpose we measured oxygen consumption using a relatively fast oxygen measuring system and made appropriate corrections for the effect of changes in the oxygen gradient. The results obtained were compared with the time course of recovery heat production as measured in pilot experiments for the same preparation.

Methods

Papillary muscles (length, 4.4–6.5 mm; diameter, 0.90–1.17 mm) were selected from the right ventricle of the rabbit heart. Experiments were performed at 20 °C in a solution containing: NaCl, 127; KCl, 2.3; $MgSO_4$, 0.6; KH_2PO_4, 1.3; $NaHCO_3$, 25; $CaCl_2$, 2.5; glucose, 5.6 (concentrations given in mM).

Measurements

a) Oxygen

A brief discussion will suffice here, because a full account of the preparation will be given somewhere else (Mast and Elzinga, submitted for publication). Isolated papillary muscles were mounted in an oxygen chamber, consisting of a glass cylinder with 2 vertical and 2 horizontal channels, which form together a closed circuit with a volume of 219 µl. The fluid in the chamber circulates rapidly, driven by a magnetically coupled spinner. The ventricular end of the muscle is fixed into a stainless steel holder at the bottom of the central channel with a short insect pin. The tendon end of the muscle is tied to a 0.2 mm diameter stainless steel wire which leaves the oxygen chamber via a glass capillary and is suspended from a force transducer.

Oxygen tension is measured with a polarographic electrode [14]. The response time of the oxygen measuring system, electrode and stirred chamber, as determined by rapid injection into the chamber of 5 µl solution with a high (or low) oxygen concentration into the chamber, was a delay of 1.0 s and a time constant of 2.1 s following that delay.

b) Heat

Temperature change during and following contraction was measured using metal-film thermopiles described in detail by Mulieri et al. [19]. The sensitivities of the two thermopiles used were 77 and 85 $\mu V \cdot K^{-1} \cdot junction^{-1}$, respectively. The rate of heat loss was determined by artificially heating the muscle with a non-stimulating 8 kHz current. It followed a single exponential time course with time constants ranging from 5.1 to 10.8 s. Correction for heat loss was done through deconvolution.

Oxygen analysis

a) Method 1

In this method the difference between two steady-state rates of oxygen consumption was studied. The assumption was made that aerobic recovery follows first order kinetics [6, 7, 15, 17]. In that case, the amount of oxygen taken up by the mitochondria during the transition from a given steady-state oxygen consumption rate to a lower one, equals the difference in steady-state rate (A) × time constant (τ) of recovery.

The quantity $A \times \tau$ cannot be measured directly, because during the rate transition some oxygen is taken up by, and stored in, the muscle.

The separation of total oxygen uptake into oxygen storage and mitochondrial oxygen consumption can be calculated on the basis of a diffusion model [11].

$$\Delta O_2 \text{ taken up} = A \times \tau + AR^2/8K. \tag{1}$$

The total amount of oxygen taken up and the steady-state rate above basal can be easily measured in the oxygen chamber. Radius R can be determined under the microscope, and K equals 6.0×10^{-4} $cm^2 \cdot min^{-1}$ for rat heart [9]. If the ratio of the total amount of oxygen taken up and the steady-state rate A is plotted as a function of $R^2/8K$, then the time constant τ is found as the intercept on the ordinate of a line through the points with slope equal to 1.

b) Method 2

A somewhat different approach was used: to separate the oxygen consumed by the mitochondria from that taken up to restore the oxygen gradient in the muscle, the diffusion

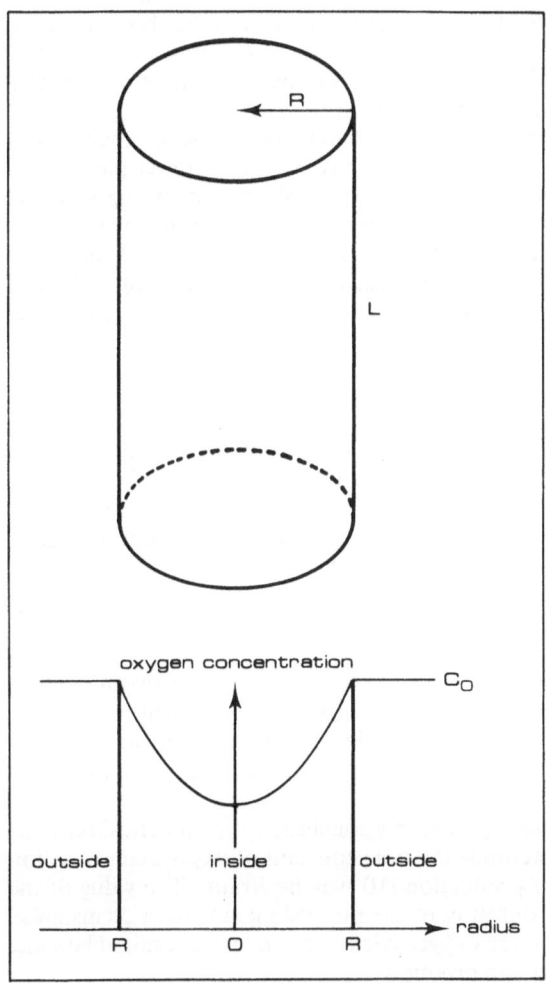

Fig. 1. Upper: Cylinder (radius R and length L) taken as model for the muscle as diffusion system; lower: Oxygen concentration in the muscle (inside) and in the surrounding fluid (outside) in the oxygen chamber. The oxygen gradient has a parabolical shape.

characteristics of the muscle were modelled. To characterize the muscle as a diffusion system it was assumed that the muscle is a circular cylinder, the radius of which is small compared to its length (Fig. 1). The diffusion constant was again taken to be 6.0×10^{-4} $cm^2 \cdot min^{-1}$, and the contribution of facilitated diffusion due to the presence of myoglobin was neglected. The function used to single out the true mitochondrial component was derived from a diffusion equation describing the time course of the oxygen concentration in the muscle when oxygen consumption changes from one steady-state level to another [2].

Results

Oxygen consumption

Force development and oxygen consumption due to a train of 120 contractions were measured. At various stimulation frequencies different steady-state rates of oxygen consumption were found. An example of a recording of stress (a) and oxygen (b) is shown in Fig. 2. To determine the oxygen uptake related to contraction, a correction has been made for the continuous decay of oxygen measured at rest. During stimulation a steady state oxygen consumption rate A is reached. With Method 1 we found, for the time constant of mitochondrial "off" kinetics: $\tau = 20 \pm 8$ s (mean $\pm$ s.d.; n = 7).

An example of the results obtained when Method 2 was used to determine the time course of mitochondrial oxygen uptake in a muscle is given in Fig. 2 (lower panel). Unfortunately the noise introduced by the deconvolution procedure made it impossible to determine the most appropriate function to fit the time course of oxygen uptake following contraction in most muscles. We therefore assumed first order kinetics and calculated the time constant from the slope of the relationship between the oxygen consumption following contraction and steady-state rate of oxygen uptake. This yielded a time constant of 26 ± 9 s (mean $\pm$ s.d.; n = 11).

Heat production

To compare the time course of oxygen consumption following contraction with the time course of recovery heat, we measured the heat production following a train of 10 contractions. An example of force and heat is given in Fig. 3. The time course of recovery heat production, corrected for heat loss, could be described adequately by a single exponential. The time constant was $\tau = 24.9 \pm 8.5$ s (mean $\pm$ s.d.; n = 9).

Discussion and conclusions

For rabbit papillary muscle at 20 °C, we found a good correspondence between the time constants of mitochondrial oxygen consumption and aerobic recovery heat production. This suggests that the close relationship between the time courses of recovery heat production and oxygen consumption, found for amphibian skeletal muscle [12], is also present in mammalian heart muscle.

However, an objection may be raised against this conclusion. For practical reasons, different train lengths were used to determine the time constant of oxygen consumption (120 twitches/train) and recovery heat production (10 twitches/train). The value of the time constant might be related to the duration of the preceeding activity. For instance, lactate formation could occur in addition to oxygen consumption, or there might be some effect due to the accumulation of other end products.

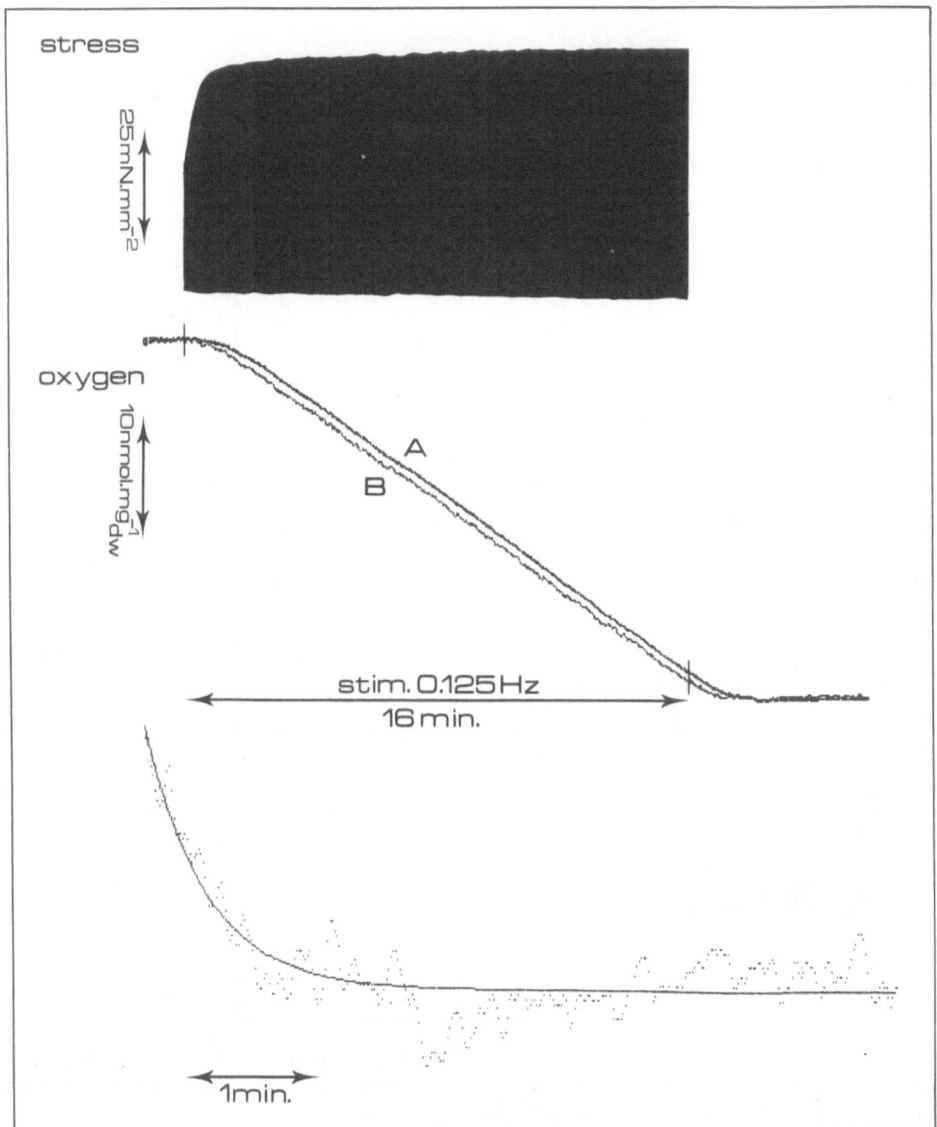

Fig. 2. Example of a recording of stress (upper panel) and oxygen (middle panel, tracing A) from a train of 120 twitch contractions. The oxygen signal has been corrected for the oxygen disappearance when the muscle is at rest. The second oxygen tracing (middle panel, tracing B) shows the mitochondrial oxygen consumption. The duration of the stimulus train is indicated on the oxygen tracings by vertical markers. In the lower panel the part of tracing (B) during recovery is shown (dashed line). The exponential curve, fitting the data, is indicated by the solid line ($\tau = 27$ s).

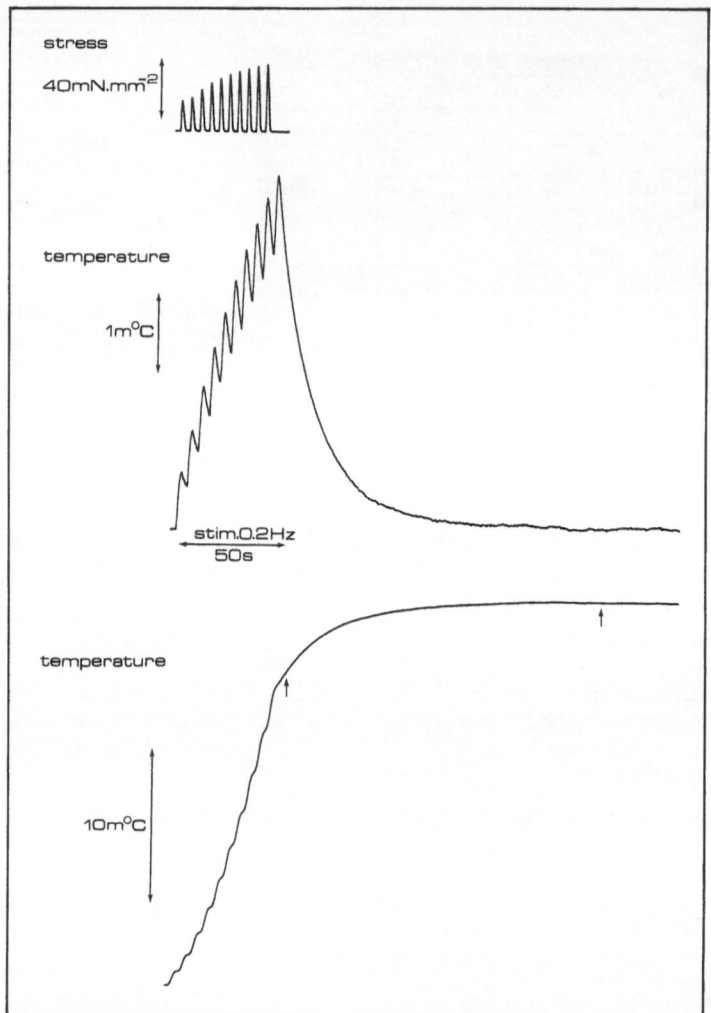

Fig. 3. Example of a recording of stress (upper panel) and temperature (middle panel) and temperature corrected for heat loss (lower panel) from a train of 10 twitch contractions. The vertical arrows indicate the part of the signal fitted with a single exponential.

Assuming that the time course of oxygen consumption after the end of a train of contractions follows first order kinetics, the time constant for aerobic recovery is 20–26 seconds. On the basis of the time constant of 20 seconds and assuming that all ATP required for a twitch is hydrolyzed immediately at the onset of contraction, and that the mitochondria are immediately activated, one can calculate that at most 10% of the oxygen required to restore ATP levels is already consumed by the mitochondria during the twitch (Fig. 4). This value is only slightly larger that the 7% contamination of initial heat with recovery heat reported by Alpert and Mulieri [1].

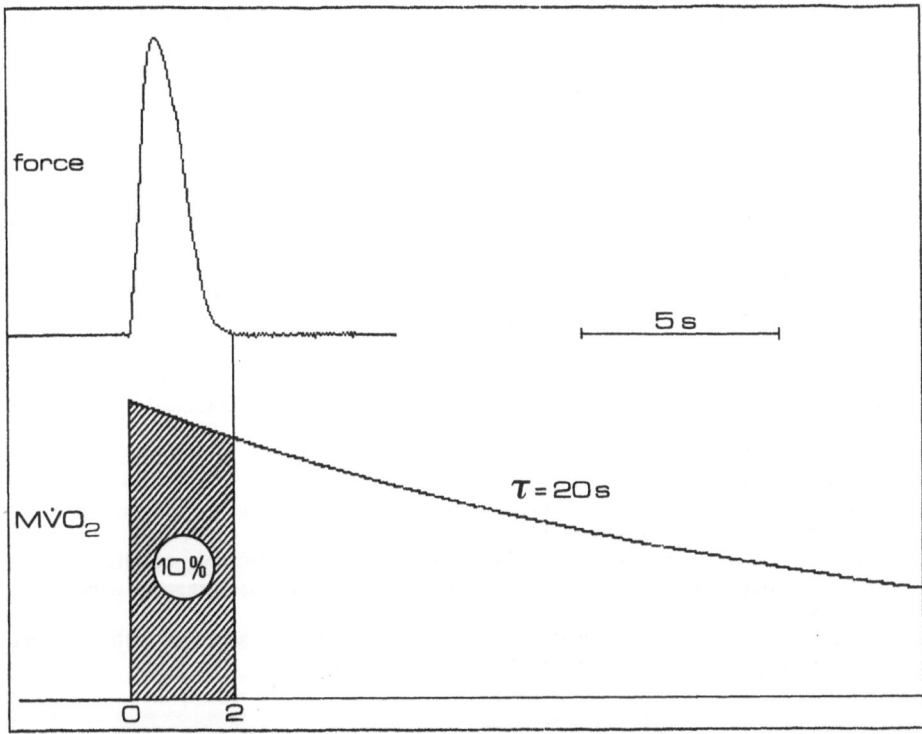

Fig. 4. Force and theoretical myocardial oxygen consumption rate $M\dot{V}O_2$ as a function of time. If the $M\dot{V}O_2$ started immediately after stimulation with an "off" time constant of 20 s, then about 10% of the total amount of oxygen consumed is taken up during the 2 s twitch.

The values of the time constant that we found correspond well with the time constant of 27.4 ± 4.6 seconds which can be derived from the $t_{1/2}$ of the "off" kinetics of NADH-fluorescence of rabbit papillary muscle at 23 °C reported by Chapman [3].

References

1. Alpert NR, Mulieri LA (1982) Increased myothermal economy of isometric force generation in compensated cardiac hypertrophy induced by pulmonary artery constriction in the rabbit. Circ Res 50:491–500
2. Carslaw HS, Jaeger JC (1959) In: Conduction of Heat in Solids. Oxford University Press, p 204
3. Chapman JB (1972) Fluorometric studies of oxidative metabolism in isolated papillary muscle of the rabbit. J Gen Physiol 59:135–154
4. Chapman JB, Gibbs CL (1974) The effect of metabolic substrate on mechanical activity and heat production in papillary muscle. Cardiovasc Res 8:656–667
5. Crow MT, Kushmerick MJ (1982) Chemical energetics of slow- and fast-twitch muscles of the mouse. J Gen Physiol 79:147–166
6. Di Prampero PE, Margaria R (1968) Relationship between O_2 consumption, high energy phosphates and the kinetics of O_2 debt in exercise. Pflugers Arch 304:11–19

7. Elzinga G, Langewouters GJ, Westerhof N, Wiechmann AHCA (1984) Oxygen uptake of frog skeletal muscle fibers following tetanic contractions at 18 °C. J Physiol 346:365–377
8. Godfraind-De Becker A (1972) Heat production and fluorescence changes of toad sartorius muscle during aerobic recovery after a short tetanus. J Physiol 223:719–734
9. Grote J, Thews G (1962) Die Bedingungen für die Sauerstoffversorgung des Herzmuskelgewebes. Pflugers Arch 276:142–165
10. Hartree W, Hill AV (1922) The recovery heat production of muscle. J Physiol 56:367–381
11. Hill AV (1965) Trails and Trials in Physiology. Edward Arnold, London
12. Hill DK (1940a) The time course of the oxygen consumption of stimulated frog's muscle. J Physiol 98:207–227
13. Hill DK (1940b) The time course of evolution of oxidative recovery heat of frog's muscle. J Physiol 98:454–459
14. Kimmich HP, Kreuzer F (1969) Catheter PO$_2$ electrode with low flow dependency and fast response. Prog Resp Res 3:100–110
15. Kushmerick MJ (1983) Energetics of muscle contraction. In: Peachey LD, Adrian RH, Geiger SR (eds) Handbook of Physiology. American Physiological Society, Bethesda, MD. Section 10, "Skeletal Muscle", pp 189–236
16. Kushmerick MJ, Paul RJ (1976) Aerobic recovery metabolism following a single isometric tetanus in frog sartorius muscle at 0 °C. J Physiol 254:693–709
17. Mahler M (1978) Kinetics of oxygen consumption after a single isometric tetanus of frog sartorius muscle at 20 °C. J Gen Physiol 71:559–580
18. Mahler M, Louy C, Homsher E, Peskoff A (1985) Reappraisal of diffusion, solubility, and consumption of oxygen in frog skeletal muscle, with applications to muscle energy balance. J Gen Physiol 86:105–134
19. Mulieri LA, Luhr G, Trefry J, Alpert NR (1977) Metal-film thermopiles for use with rabbit right ventricular papillary muscles. Am J Physiol 233:C146–C156

Authors' address:

F. Mast, Laboratory for Physiology, Free University, Van der Boechorststraat 7, 1081 BT Amsterdam, The Netherlands

Regulation of heart creatine kinase

J. S. Ingwall and J. A. Bittl

Department of Medicine, Brigham and Women's Hospital, Harvard Medical School
and Harvard Medical School NMR Laboratory, Boston, U.S.A.

Summary

Magnetization transfer nuclear magnetic resonance (NMR) provides measurement of the velocity
of the creatine kinase reaction in the intact heart. Standard one-pulse NMR spectroscopy coupled
with conventional biochemical analyses provides information about the average cytosolic concentrations of ATP, creatine phosphate (CrP), creatine (Cr) and H^+ in the heart. By combining these techniques, we tested the hypothesis that the velocity of the creatine kinase reaction in vivo was regulated
by changes in cytosolic concentrations of its substrates. We found that the reaction velocity cannot
always be predicted from its metabolite levels. We interpreted these observations as support for the
hypothesis that flux through the creatine kinase reaction is regulated by metabolite levels in microcompartments formed by localization of creatine kinase isozymes.

Introduction

In the heart, the metabolic pathways for the synthesis, transfer and utilization of the high
energy phosphate moiety are precisely integrated. Over a wide range of cardiac performance, where ATP turnover, estimated by oxygen consumption, varies by an order
of magnitude, tissue ATP content remains essentially constant. Thus, ATP synthesis via
oxidative phosphorylation (and, to a small extent, glycolysis) matches ATP utilization by
the myofibrils and ion transport sites. It is the purpose of this report to describe experiments to test the hypothesis that high energy phosphate transfer, via the creatine kinase
reaction, is also coupled to the rate of ATP synthesis by oxidative phosphorylation.

Creatine kinase (ATP; creatine N-phosphotransferase, EC 2.7.3.2; 80 kilo Dalton) is
present in high activity in heart. The enzyme catalyses the transfer of the high energy
phosphate moiety between creatine (Cr) and ADP: $H^+ + MgADP + CrP^{-2} \rightleftharpoons Cr +$
$MgATP^-$. At pH 7, the reaction has an equilibrium constant of about 170 [17]. Four elec-
trophoretically distinct dimeric creatine kinase isozymes have been identified: BB, MB,
MM (CK_3) and mitochondrial-CK (CKm). The distribution of isozymes varies with
muscle-type and species. In the rat heart, the distribution is: trace BB, 12% MB, 63%
MM and 25% CKm. Two of these isozymes are not free in solution, but are bound to
specific organelles. MM-CK is one of the components of the M-line in the sarcomere [16].
As much as one half of the total MM-CK may be associated with myofibrils and ion
transport sites in the heart [15]. CKm is located on the outside of the inner mitochondrial
membrane [6].

An historical summary of our perceptions of the role of creatine kinase in muscle is
shown in Fig. 1, taken from the monograph, Heart Creatine Kinase [7]. In 1934,
Lohmann [10] identified the reaction catalysed by creatine kinase. At that time, it was
known that ATP was the high energy phosphate compound produced by metabolism,

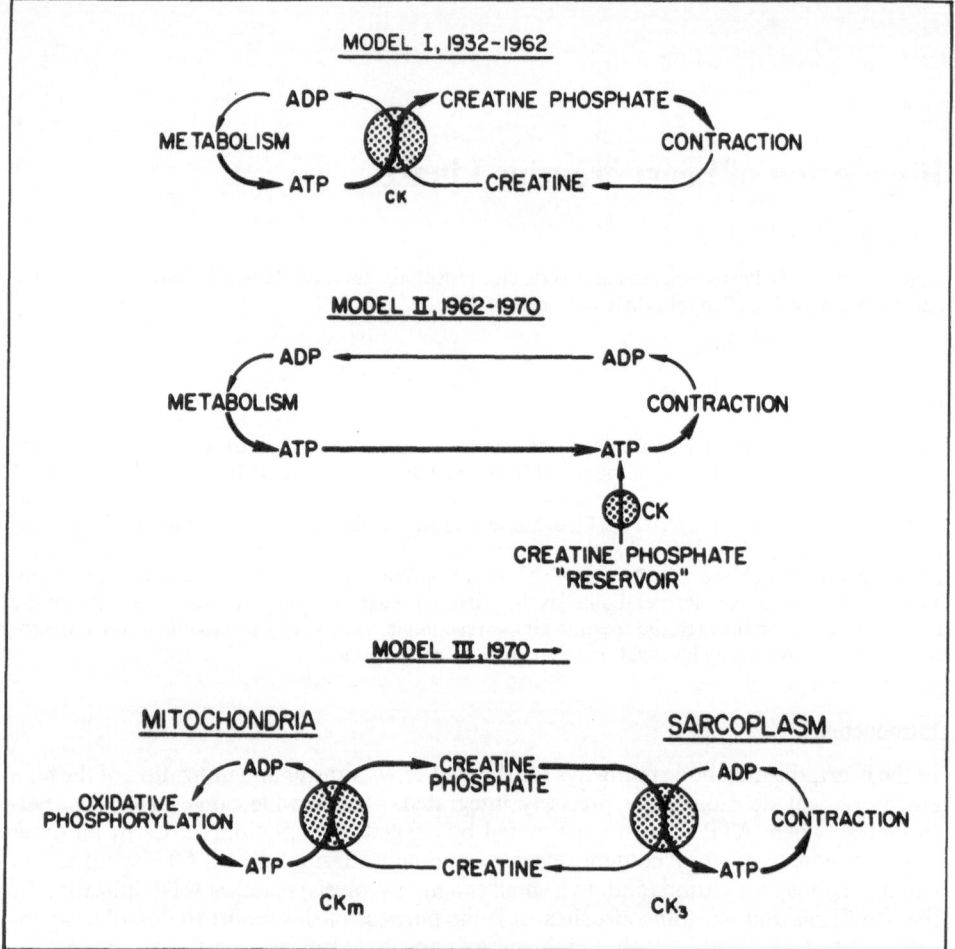

Fig. 1. Historical summary of the role of creatine kinase in energy transfer in heart. CKm and CK₃ are the mitochondrial and MM isoenzymes of creatine kinase. From reference 1.

and it was believed that ATP was the source of energy for muscular contraction. However, attempts to identify the chemical source of energy for muscular contraction in intact muscle found evidence only of creatine phosphate (CrP) hydrolysis. Thus, between 1934 and 1962, creatine kinase was thought to catalyse the transfer of a high-energy phosphate group from ATP, synthesized from metabolic processes, to form CrP to be used for muscular contraction (Model I). In 1962, Cain and Davies [3] reported their classic experiment in which they used fluorodinitrobenzene to inhibit the creatine kinase reaction in frog skeletal muscle. There was no change in CrP as the muscle twitched. Instead, they observed ATP hydrolysis and provided, for the first time, direct proof that ATP is the chemical source of energy for muscular contraction in the intact muscle. Model II was then developed identifying ATP as both the product of metabolism and the source of energy for contraction. The creatine kinase reaction was relegated to a "secondary" role, providing a reservoir of CrP to be used in times of stress, such as hypoxia or ischaemia. By the early 1970's, it became clear that Model II could not account for a number of relevant observations. One was that ATP levels did not correlate with changes in cardiac performance during normal or ischaemic conditions [12]. Another was that the localization of CKm [6] provided a direct link between the creatine kinase reaction and oxidative phosphorylation in isolated heart mitochondria. Similarly, localization of CK in the myofibrils [3] links the creatine kinase to ATP utilization. Based on these observations, Model III was developed. This model phase creatine kinase in a central position in cellular energy transport.

The central feature of Model III is that localization, rather than differences in inherent kinetic properties, of CKm on the mitochondria and MM-CK in the myofibrils and ion transport sites is important in the function of the creatine kinase system in energy transport. Although controversial, this is supported by several observations. Utilizing mitochondrial preparations, Jacobus et al. [12] measured the apparent Km of CKm for ATP with the enzyme either bound or unbound. They found that the apparent Km of bound CKm for ATP was 5- to 7-fold less than that of unbound CKm, suggesting functional coupling between the adenine nucleotide translocase and CKm. Due to this functional coupling, the creatine kinase reaction at the mitochondria would proceed primarily in the direction of CrP production. Other experiments [14] utilizing myofibrillar and sarcolemmal preparations provided evidence for functional coupling between MM-CK and ATPases. In these organelles, the creatine kinase reaction at sites of ATP degradation would proceed predominantly in the direction of ATP production. The results of these experiments suggest that when integrated in the cell, CKm and MM-CK form an energy transport system in which CrP is produced from ATP at the mitochondria and converted back to ATP at sites of utilization. The role of creatine kinase in the energy transport system of the cell has received further support from evidence that diffusion of ADP, but not creatine, could be rate limiting [8].

Thus, Model II, the equilibrium model, and Model III, the compartmentation model, describing bioenergetics of the heart, both emphasize the balance between ATP synthesis and degradation rates, but link the creatine kinase reaction to these processes differently. The equilibrium model (Model II) predicts that creatine kinase flux is regulated by changes in the average cytosolic metabolite concentration; the compartmentation model (Model III) predicts that creatine kinase flux is driven by the rate of ATP synthesis by oxidative phosphorylation, i.e., flux through the creatine kinase reaction is regulated by metabolite levels in microcompartments formed by localization of different creatine kinase isoenzymes in the mitochondria and at the myofibrils, not by average cytosolic metabolite levels.

To distinguish between these two Models, we have used the technique of magnetization transfer NMR to measure the pseudo-first-order rate constant and flux through the creatine kinase reaction in the intact heart at different levels of cardiac performance, which correspond to different rates of ATP synthesis. Based on values for cytosolic ADP concentrations calculated from the equilibrium expression

$$K_{eq} = \frac{[ATP]}{[ADP]} \frac{[Cr]}{[CrP]}$$

we compared values for the pseudo-first-order and second-order rate constants. From this comparison, we tested whether changes in the cytosolic ADP pool account for the changes in flux through the creatine kinase reaction. We found that changes in creatine kinase flux cannot be explained by changes in the cytosolic ADP levels. Rather, creatine kinase flux is coupled to cardiac performance and hence the rate of ATP synthesis by oxidative phosphorylation.

Methods

The isolated perfused rat heart

Male Sprague-Dawley rats, weighing 290–340 g, were anaesthetized with 20 mg of pentobarbital sodium intraperitoneally. The heart was removed through a median sternotomy, rinsed in ice-cold buffer, and perfused through an aortic cannula (total elapsed time 45 s). The perfusion medium contained (mEq/l): NaCl (119), KCl (4.7), CaCl$_2$ (4.0), MgSO$_4$ (2.4), ethylenediaminetetraacetate tetrasodium (0.5), NaHCO$_3$ (25), and glucose (11.0). By replacing 30 mEq/l of NaCl with KCl, some hearts were arrested. The pH was 7.4 when the medium was equilibrated with 95% O$_2$, 5% CO$_2$. The heart was perfused at a constant temperature of 36.8 °C and a constant pressure of 92 mm Hg. Securing a water-filled latex balloon in the left ventricle rendered the contraction pattern isovolumic. Heart rate and left ventricular developed pressure were recorded continuously on a Hewlett-Packard recorder. The heart was placed in a 20 mm NMR sample tube and inserted into a probe which was seated in the bore of a superconducting Oxford Instrument 360 wide-bore magnet; as previously described [5].

   Oxygen tension was measured in the perfusion medium at the level of the aortic cannula and in the coronary effluent in the right ventricle with a Clark-type electrode (Johnson Foundation Biomedical Laboratories, University of Pennsylvania). Oxygen consumption was calculated according to the formula [11]: aortic effluent pO$_2$ difference × solubility of O$_2$/mm Hg × coronary flow/dry weight in grams.

^{31}P NMR Spectroscopy

^{31}P NMR spectra were obtained on a Nicolet NT-360 spectrometer operating at 145.75 MHz and interfaced with either a Nicolet 1180 or 1280 computer. Spectra were obtained by signal averaging 80 or 96 scans of a 45 ° pulse and a 2.5 second delay. Magnetization transfer was performed by applying a low-power, narrow-band radio frequency pulse at the [γ-P]ATP resonance for 0, 0.3, 0.6, 1.2, 2.4, 3.0 or 4.8 seconds (Fig. 2). Magnetization transfer measurements of creatine kinase flux in vivo were analysed according to the two-site chemical exchange model of Forsen and Hoffman [4] as described previously [1, 2]. It is assumed that use of a two-site chemical exchange model accurately estimates flux in the forward directions since CrP participates in no reaction other than that catalysed by creatine kinase. A variance-weighted computer program for nonlinear regression was used to fit the relationship between magnetization of CrP to time of saturation at [γ-P]ATP to a single exponential function whose slope yields the value for $-1/_\tau$ where $1/_\tau = 1/T1 + k$, $T1$ is the longitudinal relaxation time for CrP, and k is the pseudo-first-order unidirectional rate constant for the forward reaction. The value of k is a pseudo-first-order rate constant, $k = [ADP]k'$ where k' is the second order rate constant: $k_f'[ADP][CrP] = k_r'[ATP][Cr]$ (ignoring pH). T1 values for CrP did not change with cardiac performance and averaged ~ 2 s.

   The content of CrP and ATP in heart was calculated from the NMR results, using a value of 25 μmol/g dry weight for heart muscle ATP [5]. Intracellular pH was calculated from the position of the Pi resonance [5]. Total tissue creatine content was measured using the method of Kammermeier [9].

   Statistical analysis was aided by the Statistics and Data Management Program of Bolt, Beranek, and Newman (VAX 11/780 computer).

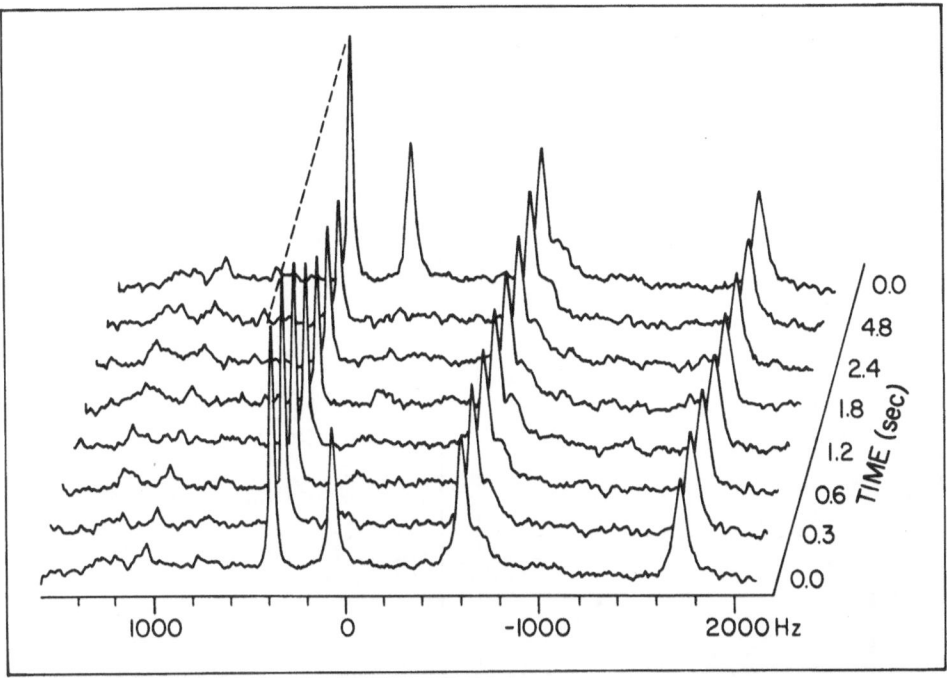

Fig. 2. Stack of ^{31}P NMR spectra illustrating the magnetization transfer method used to measure flux through the creatine kinase reaction in heart. The first and last spectra are standard one-pulse spectra obtained before and after the saturation sequence. The major resonances are assigned (from left to right) as CrP and γ-, α- and β-phosphates of ATP. The intervening spectra were obtained when the [γ-P]ATP was selectively saturated for times ranging from 0.3 to 4.8 s. Values for the pseudo-first-order rate constant of the forward creatine kinase reaction, CrP$\rightarrow$[γ-P]ATP, were calculated from the exponential defining the relationship between the magnetization (area) of the CrP resonance and time of saturation. Flux through the reaction is calculated by multiplying the pseudo-first-order rate constant and tissue CrP content. Reprinted by permission of the American Heart Association, Inc. from reference 2.

Results and discussion

Tissue content of ATP and CrP, the pseudo-first-order rate constant, k, the velocity of the creatine kinase reaction, and the rate of ATP synthesis (estimated by oxygen consumption) were measured in hearts functioning at five different levels of cardiac performance ranging from KCl-arrest to hearts whose product of heart rate and developed pressure were $\sim$45,000 mm Hg/min (Fig. 3) [1]. Changes in cardiac performance defined by the product of heart rate and developed pressure closely matched changes in ATP synthesis rates estimated by oxygen consumption (slope of the relationship between cardiac performance and oxygen consumption was 1.05). The results displayed in Fig. 3 show that as cardiac performance and hence ATP synthesis rates from oxidative phosphorylation increase, tissue ATP content remains constant and CrP content falls only slightly, while the pseudo-first-order rate constant and flux through the creatine kinase reaction increase four-fold. Thus, over this range of cardiac performance, ATP synthesis from oxidative phosphorylation increases 8-fold, while ATP synthesis via the creatine kinase reaction increases nearly 4-fold.

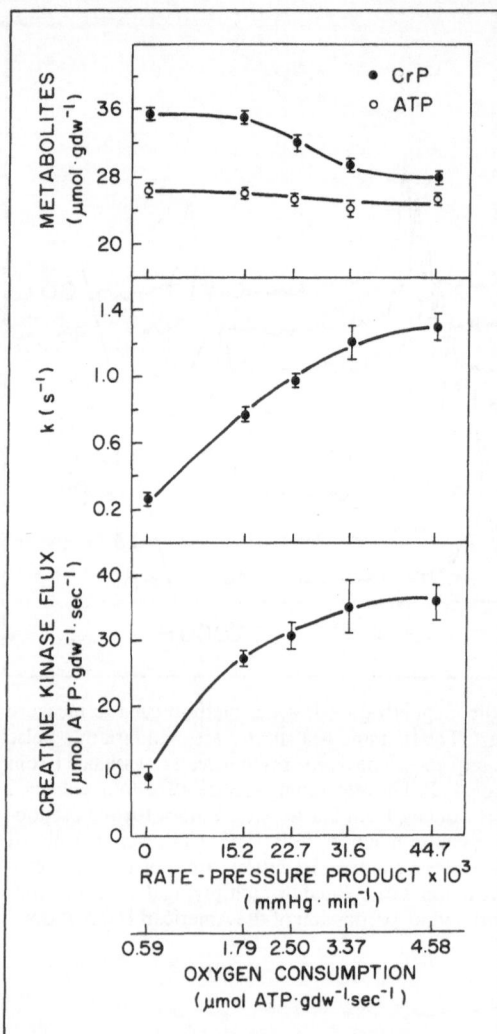

Fig. 3. Tissue content (μmol per gram dry weight) of CrP and ATP (top panel); pseudo-first-order rate constant, k (s⁻¹), for the reaction, CrP→[γ-P]ATP (middle panel); and flux through the creatine kinase reaction (μmol ATP gdw⁻¹ s⁻¹) (bottom panel) for rat hearts working at five levels of cardiac performance. Data recalculated from reference 1.

What regulates creatine kinase flux in vivo? Total creatine kinase activity did not change, nor was there a change in intracellular pH. Is the increase in flux through the creatine kinase reaction due to increased cytosolic ADP content? To answer this question, we calculated cytosolic ADP concentrations using the widely held assumption that cytosolic ADP levels can be calculated from the creatine kinase equilibrium: k'_{for} [ADP][CrP] = k'_{rev}[ATP][Cr]. We calculated cytosolic [ADP] using values for tissue levels of CrP, ATP (both obtained from ³¹P NMR spectra), free creatine (calculated from the difference between total Cr and CrP contents) and pHi (from ³¹P NMR), and using the value of 166 for the equilibrium constant, *Keq*, at pH 7.1, taken from Veech et al. [17]. Table 1 shows the measured tissue levels of ATP, CrP, free creatine and the calculated cytosolic ADP content for hearts with ATP synthesis rates ranging from 0.59 (KCl-ar-

Table 1. The creatine kinase system and ATP synthesis rates in the isovolumic rat heart

(n)	ATP synthesis rate (μmol gdw^{-1}s^{-1})	Metabolite concentrations (mM)				Pseudo 1st order rate constant (s^{-1})	Calculated 2nd order rate constant ($\times 10^3$)
		CrP	Cr	ATP	ADP		
Series I							
6	0.59±0.19	13.6	11.0	10.2	0.050	0.27±0.03	5.4
3	1.79±0.23	13.5	11.1	10.1	0.050	0.78±0.03	16
3	2.50±0.19	12.3	12.3	9.8	0.059	0.97±0.06	16
4	3.37±0.12	11.3	13.3	9.4	0.067	1.21±0.06	18
6	4.58±0.48	10.8	13.8	9.8	0.075	1.30±0.09	17
Series II							
7	0.8±0.1	14.1	10.5	9.6	0.043	0.33±0.03	7.6
5	1.0±0.1	14.2	10.4	9.6	0.042	0.80±0.08	19

Data for these calculations were taken from references 1 and 2. Cytosolic concentrations (mM) of ATP, Crp and Cr were calculated using a value of 25 μmol/g dry weight for ATP content of a beating heart, a wet to dry weight ratio of 5.2 and a value of 0.5 ml of intracellular water per g of tissue, all experimentally determined. Series I hearts encompass a wide range of cardiac performance, assessed as the product of heart rate and developed pressure, ranging from 0, for KCl-arrested hearts, to 44.600 mm Hg/min for hearts working at high levels of cardiac performance. Series II hearts were arrested with KCl, and wall stress varied by increasing end diastolic pressure from 0 to 24 mm Hg by filling a latex balloon in the left ventricle with different levels of fluid.

rest) to 4.6 μmol/gdw^{-1} s^{-1} (Series I). Also shown are values for the pseudo-first-order rate constant measured by NMR and the second-order rate constant calculated by dividing k by [ADP]. The results show that as the pseudo-first-order rate constant increases nearly five-fold, there are only small changes in metabolite levels. Cytosolic [ADP], the putative regulator of the creatine kinase reaction, increased only by 50%. Comparison of values for the second order rate constant calculated from the expression, $k = k'[ADP]$, for these five levels of cardiac performance shows that changes in metabolite levels failed to yield the expected constant value in all cases. The second order rate constant is essentially the same for beating hearts (range 16 to 18) but is 5.4 for the arrested heart.

These observations show that the average cytosolic ADP level does not always predict the velocity of the creatine kinase reaction. This suggests that the appropriate metabolite concentrations governing the reaction are not the bulk cytosolic values. This is consistent with the Compartmentation Model of creatine kinase. Based on the observation that the second-order rate constant for the arrested heart is three times lower than for the beating heart (at any level of cardiac performance), these results also suggest that high-energy phosphate transfer via the creatine kinase reaction is different in inactivated myocardium.

Unexpected variability in the calculated second-order rate constant was also observed in KCl-arrested hearts in which wall stress was increased by stretching the isovolumic balloon (Table 1, Series II) [2]. In this case, ATP synthesis increased by 20%, and the pseudo-first-order rate constant, k, and flux through the creatine kinase reaction increased more than 100%, yet no measurable changes in metabolite content could be demonstrated. Thus, the increase in the pseudo-first-order rate constant cannot be explained by increases in cytosolic ADP concentration. The calculated second-order rate constants differ by a factor of two.

Conclusions

The technique of magnetization transfer NMR is a unique tool permitting measurement of the whole organ enzymology of the creatine kinase reaction in the intact heart. Using this technique, we found that cytosolic ADP levels do not always predict the velocity of the creatine kinase reaction. Thus, the appropriate metabolite concentrations regulating the reaction are not the bulk cytosolic levels. These results provide support for the hypothesis that the velocity of the creatine kinase reaction in heart is regulated by metabolite levels in microcompartments formed by localization of creatine kinase isozymes in mitochondria and at the myofibrils, the Compartmentation Model. Our observations suggest a central role for creatine kinase in the transfer of high-energy phosphate compounds and maintenance of high ATP levels. These observations could only be made in the intact heart where pathways for ATP synthesis and degradation are integrated.

References

1. Bittl JA, Ingwall JS (1985) Reaction rates of creatine kinase and ATP synthesis in the isolated rat heart. J Biol Chem 260:3512–3517
2. Bittl JA, Ingwall JS (1986) The energetics of myocardial stretch: creatine kinase flux and oxygen consumption in the noncontracting rat heart. Circ Res 58:378–383
3. Cain DF, Davies RE (1962) Breakdown of adenosine triphosphate during a single contraction of working muscle. Biochem Biophys Res Commun 8:361–366
4. Forsen S, Hoffman RA (1963) Study of moderately rapid chemical exchange reactions by means of nuclear magnetic double resonance. J Chem Phys 39:2892–2901
5. Ingwall JS (1982) P-31 NMR spectroscopy of cardiac and skeletal muscles. Am J Physiol 242:H729–H744
6. Jacobus WE, Lehninger AL (1973) Creatine kinase of rat heart mitochondria. J Biol Chem 248:4803–4810
7. Jacobus WE, Ingwall JS (eds) (1980) Heart creatine kinase. Williams and Wilkins, Baltimore
8. Jacobus WE (1985) Theoretical support for the heart phosphocreatine energy transport shuttle based on the intracellular diffusion limited mobility of ADP. Biochem Biophys Res Commun 133:1035–1041
9. Kammermeier H (1973) Microassay of free and total creatine from tissue extracts by combination of chromatographic and fluorometric methods. Anal Biochem 56:341–345
10. Lohmann K (1934) Über die enzymatische Aufspaltung der kreatinen Phosphorsäure; zugleich ein Beitrag zum Chemismus der Muskelkontraktion. Biochem Z 271:264
11. Neely JR, Liebermeister H, Battersby EJ, Morgan HE (1967) Effect of pressure development on oxygen consumption by isolated rat heart. Am J Physiol 212:804–815
12. Pool PE, Covell JW, Chidsey CA, Braunwald E (1966) Myocardial high energy phosphate stores in acutely induced hypoxic heart failure. Circ Res 19:221–229
13. Saks VA, Kupriyanov Vv, Elizarova GV, Jacobus WE (1980) The importance of creatine kinase localization for the coupling of mitochondrial phosphorylcreatine production to oxidative phosphorylation. J Biol Chem 255:755–763
14. Saks VA, Ventura-Clapier R, Huchua ZA, Preobrazhensky AN, Emelin IV (1984) Creatine kinase in regulation of heart function and metabolism. I. Further evidence for compartmentation of adenie nucleotides in cardiac myofibrillar and sarcolemmal coupled ATPase-creatine kinase systems. Biochim Biophys Acta 803:254–264

15. Scholte HR (1973) Triple localization of creatine kinase in heart and skeletal muscle cells of the rat: evidence for existence of myofibrillar and mitochondrial isoenzymes. Biochem Biophys Acta 305:413–427
16. Turner DC, Wallimann T, Eppenberger EM (1973) A protein that binds specifically to the M-line of skeletal muscle is identified as the muscle form of creatine kinase. Proc Natl Acad Sci USA 70:702–705
17. Veech RK, Lawson JWR, Cornell NW, Krebs HA (1979) Cytosolic phosphorylation potential. J Biol Chem 254:6538–6543

Authors' address:

J.S. Ingwall, Ph.D., Department of Medicine, Brigham and Women's Hospital, Harvard Medical School, 221 Longwood Avenue, Boston, MA 02115, U.S.A.

Effect of creatine depletion on myocardial mechanics

B. Korecky and Y. Brandejs-Barry

Department of Physiology, School of Medicine, University of Ottawa, Canada

Summary

The physiological significance of energy transport by means of shuttling creatine (C) and creatine phosphate (CP) between mitochondria and the energy utilizing sites was examined in C-depleted rat hearts. Feeding a diet containing structural analogues of C [either 1% guanidinoproprionic acid (GPA) or 2% guanidinobutyric acid (GBA)] led to the decrease of total myocardial C (C + CP) by 78% and 75% after seven weeks and by 89% and 82% after ten weeks. Mechanics of isolated papillary muscles were examined under isometric conditions at different temperatures, muscle lengths and frequencies of stimulation. No differences were found in the basic characteristics of contraction and relaxation among the hearts of normal, GPA- and GBA-treated rats at low workloads. However, when paired stimulation was applied, the interval at which fusion occurred was significantly longer in C-depleted than in controls. At high workloads, the developed force (*DF*) in C-depleted decreased in the same way as in controls, but its subsequent recovery took significantly longer and the recoverd *DF* in C-depleted muscles was smaller than in controls. We conclude that C depletion has greater effects on the excitation to contraction coupling and recovery rather than on the DF during stimulation at high frequencies.

Introduction

It has been generally recognized that creatine phosphate (CP) rapidly regenerates ATP during muscular activity preventing any substantial decrease in muscle ATP [1]. It has also been proposed that CP "shuttles" energy between the mitochondria, the myofibrils and other intracellular energy consuming sites [16, 8]. However, evidence for the physiological importance of the CP shuttle has not been conclusive [13, 18] even though abundant biochemical evidence for the existence of the shuttle has been provided. In fact, experiments carried out on skeletal muscles have shown that muscles severely depleted of creatine and CP continue to function well, and some muscles become even more fatigue resistant after creatine depletion [14, 11].

Chronic creatine depletion can be attained by feeding structural analogues of creatine to experimental animals. The commonly used analogue is β-guanidinopropionic acid (GPA) which competes with creatine at the uptake sites of the sarcolemma [6], leading to maximum depletion of about 95% in skeletal muscle [14], and 90% in the myocardium [5]. GPA can accumulate in the phosphorylated form and can be used to a limited extent as a substrate for creatine kinase (CK). Another analogue, β-guanidinobutyric acid (GBA) can similarly deplete creatine in cardiac tissues but cannot be used as a substrate for CK. In this study we decided to use both analogues and compare the mechanics of isolated papillary muscles from hearts of creatine-depleted rats with those of the normal myocardium.

Methods and materials

Treatment of animals

Sprague-Dawley rats (250–300 g) were kept on a standard diet or supplemented with either 2% GBA or 1% GPA for 7 or 10 weeks. The animals were pair-fed and concentrations of total creatine, GBA and GPA determined from deep-frozen samples of myocardium, as described elsewhere [11].

Isolated papillary muscle preparation

After 7 and 10 weeks on the diet, the rats were stunned by a blow to the head, decapitated, the hearts removed, papillary muscles exposed and excised. The left ventricle was quickly frozen between aluminium blocks in liquid nitrogen. The excised papillary muscles were mounted between two acrylic clamps on isometric muscle stands and immersed in a bath with Krebs-Ringer solution of the following concentrations (in mM): NaCl 117.4, $CaCl_2$ 2.5, KCl 3.6, $MgSO_4$ 1.2, NaH_2PO_4 1.2, $NaHCO_3$ 25 and glucose 5.5. The pH was kept in the range of 7.38–7.45 by aerating the solution with a gas mixture of 95% O_2 : 5% CO_2 and the muscles were field stimulated with plate electrodes using a square-wave pulse of 5 ms duration at a supramaximal voltage. Several protocols were applied to increase the workload: 1) trains of progressively higher stimulation frequencies at low temperature, 2) paired pulsing with progressively decreasing intervals between the pair of stimuli, 3) high temperature and trains of stimuli at increasing frequencies.

Frequency-force relationship at 25 °C

The muscles of rats fed control, GBA and GPA diets for 7 and 10 weeks were left to stabilize in the bath at a temperature of 25 °C for 30 min, at a preload of 0.5 g and at a rested state stimulation rate of 3 pulses/min. The preload was then increased stepwise by 0.25 g until no further increase in force development was attained (L_{max}). The frequency-force relationship at L_{max} was examined at a stimulation rate of 3, 6, 12, 24, 48, 96 and 192 pulses/min. The rate was reduced to 3 pulses/min after each frequency step until no further increase in developed tension was observed.

Paired stimulation

The papillary muscles of rats fed control, GBA and GPA diets for 10 weeks were left to stabilize at 3 pulses/min and the L_{max} was established as above. The muscles were kept at 3 pulses/min but superimposed on that were paired-pulses with decreasing intervals of 700, 500, 300 and 200 ms. Then the paired-pulse intervals were stepwise decreased by 5 ms until fusion of the two mechanical twitches was observed. Subsequently, the muscles were stimulated at 3 pulses/min only for 5 min and then a train of single pulses at a rate of 192/min was applied for 20 s, in order to deplete the potential energy supply of the cells. This was followed by a 5 s rest period, after which a single stimulus was given to assess the recovery.

Frequency-force relationship at 31 °C

Muscles of rats fed control, GBA and GPA diets for 7 weeks were stabilized for 30 min at 3 pulses/min but at a bath temperature of 31 °C. Upon establishment of L_{max}, the muscle length was reduced to 75% of the L_{max} and the papillary muscles were stimulated at 270, 300, 330, 360 and 390 pulses/min for 85 s. Between each frequency of stimulation, the muscles were allowed to stabilize at 3 pulses/min.

Statistical analysis

All data were expressed as means and standard errors of the mean. Statistical significance was determined using one-way analysis of variance and statistically significant differences were identified using the Scheffe test.

Results

Contractile characteristics of single isometric twitches

Initial experiments were done after 7 weeks of feeding, at which time the levels of total creatine in the heart declined by 75% in by the GBA and GPA groups (Table 1). The ex-

Table 1. Concentrations of total creatine and analogues (GBA and GPA) of hearts of rats fed control, GBA and GPA diets for 7 and 10 weeks

	n	Total creatine	GBA	GPA
7 weeks				
Control	5	13.8±1.80	0.6±0.03	0.3±0.04
GBA	5	3.0±0.80	4.7±1.50	–
GPA	5	3.4±0.71	–	8.9±0.90
10 weeks				
Control	5	14.9±1.80	0.2±0.01	0.3±0.03
GBA	5	1.6±0.20	4.8±0.40	–
GPA	5	2.7±0.10	–	11.1±1.20

The values are means and standard errors of the mean expressed as µmol/g of wet weight; n refers to the number of animals

Table 2. Contractile characteristics at L_{max} of papillary muscles of rats fed control, GBA and GPA diets for 7 and 10 weeks

		DF (g/mm^2)	dT/dt (g/mm^2/s)	ΔT/Δt (g/mm^2/s)	TMT (ms)	T 0.5 R (ms)
7 weeks						
Control		3.7±0.3	37.8±2.0	23.3±0.7	179± 8	153±18
$n=$ 9	GBA-fed	4.5±0.9	45.5±9.6	23.8±4.4	186±10	190±33
$n=10$	GPA-fed	4.1±0.5	43.0±5.0	24.0±2.2	173± 9	174±23
10 weeks						
Control		3.6±0.7	33.5±3.8	19.5±2.3	192± 8	164±20
$n=10$	GBA-fed	4.0±0.5	29.8±6.3	10.0±7.1	200±15	191±14
$n=10$	GPA-feed	3.7±1.0	34.6±8.9	18.7±8.0	198± 8	192±18

The values are means and standard errors of the mean; n is the number of muscles. Five animals were used in each group. DF is the maximum development force, dT/dt ist he maximum rate of tension development, $ΔT/Δt$ is the mean rate of tension development, TMT is the time to maximal tension development, $T0.5R$ is the time to half-relaxation. Bath temperature was 25 °C

periments were repeated after 10 weeks of feeding at which time total creatine decreased
by 90% in GBA group and by 84% in GPA group. The differences in concentrations of
total creatine between GBA and GPA groups at 7 and 10 weeks were not statistically significant. Contractile characteristics were compared among control, GBA and GPA
groups after 7 weeks and 10 weeks (Table 2).

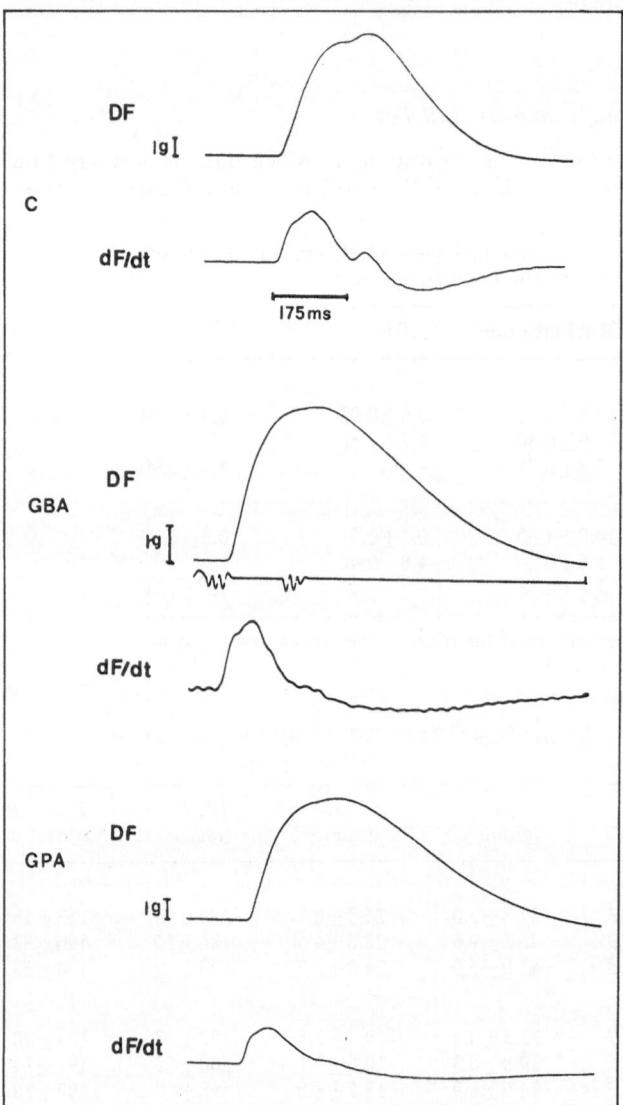

Fig. 1. A recording of a papillary muscle response to paired stimuli in control (c), GBA and GPA
fed rats for 10 weeks. The upper tracing in each case is the developed force (DF) and the lower is
the first derivative of the developed force (dF/dt). The paired-pulse interval in each case was 175 ms.
Bath temperature was 25 °C.

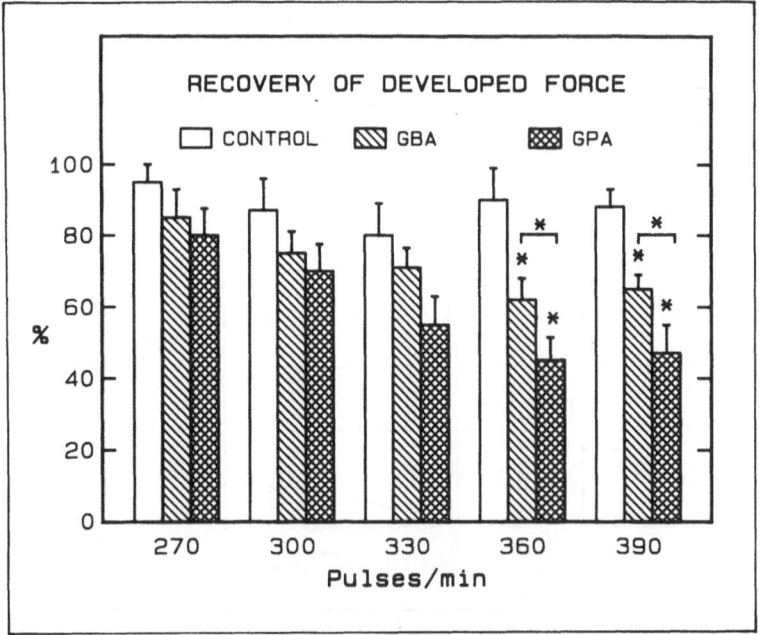

Fig. 2. Mean values and SEM of the recovered developed force at 3 pulses/min after stimulation trains of 270, 300, 330, 360 and 390 pulses/min in control, GBA and GPA rats fed for 7 weeks. The recovered developed force was taken as a percent of the developed force obtained at 3 pulses/min before the onset of the train of stimuli. The number of muscles used was 10 in the control and GPA groups and 9 in the GBA group. Five animals were used in each group. Bath temperature was 31 °C. *p<0.01.

Preload, length and cross-sectional area were similar in creatine depleted and control papillary muscles and no significant differences were observed in any of the contractile characteristics among the three groups at rested state stimulation rate.

Frequency-force relationship at 25 °C

Contractile mechanics of the papillary muscle was investigated after each stepwise increase in stimulation rate and a negative inotropic response in developed force was seen in all three groups from 6 to 192 pulses/min. No significant differences were observed at 7 or 10 weeks in any of the twitch characteristics among the three groups at any of the frequencies.

Effects of paired stimulation

The developed force (DF) of the second twitch at each paired-pulse interval was taken as a percentage of the DF obtained at 3 pulses/min, determined just before the onset of paired stimulation. In all groups, the DF of the second twitch was less than that of the first one. When DF of the second twitch was compared among the three groups, no differences were observed at any of the paired-pulse intervals. However, the fusion of the first and second twitch occurred at a longer interval in the creatine-depleted muscles than in controls. In Fig. 1 for example, the twitches of GBA and GPA groups fused at an in-

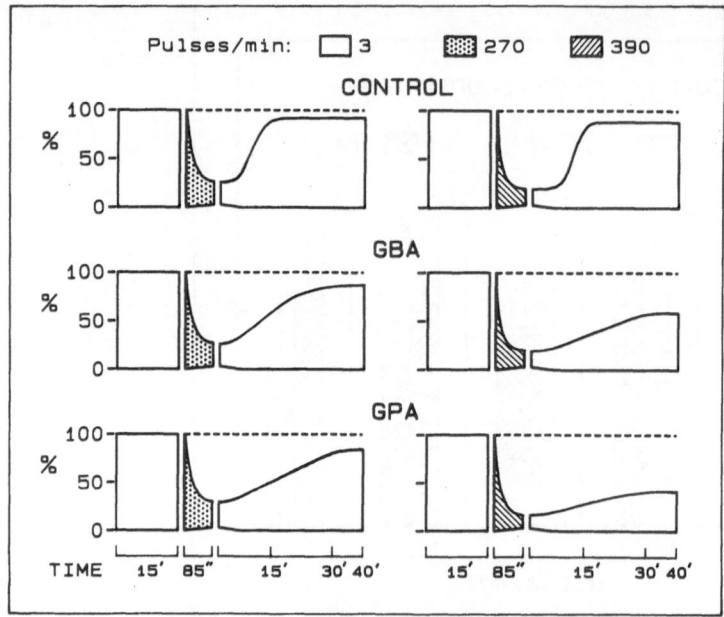

Fig. 3. Mean values of time course of recovery of the developed force during and after stimulation at 270 and 390 pulses/min in control, GBA and GPA rats fed for 7 weeks. The developed force was taken as a percent of the developed force obtained at 3 pulses/min before the onset of the stimulation frequency of 270 pulses/min. The number of muscles used was 10 in the control and GPA groups and 9 in the GBA group. Five animals were used in each group. Bath temperature was 31 °C.

terval of 175 ms while the control muscles still manifested two separate twitches. The paired-pulse interval at which fusion occurred was significantly shorter (< 0.01) for the control group, (144 ± 9 ms) than for the GBA (170 ± 11 ms) and for the GPA group (173 ± 3 ms). No significant differences were obtained between the experimental groups. After the last paired-pulse, high frequency stimulation did not result in lower DF in the creatine depleted papillary muscles than in controls and the subsequent recovery of DF was similar in all three groups.

Frequency-force relationship at 31 °C

At higher frequencies of stimulation, the DF was lower in muscles of the analogue fed rats as compared to those of control rats, however these differences in DF among the three groups were not statistically significant. After each train of stimuli lasting for 85 s, the stimulation rate was decreased to 3/min until the plateau of recovered DF was reached (Fig. 2). The recovered DF of the control group was higher than that of the GBA and GPA groups. These differences were statistically significant at 360 pulses/min ($p < 0.01$) and at 390 pulses/min ($p < 0.01$). The recovered DF of GPA group was significantly lower than the DF of the GBA group after each of these two trains of stimuli (Fig. 2). The time interval required to reach the recovered force was significantly longer ($p < 0.001$) in the creatine-depleted muscles than in controls after every train of stimuli. The control muscles recovered in about 12 min, as compared to the muscles in the GBA and GPA groups which recovered in 31 min (Fig. 3).

Discussion

The inotropic changes in the rat heart are affected by calcium movements in and out of the cell. Each action potential may liberate a small releasable fraction of intracellular bound calcium (Cr) which is taken up during recovery into another intracellular fraction (Cs) and later recycled to Cr at a slower rate [6, 12]. The cycling of calcium from the various fractions is dependent on the energy available that is ΔG ATP.

In our experiments during paired stimulation, the cycling of calcium from Cs to Cr in normal muscles became limiting when paired stimuli were less than 140 ms apart. The amount of calcium released during the first pulse was able to fully activate the contractile system but could not be retrieved into the Cr quickly enough to be released during the second pulse [10]. Therefore, the active sites remain saturated, and consequently the mechanical response to the second stimulus was not observed. When we shortened the intervals between paired stimuli in C-depleted muscles, then the fusion of the first and second twitch occurred earlier and at a longer interval than in controls, indicating that the cycling of calcium is slower. In our C-depleted rats, myocardial concentrations of total C were reduced by 75–90%. Consequently CP was most probably depleted as well, since others have shown that CP decreases by more than 90% in hearts of GPA fed rats [17, 18]. It appears then, that normal levels of CP may be essential to maintain mechanical response under extreme conditions such as short intervals between paired stimuli, but do not appear to be essential at longer intervals such as 200 ms or more.

When we imposed high workload, using trains of high frequency stimuli, and increased bath temperatures to 31 °C, C-depletion had no effect on muscle performance. However, the maximal recovered DF obtained after each train was lower and the time needed for recovery longer in the C-depleted muscles. Surprisingly, the recovery of the GPA group with somewhat lesser myocardial depletion of total C was more depressed than in the GBA group. There are several mechanisms which may account for this slow recovery. Firstly, if levels of C decrease by 75–90%, then the level of ATP may decrease as well, particularly after a burst of high frequency stimuli. This leads to a decrease in ΔG ATP and an increase of ADP, resulting in a slowing down of calcium reuptake by the sarcoplasmic reticulum [7]. Secondly, if intracellular ADP levels rise further, then the myokinase reaction will increase the formation of AMP which, after deamination to inosine, will diffuse from the myocyte [15]. Consequently, the intracellular adenine nucleotide pool will decrease, leading to failure of the adenine nucleotide cycle to restore the ATP levels. During high frequency stimulation, CP and ATP levels may transiently decline leading to a simultaneous transient increase in inorganic phosphate [9]. Increased Pi may shift the force-[Ca^{2+}] relationship to the right, leading to a decrease in the calcium sensitivity of the myofibrils [1, 9]. Third, it is generally accepted that hydrogen ions are generated during ATP hydrolysis and consumed during CP hydrolysis [4]. In normal muscles, CP buffers these hydrogen ions, while in creatine depleted hearts with low CP levels, the capacity to buffer hydrogen ions may be greatly diminished.

Conclusion

Creatine depletion does not affect the contractile mechanics of isolated papillary muscles at low workloads. The interval between paired-pulses at which fusion occurred was significantly longer in C-depleted muscles than in controls. At high workloads, DF decreased in the same way in both groups, but its subsequent recovery took significantly longer, and the recoverd DF in C-depleted muscles was smaller. From these results it appears that CP depletion has greater effects on the coupling mechanism, through recharging of the calcium fractions, than on the actual force development during stimulation at high frequencies.

Acknowledgements. This work was supported by the Medical Research Council of Canada and the Heart and Stroke Foundation of Ontario.

References

1. Allen DG, Eisner DA, Morris PG et al. (1986) Metabolic consequences of increasing intracellular calcium and force production in perfused ferret hearts. Am J Physiol 367:121–141
2. Brandejs-Barry YA, Korecky B (1987) Hemodynamic performance of isolated blood perfused working hearts of creatine depleted rats. Develop Cardiovasc Med (in press)
3. Cain DF, Davis RE (1978) Breakdown of ATP during a single contraction of working muscle. Biochem Biophys Res Commun 8:361–366
4. Curtin NA, Woledge RC (1978) Energy changes and muscular contraction. Physiol Rev 58:690–761
5. Fitch CD, Chevli R (1980) Inhibition of creatine and phosphocreatine. Accumulation in skeletal muscle and heart. Metabolism 29:686–690
6. Forrester GV, Mainwood GW (1974) Internal dependent inotropic effects in the rat myocardium and the effects of calcium. Pflugers Arch 352:189–196
7. Hasselbach W, Oetliker H (1983) Energetics and electro-genicity of the sarcoplasmic reticulum pump. Ann Rev Physiol 45:325–339
8. Jacobus WE (1985) Respiratory control and the integration of heart high-energy phosphate metabolism by mitochondrial creatine kinase. Ann Rev Physiol 47:707–725
9. Kentish J (1986) Effects of inorganic phosphate on the force-$[Ca^{++}]$ relationship of skinned muscles from rat ventricle. J Physiol 370:585–604
10. Lee SL, Mainwood GW, Korecky B (1970) The electrical and mechanical response of rat papillary muscle to period pulse stimulation. Can J Physiol Pharmacol 48:216–225
11. Mainwood GW, Alward M, Eiselt B (1982) The effects of metabolic inhibitors on the contraction of creatine-depleted muscle. Can J Physiol Pharmacol 60:120–127
12. Mainwood GW, Lee SL (1969) Rat heart papillary muscles: Actin potentials and mechanical response to paired stimuli. Science 166:396–397
13. Meyer RA, Brown RT, Kushmerick JJ (1984) CK kinetics in phosphocreatine depleted rat hearts. Biophys J 45:91
14. Petrofsky JS, Fitch CD (1980) Contractile characteristics of skeletal muscle depleted of phosphocreatine. Pflugers Arch 384:123–129
15. Rubio R, Weidmere VT, Berne RM (1972) Nucleoside phosphorylase: localisation and its role in myocardial distribution of purines. Am J Physiol 222:550–555
16. Saks VA, Rosenshtraukh LV, Smirnov VN et al. (1978) Role of creatine phosphokinase in cellular function and metabolism. Can J Physiol Pharmacol 56:691–706
17. Saks VA, Kupriyanov VN, Smirnov VN (1986) Myocardial energy: role of PCr and ATP. Proc Int Union Physiol Sci 16:66
18. Shoubridge EA, Jeffry FMH, Keogh JM et al. (1985) Creatine kinase kinetics, ATP turnover and cardiac performance in hearts depleted of creatine with the substrate analogue β-guandininopropionic acid. Biochim Biophys Acta 847:25–32

Authors' address:

Dr. B. Korecky, Dept. of Physiology, School of Medicine, University of Ottawa, 451 Smyth Road, Ottawa, Ont. KlH 8M5

ATPase activity of intact single muscle fibres of Xenopus laevis is related to the rate of force redevelopment after rapid shortening

G. J. M. Stienen [1], J. Lännergren [2] and G. Elzinga [1]

[1] Laboratory for Physiology, Free University, Amsterdam, The Netherlands
[2] Department of Physiology II, Karolinska Institutet, Stockholm, Sweden

Summary

Five different fibre types have been recognized in the iliofibularis muscle of *Xenopus laevis*. The force-velocity and histochemical characteristics of these fibres vary considerably and differences are also found in their myosin composition. In this study a comparison was made between the rate of ATP hydrolysis estimated from the stable maintenance heat rate and the mechanical performance of fibres of type 1, 2 and 3.

In the experiments, firstly the force-velocity relation of a fibre was determined, and subsequently, heat production during isometric tetanic contractions at 20 °C was measured. Force redevelopment following the fastest shortening used for the measurement of the force-velocity relationship was fitted to a single exponential.

The rate of ATP hydrolysis, estimated from the heat production, was found to be roughly proportional to the rate of force redevelopment. Crossbridge attachment rate was determined by using a simulation of a four state model of the crossbridge cycle. It appears that crossbridge attachment rate is proportional to the in vivo actomyosin ATPase activity during an isometric tetanic contraction.

Introduction

There is general agreement that the contractile mechanism in different types of muscle operates along similar lines through the cyclic interaction between actin and myosin. The variation in the mechanical characteristics such as the force-velocity relationship in different fibre types [12] suggests that different fibres have different sets of rate constants for the actomyosin interaction scheme. In order to arrive at a general model of muscle contraction, which applies to cardiac muscle as well, we studied the mechanical and energetic properties of fibres of different types. A discussion of the results will be guided mainly by a simplified scheme for the actomyosin ATP hydrolysis.

Methods

Single fibres were dissected under dark-field illumination from the iliofibularis muscle of *Xenopus laevis*. A small platinum hook was tied to each tendon end. The fibre was then transferred to a temperature controlled chamber (20 °C), filled with Ringer solution, pH = 7.0 and mounted between a force transducer and the extension of a loudspeaker coil, the position of which was servocontrolled. Sarcomere length, measured by laser diffraction, was adjusted to 2.3 μm. Fibre dimensions were measured by means of a dissection microscope. The fibre was stimulated electrically via platinum electrodes with 0.5 ms pulses at a repetition frequency of 85–100 Hz for 350–800 ms. After the maximum tetanic force (P_0) was attained, the fibre was shortened by an abrupt length change, followed by a length change at a constant velocity (v) which reduced force to a nearly constant level. The force v̄ velocity data obtained, were fitted to Hill's equation: $(P+a)(v+b)=b(P_0+a)$, in which the shortening velocity at zero load, v_{max}, equals bP_0/a [7]. After completion of these measurements the fibre was transferred to a thermopile [13b]. While kept in water-saturated air, the fibre was stimulated via the hooks in each tendon end to produce isometric tetanic contractions. Calibration of the pile, correction for heat loss and the determination of added heat capacity were carried out as described elsewhere (submitted).

Results

Different shortening velocities were applied to the tetanically contracting muscle fibres in order to construct a force-velocity curve. A tracing obtained in such a way is shown in Fig. 1 A. The fibre is stimulated electrically and after the maximum isometric tension (P_0) is reached, the fibre is shortened over a distance of 1.2 mm. Figure 1 B shows the position signal and force response in greater detail. After the length change, force recovers

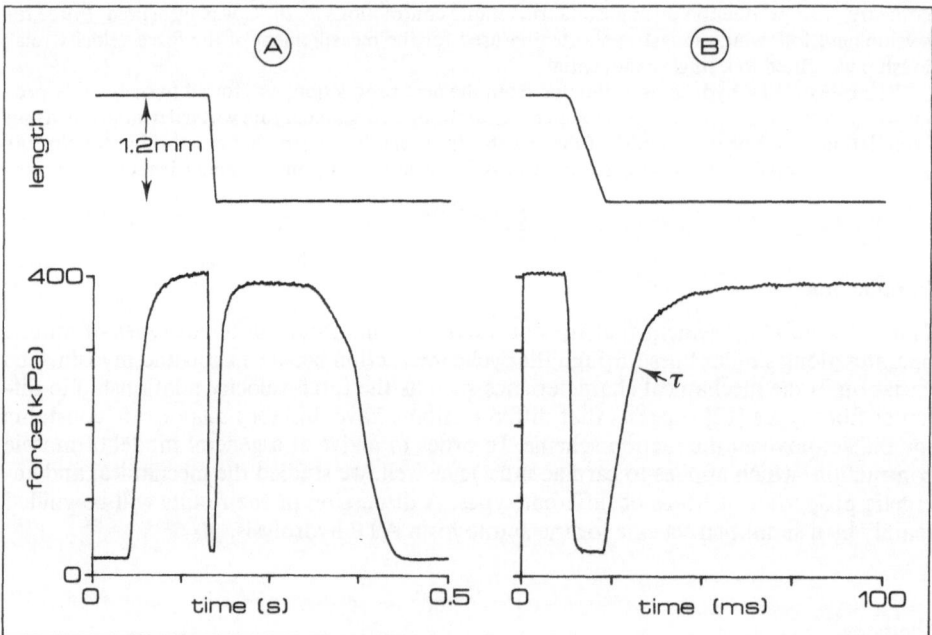

Fig. 1. In A, force and length signals are shown during a tetanic contraction. In B, these signals are shown in greater detail. The steady shortening imposed corresponded to 8.8 fibre lengths (L_0) per second. The rate of force recovery ($1/\tau$) was 150 s^{-1}. Fibre of type 1, maximum velocity of shortening 9.4 L_0/s.

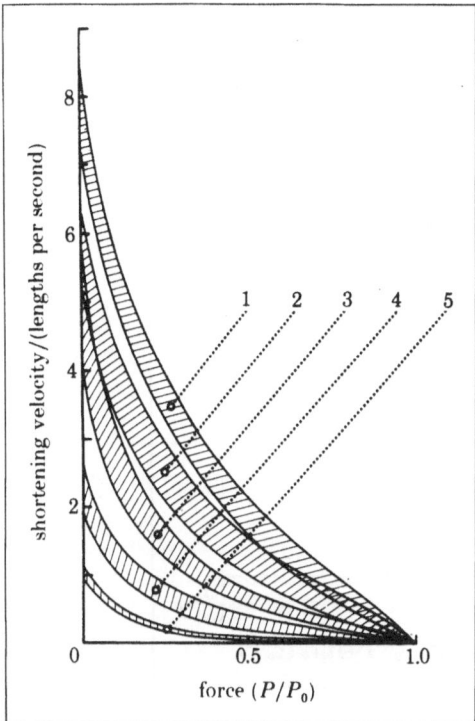

Fig. 2. Compilation of force-velocity data from the five fibre types in iliofibularis muscle of *Xenopus laevis* (adapted from [12]).

towards its original isometric level in an exponential fashion. To obtain an estimate of the rate of crossbridge attachment, force development following a shortening at about zero load was fitted with a single exponential. Although the recovery process contains a number of different rate constants, the correlation coefficient of the mono-exponential fit was in none of the cases less than 0.996. The rate of the recovery process after length changes with different velocities was somewhat dependent on the force level reached; at 50% of P_0 the rate was about 10% smaller than the rate found near zero load. The force-velocity relationships for different fibre types are illustrated in Fig. 2.

Stable maintenance heat rate (h_b) was determined from the slope of the heat production after 350 ms, where the influence of the labile heat component is negligible. Stable maintenance heat rate ranged from 0.17–0.78 W per gram dry weight. On average the heat rate was higher in type 1 than in type 2 fibres, type 3 fibres producing even less heat.

In order to relate heat production to crossbridge energetics, the heat produced by the activation process, i.e. the Ca^{2+} cycling, must be substracted from the maintenance heat. The data of Homsher and Kean [8] suggest that around 40% of the h_b is accounted for by the activation process. For the results shown in Fig. 3, relating ATP hydrolysis rate to rate of force recovery, and in the simulations presented below we assumed that this fraction is independent of fibre type. The ATP hydrolysis rate was estimated from the contribution of the AM·ATP hydrolysis in h_b (60%), taking into account the myosin concentration (0.2 mM)[1], the ratio dry weight/wet weight (0.27) and the net enthalpy change associated with ATP hydrolysis $\Delta H = -34$ kJ/mol [3]. As a starting point for the

---

[1] If a myosin concentration of 0.28 mM [4a] is used, the values obtained for the ATPase rate are 1.4 times smaller than shown in Fig. 3 and Table 1.

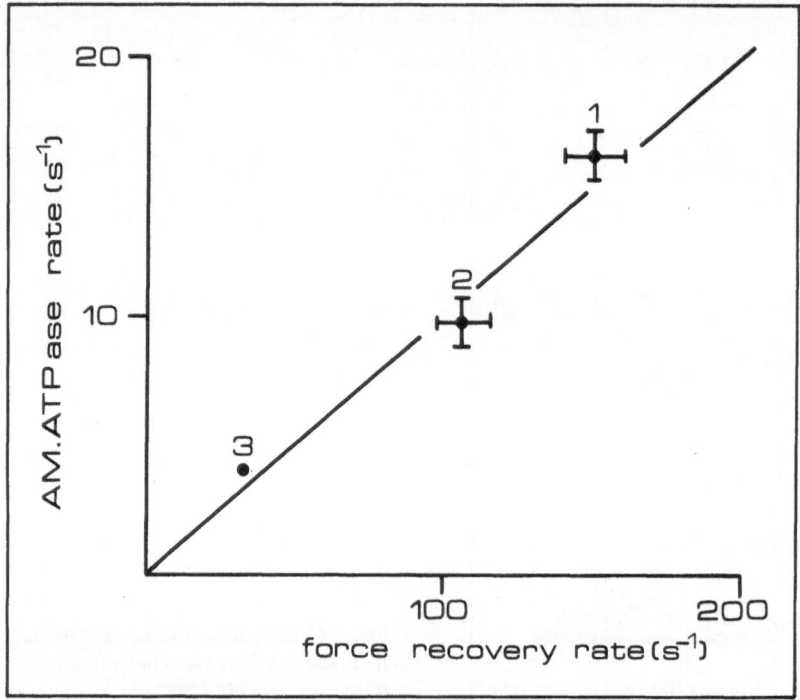

Fig. 3. Averaged values for the ATP hydrolysis rate of type 1, 2 and 3 estimated from the maintenance heat rate as a function of the rate of force recovery. The bars indicate the s.e.m.; solid line
represents the regression line forced through the origin.

simulations a four state model was used which is schematically shown in Fig. 4 [10]. For
a given set of rate constants (8 in this case), the steady state ATP hydrolysis rate can be
calculated according to the King and Altman procedure, but the general (analytical) solution of the four coupled differential equations can also be calculated and simulated by
computer. It must be emphasized that the simulations can only be used to obtain a rough
idea of the different rate constants involved in the crossbridge cycle, because a general
crossbridge model must take the distribution and rates of crossbridges with different mechanical properties into account [4b, 7]. An example of the results of a simulation is shown
in Fig. 4. The curves shown are based on a compilation of data in the literature (mainly)
for frog skeletal muscle at about 0 °C and rabbit psoas fibres (e.g. [5, 6]).

At $t=0$, the cycle is switched from a stationary isometric to an isotonic mode by a
change of CD from $1 \, s^{-1}$ to $200 \, s^{-1}$. This change in rate for the transition between attached states causes an increase in the apparent detachment rate because the subsequent
detachment step DA is very rapid. A decrease in the number of attached crossbridges rapidly takes place. A reduction in force to a low isotonic level might be due to a redistribution of crossbridges between the attached states (force as well as nonforce producing
states) and because the remaining force producing crossbridges are opposed by pushing
bridges [9]. In the simulations, the rate of increase of detached crossbridges during isotonic contraction is more rapid than the rate of decrease in the reversed mode after the
model is switched back from stationary isotonic to isometric contraction (right graph).
A change in the occupancy of any of the four states consists of the sum of three exponen-

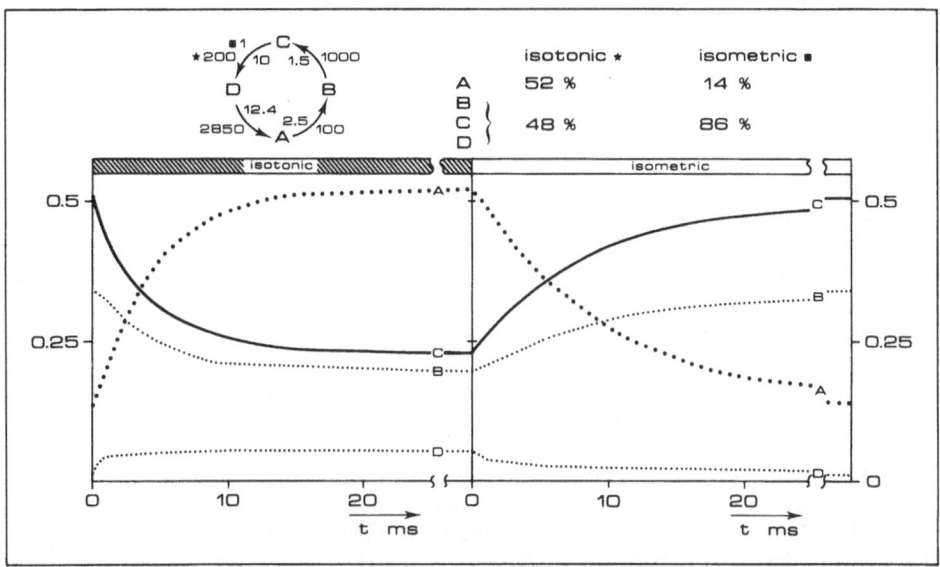

Fig. 4. Results from a simulation of a four state model for the crossbridge cycle. The detached state(s) (A), weakly attached (B), strongly attached (C) and rigor state (D) are shown. The forward rate constant chosen is given at each arrow outside the circle; the equilibrium constant chosen is indicated at the inside. The reverse rate of each transition is obtained by dividing these two values. The forward rate of the transition CD depends on the mode of contraction, isotonic ($\star$) or isometric ($\blacksquare$).

Table 1. Results of the simulations for different fibre types of the iliofibularis muscle of *Xenopus laevis*

Fibre type	1	2	3
ATPase rate (s^{-1})	15	11	3
Attachment rate AB (s^{-1})	110	77	24
Transition rate CD (s^{-1})	43	31	9

Forward rate and equilibrium constant of the transition DA were taken as $7\,500\ s^{-1}$ and 30 respectively. The remaining values used were as shown in the scheme of Fig. 4. During isometric contraction the distribution for each fibre type was, A: 28%, B: 36%, C: 35% and D: 1%.

tial contributions [1]. It appeared, however, with the set of rate constants shown, that only the contribution of the smallest rate constant was large enough to be important on the time scale of the experiments.

For the estimation of values for the attachment rate AB and the rate CD, this relaxation rate constant was assumed to be equal to the rate of force redevelopment. The turnover rate in the model was assumed to be equal to the ATP hydrolysis rate. The average results for fibres of different types are shown in Table 1. Surprisingly, the AM·ATP

---

[1] In this particular case, a value of 86% was found for the crossbridges attached during isometric contraction while during isotonic shortening 48% was attached.

hydrolysis rate, AB and CD vary in proportion even when the values for the rapid equilibrium process BC vary within an order of magnitude.

Discussion

The combined measurement of mechanical and energetic quantities in single fibres offers the opportunity to gain insight in the quantitative differences between crossbridge kinetics in different fibre types. Bárány [1] has shown for muscle tissue of different origins that the in vitro AM · ATPase activity is proportional to the maximum shortening velocity. Van der Laarse et al. [11] showed that a similar correlation is also found in different fibre types of *Xenopus laevis* when a measure of the myofibrillar ATPase is obtained by quantitative histochemistry. We found that the relation between the in vivo AM · ATPase activity estimated from heat measurements and maximum shortening velocity was nonlinear, a trend which also shows up in a summary of the results of different muscle tissue, compiled by Marston and Taylor ([13a], Table 3). A serious difficulty in explaining the relation between the in vivo AM · ATP hydrolysis rate and the maximum shortening velocity is that at least one transition rate between attached states (e.g. CD) is increased during shortening. A more elaborate crossbridge model would certainly be necessary to explain this result in detail. Eisenberg et al. [4b] argue from their model calculations that there exists a relation between the force-velocity curve and the ATPase activity via the attachment rate (AB). A similar phenomenon is present in the scheme presented here, because CD as well as AB influences the rate of ATP hydrolysis. Marston and Taylor [13a] showed that the apparent association rate of myosin S-1 with actin is equal to the rate of ATP hydrolysis for different muscle types and different actin concentrations. This has an interesting parallel with our findings of a rather linear relation between the rate of ATP hydrolysis and the rate of force recovery, which is dominated by the attachment rate. However, as Marston and Taylor indicated, the apparent association rate is the result of a series of steps. In this respect, it can be noted that Siemanskowski and White [14] indicate that at least for slow muscles, the rate constant for ADP dissociation from the AM-complex is sufficiently slow to be the rate limiting step for unloaded shortening. Recently, Brenner and Eisenberg [2] showed that there is a close correlation between the rate constant of force redevelopment in skinned rabbit psoas fibres, and the maximum actin activated ATPase activity of myosin S-1 crosslinked to actin. The data presented here indicate that a similar correlation is found in intact fibres using a different approach.

A proper test of the crossbridge model requires that the physiological properties of muscles with different contraction velocities be predicted from the sets of rate constants obtained from the corresponding actomyosin [15]. In the present paper we have indicated some of the relations between physiological properties and the kinetic biochemical data available. Unfortunately, no biochemical data are known for *Xenopus* fibres; these are needed, however, for a critical evaluation of the model.

References

1. Bárány M (1967) ATPase activity of myosin correlated with speed of muscle shortening. J Gen Physiol 50:197–218
2. Brenner B, Eisenberg E (1986) Rate of force generation in muscle: Correlation with actomyosin ATPase activity in solution. Proc Nat Acad Sci USA 83:3542–3546
3. Curtin NA, Woledge RC (1978) Energy changes and muscular contraction. Physiol Rev 58:690–761

4a. Ebashi S, Endo M, Ohtsuki I (1969) Control of muscle contraction. Q Rev Biophys 2:351–384
4b. Eisenberg E, Hill TL, Chen Y (1980) Cross-bridge model of muscle contraction. Quantitative analysis. Biophys J 29:195–227
5. Ferenczi MA, Goldman YE, Simmons RM (1984) The dependence of force and shortening velocity on substrate concentration and skinned fibres from frog muscle. J Physiol 350:519–543
6. Goldman YE, Hibberd MA, Trentham DR (1984) Initiation of active contraction by photogeneration of adenosine-5'-triphosphate in rabbit psoas muscle fibres. J Physiol 354:605–624
7. Hill AV (1938) The heat of shortening and the dynamic constants of muscle. Proc R Soc B 126:136–195
8. Homsher E, Kean CJ (1978) Skeletal muscle energetics and metabolism. Ann Rev Physiol 40:93–131
9. Huxley AF (1957) Muscle structure and theories of contraction. Prog Biophys Chem 7:255–318
10. Huxley HE, Kress M (1985) Crossbridge behaviour during muscle contraction. J Muscle Res Cell Motil 6:153–161
11. Laarse WJ van der, Diegenbach PC, Hemminga MA (1986) Calcium-stimulated myofibrillar ATPase activity correlates with shortening velocity of muscle fibres in Xenopus laevis. Histochem J 18:487–496
12. Lännergren J, Hoh JFU (1984) Myosin isoenzymes in single muscle fibres of Xenopus laevis: Analysis of five different functional types. Proc R Soc Lond, series B 222:401–408
13a. Marston SB, Taylor EW (1980) Comparison of the myosin and actomyosin ATPase mechanisms of the four types of vertebrate muscles. J Mol Biol 139:573–600
13b. Mulieri LA, Luhr G, Trefry J, Alpart NR (1977) Metal film thermopiles for use with rabbit right ventrical papillary muscles. Am J Physiol 233:C146–C156
14. Siemankowski RF, White HD (1984) Kinetics of the interaction between actin, ADP, and cardiac myosin-S1. J Biol Chem 259:8, 5045–5053
15. Taylor EW (1979) Mechanism of actomyosin ATPase and the problem of muscle contraction. CRC Crit Rev Biochem 6:103–164

Authors' address:

Dr. G.J.M. Stienen, Laboratory for Physiology, Free University Amsterdam, Van der Boechorststraat 7, 1081 BT Amsterdam, The Netherlands

II. Cardiac energetics as related to ontogenesis, chronic myocardial transformation and inotropic interventions

Developmental differences in myocardial ATP metabolism

J. W. de Jong and P. W. Achterberg

Cardiochemical Laboratory, Thoraxcenter, Erasmus University Rotterdam, The Netherlands

Summary

Little is known about postnatal changes in myocardial purine metabolism. We therefore studied how ATP catabolism was affected by hypothermia and ischaemia in neonatal and adult hearts. Hypothermia during ischaemia protected isolated adult and newborn hearts against ATP decline. Reperfusion after normothermic ischaemia resulted in higher ATP levels in newborn hearts with less release of ATP-catabolites. During normoxia adult hearts released mainly urate (80% of total purine release), while newborns released mainly hypoxanthine (64%). During early reperfusion adult and newborn hearts released mainly inosine (50–60%). The very low xanthine oxidase activity in the neonatal heart could be an important factor in the observed ATP preservation during reperfusion.

Introduction

The newborn heart copes better with hypoxia and ischaemia than the adult heart [10, 11]. Little is known, however, about postnatal changes in myocardial purine metabolism, which may be important for homeostasis of the myocardial ATP-pool after ischaemia and reperfusion, or for the extent of free-radical induced reperfusion damage, e.g. via xanthine oxidase [3]. We studied adenine nucleotide breakdown and ATP-catabolite release in adult and newborn rat hearts, subjected to normothermic and hypothermic ischaemia followed by reperfusion. The newborn heart differed in normoxic and post-ischaemic patterns of purine release from the adult heart. This coincided with a better capacity of the neonatal heart to maintain its post-ischaemic adenine nucleotide content.

Methods

Isolated rat hearts were retrogradely perfused with a modified Tyrode buffer (10 mM glucose) as described elsewhere [1]. A pressure of 50 mm Hg was used; pacing frequency was 300 beats/min. After a 20 min stabilization period, hearts were made ischaemic for 15 min by complete arrest of flow. Ischaemic hearts were submerged into perfusion buffer kept at either 37, 30 or 23 °C. Pacing was continued during 20 min of reperfusion (37 °C), then the hearts were quickly removed from the perfusion apparatus, freeze-clamped, and analysed for adenine nucleotides by HPLC [8]. Perfusate samples were used to determine the purines (urate, xanthine, hypoxanthine, inosine and adenosine) by HPLC, as described before [2].

Results

At 50 mm Hg perfusion pressure both adult and newborn hearts demonstrated stable biochemical characteristics between 20 and 55 min of control perfusion. The myocardial

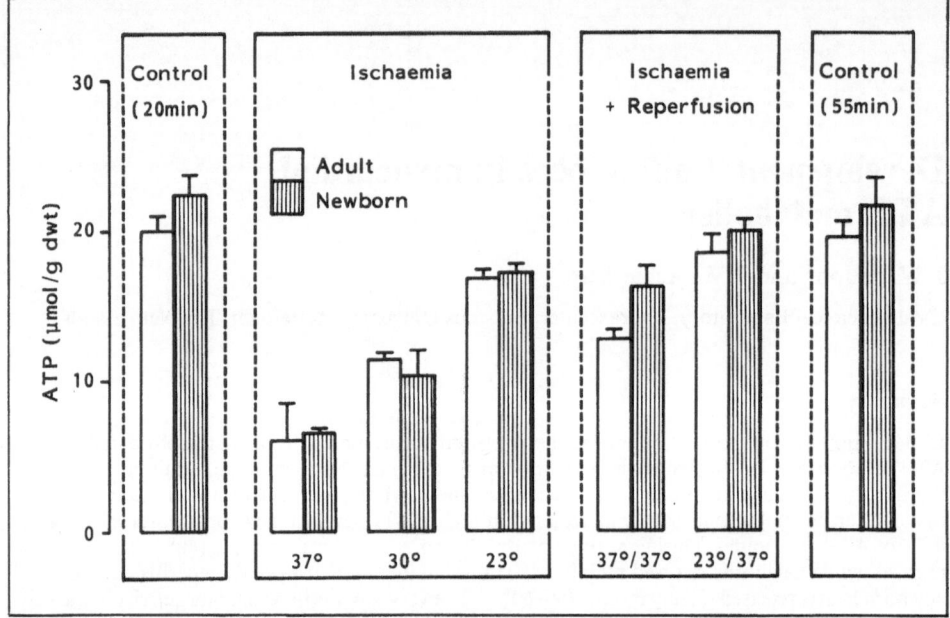

Fig. 1. Myocardial ATP content after normoxic control perfusions of 20 and 55 min; after 20 min of control perfusion at 37 °C plus 15 min of ischaemia (at 37, 30 or 23 °C; and after a similar period of ischaemia at 37 or 23 °C with 20 min reperfusion at 37 °C. Means ± SEM (n = 4).

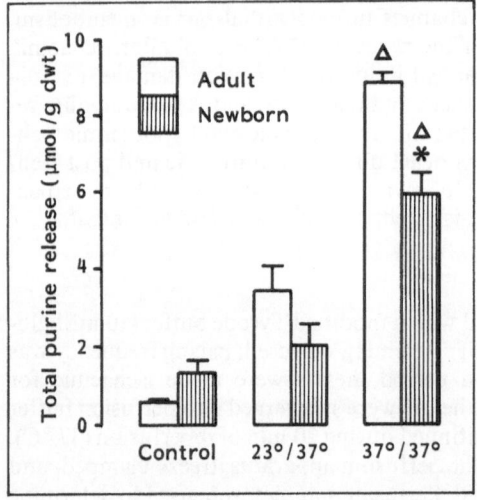

Fig. 2. Release of adenosine, inosine, hypoxanthine, xanthine and urate after 15 min of ischaemia at 23 or 37 °C, plus 20 min of reperfusion at 37 °C. Control data from hearts with no ischaemia. Means ± SEM (n = 4): $p < 0.05$ vs hypothermia ($\triangle$); $p < 0.05$ vs adult (*).

ATP content remained high (Fig. 1). Adult hearts released half as much purine at the end of a 55 min normoxic period as newborns did (data not shown). Coronary flow was also lower in adults than in newborns (38 ± 6 vs 56 ± 9 ml/min per g dry wt; $p < 0.05$).

Total arrest of flow with simultaneous immersion of the hearts in the perfusion medium at 37, 30 or 23 °C for 15 min resulted in a decline of ATP in both adult and newborn

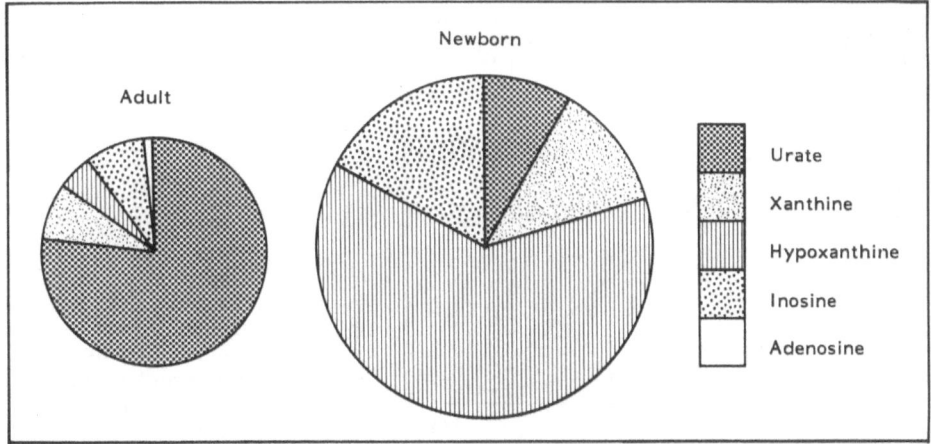

Fig. 3. Relative amounts of purines produced during normoxia.

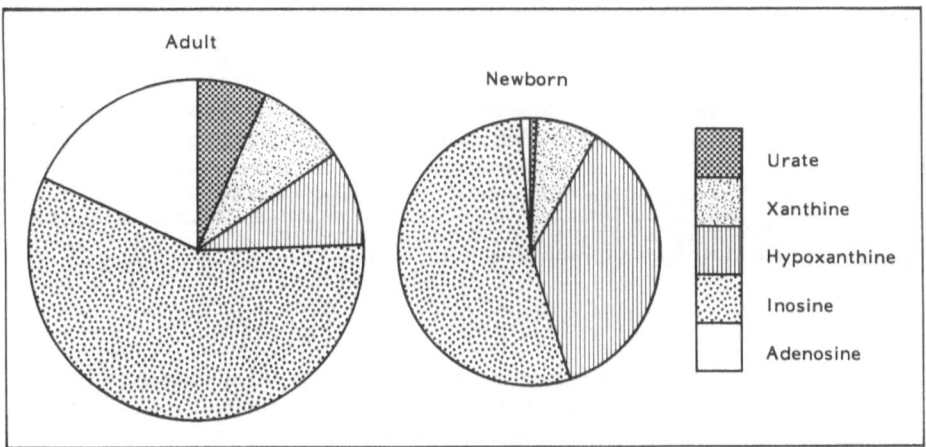

Fig. 4. Relative amounts of purines produced during reperfusion after ischaemia at 37 °C.

hearts (Fig. 1). The changes were much less at 23 °C, with intermediate levels at 30 °C, in both adult and newborn hearts (Fig. 1). After normothermic ischaemia and reperfusion, ATP levels were higher ($p = 0.06$) in newborn than in adult hearts. The decreased ATP catabolism was supported by the finding of significantly less purine release from newborn hearts after normothermic ischaemia followed by reperfusion (Fig. 2). The protective effect of hypothermia was also clearly reflected in post-ischaemic purine release, which was significantly lower after hypothermic ischaemia in both adult and newborn hearts (Fig. 2).

Purine release during the last 5 min of control perfusion amounted to 84 ± 10 nmol/5 min per g, of which 77% was urate (Fig. 3). Newborn hearts released 188 ± 42 nmol/5 min per g ($p < 0.05$), which consisted mainly of hypoxanthine (64%) and inosine (15%).

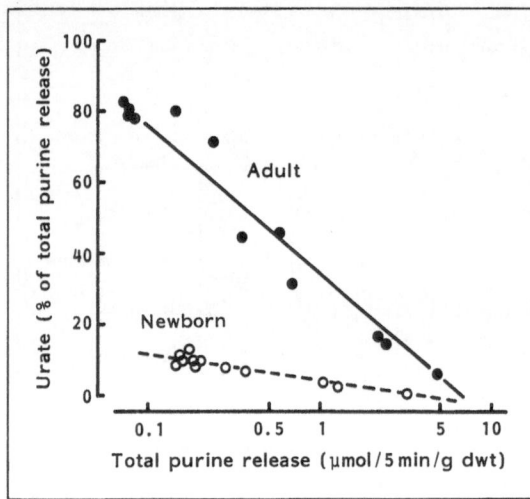

Fig. 5. Urate release as a percentage of total purine release. Perfusates were collected over four consecutive 5 min periods during reperfusion, after hypothermic or normothermic ischaemia, or during prolonged control perfusion. Each point is the average of four experiments.

During the first 5 min of reperfusion after normothermic ischaemia, purine release from adult heart amounted to 4.9 ± 0.4 µmol per g, while newborns released 3.1 ± 0.3 µmol/5 min per g over the same period ($p < 0.05$). Inosine was in both cases the major purine released (58% of the total in adults and 53% in newborns; Fig. 4). The relative contribution of urate to total purine release from adult and newborn hearts is plotted in Fig. 5 as a function of the total amount of purines released during reperfusion. Under conditions of similar total purine release, newborn hearts released much less urate (Fig. 5), but much more hypoxanthine than adult hearts.

Discussion

Good evidence exists that purine release is a valuable indicator of an imbalance between ATP formation and ATP breakdown [1, 4, 7]. During the first 5 min of reperfusion after hypothermic ischaemia, purine release from newborn hearts was significantly increased, as compared to release in controls (Fig. 2). However, only a minor insignificant decline in ATP was found at the end of hypothermic ischaemia (Fig. 1). This points to the relatively high sensitivity of purine release as an indicator of ATP breakdown.

Detailed knowledge of the pathways of myocardial purine metabolism has become increasingly important, because several studies have indicated that restoration of the myocardial ATP-pool after ischaemia can be achieved by giving purines or purine precursors or by preventing washout or breakdown of endogenously formed purines [5, 6, 9]. No major efforts have been undertaken, however, to characterize developmental differences in myocardial purine metabolism, which could lead to a different approach in the protection of newborn versus adult hearts.

Newborn hearts preserved myocardial ATP levels better than adult hearts after normothermic ischaemia followed by reperfusion ([10, 11]; Fig. 1). Moreover, the newborn hearts had a lower total purine release over the 20 min reperfusion period (4.7 vs 8.3 µmol/g dry wt in adults). Again purine release appears to be a powerful indicator for determining the ATP balance non-invasively; an approach comparable only to NMR techniques, but much more sensitive.

The observed pattern of purine release from adult hearts at varying rates of total purine release has been described before [2]. The different patterns of purine release in newborn and adult hearts presumably reflect age-dependent differences in the enzymes of purine metabolism.

Apparent maximal xanthine oxidase/dehydrogenase activities can be computed from the release of urate and xanthine (2 × urate + xanthine = xanthine oxidase activity). For adult hearts this calculation gives a maximal xanthine oxidase activity of 31 ± 3 nmol/min per g wet wt, which accurately resembles the xanthine oxidase activity determined by us in adult rat heart homogenates [13]. The same calculation for 10 day old rat hearts gives several fold lower activities of this enzyme (6 ± 2 nmol/min per g). Xanthine oxidase activity is virtually absent in fetal rat-heart homogenates (B. Schoutsen, personal communication). The absence of xanthine/dehydrogenase could be beneficial to the newborn hearts, because: I) it could result in higher levels of hypoxanthine in the heart and blood. This purine can be reincorporated into the myocardial ATP-pool, as we showed for adult rat hearts [9]; II) it has been reported that myocardial xanthine dehydrogenase, the native form of the enzyme, can be converted to the oxidase-form during ischaemia [3]. During reperfusion the oxidase can cause free-radical formation and contribute to reperfusion damage in the adult heart [12].

Our perfusion conditions appear to put a higher strain on newborn than on adult hearts (as reflected by purine release during normoxia; Fig. 2). Nevertheless, the effects of ischaemia and reperfusion were better tolerated by the newborn heart. Despite this better tolerance to normothermic ischaemia and reperfusion, it is clear that hypothermia will significantly contribute to the protection of the newborn heart against ischaemic myocardial ATP depletion. The major differences with regard to ATP preservation become clear during the reperfusion period.

We conclude that age-dependent differences in rat heart purine metabolism may be responsible for the better capability of the newborn heart to preserve its ATP-pool during ischaemia and reperfusion.

Acknowledgements. The authors gratefully acknowledge the secretarial assistance of Mirjam Zweserijn, the technical help of Selma Nieukoop, and the financial support by the Netherlands Heart Foundation.

References

1. Achterberg PW, Nieukoop AS, Schoutsen B, De Jong JW (1987) Different purine metabolism in newborn heart coincides with better ATP preservation after ischemia and reperfusion. Am J Physiol, submitted
2. Achterberg PW, Stroeve RJ, De Jong JW (1986) Myocardial adenosine cycling rates during normoxia and under conditions of stimulated purine release. Biochem J 235:13–17
3. Chambers DE, Parks DA, Patterson G, Roy R, McCord JM, Yoshida S, Parmley LF, Downey JM (1985) Xanthine oxidase as a source of free radical damage in myocardial ischemia. J Mol Cell Cardiol 17:145–152
4. De Jong JW, Huizer T, Tijssen JGP (1984) Energy conservation by nisoldipine in ischaemic heart. Br J Pharmacol 83:943–949
5. Ely SW, Mentzer RM, Lasley RD, Lee BK, Berne RM (1985) Functional and metabolic evidence of enhanced myocardial tolerance to ischemia and reperfusion with adenosine. J Thorac Cardiovasc Surg 90:549–556
6. Foker JE, Einzig S, Wang T, Anderson RW (1980) Adenosine metabolism and myocardial preservation. J Thorac Cardiovasc Surg 80:506–516

7. Harmsen E, De Jong JW, Serruys PW (1981) Hypoxanthine production by ischemic heart demonstrated by high pressure liquid chromatography of blood purine nucleosides and oxypurines. Clin Chim Acta 115:73–84
8. Harmsen E, De Tombe PP, De Jong JW (1982) Simultaneous determination of myocardial adenine nucleotides and creatine phosphate by high-performance liquid chromatography. J Chromatogr 230:131–136
9. Harmsen E, De Tombe PP, De Jong JW, Achterberg PW (1984) Enhanced ATP and GTP synthesis from hypoxanthine or inosine after myocardial ischemia. Am J Physiol 246:H37–H43
10. Jarmakani JM, Nagatomo T, Nakazawa M, Langer GA (1978) Effect of hypoxia on myocardial high-energy phosphates in the neonatal mammalian heart. Am J Physiol 235:H475–H481
11. Nishioka K, Jarmakani JM (1982) Effect of ischemia on mechanical function and high-energy phosphates in rabbit myocardium. Am J Physiol 242:H1077–H1083
12. Schlafer M, Kane PF, Kirsh MM (1982) Superoxide dismutase plus catalase enhances the efficacy of hypothermic cardioplegia to protect the globally ischemic, reperfused heart. J Thorac Cardiovasc Surg 83:830–839
13. Schoutsen B, De Jong JW, Harmsen E, De Tombe PP, Achterberg PW (1984) Myocardial xanthine oxidase/dehydrogenase. Biochim Biophys Acta 762:519–524

Authors' address:

Dr. J. W. de Jong, Cardiochemical Laboratory, Thoraxcenter, Erasmus University Rotterdam, P.O. Box 1738, 3000 DR Rotterdam, The Netherlands

The effects of acute and chronic inotropic interventions on tension independent heat of rabbit papillary muscle

E. M. Blanchard, L. A. Mulieri, and N. R. Alpert

University of Vermont College of Medicine, Department of Physiology and Biophysics, Given Building, Burlington, Vermont U.S.A.

Summary

We have used the myothermal method to noninvasively monitor the amount of calcium cycled during a single isometric twitch of rabbit papillary muscle. Experiments were designed to test the working hypothesis that changes in peak twitch tension caused by pharmacological agents or changing haemodynamic conditions are accompanied by parallel changes in the tension independent heat (TIH) signal associated with Ca^{2+} cycling. We isolated the TIH signal by eliminating the tension dependent component of initial heat with a hyperosmotic Krebs solution containing 2,3-butanedione monoxime. Contrary to the working hypothesis, positive or negative inotropic effects on twitch tension caused by pressure overload hypertrophy, thyrotoxic hypertrophy, isoproterenol, and $UDCG_{115}$ were not accompanied by parallel changes in TIH. Alternative explanations for the relation between peak twitch tension and TIH are explored.

Introduction

Pharmacological agents and changes in haemodynamic conditions can produce distinct changes in peak twitch force (Ptw), maximum rate of force development (dP/dt max) and tension time integral ($\int$ Pdt) of cardiac papillary muscle at a fixed muscle length. Biochemical analysis of isolated components of the excitation-contraction coupling (EC coupling) system in cardiac muscle strongly suggests that acute or chronic changes in the cycling of Ca^{2+} can account for some aspects of the altered mechanical response. For example, conditions that increase the amount of Ca^{2+} cycled per twitch because of effects of the slow inward Ca^{2+} current on the loading and release of Ca^{2+} from the sarcoplasmic reticulum could potentiate the peak twitch tension or dP/dt by increasing the free Ca^{2+} available to activate the contractile apparatus [5]. Thus, one working hypothesis is that changes in peak twitch tension or dP/dt are accompanied by parallel changes in the total amount of Ca^{2+} cycled and the peak free Ca^{2+} transient per twitch. It is difficult to test the hypotheses generated by in vitro experiments that explain changes in the inotropic state of intact cardiac tissue, since there are few methods available to monitor the dynamics of EC coupling noninvasively. The free Ca^{2+} concentration can be monitored by various dyes or aequorin bioluminescence and inferences about the strength and duration of Ca^{2+} activation can be drawn. The myothermal technique is an alternative noninvasive way to monitor Ca^{2+} cycling in intact cardiac tissue.

The biochemical reactions causing tension production of an isometric papillary muscle produce heat when a stimulus initiates activity and this heat signal can be measured by sensitive thermopiles (Fig. 1). This heat output of the isometrically contracting

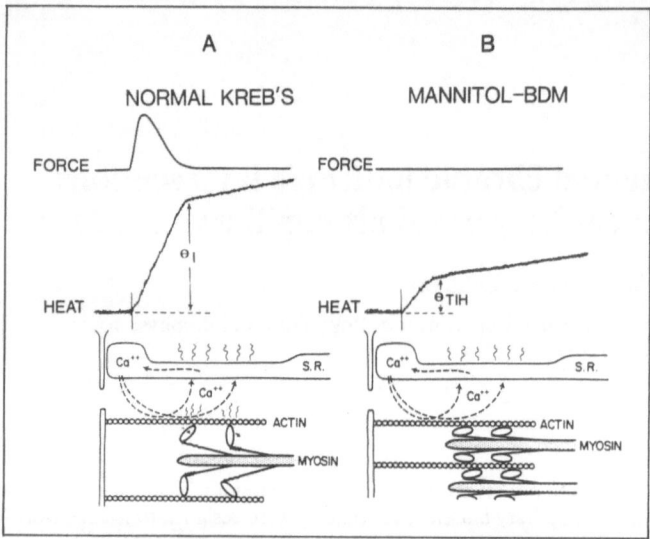

Fig. 1. An illustration of initial heat production after incubation of the papillary muscle in normal Krebs (A) and tension independent heat production after incubation in mannitol-BDM Krebs (B). Below each heat record (corrected for heat loss) are schematics of the sources of the heat.

papillary muscle can then be used to monitor the extent and kinetics of these biochemical reactions. The heat evolved within the duration of the twitch is largely caused by the exothermic reactions of crossbridge cycling (tension dependent heat, TDH) and the cycling of Ca^{2+} between storage sites and the myofilament space (tension independent heat, TIH) (Fig. 1 A). The aim of this paper is to explain how the myothermal method can be used to noninvasively determine whether or not changing haemodynamic conditions and pharmacological agents alter the total amount of Ca^{2+} cycled in a single twitch of a papillary muscle. According to the working hypothesis, the amount of TIH at the time of complete mechanical relaxation should parallel changes in peak twitch tension.

In order to derive information about Ca^{2+} cycling, the component of initial heat caused by crossbridge cycling must be eliminated (Fig. 1 B). The tension dependent heat caused by crossbridge cycling can be eliminated in some skeletal muscle preparations by stretching the muscle to a length where no overlap of the thick and thin filaments occurs [12, 22]. The tension independent heat signal remaining at zero overlap is considered the thermal equivalent of calcium cycling. The stretching method cannot be used to isolate tension independent heat in cardiac muscle as the heat output continues to rise while force declines when papillary muscles are stretched beyond optimal length [10] suggesting that the muscle is being altered in a non-simple manner by stretch [21]. An alternative approach to estimating the amount of tension independent heat occurring during a papillary muscle twitch has been to shorten the muscle from the optimal length for force production to very short lengths [6, 7]. Tension and the tension dependent heat component of initial heat decline as the muscle is shortened and the resulting heat-tension relation is extrapolated to zero tension to estimate tension independent heat. Although the shortening method has been useful in assessing the effects of excitation-contraction coupling modulators on tension independent heat (TIH), the quantitative description of

Ca^{2+} cycling is probably confounded by the length-dependence of Ca^{2+} activation [1] and by tension dependent heat contamination of TIH at very short lengths.

We have used a method to isolate TIH that does not involve stretching or shortening papillary muscles from the optimal length for tension production. The muscles are soaked in a modified Krebs-Ringer solution that contains either mannitol to make the solution hyperosmotic or both mannitol and the compound 2,3-butanedione monoxime (BDM). BDM [16] and the increased intracellular ionic strength [13] caused by the mannitol act to suppress crossbridge cycling and the accompanying tension dependent heat. Even though tension is usually eliminated with this treatment, a triggerable heat output (TIH) is obtained following each stimulus. TIH isolated in this manner reveals important information about the effects of changing haemodynamic conditions and positive inotropic drugs on Ca^{2+} cycling in papillary muscles.

Methods

Animals and muscle preparation

Male rabbits weighing about 2.0 kg were sacrificed by cervical dislocation and the hearts rapidly removed and washed free of blood by serial immersion in Krebs-Ringer solution at 37 °C. Silk ligatures (4-0) were tied to both ends of a right ventricular papillary muscle and the muscle was mounted on a thermopile (Fig. 2). The ligature at the tendonous end was attached to a stationary hook (B) while the ligature at the other end (C) of the muscle

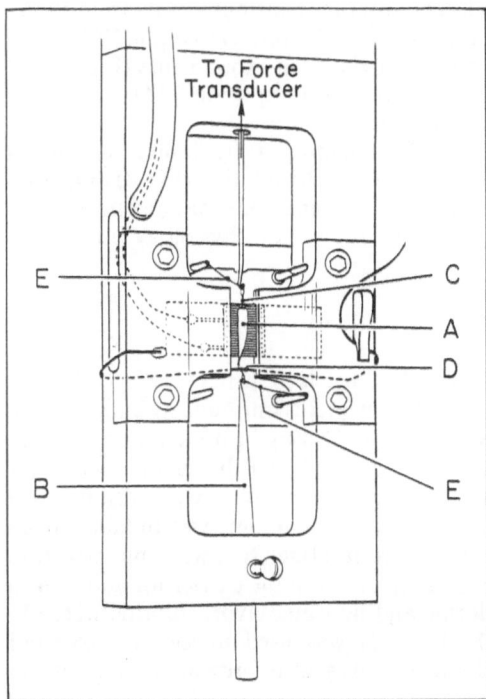

Fig. 2. Diagram of the muscle and thermopile assembly. A, papillary muscle; B, stationary glass hook; C, ligature and glass rod connected to force transducer; D, tether; E, stimulating electrode wires embedded in silk ligatures.

was attached to a glass rod connected to a capacitance-type force transducer. The muscle was mounted over the centre of the thermopile (A). A tether (D) was used to make sure that the muscle was in close contact with the thermopile. Platinum wires embedded in the ligatures delivered the stimulus to the muscle. The apparatus of Fig. 2 was enclosed in a muscle chamber which was submerged in a 70 litre water bath at 21 °C. After a 2 hr equilibration period (stimulus frequency 0.2 Hz, stimulus strength 10% > threshold), the muscle was stretched to L_o for maximum twitch tension. Heat measurements were made after draining the chamber.

Pressure overload hypertrophy was induced by placing a spiral Monel metal clip around the pulmonary artery [9].

Thyrotoxic hypertrophy was induced by daily intramuscular injections of 0.2 mg/kg of L-thyroxine [3].

Thermal measurements

The muscle temperature was measured with antimony-bismuth thermopiles [20] which consisted of junctions between the two metals sandwiched between two thin mica sheets. Reference junctions were in thermal contact laterally with the brass jaws clamping the mica. The measuring junctions were in thermal contact with the muscle in the central zone. An isometric muscle paced at 0.2 Hz exhibits a steady-state temperature oscillation. TIH data were corrected for heat loss (displayed in Figs. 3, 4 and 5) from the individual temperature data by the single time constant method described by Hill [10]. The heat loss coefficient, cool-off time constant, and thermal capacity of the muscle and adhering solutions were used to convert corrected temperature records to heat units (mcal/g) before the normalizations shown were calculated. The cool-off time constant and thermal capacity were determined by a modification [20] of the method of Kretzschmar and Wilkie [15] for the data in Fig. 3. The cool-off time constant for the drug experiments (Figs. 4 and 5) was determined by following the cool-off after Joule heating using 1 MHz sinewave current delivered through the stimulating electrodes [8].

We partitioned initial heat into the TDH and TIH components by eliminating tension and TDH with either a strongly hyperosmotic ($2.5 \times N$) Krebs solution (Fig. 3) or a moderately hyperosmotic Krebs solution containing 2,3-butanedione monoxime (BDM) (Figs. 4 and 5) [details in manuscript submitted to the Journal of Physiology (London)].

Solutions and chemicals

Krebs-Ringer solution contained in mM: Na, 152; K, 3.6; Cl, 135; HCO_3^-, 25; Mg^{2+}, 0.6; $H_2PO_4^-$, 1.3; SO_4^-, 0.6; Ca^{2+}, 2.5; glucose, 5.6. Solutions were continuously bubbled with 95% O_2—5% CO_2 during dissection and equilibration. During heat measurements, the same moisturized and temperature equilibrated gas was bubbled through a 2 mm layer of Krebs-Ringer solution at the bottom of the thermopile chamber when drained. All salts were Mallincrodt analytical grade. 2,3-butanedione monoxime was obtained from Sigma and dextrose was prescription-grade Mallincrodt. $UDCG_{115}$ was obtained from Holubarsch (University of Freiburg, West Germany) in a soluble carrier for addition to the Krebs solution. Isoproterenol hydrochloride and lidocaine hydrochloride were obtained from Abbott & Astra. Lidocaine (1×10^{-4} M) was used in the isoproterenol experiment to prevent aftercontractions. Lidocaine was also present in the control solutions for the isoproterenol experiments.

Results

Hypertrophy models

Isometric twitch force for pressure overload (PO) muscles was slightly less (13%) than normal, although the difference is not significant. Thyrotoxicosis caused a significant reduction in peak twitch force (28%). Our working hypothesis predicts that the same amount of Ca^{2+} is cycled in the PO muscle while significantly less Ca^{2+} is cycled in the thyrotoxic muscle. However, representative TIH records corrected for heat loss (Fig. 3 B) from normal (con), pressure overload (PO), and thyrotoxic (thyro) rabbits show that the predicted result did not occur.

Differences in TIH within the duration of the respective twitches (indicated by arrows) are evident when the three TIH records are plotted relative to the control record (Fig. 3 B). Total TIH of PO muscles was only 50% of control values suggesting that 50% less Ca^{2+} was cycled during the twitch. Although total TIH of thyrotoxic muscles on average was less than control, the decrease in TIH (20%) was not significant, suggesting that similar amounts of Ca^{2+} were cycled in these two preparations. In all three records TIH evolution is largely finished within the duration of the twitch. The TIH records in Fig. 3 B suggest that the transport systems removing Ca^{2+} from the myofilament space are fastest in the case of thyrotoxicosis, since the time required for TIH to approach a plateau is shortest for the thyrotoxic muscle.

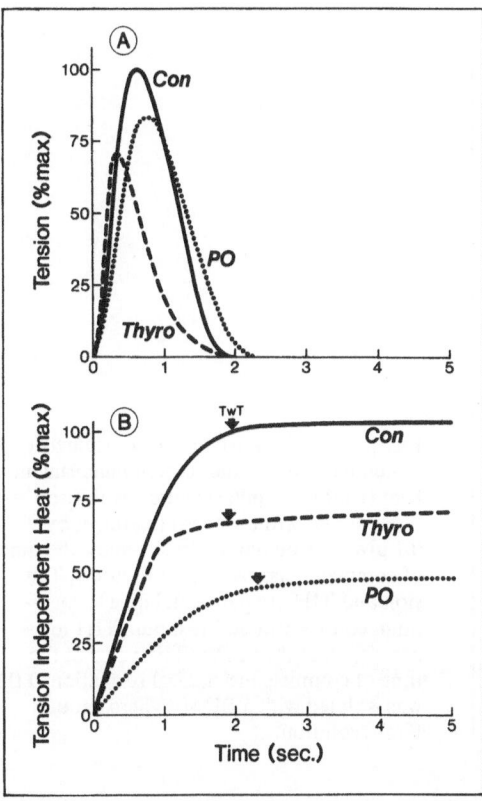

Fig. 3. Representative records of isometric twitch tension (A) and tension independent heat (B) for papillary muscles from control (con), pressure overload (PO), and thyrotoxic (thyro) rabbits. The arrows on the TIH graph labelled TWT indicate the time at which tension has relaxed completely to baseline levels. PO and thyro records are plotted relative to the control record. TIH was measured after incubation of the muscles in $2.5 \times N$ hyperosmotic Krebs solution.

Positive inotropic drugs

The results of the experiments with isoproterenol and UDCG$_{115}$ also did not support the working hypothesis that the amount of TIH parallels changes in peak twitch tension induced by the drugs. Isoproterenol (0.1 µM) potentiated peak active tension by 24% and shortened twitch duration by 37% (Fig. 4A). The predicted increase in TIH by the time of complete mechanical relaxation did not occur. On average, TIH production was not different from control at any time in the interval between stimuli (5 s). It was also surprising that no increase in TIH rate was apparent with isoproterenol present. β-adrenergic stimulation is generally thought to cause an increase in the intrinsic pumping rate of the sarcoplasmic reticulum by cAMP-dependent phosphorylation of phospholamban [25].

UDCG$_{115}$ (200 µM) also potentiated peak twitch tension (33%) but did not cause an increase in TIH evolved at any time in the interval between stimuli. TIH was actually significantly less than control at the time of complete mechanical relaxation. The time taken to complete mechanical relaxation was not affected by UDCG$_{115}$ (Fig. 5).

In summary, the results suggest that an alternative working hypothesis is necessary to explain the relation between peak twitch tension and the total amount of Ca^{2+} cycled within the duration of the isometric twitch.

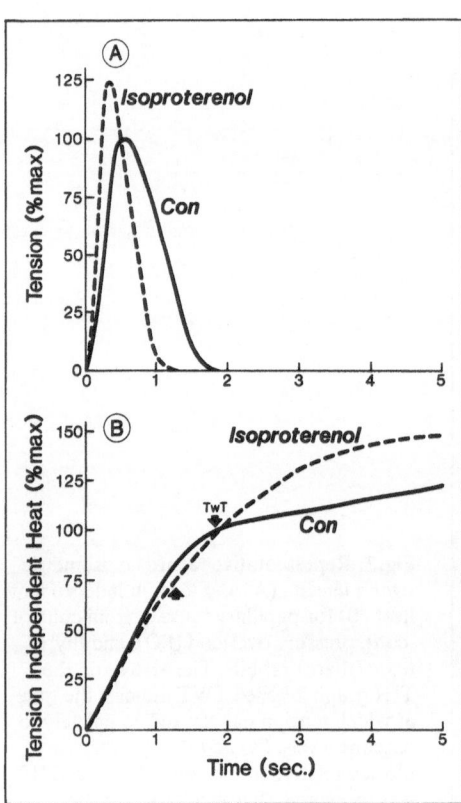

Fig. 4. Representative records of isometric twitch tension (A) and tension independent heat (B) for a papillary muscle with (isoproterenol) and without (con) treatment by 0.1 µM isoproterenol. Arrows mark the time of complete relaxation as in Figure 3. Tension and TIH are plotted relative to maximum control values. Maximum TIH for control was taken as the TIH value at the time of complete mechanical relaxation. TIH was isolated with a BDM-hyperosmotic Krebs solution.

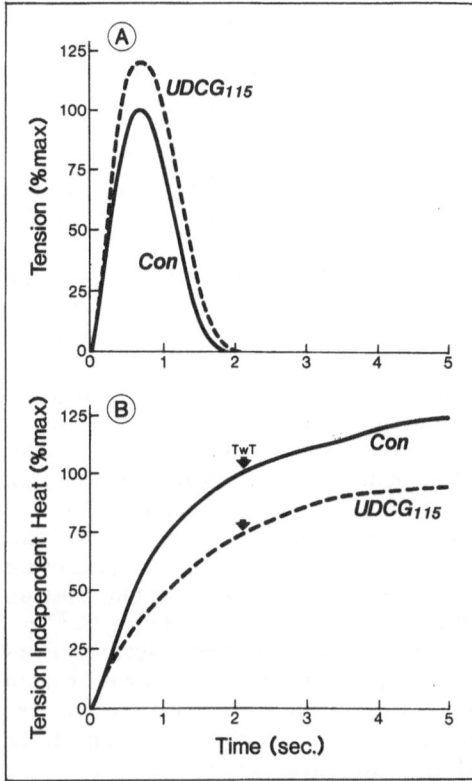

Fig. 5. Representative records of isometric twitch tension (A) and tension independent heat (B) for a papillary muscle with (UDCG$_{115}$) and without (con) treatment by 200 μM UDCG$_{115}$. The normalizations, arrows, and TIH isolation are as described for Figure 4.

Discussion

This paper shows how the analysis of tension independent heat data from papillary muscles soaked in a solution to eliminate crossbridge cycling and tension dependent heat reveals important information about the effects of changing haemodynamic conditions and inotropic compounds on Ca^{2+} cycling in cardiac muscle. However, the relation between peak twitch tension and the amount of Ca^{2+} cycled appears to be complex. The TIH data for pressure overload and UDCG$_{115}$ treatments clearly demonstrate differences in the amount of Ca^{2+} cycled compared to controls without parallel changes in peak twitch tension. On the other hand, the amount of Ca^{2+} cycled was not different with thyrotoxicosis and isoproterenol treatments even though twitch tension was decreased and increased, respectively. One way in which the degree of contractile activation can be altered without a parallel change in the amount of Ca^{2+} cycled, occurs when the intervention causes a change in the Ca^{2+} sensitivity of troponin C. Such a mechanism is unlikely, though, to account for the effects of thyrotoxicosis, pressure overload, and isoproterenol on peak twitch tension since Ca^{2+} sensitivity is unaffected by the first two conditions [17, 18] and is reduced by β-adrenergic antagonists [11]. A substantial increase in the Ca^{2+} sensitivity of the contractile apparatus may explain the twitch potentiation caused by UDCG$_{115}$, even though the amount of Ca^{2+} cycled was lower than control since the parent compound ARL 115 BS has been shown to act by this mechanism [24].

In the absence of changes in the affinity of troponin C for Ca^{2+}, peak twitch tension could be altered without parallel changes in the total amount of Ca^{2+} cycled if the time that released Ca^{2+} is available for activation is affected by the intervention. For example, although much less Ca^{2+} is cycled in the PO muscle, twitch tension may be near normal if the duration of Ca^{2+} activation is prolonged by slow sequestration of Ca^{2+} by storage sites. Similarly, rapid sequestration of released Ca^{2+} in the thyrotoxic muscle could attenuate the active state and lead to lower peak twitch tension. The TIH records support this idea since the time for TIH to reach the final value appears longer than control for the PO muscle and shorter than control for the thyrotoxic muscle.

The effect of isoproterenol on the relation between twitch tension and TIH is difficult to explain since a decrease in Ca^{2+} sensitivity and the duration of Ca^{2+} activation are expected, due to covalent phosphorylation of troponin I [23] and the sarcoplasmic reticulum [14], respectively. These two consequences of phosphorylation reactions initiated by β-adrenergic stimulation would be expected to decrease peak twitch tension unless a higher than normal free Ca^{2+} transient also occurred during the contractile phase of the twitch. The results of experiments with the photoprotein aequorin injected into papillary muscles confirms that β-adrenergic antagonists potentiate the free calcium transient [19]. Thus, a dissociation between changes in the free Ca^{2+} transient and total Ca^{2+} cycled is suggested by the isoproterenol data in this study. Such a dissociation between the two measures of Ca^{2+} flux might occur in conditions where the kinetics of Ca^{2+} release are altered. For instance, a normal amount of Ca^{2+} released very rapidly could cause an elevated free Ca^{2+} transient and higher peak twitch tension if the circulating Ca^{2+} arrived at aequorin and troponin C sites before the sequestering systems effectively competed with these binding sites for the Ca^{2+}. The kinetics of the aequorin light signal in the presence of isoproterenol support the idea that the rate of Ca^{2+} release is enhanced by the drug [4]. An alternative explanation for the lack of effect of isoproterenol on the extent and kinetics of TIH is that the BDM-mannitol Krebs solution (used to isolate TIH) interfered with the phosphorylation reactions expected in the presence of isoproterenol. We are currently testing this last hypothesis.

In conclusion, the myothermal method has great potential for providing information about Ca^{2+} cycling in cardiac muscle. We emphasize however, that the information available from TIH analysis is different from the information available from methods such as aequorin bioluminescence which monitor the free Ca^{2+} concentration early in the twitch. Myothermal records of TIH only monitor total Ca^{2+} as it is removed from the myoplasm and thus provides little information about the nature of the Ca^{2+} release process. Monitoring free Ca^{2+} concentration does have the advantage of following the kinetics of Ca^{2+} release but has the disadvantage of not being able to distinguish whether or not a change in the free Ca^{2+} transient is caused by a change in the total amount of Ca^{2+} cycled, or by other mechanisms such as alteration in the rate of Ca^{2+} release. Simultaneous measurements of free Ca^{2+} concentration and TIH are clearly needed for an integrated view of the fate of Ca^{2+} during cardiac excitation-contraction coupling.

References

1. Allen DG, Kentish JC (1985) The cellular basis of the length-tension relation in cardiac muscle. J Mol Cell Cardiol 17:821–840
2. Alpert NR, Mulieri LA (1982) Increased myothermal economy of isometric force generation in compensated cardiac hypertrophy induced by pulmonary artery constriction in the rabbit. Circ Res 50:491

3. Banarjee SK, Flink IL, Morkin E (1976) Enzymatic properties of natural and N-ethylmaleimide modified cardiac myosin from normal and thyrotoxic rabbits. Circ Res 34:319–326
4. Fabiato A (1981) Effects of cyclic AMP and phosphodiesterase inhibitors on the contractile activations and the Ca^{2+} transient detected with aequorin in skinned cardiac cells from rat and rabbit ventricles. J Gen Physiol 78:15a–16a
5. Fabiato A (1983) Calcium-induced release of calcium from the cardiac sarcoplasmic reticulum. Am J Physiol 245:C1–C14
6. Gibbs CL (1967) Changes in cardiac heat production with agents that alter contractility. Aust J Exp Biol Med Sci 45:379–392
7. Gibbs CL (1982) Modification of the physiological determinants of cardiac energy expenditure by pharmacological agents. Pharmacol Ther 18:133–157
8. Gibbs CL, Mommaerts WFHM, Ricchiuti NV (1967) Energetics of cardiac contractions. J Physiol 191:25–46
9. Hamrell BB, Alpert NR (1979) The mechanical characteristics of hypertrophied rabbit cardiac muscle in the absence of congestive heart failure. Circ Res 40:20–25
10. Hill AV (1939) Recovery heat in muscle. Proc R Soc Lond (Biol) 127:297–307
11. Holroyde MJ, Potter JD, Solaro RJ (1979) Modification of calcium requirements for activation of cardiac myofibrillar ATPase by cyclic AMP dependent phosphorylation. Biochim Biophys Acta 586:63–69
12. Homsher E, Mommaerts WFHM, Ricchiuti NV, Wallner A (1972) Activation heat, activation metabolism and tension-related heat in frog semitendinosus muscles. J Physiol 220:601–625
13. Kentish JC (1984) The inhibitory effects of monovalent ions on force development in detergent-skinned ventricular muscle from guinea pig. J Physiol 352:353–374
14. Kranias EG, Garvey JL, Srivastava RD, Solaro RJ (1985) Phosphorylation and functional modifications of sarcoplasmic reticulum and myofibrils in isolated rabbit hearts stimulated with isoprenaline. Biochem J 226:113–121
15. Kretzschmar KM, Wilkie DR (1972) A new method for absolute heat measurements utilizing the Peltier effect. J Physiol 224:18P–20P
16. Li T, Sperelakis N, Teneick RE, Solaro RJ (1985) Effects of diacetyl monoxime on cardiac excitation-coupling. J Pharmacol Exp Ther 232:688–695
17. Litten RZ, Martin BJ, Howe ER, Alpert NR, Solaro RJ (1981) Phosphorylation and adenosine triphosphatase activity of myofibrils from thyrotoxic rabbit hearts. Circ Res 48:498–501
18. Maughan D, Low E, Litten R, Brayden J, Alpert N (1979) Calcium-activated muscle from hypertrophied rabbit hearts. Circ Res 44:279–287
19. Morgan JP, Blinks JR (1982) Intracellular Ca^{2+} transients in the cat papillary muscle. Can J Physiol Pharmacol 60:524–528
20. Mulieri LA, Luhr G, Trefry J, Alpert NR (1977) Metal-film thermopiles for use with rabbit right ventricular papillary muscles. Am J Physiol 233:C146–C156
21. Mulieri LA, Alpert NR (1982) Activation heat and latency relaxation in relation to calcium movement in skeletal and cardiac muscle. Can J Physiol Pharmacol 60:529–541
22. Smith ICH (1972) Energetics of activation in frog and toad muscle. J Physiol 220:583–599
23. Solaro RJ, Robertson SP, Johnson JD, Holroyde MJ, Potter JD (1981) Troponin-I phosphorylation: A unique regulator of the amounts of calcium required to activate cardiac myofibrils. In: Protein Phosphorylation. Cold Spring Harbor Conferences on Cell Proliferation 8:901–911
24. Solaro RJ, Ruegg JC (1982) Stimulation of calcium binding and ATPase activity of dog cardiac myofibrils by ARL 115 BS, a novel cardiotonic agent. Circ Res 51:290–294
25. Tada M, Kirchberger MA, Katz AM (1975) Phosphorylation of a 22,000-dalton component of the cardiac sarcoplasmic reticulum by adenosine 3′:5′-monophosphate-dependent protein kinase. J Biol Chem 250:2640–2647

Authors' address:

Dr. Norman R. Alpert, Dept. Physiology & Biophysics, University of Vermont, College of Medicine, Given Building, Burlington, VT 05405

Chronic cardiac reactions. I. Assessment of ventricular and myocardial work capacity in the hypertrophied and dilated ventricle *

R. Jacob, M. Vogt, and K. Noma

Physiologisches Institut II, Universität Tübingen, F.R.G.

Summary

The endsystolic and end-diastolic pressure-volume or stress-length curves define the margins of the various conceivable courses of pressure-volume or stress-length loops. Although the endsystolic pressure-volume and stress-length relations of isovolumetric and afterloaded contractions are not entirely identical, the area between isovolumetric maxima- and enddiastolic minima curves in the pressure-volume or stress-length diagram can be taken as a measure of potential ventricular and myocardial work under different yet defined mechanical conditions.

The normalized stress-length area, as derived from the left ventricular pressure-volume diagram and myocardial mass, renders a rational basis for global quantitative evaluation of myocardial work capacity. The area obtained is independent of ventricular mass and size and as such is invaluable for assessing hypertrophied and/or dilated hearts, and thus interindividual comparison of myocardial contractile capability based on physical principles. However, this measure should be supplemented by considering time dependent parameters (e.g. maximum rate of stress development as a function of end-diastolic stress).

The principle set here for evaluating ventricular and myocardial performance should always be borne in mind, especially when referring to more empirical parameters.

Introduction

Evaluation of cardiac performance in general, especially in studies dealing with energetics, requires a rational basis of assessing ventricular and myocardial function.

Under physiological, pathophysiological and pharmacological conditions, great variability of the myocardial contraction process is found with regard to amplitude and time course of the contraction cycle. Isometric tension and unloaded shortening velocity can even change in different directions, e.g. in pressure induced hypertrophy [5, 6, 7]. Therefore, an extensive description and evaluation of the contraction process is impossible solely on the basis of single parameters.

Global evaluation of ventricular and myocardial function on the basis of physical principles should be made with reference to potential pressure-volume and stress-length work as well as mechanical power under defined mechanical conditions (end-diastolic pressure, aortic pressure, end-diastolic wall stress, systolic wall stress) [4, 8].

The isovolumetric maxima curves and the end-diastolic minima curves circumscribe the extreme limits of potential pressure and volume changes, or stress and length changes,

---

* Supported by the Deutsche Forschungsgemeinschaft.

respectively [1]. Thus, the pressure-volume area between the curves is a reasonable measure of ventricular work capacity. Furthermore, the normalized stress-length area, being a measure of myocardial work capacity, can be used to assess myocardial function under considerable changes in ventricular geometry [5, 9]. Since calculation of stress-length values from pressure-volume relations implies considerable simplifications, the applicability of this concept is investigated in the present study using the example of cardiac failure in hypertensive rats.

Theoretical basis

Although the heart functions in the form of single twitches, and the isovolumic maxima curves are not entirely identical with the end-systolic pressure volume (or stress-length) relationships derived under afterloaded conditions [2, 4, 11], the area between isovolumic maxima and diastolic minima curve (related to a given end-diastolic pressure) can be taken as a quantitative measure of potential ventricular work per beat under defined mechanical conditions. Assuming constant contractility, the course of a given pressure-volume loop falls within the margins of isovolumetric maxima and end-diastolic minima curves, i.e. the area between these two curves (up to a defined end-diastolic pressure) includes all conceivable pressure-volume loops.

The same applies to the stress-length diagram. As shown in Fig. 1 C, the area within which the individual stress-length diagrams (or pressure-volume diagrams) are situated, is rather overestimated when related to isovolumetric maxima and underestimated when related to endsystolic stress-length relations (or pressure-volume relations), obtained by increasing preload under afterloaded conditions.

However, it must be emphasized that the values of work do not take into consideration the time course involved and therefore velocity of the contraction process. Thus, it would be appropriate to relate the pressure-volume or stress-length area also to duration of systole in order to obtain a measure for ventricular and myocardial power capacity. Such a procedure, however, is only possible on condition, since the mechanical parameters per se influence the duration of systole.

Methods

Animals, experimental procedures and calculations

A representative analysis if presented using a 14-month-old spontaneously hypertensive rat (SHR) with left ventricular hypertrophy and dilatation as well as pleural effusion, ascites and skin oedema. The data were compared to that of a normotensive age-matched control. Haemodynamic measurements were performed on open-chest animals. The maxima curves were obtained from isovolumetric beats performed by intermittent clamping of the aorta ascendens. The data of volume and pressure were used to calculate midwall circumference (length) and midwall circumferential stress (σ_R), assuming a spherical shell.

$$\sigma_R = \frac{P_{r_i}^3}{r_0^3 - r_i^3}\left(\frac{r_0^3}{2R^3}+1\right) = \frac{V \cdot P}{W}\cdot\left(1+\frac{4(V+W)}{(V^{1/3}+(V+W)^{1/3})^3}\right). \tag{3}$$

Pressure-volume area and length-stress area were assessed as already described. The measured part of the isovolumic pressure-volume curve and stress-length curve were extrapolated as shown in Fig. 1. The pressure-volume area can be related to ventricular weight in analogy to methods used in technological applications (e.g. power to weight ratio).

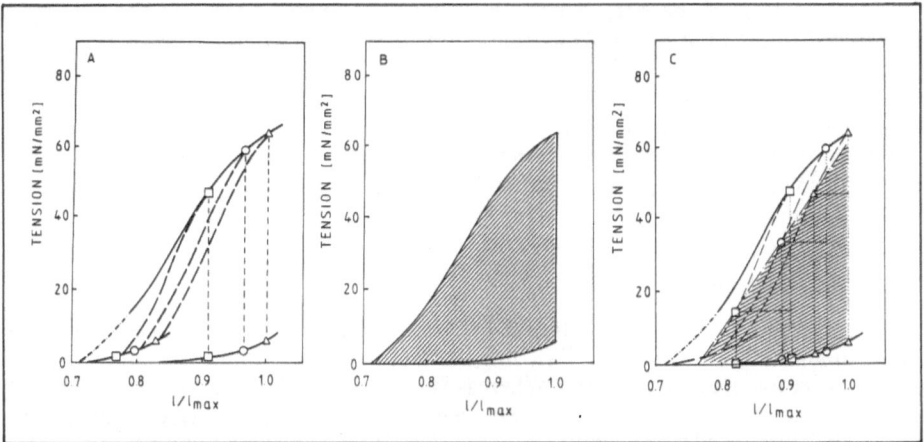

Fig. 1. Length-tension area as a measure of myocardial work capacity based on isometric and after-loaded contractions: length-tension diagrams of the isolated rat ventricular muscle (according to measurements of Gülch [2]). A) Curve of isometric maxima (total tension), resting tension curve and 3 maximum curves of afterloaded contractions; B) Area between the isometric maximum curve and resting tension curve (to the left of l_{max}); C) Area between resting tension curve and "endsystolic length-tension curve" (———).

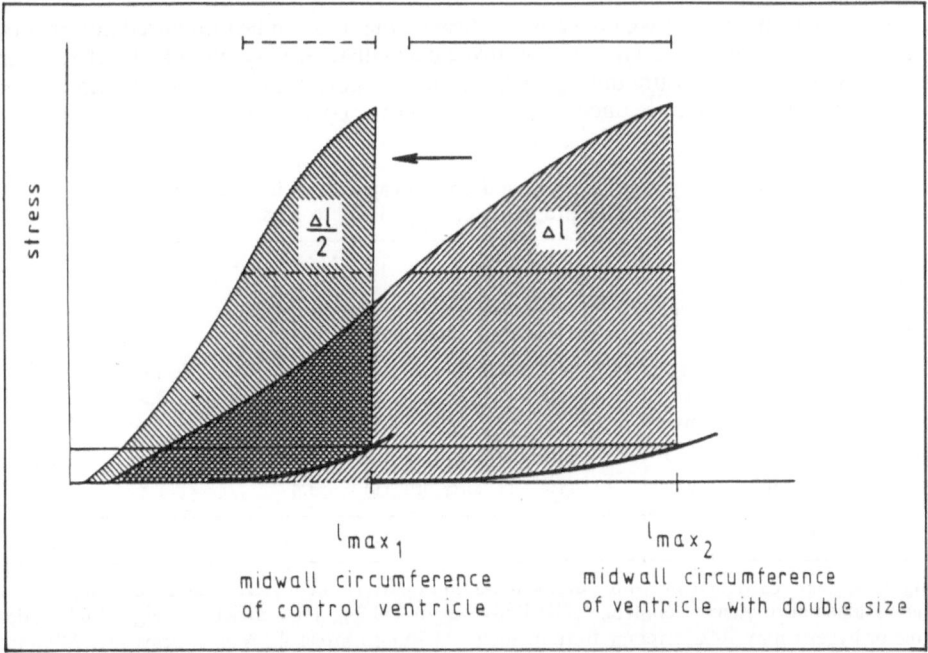

Fig. 2. Normalization of the stress-length area by relating the "degree of shortening" (more precisely the horizontal dimensions of area) to end-diastolic midwall circumference of control ventricle.

Normalization of the stress-length area was performed by relating the "degree of shortening" (more precisely the horizontal dimensions of area) to initial length (end-diastolic midwall ventricular circumference) of normotensive controls, as illustrated in Fig. 2. An absolut measure independent of heart size and muscle mass can be obtained by relating area to unit midwall circumference.

For the correlation between ejection fraction (or circumferential shortening velocity) and systolic wall stress (Figs. 5A, 5B) systolic ventricular pressure was varied by partial aortic clamping.

Results and Discussion

Figures 3A and 4A show representative left ventricular pressure-volume and stress-circumferential relationships of an SHR in a state of congestive failure as compared to a normotensive age-matched control. Whereas ventricular work capacity (Fig. 3B) is only moderately reduced, there is a 55% decrease in myocardial work capacity, evaluated on the basis of the normalized stress-length area (Fig. 4B). By relating area to duration of systole (the limitations of this procedure as discussed under Methods) a measure of myocardial power capacity is obtained which, in the present example, is reduced by 60%.

The results based on this concept are compared in Table 1 with other parameters (Figs. 5, 6) which are indicative of myocardial work capacity or power capacity, respectively. As would be expected, the difference (expressed in %) between insufficient hearts and control hearts is generally greater for time-dependent parameters than for stress and work parameters. In the given context, it should be borne in mind that the relative differences between the two curves in Fig. 6 depends on the preload or diastolic stress taken as the basis.

Of course, all data derived from whole heart dynamics can be influenced not only by changes in the spread of excitation and myocardial distensibility but also by changes in coronary perfusion pressure during the procedure of aortic clamping. Furthermore, the left ventricular performance under experimental aortic coarctation is not completely iso-

Table 1. Comparison of myocardial work (and power) capacity with other parameters which are indicative for myocardial contractile properties corresponding to Figs. 4–6

Basis of evaluation	Dimension	Normotensive control	SHR cong. failure	$\Delta\%$
Myocardial working capacity	$(10^{-5}\,\mathrm{Nm/mm^2 \cdot cm})$	4.13	1.84	−55
Isovolumetric systolic wall stress σ	$(10^{-4}\,\mathrm{N/mm^2})$	381.0	254.0	−33
Max. rate of stress rise $d\sigma/dt$ (at $\sigma_{diast}=10 \cdot 10^{-4}\,\mathrm{N/mm^2}$)	$(10^{-1}\,\mathrm{N/mm^2 \cdot s})$	15.3	7.2	−53
Stress-velocity area	$(10^{-2}\,\mathrm{N \cdot circ/mm^2 \cdot s})$	7.6	3.9	−49
Maximal momentary power	$(10^{-2}\,\mathrm{N \cdot circ/mm^2 \cdot s})$	3.4	1.8	−47

Fig. 3. A) Left ventricular pressure volume diagram of a SHR in congestive failure as compared to age-matched normotensive control. SHR: Body weight 450.0 g; left ventricular weight 1.625 g; degree of hypertrophy 70%; average fibre diameter 23.7 μm; fibrosis 4.1 Vol%; heart rate 220/min. Control rat: Body weight 410.0 g; left ventricular weight 0.96 g; heart rate 250/min. B) Left ventricular pressure-volume area (between isovolumetric maximum and end-diastolic pressure-volume relations (to the left of the end-diastolic pressure of $2 \cdot 10^3\mathrm{N/m^2}$).

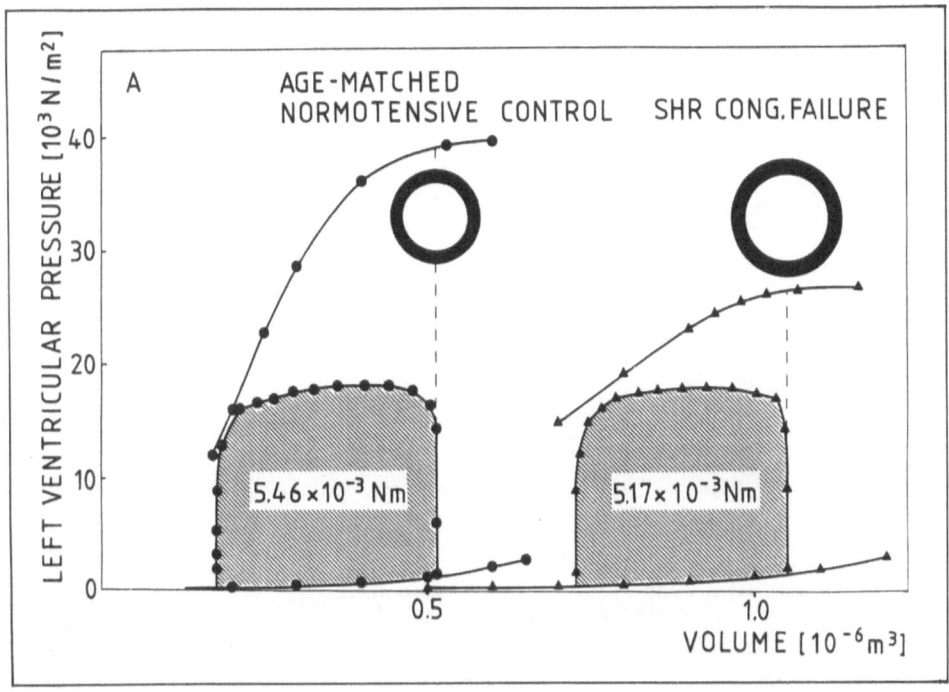

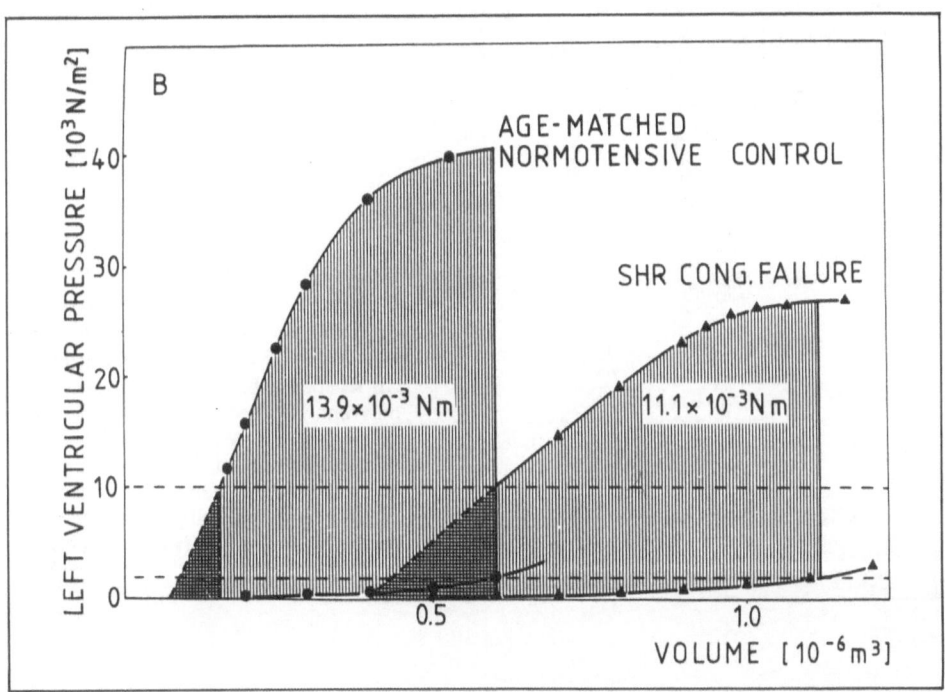

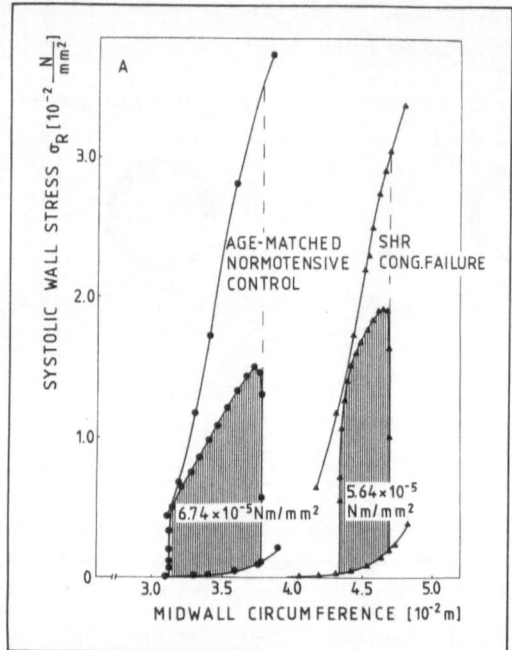

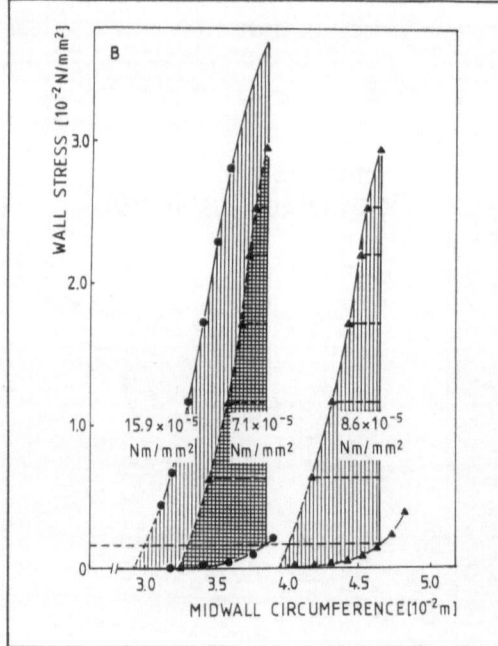

Fig. 4. A) Midwall stress-length (stress circumference) diagrams, calculated from Fig. 3 A. B) Evaluation of myocardial work capacity based on the stress-length diagram. Hatched area: The stress-length area of the failing heart (right) is related to circumference of the control ventricle. Normalization of both areas by relating areas to 1 cm midwall circumference results in $4.13 \cdot 10^{-5}$ (Nm/mm² · cm) (control); $1.84 \cdot 10^{-5}$ (Nm/mm² · cm) (SHR, congestive failure).

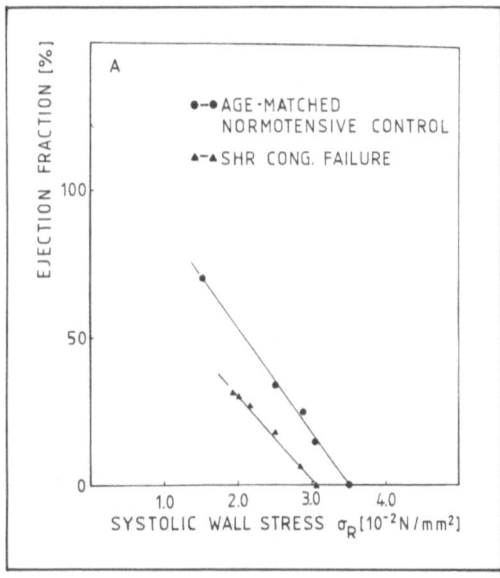

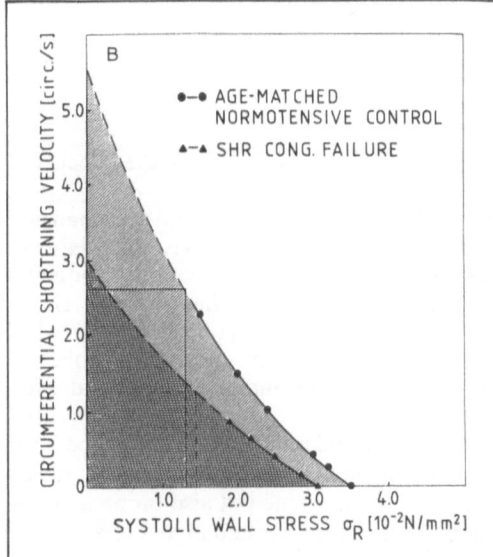

Fig. 5. A) Ejection fraction as a function of systolic (peak) midwall stress (the same specimen as in Fig. 3 and Fig. 4). B) Calculated stress-shortening velocity relation as derived from A). The area under the latter curve is considered representative of myocardial power capacity. The data were obtained under the conditions of high end-diastolic pressure (ca. 15 mm Hg) in order to guarantee similar end-diastolic sarcomere length in all measurements. Values in $N \cdot circ/mm^2 \cdot s$: Stress-velocity area: $7.62 \cdot 10^{-2}$ (control); $3.94 \cdot 10^{-2}$ (SHR, cardiac failure). Maximal momentary power: $3.41 \cdot 10^{-2}$ (control); $1.84 \cdot 10^{-2}$ (SHR, cardiac failure).

volumetric. Nevertheless, this approach facilitates quantitative evaluation of myocardial performance independent of ventricular configuration and mechanical conditions and thus allows for assessment of myocardial function and individual comparisons based on physical principles. This study demonstrates that despite the use of simple models, valuable results can be obtained. At any rate, the present concept of evaluation is preferable to those based solely on the slope of the maxima curves of afterloaded contractions (or the end-systolic pressure volume relationships [10, 12, 13] and the interception of these curves

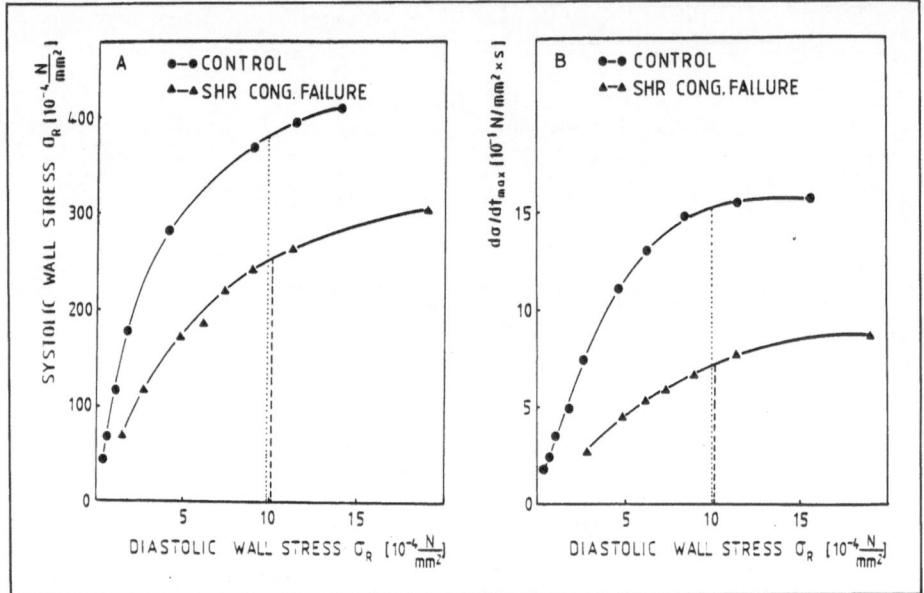

with the volume abscissa [12], respectively, since the values obtained have the dimensions of work.

Potential myocardial work is described for all the conceivable pre- and afterload conditions at a given state of "contractility", independent of ventricular mass and diameter. In contrast to using various "indices of contractility", a global evaluation of potential ventricular and myocardial work is made possible, which, needless to say, should be supplemented by velocity parameters. A "power index" [14], however, which is determined from the quotient of stroke power per wall volume and end-diastolic wall stress, is apparently a less accurate method with regard to normalization (preload, afterload).

Our principle of evaluating ventricular and myocardial performance should be borne in mind also when referring to more empirical parameters, e.g. ejection fraction, the absolute values of which not only depend on the contractile state but also on preload and afterload, not to mention geometric conditions (ventricular size and myocardial mass).

References

1. Frank O (1898) Die Wirkung von Digitalis (Helloborein) auf das Herz. Sitz ber Ges f Morph u Physiol 14:14
2. Gülch RW (1985) The "end-systolic" length-tension relation in mammalian myocardium. Basic Res Cardiol 80:636–641
3. Hepp A, Hansis M, Gülch R, Jacob R (1974) Left ventricular isovolumetric pressure-volume relations, "diastolic tone", and contractility in the rat after physical training. Basic Res Cardiol 69:516–532

4. Jacob R (1966) Die Dynamik des linken Ventrikels im natürlichen Kreislauf und ihre muskel-physiologischen Grundlagen. Habilitationsschrift (Würzburg)
5. Jacob R (1986) Cardiac responses to experimental chronic pressure overload. In: Zanchetti A, Tarazi RC (eds) Handbook of hypertension, vol 7: Pathophysiology of hypertension – cardiovascular aspects. Elsevier Science Publishers BV, pp 59–83
6. Jacob R, Kissling G, Ebrecht G, Holubarsch C, Medugorac I, Rupp H (1983) Adaptive and pathological alterations in experimental cardiac hypertrophy. In: Chazov E, Saks V, Rona G (eds) Advanc Myocardiol 4. Plenum Publishing Corporation, New York, pp 55–77
7. Jacob R, Ebrecht G, Kissling G, Rupp H, Takeda N (1986) Functional consequences of cardiac myosin isoenzyme redistribution. In: Rupp H (ed) The regulation of heart function. Thieme, Stuttgart New York, pp 305–326
8. Jacob R, Gülch RW, Holubarsch C, Kissling G (1975) Die Beurteilung der myokardialen Leistungsfähigkeit. Med Klinik 70:1347–1365
9. Jacob R, Vogt M, Rupp H (1986) Physiological and pathological hypertrophy. In: Dhalla NS, Singal PK, Beamish RE (eds) Pathophysiology of heart disease. Martinus Nijhoff Publishing, Boston pp 39–56
10. Jacob R, Weigand KH (1966) Die endsystolischen Druck-Volumenbeziehungen als Grundlage einer Beurteilung der Kontraktilität des linken Ventrikels in situ. Pflügers Arch 289:37–49
11. Kissling G, Takeda N, Vogt M (1985) Left ventricular end-systolic pressure-volume relationships as a measure of ventricular performance. Basic Res Cardiol 80:594–607
12. Sagawa K (1981) The endsystolic pressure-volume relation of ventricular definition modifications and clinical use. Circulation 63:1117–1223
13. Suga H, Igarashi Y, Yamada O, Goto Y (1986) Cardiac oxygen consumption and systolic pressure volume area. Basic Res Cardiol [Suppl 1] 81:39–50
14. Unterberg RH, Körfer R, Pölitz B, Schmiel FK, Spiller P (1984) Assessment of left ventricular function by a power index: an intraoperative study. Basic Res Cardiol 79:423–431

Authors' address:

Prof. Dr. R. Jacob, Physiologisches Institut II, Universität Tübingen, Gmelinstraße 5, D-7400 Tübingen

Chronic cardiac reactions. II. Mechanical and energetic consequences of myocardial transformation versus ventricular dilatation in the chronically pressure-loaded heart

M. Vogt, R. Jacob, G. Kissling, H. Rupp

Physiologisches Institut II, Universität Tübingen, F.R.G.

Summary

Mechanical and energetic consequences of myocardial transformation and of ventricular configuration on the other were separately analysed. The considerations were realized on representative samples of normotensive rats and spontaneously hypertensive rats (SHR) in compensated stages, as well as in SHR in a state of congestive cardiac failure. Cardiac dynamic measurements were performed under Urethane anaesthesia and open chest conditions. Myosin isoenzyme pattern was determined by pyrophosphate gel electrophoresis. Energetic calculations were based on oxygen consumption data, measured in a specified heart-lung model [19].

In the compensated stage of SHR the concentric type of left ventricular hypertrophy with renormalized systolic auxotonic wall stress predominated. The process of cardiac hypertrophy was associated with a shift in the myosin isoenzyme pattern towards the "slow" VM-3. Myocardial transformation did not significantly reduce myocardial performance and pumping ability, but caused a decrease in oxygen consumption as related to developed stress and LV weight. Thus, the efficiency of the hypertrophied ventricle of SHR was improved. However, due to the moderate effect of isoenzyme pattern redistribution for total energy turnover and the limited adaptive reserve of normotensive controls, the extent of improvement was small.

In SHR with congestive heart failure, myocardial contractility was severely impaired, when structural dilatation of the left ventricle had set in. Reduced myocardial contractility could not be explained solely on the basis of a shift in the myosin isoenzyme pattern. Both impaired myocardial contractility and structural dilatation contributed to reduced ventricular performance. Myocardial transformation, along with its energy economizing effect, failed to compensate for unfavourable energetic consequences of structural dilatation and therefore the reduced ventricular efficiency is assumed to be another deleterious factor in the dilated failing heart.

Introduction

A multitude of morphological and functional alterations with diverse manifestations at various levels of the organ, occur during cardiac adaptation to chronic haemodynamic overload and later on in the development of cardiac insufficiency.

In our present paper, special reference should be given to myocardial transformation and to ventricular hypertrophy and configuration. Changes in biochemical structures at the myofibrillar level are well investigated in the rat, revealing three different types of myosin isoenzymes [10, 28]. Pressure-induced ventricular hypertrophy leads to redistribution towards the "slow" isoenzyme VM-3 [22, 28], which is associated with reduced myofibrillar ATPase activity [22, 28] and a decrease in normalized oxygen consumption [19].

In the compensated stage of chronic pressure overload the concentric type of left ventricular hypertrophy predominates, whereas congestive cardiac failure is usually accompanied by structural dilatation. Due to geometric conditions, resulting from dilatation, the systolic wall stress increases and thus the degrees of myocardial shortening and mechanical efficiency are reduced [21].

The purpose of the present study was to investigate, in the model of SHR, the effects of changes in myosin isoenzyme pattern, as well as in ventricular configuration, on cardiac mechanics and energetics. In particular, the ability of myocardial transformation towards a "slower" muscle to compensate for unfavourable consequences of structural dilatation on cardiac energetics should be evaluated.

Methods

Experimental protocol

The investigations were carried out on male spontaneously hypertensive rats (SHR, Aoki-Okamoto strain) in the compensated stage, as well as in SHR in congestive cardiac failure. Normotensive age-matched Wistar rats served as a control.

Cardiac dynamic measurements were performed under Urethane anaesthesia (1.2 g/kg b.w. intra-abdominally) in open chest conditions. The left ventricle was pierced at the apex with a steel cannula No. 1 connected to a fluid-filled pressure transducer. An electromagnetic flow probe was placed around the ascending aorta. Left ventricular pressure (LVP) amplitude, LV end-diastolic pressure (LVEDP; high amplification), rate of pressure rise (dP/dt) and aortic flow (dV/dt) were recorded simultaneously. Variation of afterload and isovolumetric beats were performed by gradually clamping the ascending aorta.

The end-diastolic volume was derived from end-diastolic pressure values and the end-diastolic pressure: volume relationships registered at the end of each experiment (modified method according to Ullrich et al. [35]).

Circumferential wall stress was calculated for the midwall region assuming a thick-walled spherical shell, according to the formula:

$$\sigma_R = \frac{P \cdot r_i^3}{r_0^3 - r_i^3} \cdot \frac{r_0^3}{2R^3} = \frac{P \cdot V}{W} \cdot \left(1 + \frac{4(V+W)}{(V^{1/3} + (V+W)^{1/3})^3}\right). \qquad (9, 24)$$

(V = inner volume; W = left ventricular wall volume)
Non-dissociating gel electrophoresis in the presence of pyrophosphate was carried out to determine the myosin isoenzyme pattern according to [10, 28].

Evaluation of efficiency

The evaluation of efficiency was based on the assumption that peak tension, or peak systolic wall stress yields a reliable measure of muscle energy expenditure. Data of oxygen consumption, as related to peak wall stress and to myosin isoenzyme pattern, were based on the measurements of Kissling [19] in a specified heart-lung model.

Since the heart beats nearly isovolumetrically in the above mentioned model and since more energy is liberated in afterloaded contractions than in "equivalent" isometric contractions, the amount of external work has to be considered [5]. However, an additional amount for shortening was neglected:

$$\text{efficiency} = \frac{EW}{OC_T + EW}$$

EW = external work (mJ/g); OC_T = oxygen consumption as related to peak wall stress (mJ/g · N/cm^2).

The mechanical equivalent of the consumed oxygen was calculated assuming a respiratory quotient of 0.94 and a respiratory equivalent of 4970 cal/l O_2. In addition, changes in efficiency were evaluated by taking tension-time integral as an index of muscle energy expenditure.

Results and Discussion

A. Mechanical consequences of myocardial transformation as well as of ventricular hypertrophy and configuration

Spontaneously hypertensive rats (SHR) originally developed by Aoki-Okamoto [26] represent a model for slowly proceeding left ventricular pressure overload due to severe systemic hypertension. In response to the sustained functional demands, left ventricular myocardial mass increases. During the process of hypertrophy, beside others, changes in ventricular configuration and alterations at the myofibrillar level occur. The purpose of this section is to elucidate the relations between myocardial transformation, myocardial contractile capability, geometric state and ventricular performance, as well as to evaluate their relative significance for the pumping function of the left ventricle.

1. Left ventricular pressure volume relations

In the compensated stage of SHR, concentric left ventricular hypertrophy occurred. As could be expected from the increase in muscle mass, the hypertrophied left ventricle of

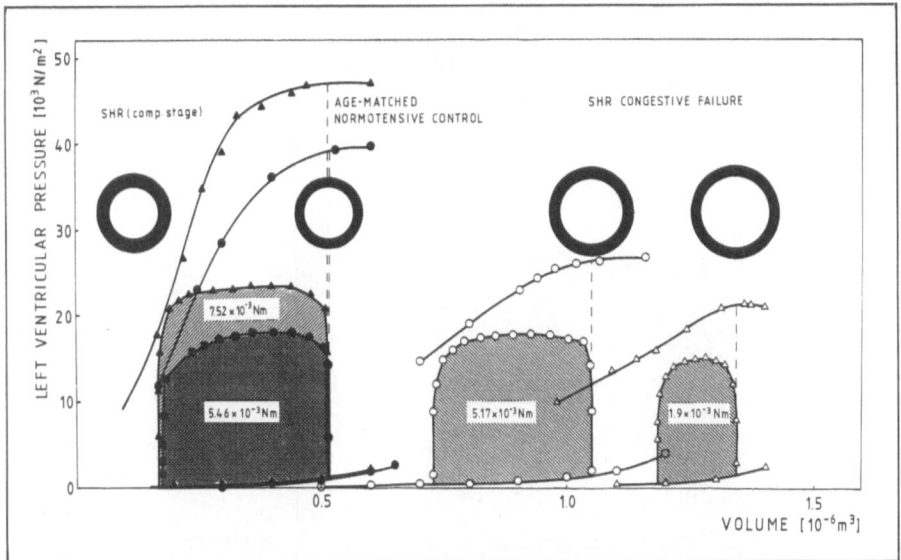

Fig. 1. Left ventricular isovolumetric pressure volume relationships and pressure volume loops of a normotensive rat, an SHR in compensated stage (on the left) and SHR in congestive cardiac failure (two curves to the right). The stroke work (hatched area) is shown. Left ventricular weight: normotensive control, 0.96 g; SHR compensated stage, 1.329 g; SHR in congestive failure, 1.625 g and 1.68 g.

SHR developed a higher isovolumetric pressure at a given end-diastolic pressure in comparison to normotensive controls. Ventricular working capacity, indicated by the area between end-diastolic pressure volume curve and isovolumic maxima as related to a defined end-diastolic pressure [17], was markedly enhanced (Fig. 1).

In SHR in congestive cardiac failure (peripheral oedema, pleural effusion and ascites) structural dilatation has appeared with a significant rightward shift in pressure volume relations. Ventricular diastolic size was markedly increased. However, due to a severe decrease in isovolumic maxima, ventricular working capacity was severely reduced (Fig. 1).

2. Myocardial contractile capability

The area between end-diastolic and systolic isovolumetric length stress curve – related to a given end-diastolic stress and to control ventricular circumference – yields an appropri-

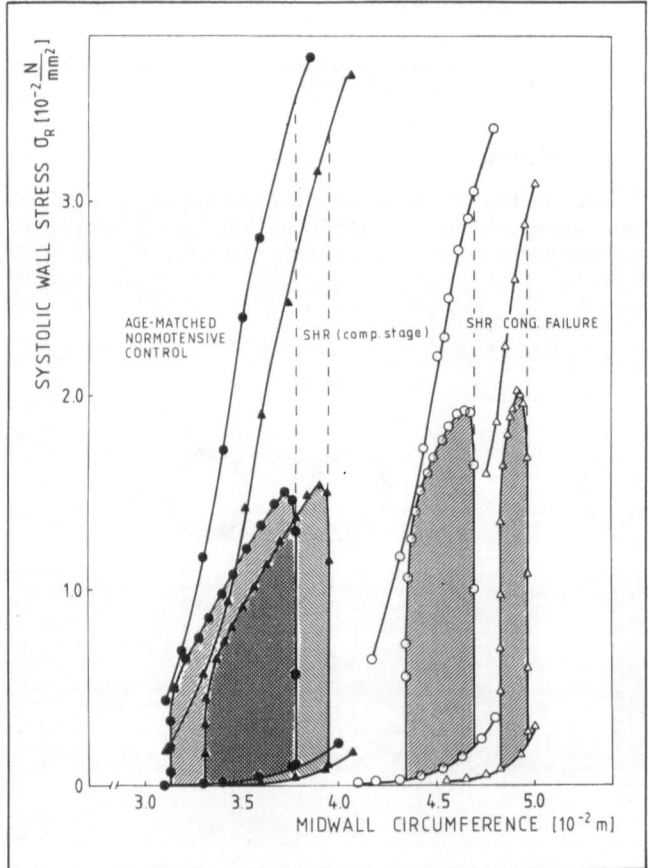

Fig. 2. Left ventricular length-stress diagrams, calculated from pressure volume relations of Fig. 1. Circumferential wall stress was calculated for the midwall region. Note the increase in peak auxotonic systolic wall stress in the structurally dilated ventricles of SHR in congestive failure.

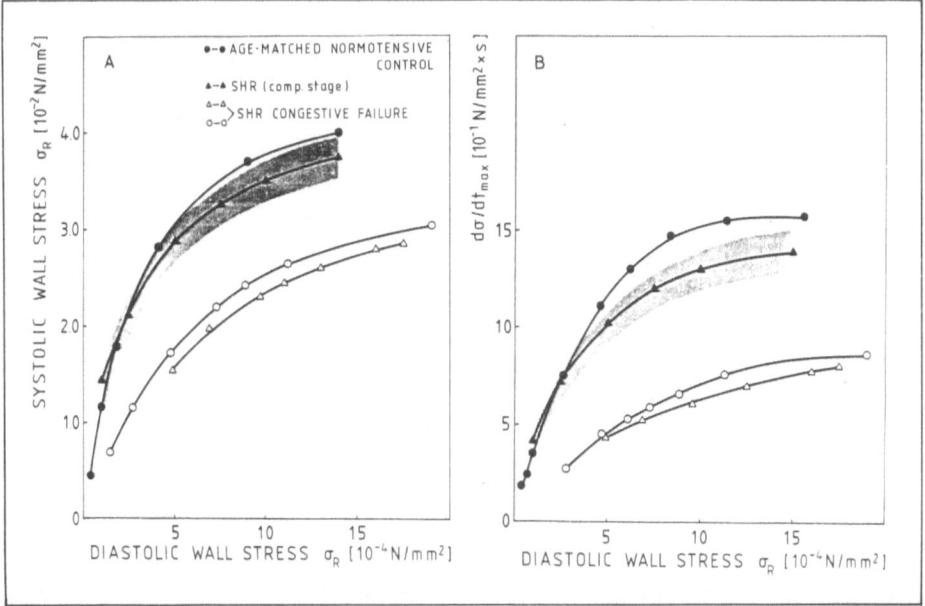

Fig. 3. Myocardial function in SHR in the compensated stage and in the stage of congestive cardiac failure as compared to normotensive rats. A) Isovolumetric circumferential systolic wall stress as a function of diastolic wall stress. B) Maximum rate of rise in systolic wall stress as a function of diastolic wall stress.

ate measure of myocardial working capacity [17]. For SHR in the compensated stage, myocardial working capacity was normal or even slightly increased, and therefore the enhanced ventricular working capacity was due to the increase in muscle mass. On the other hand the calculations revealed a markedly reduced myocardial working capacity of the dilated ventricles (Fig. 2). This would be even more pronounced when the length-stress area is normalized, i.e. shortening is related to control midwall ventricular circumference.

Former investigations have revealed that a shift in myosin isoenzyme pattern towards the "slow" VM-3 induced by pressure-overload, hypothyreosis or intermittent feeding did not lead to a significant reduction in developed isovolumic wall stress or isometric tension development at the myofibrillar level [4, 15, 36]. This can be explained on the basis of current concepts of cross-bridge cycling. Supported by the findings of a reduced heat release per unit developed tension in isolated myocardium with prevailing isoenzyme VM-3, it was suggested that reduced myofibrillar ATPase activity is associated with a prolongation of the force-generating state of cross-bridges under the overall reduced rate of the cross-bridge cycle [1]. In native heart muscle alterations in excitation and excitation-contraction coupling in the whole ventricle, the arrangement of fibres, spread of excitation, synchronization of fibre contractions and changes on the tissue level are additionally superimposed onto these myofibrillar effects. Thus, based on the severely reduced systolic wall stress in SHR with congestive failure (Fig. 3A), the reduction in myocardial contractile capability was not regarded as mainly due to myocardial transformation but rather to other above mentioned alterations. As would be expected from alter-

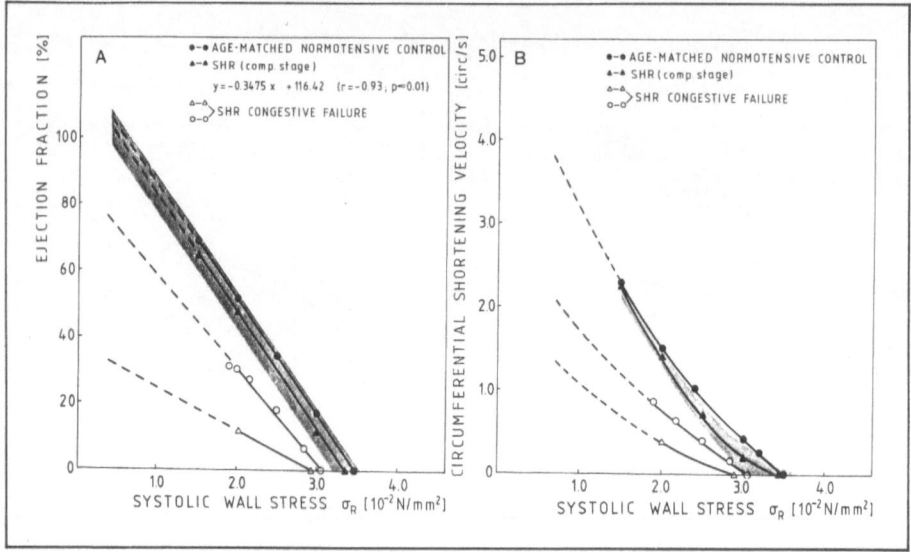

Fig. 4. Myocardial function in SHR in the compensated stage and in the stage of congestive cardiac failure as compared to normotensive rats. A) Ejection fraction-afterload relationship. Ejection fraction as a function of systolic wall stress. Afterloaded contractions were performed at high diastolic ventricular filling in order to minimize the influence of differences in sarcomere lengths. B) Normalized circumferential shortening velocity as related to systolic wall stress. Both parameters were calculated for the midwall region. Same experimental conditions as in A).

ations in cross-bridge kinetics time-dependent parameters, i.e. maximum rate of stress development (Fig. 3 B) or circumferential shortening velocity (Fig. 4 B) tended to be more reduced than stress development or relative shortening in SHR with redistribution towards VM-3 as compared to normotensive controls. Again, the reduction in the rate of wall stress rise of SHR in congestive failure exceeded the value predicted by the proportion of VM-3.

Deduced from basic principles of muscle physiology – i.e. the force shortening relationship – the myocardial contractile state can be evaluated on the basis of the ejection fraction : afterload relation. A significant inverse linear relationship between ejection fraction and peak auxotonic systolic wall stress could be obtained in the acute experiments by variation of afterload in ejecting beats. Since this relation depends on preload, the experiments were performed at high diastolic ventricular filling to minimize the influence of differences in sarcomere lengths. Although there was a shift in myosin isoenzyme pattern towards VM-3 in SHR as compared to normotensive controls no significant difference occurred in the ejection fraction – afterload relation during the compensatory stage. As can be seen in Fig. 4 A the reduction in ejection fraction of SHR in congestive failure exceeded the extent predicted by the increase in systolic wall stress. A difference in axial intercepts and slope of the ejection fraction – afterload relation could be demonstrated, indicating impaired myocardial contractile state. Thus, in our investigation, both geometric state and impaired myocardial contractility contributed to the reduction in ejection fraction which per se, however, did not give evidence for absolute values of stroke volume. Instead of the hybrid plot ejection fraction versus afterload, the evaluation of shortening as a function of wall stress should be preferable, particularly since in the case of considerably hypertrophied ventricle, ejection fraction is no longer a precise measure of circumferential shortening in the midwall region. With increasing wall thickness the latter is overestimated when evaluated on the basis of the ejection fraction.

3. Ventricular performance

As shown by the pressure volume loops (Fig. 1) the pressure loaded heart ejected a stroke volume that corresponded to control values, although the afterload on the whole ventricle was increased due to systemic hypertension. Thus, in accordance with other investigators [3, 27] we found normal pumping ability of the hypertrophied left ventricle of SHR in the compensated stage, which was also described for Goldblatt rats [18]. The stroke work was enhanced. The increase in myocardial mass and the decrease in radius-to-wall thickness ratio are advantageous in transforming developed wall stress into pressure and result in renormalized systolic auxotonic wall stress for SHR in the compensated stage (Fig. 2). Thus, on the basis of the force-shortening relationship, the degree of myocardial shortening can be expected to be normal; all the more so as a conceivable increase in frictional or deformational resistance due to increased myocardial mass could not be ascertained [13]. In addition, considering that an increase in activation duration, indicated by prolongation of action potential [8], and consequently prolonged ejection time may compensate for a reduced shortening velocity, the significance of a shift in myosin isoenzyme pattern towards the "slow" VM-3 alone for the pumping function of the ventricle seems to be limited, at least under resting conditions. However, when high heart rates are required, the reduction in ATPase activity and shortening velocity is unfavourable.
    Increase in cavitary size with inadequate increase in left ventricular mass and thus an increased radius-to-wall thickness ratio caused an increase in systolic auxotonic wall stress in SHR with congestive cardiac failure (Figs. 1 and 2). Despite enhanced end-diastolic pressure, stroke volume and stroke work were reduced. Based on the force shortening relationship a diminution in external myocardial shortening had to occur with the increased wall stress. However, deduced from theoretical considerations [16], a moderate dilatation with already declined ejection fraction should lead to an augmentation of the absolute value of stroke volume, whereas only extreme dilatation would, in principle, cause cardiac pumping failure in the absence of any impairment of myocardial "contractility". In our experiments, however, reduced myocardial shortening was not only due to increased afterload, but also a result of impaired myocardial contractile capability, as could be seen in Figs. 3A, 4A and 4B. Impaired cardiac performance and an inverse relationship between ejection fraction and ventricular systolic wall stress is also found under clinical conditions [7, 32] and in chronic animal experiments [3, 16, 25].

B. Energetic consequences of myocardial transformation, ventricular hypertrophy and configuration

1. Consequences of cardiac hypertrophy and ventricular configuration on cardiac energetics

A variety of controversial concepts are regarded as the predominant causes of cardiac failure. At the cellular level, an energy deficit ensuing from inadequate oxygen and substrate supply along with poor ATP production is held responsible by many authors (for review see 16).

With regard to this topic, the energetic consequences of structural and functional alterations that occur during chronically increased haemodynamic load of the heart, as well as the question of the extent to which the hypertrophy process itself leads to a reduction in energy supply of the contractile apparatus, are of special interest.

As a rule, below critical weight, heart enlargement reflects the growth of individual cells [20]. Along with a diminution of coronary capillaries per unit mass this results in an increase in the diffusion distance for oxygen [12, 20].

An unfavourable balance between energy-producing and contractile units originates from a significant decrease in the mitochondrial fractional volume and is the mitochondrial: myofibril ratio, which cannot be compensated for by an increase in the respiratory activity of mitochondria [6, 38]. Also, a decrease in oxidation-phosphorylation coupling in the mitochondria of hypertrophied hearts and thus the decreased effectiveness of oxygen consumption was proposed [23]. During pressure-induced hypertrophy, fibrosis occurs, not only due to hypertensive vasculopathy or ischaemic cell injury, and may act as a diffusion barrier [14, 37]. In addition, coronary reserve is described as reduced in hypertrophied ventricles [23, 33] and, when structural dilatation sets in, mechanical efficiency of the ventricles is reduced [21].

In the light of these alterations, endangering the energy supply of the contractile apparatus, energy saving adjustments are of particular significance in the hypertrophied ventricle. As expected on the basis of reduced myofibrillar ATPase activity, a reduction in energy turnover, ascertained by means of measurements of oxygen consumption and heat liberation, could be demonstrated for hearts with pressure-induced hypertrophy and in the state of hypothyreosis [1, 11, 19]. These events with a greater economy of tension development were associated with a shift in the myosin isoenzyme pattern towards VM-3. Thus myocardial transformation towards a "slower" muscle represents such an energy-saving adjustment.

2. Does myocardial transformation contribute decisively to the improvement of ventricular efficiency?

Up to now no general agreement has been obtained concerning the best index of energy expenditure of cardiac muscle. A variety of different concepts have been proposed. Peak systolic wall stress, tension-time integral, velocity of contraction, systolic pressure-volume area and more comprehensive equations were emphasized as most suitable for predicting oxygen consumption of the heart [2, 29, 30, 31, 34]. Disregarding differences in experimental conditions, methodological problems in measurements of heat production and oxygen consumption, and obstacles in the choice of an adequate mathematical approach, it may be concluded that there is no single parameter that holds for all conditions in predicting oxygen consumption, but rather that the best predictor depends on the mode of contraction and on other physiological circumstances.

In our theoretical consideration we regarded peak systolic wall stress as an adequate measure of energy expenditure. Our calculations were based on the finding that myocardial transformation towards a "slower" muscle was associated with a reduction in oxygen consumption as related to ventricular weight and peak systolic wall stress (Fig. 5 A). The difference in oxygen consumption between homogeneous VM-1 and homogeneous VM-3 myocardium was 31%.

In the pressure-loaded heart of SHR peak systolic wall stress is renormalized due to ventricular hypertrophy. Neglecting the energy saving effect of myocardial transformation and assuming the same values of normalized oxygen consumption for both normotensive and spontaneously hypertensive rats, the calculated efficiency for SHR was not

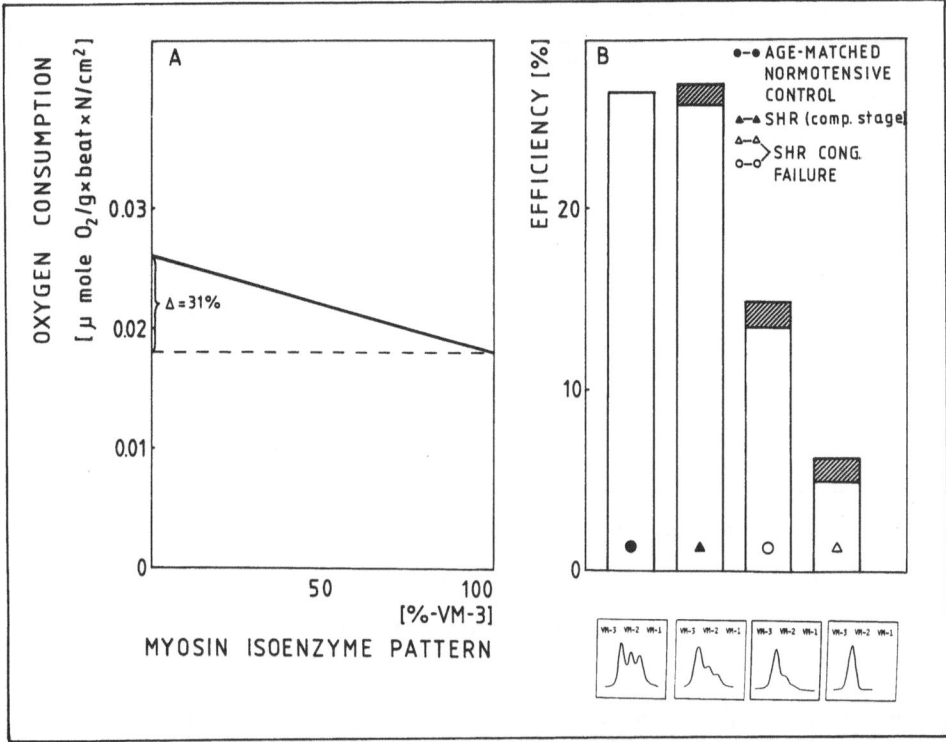

Fig. 5. Influence of myosin isoenzyme pattern on cardiac energetics. A) Correlation of myocardial oxygen consumption per gram of left ventricular weight, and beat, normalized to peak systolic wall stress with the percentage proportion of VM-3. Data were obtained in a specified heart-lung model [4]. B) Ventricular efficiency for a normotensive control, an SHR in the compensated stage and two SHR in congestive cardiac failure. Open columns = ventricular efficiency assuming the same normalized oxygen consumption for all 4 specimens. Hatched areas = the improvement in efficiency, taking energy saving effect of myocardial transformation into account. Calculations based on data of Fig. 5 A. Peak systolic wall stress was assumed to be the best index of energy expenditure of cardiac muscle. Corresponding myosin isoenzyme profiles are given below the figure.

significantly altered since stroke work, as related to LV weight, lay within the range of normotensive controls. When the myocardial transformation (normotensive, 40% VM-3; SHR, 55% VM-3) was taken into account, ventricular efficiency increased slightly from 25.7 to 26.75%. Thus, under auxotonic conditions, the often proposed improvement in ventricular efficiency due to more economic tension development of the hypertrophied myocardium with a higher proportion of VM-3 was rather small (Fig. 5 B; Table 1).

In the SHR in congestive cardiac failure structural dilatation with increased peak auxotonic systolic wall stress caused an increase in oxygen demand, whereas the stroke work was diminished. Consequently ventricular efficiency was severely reduced. When the coincident shift in the myosin isoenzyme pattern was taken into account (SHR congestive failure, 70% VM-3 and 100% VM-3, respectively), the ventricular efficiency was slightly improved but still markedly below normal (Fig. 5 B and Table 1).

Table 1. Data for evaluation of ventricular efficiency

		Normoten-sive control	SHR comp. stage		SHR cong. failure			
Stroke work	($\cdot 10^{-3}$ Nm)	5.46	7.52		5.17		1.90	
Stroke work/g LVW	($\cdot 10^{-3}$ Nm/g)	5.71	5.66	-0.8%	3.18	-44.3%	1.13	-80.2%
Peak auxotonic wall stress	(N/cm^2)	1.50	1.54	$+2.6\%$	1.92	$+28.0\%$	2.03	$+35.3\%$
Tension time integral	(N$\cdot$s/cm^2)	0.137	0.142	$+3.6\%$	0.226	$+65.0\%$	0.214	$+56.2\%$
Myosin isoenzyme	(% VM-3)	40.0	55.1		69.6		100.0	
Efficiency	(%)	26.38	25.70		13.49		4.99	
Efficiency (corrected)	(%)	26.38	26.75		14.84		6.24	

Efficiency: Calculated, assuming the same normalized oxygen consumption for all specimens. Corrected efficiency: Oxygen consumption was corrected according to the myosin isoenzyme pattern (Fig. 5A). Note: Ventricular efficiency would be even more decreased for SHR in congestive failure, when calculations were based on tension-time integral (see also Fig. 6).

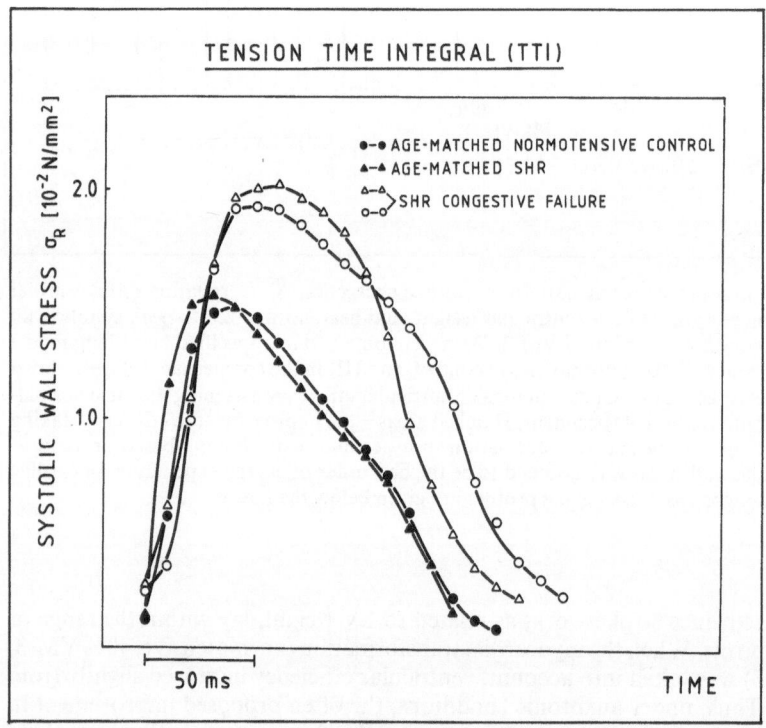

Fig. 6. Time course of systolic wall stress during an ejecting beat. Note the marked increase in peak systolic wall stress and in stress time integral for SHR in congestive failure. For values of TTI see Table 1.

If myocardial transformation was able to compensate for unfavourable effects of structural dilatation in the case of the most dilated ventricle in our experiments, the difference in normalized oxygen consumption for the marginal values of VM-1 and VM-3 should be 80–90%, instead of 31%, and thus the value of oxygen consumption in the case of homogeneous VM-3 should be diminished to $^1/_5$ of its actual value. The limited adaptive reserve capacity, arising from the energy saving effect of myocardial transformation, is additionally lessened by the fact that normotensive controls of comparable age already have a considerable proportion of the isoenzyme VM-3. Thus by virtue of its energy saving effect, the redistribution towards VM-3 may be regarded as an adaptive process; however, the limits will soon be reached when structural dilatation or impairment of energy supply occur.

In summary, ventricular and myocardial working capacity are hardly affected by myocardial transformation. Reduction in time-dependent parameters is more pronounced. In SHR in congestive failure, ventricular performance is impaired by both geometric state and reduced myocardial "contractility". Redistribution towards the isoenzyme VM-3 is not regarded as the main cause of the development of cardiac failure. Myocardial transformation failed to compensate for unfavourable consequences of structural dilatation on cardiac energetics. The significance of myocardial transformation in the impairment of ventricular performance and the improvement of ventricular efficiency seems to have been overestimated in the literature.

References

1. Alpert NR, Mulieri LA (1981) The utilization of energy by the myocardium hypertrophied secondary to pressure overload. In: Strauer BE (ed) The heart in hypertension. Springer, Berlin Heidelberg New York, pp 153–163
2. Bretschneider HJ (1972) Die hämodynamischen Determinanten des myokardialen Sauerstoffverbrauchs. In: Dengler HJ (ed) Die therapeutische Anwendung β-sympathikolytischer Stoffe. Springer, Stuttgart New York, pp 45–60
3. Bürger SB, Strauer BE (1981) Left ventricular hypertrophy in chronic pressure load due to spontaneous essential hypertension. I. Left ventricular function, left ventricular geometry, and wall stress. In: Strauer BE (ed) The heart in hypertension. Springer, Berlin Heidelberg New York, pp 13–36
4. Ebrecht G, Rupp H, Jacob R (1982) Alterations of mechanical parameters in chemically skinned preparations of rat myocardium as a function of isoenzyme pattern of myosin. Basic Res Cardiol 77:220–234
5. Gibbs CL, Gibson WR (1970) Energy production in cardiac isotonic contractions. J Gen Physiol 56:732–750
6. Goldstein MA, Sordahl LA, Schwartz A (1974) Ultrastructural analysis of left ventricular hypertrophy in rabbits. J Mol Cell Cardiol 6:265–273
7. Grossman W, Carabello BA, Ganter S, Fifer MA (1983) Ventricular wall stress and the development of cardiac hypertrophy and failure. In: Alpert NR (ed) Myocardial hypertrophy and failure. Raven Press, New York, pp 1–18
8. Gülch RW (1980) The effect of elevated chronic loading on the action potential of mammalian myocardium. J Mol Cell Cardiol 12:415–420
9. Hepp A, Hansis M, Gülch R, Jacob R (1974) Left ventricular isovolumetric pressure-volume relations, "diastolic tone" and contractility in the rat heart after physical training. Basic Res Cardiol 69:516–532
10. Hoh JFY, McGrath PA, Hale PT (1978) Electrophoretic analysis of multiple forms of rat cardiac myosin: Effects of hypophysectomy and thyroxine replacement. J Mol Cell Cardiol 10:1053–1076

11. Holubarsch C, Alpert NR, Goulette R, Mulieri LA (1981) Isometric force generation in hypo-thyrotic rat myocardium. Mechanical and energetic changes. Pflügers Arch [Suppl] 389:R7

12. Honig CR, Bourdeau-Martini J (1974) Extravascular component of oxygen transport in normal and hypertrophied hearts with special reference to oxygen therapy. Circ Res [Suppl II] 34/35:97–115

13. Jacob R, Kissling G (1981) Left ventricular dynamics and myocardial function in Goldblatt hypertension of the rat. Biochemical, morphological and electrophysiological correlates. In: Strauer BE (ed) The heart in hypertension. Springer, Berlin Heidelberg New York, pp 89–107

14. Jacob R, Kissling G, Ebrecht G, Holubarsch C, Medugorac I, Rupp H (1983) Adaptive and pathological alterations in experimental cardiac hypertrophy. In: Chazov E, Saks V, Rona G (eds) Advances in myocardiology, vol 4. Plenum Publishing Corporation, pp 55–77

15. Jacob R, Kissling G, Ebrecht G, Jörg E, Rupp H, Takeda N (1984) Cardiac alterations at the myofibrillar level: Is a redistribution of the myosin isoenzyme pattern decisive for cardiac failure in haemodynamic overload? Eur Heart J [Suppl F] 5:13–26

16. Jacob R, Vogt M, Rupp H (1986) Pathophysiological mechanisms in cardiac insufficiency induced by chronic pressure overload. Basic Res Cardiol [Suppl 1] 81:203–216

17. Jacob R, Vogt M (1986) Chronic cardiac reactions. I. The normalized length-stress area as a basis of assessment of myocardial work- and power-capacity in the hypertrophied and dilated ventricle. This issue.

18. Kissling G, Wendt-Gallitelli MF (1977) Dynamics of the hypertrophied left ventricle in the rat. Basic Res Cardiol 72:178–183

19. Kissling G, Rupp H, Malloy L, Jacob R (1982) Alterations in cardiac oxygen consumption under chronic pressure overload. Significance of the isoenzyme pattern of myosin. Basic Res Cardiol 77:255–269

20. Linzbach AJ (1948) Herzhypertrophie und kritisches Herzgewicht. Klin Wschr 26:459–463

21. Linzbach AJ (1960) Heart failure from the point of view of quantitative anatomy. Am J Cardiol 5:370–382

22. Lompré AM, Schwartz K, d'Albis A, Lacombe G, van Thiem N, Swynghedauw B (1979) Myosin isoenzyme redistribution in chronic heart overload. Nature 282:105–107

23. Meerson FZ (1976) Insufficiency of hypertrophied heart. Basic Res Cardiol 71:343–354

24. Mirsky I, Parmley WW (1973) Assessment of passive elastic stiffness for isolated heart muscle and the intact heart. Circ Res 33:233–243

25. Mirsky I, Pfeffer JM, Pfeffer MA, Braunwald E (1983) The contractile state as the major determinant in the evolution of left ventricular dysfunction in the spontaneously hypertensive rat. Circ Res 53:767–778

26. Okamoto K, Aoki K (1963) Development of a strain of spontaneously hypertensive rats. Jpn Circ J 27:282–293

27. Pfeffer MA, Pfeffer JM, Frohlich ED (1976) Pumping ability of the hypertrophying left ventricle of the spontaneously hypertensive rat. Circ Res 38:423–429

28. Rupp H, Jacob R (1982) Response of blood pressure and cardiac myosin polymorphism to swimming training in the spontaneously hypertensive rat. Can J Physiol Pharmacol 60:1098–1103

29. Sarnoff SJ, Braunwald E, Welch GH, Case RB, Strainsley WM, Macruz R (1958) Hemodynamic determinants of oxygen consumption of the heart with special reference to the tension-time index. Am J Physiol 192:148–156

30. Sonnenblick EH, Ross J Jr, Covell JW, Kaiser GA, Braunwald E (1965) Velocity of contraction as a determinant of myocardial oxygen consumption. Am J Physiol 209:919–927

31. Strauer BE, Beer K, Heitlinger K, Höfling B (1977) Left ventricular systolic wall stress as a primary determinant of myocardial oxygen consumption: Comparative studies in patients with normal left ventricular function, with pressure and volume overload and with coronary heart disease. Basic Res Cardiol 72:306–313

32. Strauer BE (1983) Das Hochdruckherz. Springer Verlag, Berlin Heidelberg New York Tokyo, p 147

33. Strauer BE (1985) Progression und Regression der Herzhypertrophie beim arteriellen Bluthochdruck: Pathophysiologie und Klinik. Z Kardiol [Suppl 7] 74:171–178

34. Suga H (1979) Total mechanical energy of a ventricle model and cardiac oxygen consumption. Am J Physiol 236:H498–H505
35. Ullrich KJ, Riecker G, Kramer K (1954) Das Druckvolumendiagramm des Warmblüterherzens. Pflügers Arch 259:481–498
36. Vogt M, Onegi B, Noma K, Rupp H, Jacob R (1986) Significance of myosin isoenzyme pattern and cardio-adrenergic drive in the generation of cardiac failure in spontaneously hypertensive rats. Pflügers Arch [Suppl] 406:R36
37. Vogt M, Noma K, Onegi B, Rupp H, Jacob R (1986) Chronic cardiac reactions. III. Factors involved in the development of structural dilatation. This issue.
38. Wendt-Gallitelli MF, Ebrecht G, Jacob R (1979) Morphological alterations and their functional interpretation in the hypertrophied myocardium of Goldblatt hypertensive rats. J Mol Cell Cardiol 11:275–287

Authors' address:

Dr. M. Vogt, Physiologisches Institut II, Universität Tübingen, Gmelinstraße 5, D-7400 Tübingen

Chronic cardiac reactions.
III. Factors involved in the development
of structural dilatation *

M. Vogt, R. Jacob, K. Noma, B. Onegi, H. Rupp

Physiologisches Institut II, Universität Tübingen, F.R.G.

Summary

The significance of various factors for the development of structural dilatation in the chronically pressure-loaded and failing heart were evaluated. The investigations were performed on male rats with renal (Goldblatt II) and spontaneous (Aoki-Okamoto) hypertension at different stages of haemodynamic overload. Two groups of SHR were submitted to intermittent feeding (SHR IF); one group received additionally the β-blocking agent atenolol (50 mg/kg b.w.; SHR IF + βBl.). Haemodynamic measurements were carried out under open chest conditions. Myosin isoenzyme pattern, hydroxyproline concentration and circulating blood volume were determined.

Transformation to slower myocardium per se, induced by IF, did not lead to significant change in ventricular configuration. After additional blockade of β-adrenergic receptors there were indications of unfavourable development of left ventricular configuration. Inhibition of hypertrophic mass increase due to curtailed adrenergic stimulation could be an influential factor in the development of dilatation. Further investigations, however, are required to establish the relationship between the adrenergic system, on the one hand, and degree of hypertrophy as well as structural dilatation of the ventricle, on the other hand. The established marked increase in hydroxyproline concentration of the dilated ventricle of SHR in congestive failure is consistent with the assumption of a causal link between the degree of fibrosis and structural dilatation. Observations on rats with aorto-caval shunt and Goldblatt II rats with eccentric hypertrophy and corresponding increase in filling potential or circulating blood volume indicate a correlation between the latter and ventricular size.

Thus, we assume that curtailed protein synthesis, fibrosis and regulatory processes related to water and electrolyte balance, but not myocardial transformation per se, play a role in the development of structural dilatation. The relative contribution of each factor, however, may depend on the experimental model that is used.

Introduction

During chronically increased haemodynamic loading of the heart, alterations occur at various levels of the organ. With respect to the pumping function of the heart, changes in myocardial mass and ventricular configuration are of particular significance. As is well known, the pressure hypertrophied ventricle has concentric configuration, while a more eccentric hypertrophy is a characteristic of the volume-overloaded ventricle. A frequent event in the final state of prolonged haemodynamic overload is structural dilatation [38] which is characterized by an insufficient increase in wall thickness relative to ventricular radius [22]. The importance of structural dilatation in the genesis of cardiac failure is dis-

---

* Supported by the Deutsche Forschungsgemeinschaft.

cussed controversially [16, 17, 22]. Nevertheless, an unfavourable geometrical state may contribute to cardiac failure by necessitating an increase in systolic wall stress and thus reducing the degree of myocardial shortening and mechanical efficiency. In view of therapeutic problems in haemodynamic overload, including the choice of the optimum time for surgical interventions, the factors determining ventricular configuration are of special interest. The purpose of our present study was to evaluate some of the factors involved in the development of structural dilatation using the models of Goldblatt II and spontaneous hypertension in rats.

Methods

The investigations were performed on male rats with renal (Goldblatt II) and spontaneous (Aoki-Okamoto strain) hypertension at different stages of haemodynamic overload. Normotensive, age-matched Wistar rats served as controls.

One group of SHR was submitted to an intermittent feeding schedule (24 h food ad libitum, 24 h fasting, alternately, for a total duration of 10 weeks) with the purpose of additional redistribution of the myosin isoenzyme pattern towards the "slow" isoenzyme VM-3 (SHR IF). After 6 weeks of intermittent feeding a further group received the β-blocking agent atenolol (50 mg/kg b.w.xdie, administered through drinking water) (SHR IF + β-Bl.). At the end of this procedure the animals were 10 to 11 months old.

Cardiac dynamic measurements (open chest) were carried out under urethane anaesthesia (1.2 g/kg b.w. intra-abdominally). Left ventricular pressure (LVP), left ventricular enddiastolic pressure (LVEDP; high amplification), rate of pressure rise (dP/dt) and aortic flow were simultaneously recorded. At the end of each experiment left ventricular enddiastolic pressure-volume relations and wet weight of the left ventricle were determined. Circulating blood volume was ascertained by the dye dilution technique (Evans-blue). To determine myosin isoenzyme pattern nondissociating gel electrophoresis in the presence of pyrophosphate was carried out according to Hoh [11] and Rupp [30]. Collagen content of ventricular tissue could be evaluated by means of hydroxyproline concentration according to the method of Stegemann [37].

Assuming a thick-walled spherical shell, circumferential stress θ_R at the midwall region was calculated, based on data of left ventricular pressure (P), ventricular inner volume (V) and ventricular wall volume (W):

$$\sigma_R = \frac{P \cdot r_i^3}{r_0^3 - r_i^3}\left(\frac{r_0^3}{2R^3} + 1\right) = \frac{V \cdot P}{W}\left(1 + \frac{(V+W)}{(V^{1/3} + (V+W)^{1/3})^3}\right)$$

according to Gülch (10) and to [27, 40].
Midwall differential elastic modulus E_R was evaluated using the following equation:

$$E_R = d\sigma_R/d\varepsilon = a \cdot b \cdot e^{b \cdot \varepsilon} = b(\sigma_R - c)$$

$$= 3 \cdot \left[\frac{V \cdot P}{W} - \sigma_R + \left(\frac{\sigma_R}{V} + \frac{W \cdot \sigma_R - V \cdot R}{W(V+W)} + \frac{\sigma_R}{P} \cdot \frac{dP}{dV}\right)\right.$$
$$\left. \cdot \frac{V^{1/3} + (V+W)^{1/3}}{V^{-2/3} + (V+W)^{-2/3}}\right].$$

The stiffness constant b is considered to be an index of the intrinsic elasticity of cardiac muscle [27, 40] and was calculated as the slope of the R_E versus σ_R plot by linear regression analysis:

$$b = \frac{E_R}{(\sigma_R - c)} \quad \text{(stiffness constant)}.$$

Multiple statistical comparisons were performed using the rank test of Kruskal and Wallis. Statistical significance was assumed at $p < 0.05$.

Results and Discussion

The significance of myocardial transformation for the development of structural dilatation

In rat ventricles, changes in the myosin isoenzyme population arise from altered expression of genes coding for myosin heavy chains [11] and are influenced by a variety of

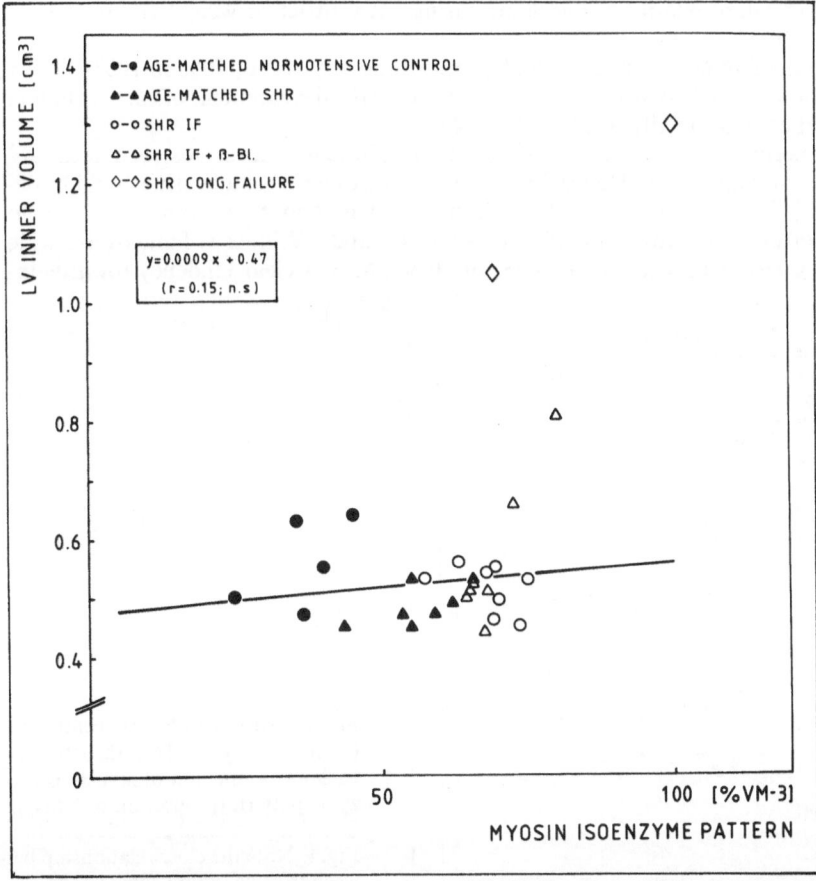

Fig. 1. Left ventricular inner volume at a defined filling pressure (LVEDP = 10 mm Hg) as a function of myosin isoenzyme pattern (proportion of VM-3). Note the striking increase in ventricular size with the proportion of VM-3 exceeding a value of 70%.

factors such as haemodynamic overload [25, 31], physical exercise [30], emotional stress [31], endocrine status [5, 11, 16], and ontogenetic development or maturation [25, 32]. An altered myosin isoenzyme population is combined with changes in myofibrillar ATPase activity [30], in the velocity of shortening [6], and in the efficiency of stress generation [1, 19].

A close correlation exists between left ventricular mass and the proportion of ventricular myosin VM-3 [32]. However, this relationship does not hold for all conditions. In our experiments intermittent feedings tends to reduce left ventricular weight and after additional β-blockade left ventricular weight was significantly reduced, while both interventions were accompanied by a significant shift of myosin isoenzyme towards VM-3, the isoenzyme with the lowest ATPase activity [left ventricular weight (g): SHR Co., 1.21 ± 0.15; SHR IF, $1.13 \pm 0,08$, n.s.; SHR IF + β-bl., 1.06 ± 0.13, $p < 0.05$; proportion of VM-3 (%): SHR Co., 53.0 ± 7.0; SHR IF, 68.2 ± 5.7, $p < 0.01$; SHR IF + β-Bl., 68.5 ± 5.2, $p < 0.01$].

Also observations in hyperthyreosis [16], hypothyreosis [16] and in the early postnatal period [25] showed deviation from the above mentioned relationship and therefore contradict a direct causal relation between increasing left ventricular weight and increasing proportion of VM-3.

As a rule the failing hearts reveal high proportions of VM-3 of 70% to 100%. However, the significance of myosin isoenzyme pattern for the development of cardiac failure was discussed controversially [16, 17, 23, 41, 43].

In our experimental models, congestive cardiac failure was never observed in the absence of structural dilatation. Hence the question arose of whether a bearing could be detected between the proportion of VM-3 and the ventricular configuration. There was no significant correlation between the fraction of VM-3 and LV inner volume over a wide range (Fig. 1). However, exceeding values of 70% VM-3, a clear tendency towards in-

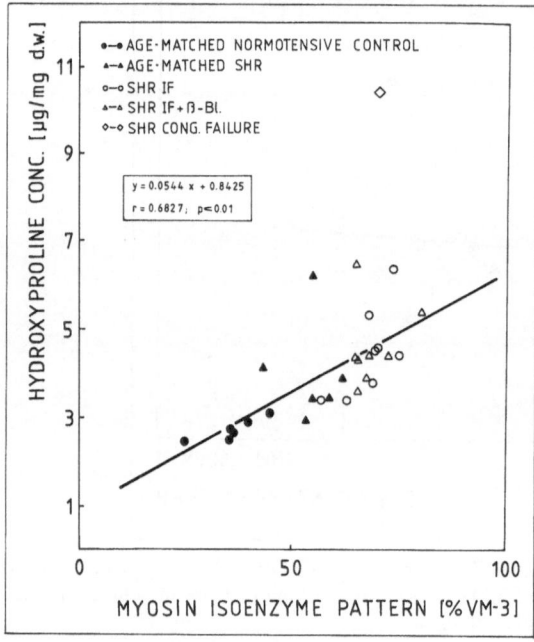

Fig. 2. Hydroxyproline concentration (related to dry weight of the left ventricle) as a function of myosin isoenzyme pattern (proportion of VM-3). Same experimental conditions as in Fig. 1. Note the close relationship between the two parameters and the overproportional increase in hydroxyproline concentration for SHR in congestive failure

creased ventricular volumes was evident, which indicates that additional factors have to be involved and that myosin isoenzyme pattern per se was not the main cause for dilatation. Although a close relationship between hydroxyproline concentration and proportion of VM-3 could be found, the SHR in congestive failure showed deviating behaviour with an overproportional increase in hydroxyproline concentration (Fig. 2). The latter suggests that this close relationship was not causal but rather concomitant in nature.

In conclusion, our results indicate that a change in myosin population per se was not a main factor for the development of structural dilatation.

On the other hand, since a severe reduction of noradrenaline content of the pressure loaded ventricle was found at the same time when the proportion of VM-3 had reached a certain level (about 70%) [32], the proportion of VM-3 could mark the onset of deterioration in cardiac adaptational process. However, this coincidence in time does not necessarily imply a causal mutual relationship.

Role of protein biosynthesis for the development of structural dilatation

Ventricular myocardium reacts to prolonged increased pressure load with hypertrophy of the myocytes in order to normalize wall stress. Within hours after the onset of an elevated work-load there is an increased adenylcyclase activity associated with rises in ribonucleic acid synthesis followed by an increased myosin biosynthesis [33, 34]. The nucleotide c-AMP has been shown to control the synthesis of selective proteins in several types of tissues [42]. Also, in heart muscle there is a great deal of evidence that c-AMP and thus the sympathetic adrenergic system represent an important regulating factor in the hypertrophy process [28, 35].

In our experiments blood pressure was reduced in both SHR IF and SHR IF + β-Bl., but still remained in an unequivocally hypertensive range. However, systolic auxotonic wall stress was slightly increased in SHR IF and even significantly increased in SHR IF + β-Bl. (systolic auxotonic wall stress [$\times 10^{-4}$ N/mm^2]: SHR control, 160.06 ± 12.32; SHR IF, 165.35 ± 12.11, n.s.; SHR IF + β-Bl., 175.18 ± 19.41, p < 0.05). The increase in radius-to-wall thickness ratio (r/h-ratio) at a defined filling pressure, which can be regarded as a measure for ventricular configuration, was significantly greater in SHR IF + β-Bl. than in SHR IF (p < 0.05). From a functional point of view, the plot of systolic auxotonic wall stress versus r : h-ratio yields information by which one can distinguish the eccentric type of hypertrophy from dilatation. Thus, the increase in systolic auxotonic wall stress as well as in r : h-ratio seem to indicate the onset of structural dilatation in SHR IF + β-Bl. (Fig. 3). Since there was no difference in the proportion of VM-3 and hydroxyproline concentration between SHR IF and SHR IF + β-Bl. this might be the consequence of an inhibition of hypertrophic mass increase, due to curtailed adrenergic stimulation. However, these results require further verification using larger groups of animals and including various models of haemodynamic overload. In principle, a restrained protein synthesis is one factor which promotes development and structural dilatation. In the prolonged pressure-overload of the heart the curtailment of the process of hypertrophy may arise from different events besides an "exhaustion" of protein synthesis. In heart failure, sympathetic nervous activity and circulating catecholamines from the adrenals are increased. Subsequently, "down-regulation" of β-adrenoceptor population in the failing heart occurs, i.e. a decrease in receptor density or sensitivity [3, 21]. In the final state a decrease in the noradrenaline content comes into play due to reduced synthesis of noradrenaline and depletion of the stores [32, 36]. These events may summate in reducing protein synthesis rate and may promote structural dilatation. Indications for the significance of cur-

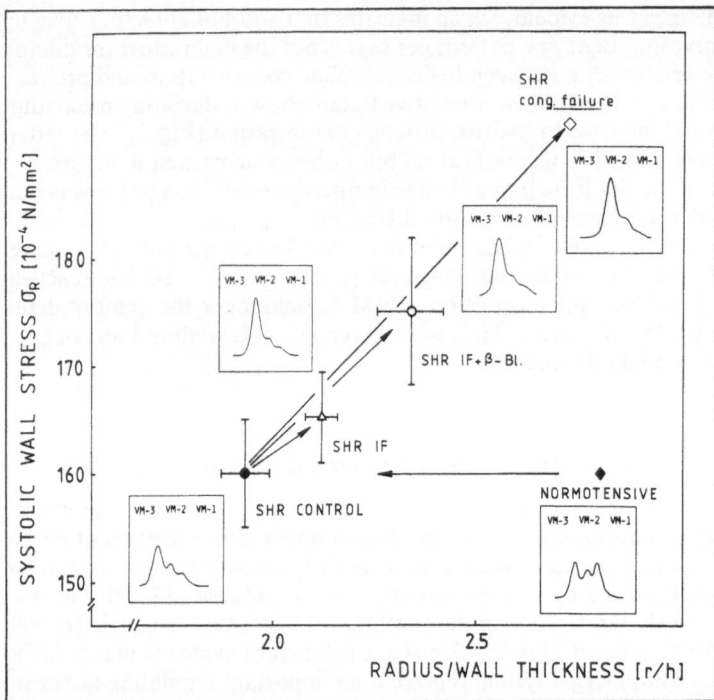

Fig. 3. Influence of geometric conditions on peak auxotonic systolic wall stress. Representative myosin isoenzyme profiles are depicted. Dilatation of the left ventricle was indicated by a simultaneous increase in systolic wall stress and in the r:h-ratio. Note the onset of structural dilatation in SHR IF + β-Bl.

tailed protein synthesis are also given by findings on the antineoplastic agent Adriamycin, which inhibits RNAse activity also in the myocardial cell and which frequently leads to dilative cardiomyopathy in a dose-dependent manner [26].

The relationship between tissue alterations and structural dilatation

According to Linzbach [22] a principle cause of structural dilatation is ischemic cardiac cell injury, which leads to multiple necroses and scars and consequently results in fibre slippage. From this the question arises as to whether there is a causal link between alterations on the tissue level, on the one hand, and structural dilatation on the other. In our experiments an increase in hydroxyproline concentration, being a measure of collagen content and degree of fibrosis, with increasing left ventricular weight was evident, when comparing SHR with normotensive controls (hydroxyproline conc. [µg/mg dry weight]: Normotensive control, 2.70 ± 0.27; SHR control, 4.02 ± 1.14, $p < 0.05$). The correlation between hydroxyproline concentration and left ventricular weight was significant. SHR in congestive failure, however, revealed an overproportional increase in hydroxyproline concentration (Fig. 4). Fibrosis, which sets in even earlier compared to SHR, was also described in the Goldblatt model [14, 17]. The increase in connective tissue during the pro-

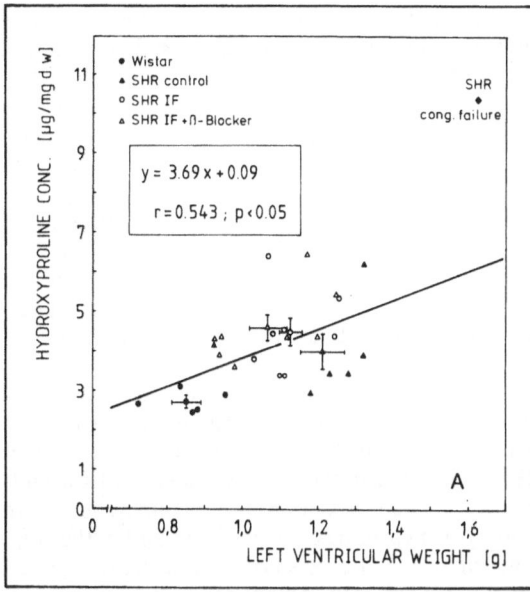

Fig. 4. Hydroxyproline concentration as a function of left ventricular weight. There is a significant correlation between the two parameters. In SHR (congestive failure) hydroxyproline concentration was markedly higher than predicted by left ventricular weight.

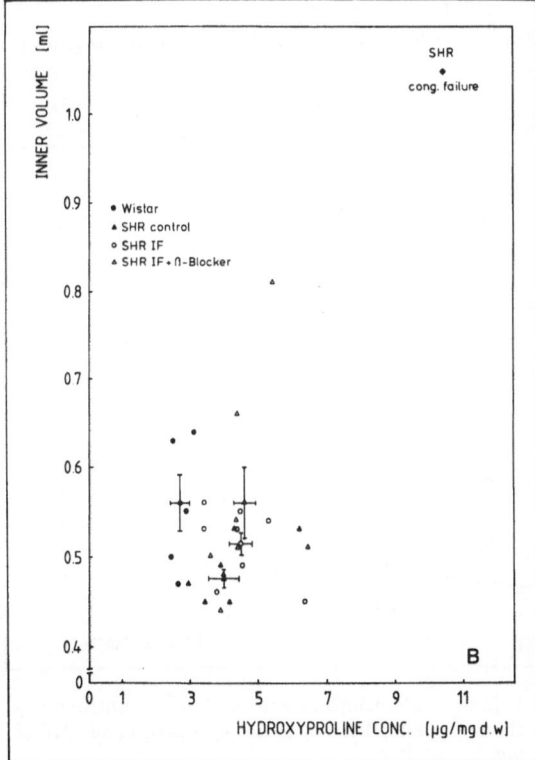

Fig. 5. Left ventricular size (inner volume at the enddiastolic pressure of 10 mm Hg) as a function of hydroxyproline concentration. Note both the high degree of scattering over a wide range and the extreme values of SHR in congestive failure.

cess of hypertrophy seems to depend on the rapidity of increase in blood pressure; however, the exact stimulus for collagen synthesis is still enigmatic. It has been described that hypoxic conditions can serve as such a stimulus [2]. Thus myocardial hypoxia, occuring subsequently to increased ventricular wall stress [13] may be involved in the growth of connective tissue at the onset and early phase of pressure overload and particularly during dilatation. In later stages, hypertensive vasculopathy with ischaemic cell injury and scar formation additionally occur.

Due to a large degree of scattering, when plotting inner volume as a function of hydroxyproline concentration, no significant correlation was obtained (Fig. 5). However, the findings in extremely dilated ventricles do not disprove the assumption that a causal relationship exists between fibrosis and structural dilatation. To obtain direct evidence, measurements of individual fibre length are required, but may pose great methodological problems.

Fibrosis not only has consequences for systolic function by acting as a diffusion barrier, but also leads to an absolute or relative decrease of contractile material. The impairment of cardiac function by fibrosis can also be mediated by disturbance of diastolic function, e.g. alterations in myocardial distensibility. The drastic rightward shift of the end-diastolic pressure volume relations in the dilated ventricles of SHR in congestive failure was accompanied by a marked increase in hydroxyproline concentration and in the stiffness constant b (Figs. 6 and 7), corresponding to the "fibrosis type" of decreased distensibility [12, 14]. However, considerable increase in collagen content can occur without any significant change in myocardial elasticity Figs. 6 and 7).

A comparable increase in hydroxyproline concentration resulted in a more marked increment of the stiffness constant b in Goldblatt rats, as compared with SHR [14]. Thus it can be concluded that not only the total amount of collagen, but als the arrangement of connective tissue is significant.

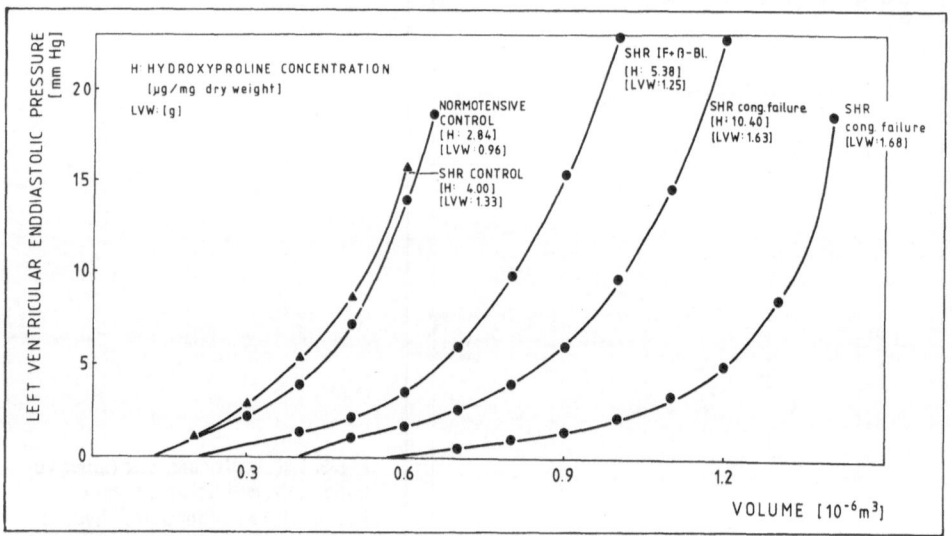

Fig. 6. Left ventricular enddiastolic pressure-volume relationships of a normotensive control and a SHR in compensated stages (left), SHR in a state of preinsufficiency (SHR IF + β-Bl.) and SHR in congestive cardiac failure (two curves to the right).

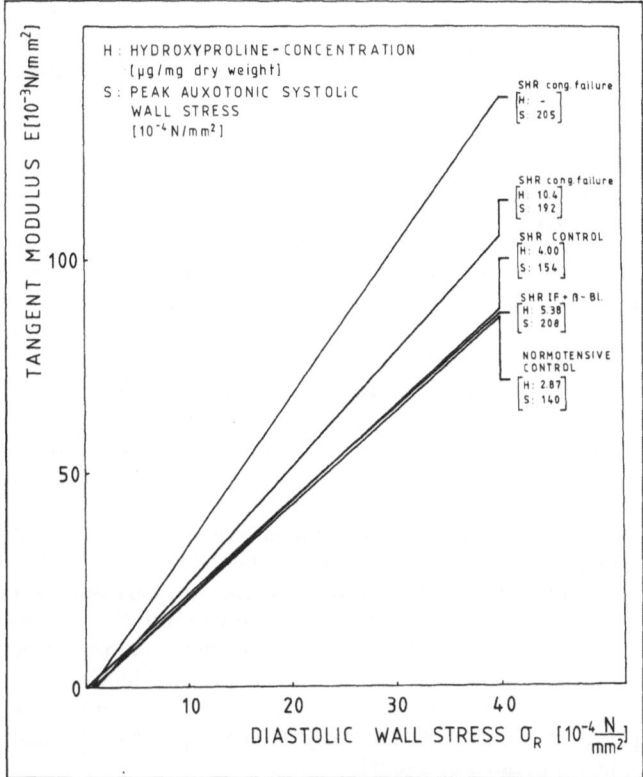

Fig. 7. Tangent modulus E (midwall differential elastic modulus) as a function of diastolic wall stress. Stiffness constant *b* is represented by the slope of the plot and is considered to be an index of the intrinsic elasticity of cardiac muscle. Note: Structural dilatation of the left ventricle led to a marked increase in the stiffness constant *b*. Considerable concentrations of hydroxyproline can occur before an increase in stiffness constant *b* appears. (The same specimen as in Fig. 4.)

Circulating blood volume

In our experimental models congestive cardiac failure with skin oedema, pleural effusion and ascites was never observed independently of structural dilatation. Thus, dilatation had already set in before symptoms of heart failure became manifest, at least under resting conditions. Simultaneously, an impairment of myocardial contractile capability was found when a marked degree of dilatation was reached, indicating that a reduction in myocardial "contractility" preceded structural dilatation. However, due to cardiac reserve mechanisms, including neuroendocrine reactions, cardiac output at rest is restored. Many of the adjustments to heart failure are similar to the homeostatic mechanisms utilized by the body in response to circulatory failure from any cause, e.g. acute blood loss [9].

One important adjustment to heart failure is the increased sympathetic activity in combination with increased plasma concentrations of norepinephrine, which leads to enhanced myocardial contractility, generalized arterial vasoconstriction and an increase in venous tone [39]. Another important compensatory mechanism is the increase in ventricular filling pressure produced by an increase in plasma volume as the result of salt and water retention by the kidneys. Although the precise mechanisms for the initial changes in the kidneys are still not clear, the increased sympathetic activity with increased renal vascular resistance and reduced total renal blood flow, as well as a stimulation of the renin-angiotensin-aldosterone system are considered to be of particular significance [4, 7]. Thus, the mechanism leading to an increase in plasma volume may ultimately contribute to the development of structural dilatation. The significance of circulating blood volume for ventricular cavity size was emphasized by findings at the 4-week stage in Goldblatt II rats [18]. Certainly, due to a significant increase in the circulating blood volume, a transient augmentation of the diastolic ventricular size was found in this early stage. Furthermore, an influence of venous filling potential and cardiac output on ventricular size is consistent with observations by other authors on experimental infarction [8] or on rats with aorto-caval shunt [20]. However, it should be mentioned that clinical experience has shown that a close correlation between the degree of failure, increase in enddiastolic volume and increase in diastolic filling pressure is not always found [29]. Thus, an increased filling pressure or increased filling potential may not be generally accepted unequivocally as a factor for the development of structural dilatation, but seem to be involved in the dependence on the degree of activation of neuroendocrine compensatory mechanisms.

In summary, our investigations led to the following conclusions. Myocardial transformation towards a "slower muscle" per se has no significant effect on dilatation of the left ventricle. The findings in extremely dilated hearts are consistent with the assumption that a causal relationship exists between fibrosis and structural dilatation. Curtailment of protein synthesis in the late stages of chronic haemodynamic overload and regulatory processes related to water and electrolyte balance, and thus to circulating blood volume, contribute to structural dilatation. The effects of adrenergic blockade require further investigation, particularly in view of the clinical relevance. The relative significance of each factor may depend on the applied experimental model.

References

1. Alpert NR, Mulieri LA (1981) The utilization of energy by the myocardium hypertrophied secondary to pressure overload. In: Strauer BE (ed) The heart in hypertension. Springer, Berlin Heidelberg New York, pp 153–163
2. Bartosova D, Chuapic M, Korecky B, Poupa O, Rakusan K, Turek Z, Vizek M (1969) The growth of the muscular and collagenous parts of rat heart in various forms of cardiomegaly. J Physiol 200:285–295
3. Bristow MR, Ginsburg R, Minobe WBS, Cubicioki RS, Sageman WS, Lurie K, Billingham ME, Marrison DC, Stinson EB (1982) Decreased catecholamine sensitivity and β-adrenergic-receptor density in failing human hearts. New Engl J Med 307:205–211
4. Cohn JM (1980) Progress in vasodilatator therapy for heart failure. New Engl J Med 302:1414
5. Dillmann WH (1980) Diabetes mellitus induces changes in cardiac myosin of the rat. Diabetes 29:579–582
6. Ebrecht G, Rupp H, Jacob R (1982) Alterations of mechanical parameters in chemically skinned preparations of rat myocardium as a function of isoenzyme pattern of myosin. Basic Res Cardiol 77:220–234
7. Eichna LW, Farber SJ, Berger AR, Earle DP, Rader B, Pellegrino E, Albert RE, Alexander JD, Taube H, Youngwirth S (1953) Cardiovascular dynamics, blood volumes, renal functions, and electrolyte excretions in the same patients during congestive heart failure and after recovery of cardiac decompensation. Circulation 7:674–686
8. Gaudron PJ, Pfeffer JM, Pfeffer MA (1986) Chronische NaCl-Restriktion mindert strukturelle Ventrikeldilatation bei Ratten mit Myokardinfarkt. Z Kardiol [Suppl 1] 75:114

9. Harris P (1983) Evolution and the cardiac patient. Cardiovasc Res 17/6:313–319; 17/7:373–378; 17/8:437–445
10. Hepp A, Hansis M, Gülch R, Jacob R (1974) Left ventricular isovolumetric pressure-volume relations, "diastolic tone", and contractility in the rat heart after physical training. Basic Res Cardiol 69:516–532
11. Hoh JFY, McGrath PA, Hale PT (1978) Electrophoretic analysis of multiple forms of rat cardiac myosin: Effects of hypophysectomy and thyroxine replacement. J Mol Cell Cardiol 10:1053–1076
12. Holubarsch Ch, Jacob R (1979) Evaluation of elastic properties of myocardium. Experimental models of fibrosis and contracture in heart muscle strips. Z Kardiol 68:123–127
13. Honig CR, Bourdeau-Martini J (1974) Extravascular component of oxygen transport in normal and hypertrophied hearts with special reference to oxygen therapy. Circ Res [Suppl II] 34/35:97–115
14. Jacob R, Kissling G (1981) Left ventricular dynamics and myocardial function in Goldblatt hypertension of the rat. Biochemical, morphological and electrophysiological correlates. In: Strauer BE (ed) The heart in hypertension. Springer, Berlin Heidelberg New York, pp 89–106
15. Jacob R, Kissling G, Ebrecht G, Jörg E, Rupp H, Takeda M (1984) Cardiac alterations at the myofibrillar level: Is a redistribution of the myosin isoenzyme pattern decisive for cardiac failure in haemodynamic overload. Eur Heart J [Suppl F] 5:13–26
16. Jacob R, Vogt M, Rupp H (1986) Pathophysiological mechanisms in cardiac insufficiency induced by chronic pressure overload. Basic Res Cardiol [Suppl 1] 81:203–216
17. Jacob R, Vogt M, Rupp H (in press) Physiological and pathological hypertrophy. In: Dhalla NS, Singal PK, Beanish RE (eds) Pathophysiology of heart disease. Martinus Nijhoff Publishing, Boston
18. Kissling G, Gassenmaier T, Wendt-Gallitelli MF, Jacob R (1977) Pressure volume relations, elastic modulus and contractile behaviour of the hypertrophied left ventricle of rats with Goldblatt II hypertension. Pfluegers Arch 369:213–221
19. Kissling G, Rupp H, Malloy L, Jacob R (1982) Alterations in cardiac oxygen consumption under chronic pressure overload. Significance of the isoenzyme pattern of myosin. Basic Res Cardiol 77:255–269
20. Kissling G, Takeda N, Vogt M (1985) Left ventricular end-systolic pressure-volume relationships as a measure of ventricular performance. Basic Res Cardiol 80:594–607
21. Lehmann M, Rühle K, Schmid P, Klein H, Matthys K, Keul J (1983) Hemodynamic values, plasma catecholamines, and β-adrenergic receptors on intact polymorphonuclear leucocytes in trained and untrained subjects and patients with cardiac insufficiency. Z Kardiol 72:529–536
22. Linzbach AJ (1960) Heart failure from the point of view of quantitative anatomy. Am J Cardiol 5:370–382
23. Lompré AM, Schwartz K, d'Albis A, Lacombe G, van Thiem N, Swynghedauw B (1979) Myosin isoenzyme redistribution in chronic heart overload. Nature 282:105–107
24. Meerson FZ (1976) Insufficiency of hypertrophied heart. Basic Res Cardiol 71:343–354
25. Mercadier JJ, Lompré AM, Bouveret P, Samuel JL, Rappaport L, Swynghedauw B, Schwartz K (1983) Myosin isoenzymic distribution in hypertrophied rat and human hearts. In: Jacob R, Gülch RW, Kissling G (eds) Cardiac adaptation to hemodynamic overload; training and stress. Steinkopff Verlag, Darmstadt, pp 104–112
26. Minow RA, Benjamin RS, Gottlieb JA (1975) Adriamycin (MCS-123 127) cardiomyopathy. An overview with determination of risk factors. Cancer Chemother Rep 6:195
27. Mirsky I, Parmley WW (1973) Assessment of passive elastic stiffness for isolated heart muscle and the intact heart. Circ Res 33:233–243
28. Östman-Smith I (1981) Cardiac sympathetic nerves as a final common pathway in the induction of adaptive cardiac hypertrophy. Clin Sci 61:265–272
29. Reindell H, Musshoff K, Klepzig H (1960) Physiologische und pathologische Grundlagen der Größen- und Formänderungen des Herzens. In: Schwiegk H (ed) Handbuch der Inneren Medizin, vol IX/1. Springer, Berlin Göttingen Heidelberg
30. Rupp H, Jacob R (1982) Response of blood pressure and cardiac myosin polymorphism to swimming training in the spontaneously hypertensive rat. Can J Physiol Pharmacol 60:1098–1103

31. Rupp H, Felbier HR, Bukhari A, Jacob R (1984) Modulation of myosin isoenzyme populations and activities of monoamine oxidase and phenylethanolamine-N-methyltransferase in pressure-loaded and normal rat heart by swimming exercise and stress arising from electrostimulation in pairs. Can J Physiol Pharmacol 62:1209–1218

32. Rupp H, Jacob R (1986) Correlation between total catecholamine content and redistribution of myosin isoenzymes in pressure loaded ventricular myocardium of the spontaneously hypertensive rat. Basic Res Cardiol [Suppl 1] 81:147–155

33. Schreiber SS, Oratz M, Rothschild MA (1966) Protein synthesis in the overloaded mammalian heart. Am J Physiol 211:314–318

34. Schreiber SS, Klein IL, Oratz M, Rothschild MA (1971) Adenyl cyclase activity and cyclic AMP in acute cardiac overload: A method for measuring cyclic AMP production based on ATP specific activity. J Mol Cell Cardiol 2:55–65

35. Sen S, Tarazi RC (1983) Regression of myocardial hypertrophy and influence of adrenergic system. Am J Physiol 244:H97–H101

36. Spann JF, Chidsey CA, Pool PE, Braunwald E (1965) Mechanism of norepinephrine depletion in experimental heart failure produced by aortic constriction in the guinea pig. Circ Res 17:312–321

37. Stegemann H (1958) Mikrobestimmung von Hydroxyprolin mit Chloramin-T und p-Dimethyl-amino-benzaldehyd. Hoppe Seyler's Z Physiol Chem 311:41

38. Strauer BE (1983) Das Hochdruckherz. Springer Verlag, Berlin Heidelberg New York Tokyo, p 147

39. Thomas JA, Marks BH (1978) Plasma-norepinephrine in congestive heart failure. Am J Cardiol 41:233–242

40. Vogt M, Jacob R (1985) Myocardial elasticity and left ventricular distensibility as related to oxygen deficiency and right ventricular filling. Analsis in rat heart model. Basic Res Cardiol 80:537–547

41. Vogt M, Onegi B, Noma K, Rupp H, Jacob R (1986) Significance of myosin isoenzyme pattern and cardioadrenergic drive in the generation of cardiac failure in spontaneously hypertensive rats. Pflügers Arch [Suppl] 406:R36

42. Wicks WD (1974) Regulation of protein synthesis by cyclic AMP. In: Greengard P, Robinson GA (eds) Advances in cyclic nucleotide research, vol 4. 335–415

43. Wikman-Coffelt J, Parmley WW, Mason DT (1979) The cardiac hypertrophy process. Analyses of factors determining pathological versus physiological development. Circ Res 45:697–707

Author's address:

Dr. M. Vogt, Physiologisches Institut II, Universität Tübingen, Gmelinstr. 5, D-7400 Tübingen

Chronic cardiac reactions.
IV. Effect of drugs and altered functional loads on cardiac energetics as inferred from myofibrillar ATPase and the myosin isoenzyme population

H. Rupp, R. Wahl* and R. Jacob

Physiologisches Institut (II) and *Medizinische Klinik (IV), Universität Tübingen, F.R.G.

Summary

A major determinant of myocardial energetics is the ATPase activity of myofibrils. In order to account for chronic changes in myofibrillar ATPase, the state equation of the intertropomyosin-interaction model of Tawada et al. [35] was extended by introducing the rates of cross-bridge cycling of myofibrils composed of V-1 or V-3 and the concentration of the myosin isoenzymes. Cross-bridge cycling rates of 1.0 or 0.7 were derived for myofibrils composed of V-1 or V-3, respectively. Ca^{2+} responsiveness and positive co-operativity were not significantly affected by the myosin isoenzymes. Redistribution of the myosin isoenzyme population and thus altered myocardial energetics was observed following administration of various drugs and as a result of different functional loads. Besides thyroid hormones, catecholamines had a marked influence on myosin. Reducing the adrenergic drive by administration of atenolol, guanethidine or reserpine led to a shift in the direction of V-3. Since serum T_3 levels were not significantly reduced by these interventions, the drugs act most probably at the organ level. The functional states responsible for the increase in the proportion of V-3 (pressure load, intermittent feeding, schedule-induced stress) also did not affect circulating T_3 in a manner that could entirely explain the redistribution. Hypertrophy-induced dilution of sympathetic nerve fibres or reduced adrenergic responsiveness most likely play a role in the redistribution. An increase in the proportion of V-1 was observed following swimming exercise but not, however, after spontaneous or enforced running. In the swim-exercised rats, T_3 was markedly reduced. Thus, the trigger reactions linked most probably to the high adrenergic drive during swimming have to overcome the lower T_3 level. It is concluded that myocardial energetics can be decisively altered by a variety of drugs and functional loads, whereby the trigger reactions leading to an altered gene expression of myosin cannot be accounted for entirely by altered circulating T_3 but most probably involve the adrenergic system.

Introduction

The major determinant of myocardial energetics is the rate of cross-bridge cycling in myofilaments. Besides effects of Ca^{2+} or phosphorylation of myosin light chain [7, 15, 34], cross-bridge cycling depends on the myosin isoenzyme population [1, 10, 11, 19]. Although the accessible range of the myosin isoenzyme population in the direction of V-3 is particularly pronounced in small mammals, redistribution of the isoenzyme population has been demonstrated also for large mammals [36]. Thus we considered it important, not only to focus on elementary steps in the cross-bridge mechanism for a given myosin isoenzyme population, but also to define the trigger reactions which lead to an altered isoenzyme population and to changes in myocardial energetics. Great progress has been made in elucidating the organization of the myosin genes [13]; nevertheless little is known about the physiological or pathophysiological stimuli in the intact organism which give rise to altered gene expression. The present approach outlines the effects of various drugs, altered neuro-endocrine status of the periphery and different functional demands imposed on the heart. In particular, the nature of trigger mechanisms involved in the redistribution of the myosin isoenzyme population was investigated.

Methods

Myofibrils from left ventricles were prepared as described previously [19]. The reaction mixture for the assay of myofibrillar ATPase contained 3.16 mM MgATP, 0.32 mM free Mg^{2+}, 40 mM imidazole, 5 mM NaN_3, 4 mM EGTA and a variable amount of free Ca^{2+}; ionic strength was 0.12 and pH was 7.0. The free metal and chelate concentrations were calculated using the following apparent stability constants $4.9 \cdot 10^6$ M^{-1} CaEGTA, 40 M^{-1} MgEGTA, $5 \cdot 10^3$ M^{-1} CaATP, $11.4 \cdot 10^3$ M^{-1} MgATP. The myosin isoenzyme population was determined using pyrophosphate gel electrophoresis as given previously [18, 22, 23].

Triiodothyronine (T_3) in serum was measured using an immuno assay (Immophase, Corning, U.S.A.). 6–8-month-old male Wistar/WU rats and spontaneously hypertensive rats (SHR) were obtained from Ivanovas (Kissleg, F.R.G.). They were fed either ad libitum or intermittently [20] (24 h free access to food followed by 24 h fasting). In "schedule-induced stress" [6] rats were given 35 mg food pellets every 80 s using a food pellet dispenser. The feeding schedule lasted 8 h/day. In the exercise groups, rats ran spontaneously up to 20 km/day in activity wheels attached to their home cages, or were forced to run at 10 m/min for max 2×3 h/day in a rotating drum, or swam in either 35 °C or 31 °C water (max. 2×1.5 h/day). Types and mode of drug treatment are specified in Table 1. All experimental routines lasted 2–4 weeks.

Results and discussion

Myofibrillar ATPase and the myosin isoenzyme population

To quantitatively describe myofibrillar ATPase activity the model proposed by Tawada et al. [35] was used. In this approach, the ATPase activity depends on the apparent association constant for Ca^{2+}-binding to troponin and on terms accounting for the positive co-operativity attributed to an interaction between tropomyosin molecules:

$$A = \left(1 + \lambda A^4 \left(\frac{[Ca^{2+}]}{k + [Ca^{2+}]}\right)^2\right) / (1 + \lambda A^4),$$

where A is the ATPase activity; $1/k$ is the apparent association constant for Ca^{2+}-binding to troponin; $\lambda = \exp(\varepsilon/KT)$; $A = \exp(-\eta/KT)$; ε is the energy required to promote one unit of tropomyosin from the resting state to the contracting state; η is a constant, if there is no co-operativity then $\eta = 0$; K is the Boltzmann constant.

The model does not account for other possible regulatory mechanisms such as phosphorylation of myosin light chain-2, the interaction between myosin heads, diverse protein-protein interactions or altered ratio of Ca^{2+} or Mg^{2+} bound to myosin. Furthermore, it should be noted that the nature of the actual trigger reaction responsible for at-

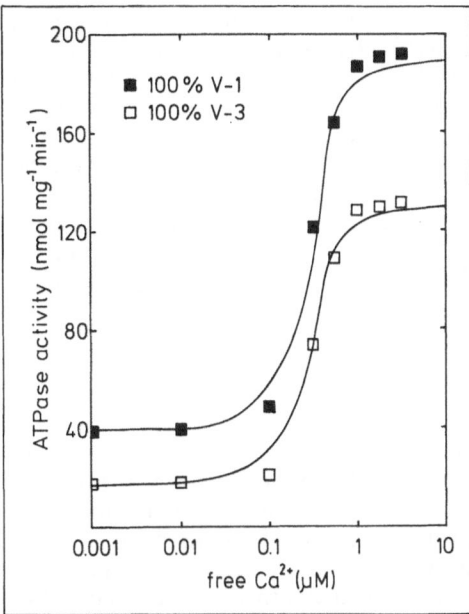

Fig. 1. Activation of myofibrillar ATPase by Ca^{2+}. Activities are given for myofibrils of homogeneous V-1 (5-week-old Wistar/WU rats) and V-3 (Wistar/WU rats treated with 6-n-propyl-2-thiouracil). The experimental data points were fitted to the extended state equation with Newton's algorithm after subtraction of basal ATPase activities at 0.001 and 0.01 μM free Ca^{2+} from all values. Fit parameters were: association constant $4.8 \times 10^6\,M^{-1}$, co-operation term $\Lambda\,1.9 \times 10^{-3}$, $\lambda\,100$, cross-bridge cycling rate parameters $r_\alpha\,1.0$ and $r_\beta\,0.7$. Following fitting of the data points, basal ATPase activities were again added to the data points and the theoretical curves. Note, Ca^{2+} responsiveness and co-operativity were not significantly different for myofibrils composed of V-1 or V-3.

tachment of cross-bridges is still controversial. Nevertheless the model is a useful approach which quantitatively describes the Ca^{2+}-dependency of myofibrillar ATPase.

In order to account for myofibrillar ATPase activity also in terms of the myosin isoenzyme population, the model of Tawada et al. was extended [19] to include the rates of cross-bridge cycling of myosins containing only one type of heavy chain. Under the present conditions, ATPase activity of myofibrils composed of homogeneous V-1 was 192 nMol P mg^{-1} min^{-1} and that of myofibrils with homogeneous V-3 was 131 nMol P mg^{-1} min^{-1} (Fig. 1). Rate of cross-bridge cycling can thus be approximated by using rate parameters of 1.0 for V-1 and 0.7 for V-3. This approach is corroborated by the fact that unloaded velocity of shortening of skinned fibres, for which the myofilament Ca^{2+} concentration is well defined, varied over a comparable range. Since V-2 is a heterodimer composed of one alpha- and one beta-heavy chain, the proportion of V-2 was halved and added to the homodimeric myosins V-1 and V-3, thus reducing the experimental isoenzyme population to a hypothetical population consisting of only 2 isoenzymes composed of either 2 alpha- or 2 beta-heavy chains. The overall cross-bridge cycling rate of myofibrils, containing a mixed population of myosins with either 2 alpha- or 2 beta-heavy chains, was assumed to arise in an additive manner from the activities of myofibrils composed of only one type of heavy chain, taking into account their respective amounts in the population of unknown ATPase activity. The additional term R which accounts for different cross-bridge cycling rates is given by the following equation:

$$A = \left[\left(1 + \lambda\Lambda^A\left(\frac{[Ca^{2+}]}{k + [Ca^{2+}]}\right)^2\right) \Big/ (1 + \lambda\Lambda^A)\right] \cdot R,$$

$$R = (r_\alpha c_\alpha + r_\beta(1 - c_\alpha))M_{0,\alpha}$$

where c_α is the concentration of myosin alpha-heavy chains in a given isoenzyme population and r_α, r_β are cross-bridge cycling rate parameters derived from myofibrils composed of either 2 alpha- or 2 beta-heavy chains. The parameter $M_{o,\alpha}$ represents the activity of myofibrils composed of only alpha-heavy chains measured under standard conditions. When the activity of myofibrils was assayed at 30 °C in the presence of 3.16 mM MgATP, 0.32 mM free Mg^{2+}, 40 mM imidazole, 5 mM NaN_3, ionic strength 0.12, pH 7.0, then $M_{o,\alpha}$ was 192 nMol P mg^{-1} min^{-1}. Using this approach, it was possible to derive myofibrillar ATPase activity for a given free Ca^{2+} concentration by determining the myosin isoenzyme population. Because the reproducibility of determinations of the myosin isoenzyme population is much higher than that of myofibrillar ATPase activity, we preferred to determine the isoenzyme population and to deduce the corresponding changes in myofibrillar ATPase activity from the above equations.

Drugs affecting the myosin isoenzyme population

Since thyroid hormones are required for the expression of genes coding for alpha-heavy chains [2, 12, 13, 30], any change in circulating thyroid hormones necessarily influences the proportion of V-1. If the serum concentration of T_3 was below approximately 35 ng/dl, only beta-heavy chains were synthesized (Table 1). Thyroid hormone analogs such as D-thyroxine or 3,3',5-triiodothyroacetic acid were ineffective in the euthyroid rat in inducing an enhanced biosynthesis of V-1 (Table 1). Other trigger reactions which, at the present state of knowledge, are still rather ill-defined superimpose on the regulation by thyroid hormones. Best characterized is the effect of catecholamines. A reduction in sympathetic drive of heart due to chemical sympathectomy with 6-hydroxydopamine [26] or norepinephrine depletion at the nerve terminals using guanethidine or reserpine resulted in a higher proportion of V-3 (Table 1). Treatment with the $beta_1$-selective adrenoreceptor blocking drug atenolol also increased the proportion of V-3 (Table 1). The effect of the non-selective blocking agent propranolol has been controversial, leading either to a higher proportion of V-3 [17] or no change [32]. In a low dose, the beta-agonists isopro-

Table 1. Effects of drugs on the myosin isoenzyme population and serum T_3 in Wistar/WU rats. For the sake of clarity, the change in alpha-heavy chain content (%) and T_3 (ng/dl) in experimental rats is given relative to the respective control rats

	Δ alpha-chain	ΔT_3
L-thyroxine, 0.1 mgkg^{-1}d^{-1}	+21.8[b] (max. +100)	+100.6[b]
D-thyroxine, 0.1 mgkg^{-1}d^{-1}	+ 2.4[NS]	+ 6.6[a]
3,3',5-triiodothyroacetic acid, 0.1 mgkg^{-1}d^{-1}	+ 3.5[NS]	− 5.3[a]
6-n-propyl-2-thiouracil, 13 mgkg^{-1}d^{-1}	−78.2[b] (max. −100)	− 60.7[b]
atenolol, 50 mgkg^{-1}d^{-1}	− 8.7[b]	+ 9.2[NS]
guanethidine, 60 mgkg^{-1}d^{-1}	−19.6[b]	+ 27.7[b]
reserpine, 0.1 mgkg^{-1}d^{-1}	−27.0[b]	− 4.8[NS]
isoproterenol, 5 mgkg^{-1}d^{-1}	− 9.3[a]	− 2.8[NS]
alloxan, 135 mgkg^{-1}d^{-1}	−69.1[b]	− 48.7[b]

L-thyroxine, D-thyroxine, 3,5',5-triiodothyroacetic acid, reserpine were injected i.p. daily; alloxan was injected i.p. once; isoproterenol was administered by Alzet osmotic minipumps; 6-n-propyl-2-thiouracil, atenolol, guanethidine were given in the drinking water. Statistical comparisons between the experimental and the respective control groups were performed using Student's t-test; NS $P > 0.05$, [a] $P < 0.05$, [b] $P < 0.01$.

terenol [21, 33] and dobutamine [26] resulted in a redistribution in favour of V-1. However, when isoproterenol was chronically infused using osmotic minipumps, the proportion of V-3 was increased (Table 1). In order to show that the observed changes in myosin isoenzyme population can be attributed to a direct action of catecholamines and not secondarily to an altered thyroid status, serum T_3 levels were determined (Table 1). Although in the reserpinized rats, T_3 concentration was slightly reduced there was no evidence that in the case of guanethidine or atenolol the increased proportion of V-3 could be due to reduced circulating T_3. In isoproterenol infused rats, the T_3 concentration remained unaltered, suggesting again a direct action, probably involving in this case down-regulation of ventricular beta-adrenergic receptors.

In contrast to the drugs which exert their main effect at the organ level, some other interventions not only diminish the influence of the sympathetic system on the heart, but also reduce the concentration of serum T_3. An example is experimentally-induced diabetes, where the number of beta-adrenergic receptors is reduced [8] and the proportion of V-3 is increased [4]. In diabetes induced by alloxan (Table 1), the proportion of V-1 was positively correlated with serum T_3 and inversely with serum glucose (not shown). Although this correlation would suggest that thyroid hormones are mainly responsible for the altered myosin isoenzyme population, it cannot be inferred that the reduced T_3 level is the final trigger for the altered gene expression. Evidence has indeed been obtained indicating that changes in substrate consumption in the diabetic heart are of particular relevance in mediating the altered myosin isoenzyme population [3].

Changes in myosin isoenzyme population following different functional loads

The drug-induced redistributions in myosin isoenzyme population demonstrate that both thyroid hormones and adrenergic activity should be considered in the analysis of an altered gene expression in different functional states. Because a markedly altered sympathetic drive of the heart is expected to affect myocardial norepinephrine stores, the content of norepinephrine was determined in those ventricles in which the myosin isoenzyme population was studied. One of the best documented examples of a redistribution in the myosin isoenzyme population as a result of chronically altered functional demands is pressure overload. In longstanding hypertension of the SHR, the proportion of V-3 can markedly increase, reaching the limiting state of homogeneous V-3 (Table 2). The proportion of V-3 was closely correlated with ventricular mass, where SHR and Wistar/WU rats seemed to follow a common relationship. Wistar/WU rats exhibited, at an age where ventricular mass was not significantly different from SHR, a similar isoenzyme population as SHR [26].

However, when the high blood pressure of SHR was reduced towards normal levels by spontaneous running in activity wheels, regression of cardiac hypertrophy was not observed, neither was there any reduction in V-3 [24] (Table 2). This demonstrates that the correlate of the increased proportion of V-3 in hypertrophied ventricles of SHR should be seen in the greater mass and not in the pressure load. An enforced running routine involving an activity pattern untypical of spontaneous running neither reduced blood pressure nor ventricular mass and the proportion of V-3 was unchanged (Table 2).

The increased proportion of V-3 in the hypertrophied ventricles cannot be attributed to a lower T_3 concentration in SHR when compared with Wistar/WU rats. It would also seem improbable that circulating thyroid hormones changed fortituously in a manner resulting in the close correlation between the proportion of V-3 and ventricular mass. Also in the case of pressure overload due to abdominal aortic constriction in adult rats, the plasma T_4 level was not reduced [14]. Even when serum T_4 was reduced as in the case

Table 2. Effects of different functional states on the myosin isoenzyme population and serum T_3. For the sake of clarity, the change in alpha-heavy chain content (%) and T_3 (ng/dl) in experimental rats is given relative to the respective control rats

	Δ alpha-chain	ΔT_3
SHR vs. Wistar/WU (6-months-old)	−33.7[b]	+ 7.4[NS]
spontaneous running, post 18 h		
Wistar/WU	+ 2.1[NS]	− 8.6[a]
SHR	+ 3.2[NS]	− 8.3[a]
enforced running, post 18 h		
Wistar/WU	− 4.9[NS]	−15.7[a]
SHR	− 0.5[NS]	− 8.8[a]
swimming exercise, Wistar/WU		
35 °C, 1×2 h, post 18 h	+11.2[b]	− 2.8[NS]
35 °C, 2×1.5 h, post 18 h	+20.4[b]	−17.0[a]
35 °C, 3×2 h, post 18 h	+13.8[b]	+24.0[b]
31 °C, 1×2 h, post 18 h	+ 8.3[a]	n.d.
35 °C, 2×1.5 h, post 0 h	+19.9[b]	− 1.7[NS]
swimming exercise, SHR		
35 °C, 2×1.5 h, post 0 h	+17.3[b]	−17.1[b]
35 °C, 2×1.5 h, post 18 h	+17.9[b]	− 1.9[NS]
intermittent feeding		
Wistar/WU, post 0 h	−27.2[b]	−6.8[NS]
Wistar/WU, post 18 h	−15.2[b]	− 6.4[a]
SHR, post 18 h	− 5.7[a]	+ 3.3[NS]
"schedule-induced stress"		
Wistar/WU, post 18 h	−16.8[b]	− 9.8[NS]
intermittent feeding and swimming exercise (35 °C, 2 × 1.5 h)		
Wistar/WU, post 18 h	+11.3[b]	−22.9[b]
SHR, post 18 h	+16.2[b]	−13.0[a]

For the swimming routines, water temperature and max. swimming time per day are given. Rats were sacrificed either approx. 18 h or immediately after finishing a given experimental routine. The routines extended over approx. 4 weeks. Statistical comparisons between the experimental and the respective control groups were performed using Student's t-test; NS $P > 0.05$, [a] $P < 0.05$, [b] $P < 0.01$; n.d., not determined.

of right ventricular hypertrophy induced by ingestion of Crotalaria spectabilis seeds, the proportion of V-3 was still higher in the loaded right ventricle compared with the left ventricle [27]. This suggests that other mechanisms which superimpose on the effect of reduced thyroid hormones operate to change gene expression in favour of V-3.

Only in old SHR with nearly homogeneous V-3, the norepinephrine content was markedly reduced, demonstrating that the myosin isoenzyme population can be redistributed in the direction of V-3 at least to 60% without concomitant reduction in norepinephrine content [25]. An increase in the proportion of V-3 is thus not necessarily associated with an impairment of myocardial performance to the extent that a sustained increase in sympathetic drive ensues leading to norepinephrine depletion. It should, however, be noted that despite unchanged total norepinephrine content, norepinephrine concentration was inversely correlated with ventricular mass (not shown). In view of the finding that injections of catecholamines can induce a higher proportion of V-1, one might conclude that the hypertrophy-induced dilution of sympathetic fibres could be linked to an increasingly smaller sympathetic influence on myocytes located between nerve fibres.

This could contribute also to the heterogeneous nature of myocytes with respect to their isoenzyme distribution [29]. One would postulate that myocytes adjacent to nerve fibres exhibit a higher proportion of V-1 compared to those were diffusion distances for catecholamines are greater.

In contrast to spontaneous running which did not lower the proportion of V-3 despite a greatly normalized blood pressure, swimming exercise reduced the proportion of V-3 in Wistar rats [16, 18, 31], SHR [22, 23] and renal hypertensive rats [31]. The duration of the daily exercise routine proved to be critical for the extent of redistribution. Increasing the swimming time from 2 h/day to 2×1.5 h/day led to a higher proportion of V-1. A further increase to 3×2 h/day reduced, however, the redistribution. Lowering the water temperature from 35 to 31 °C did not significantly affect the isoenzyme population, indicating that heat loss during swimming was not a decisive factor. Furthermore, emotional stress was probably also not important [22]. In swim-exercised SHR, ventricular mass was not reduced to values typical of exercised Wistar/WU rats although blood pressure was markedly diminished [23]. The trigger involved in the expression of a higher proportion of alpha-heavy chains in SHR is thus not related to the reduced pressure load and operates independently of ventricular mass.

Serum T_3 was not affected in a way which could explain the higher proportion of V-1 (Table 2). In rats exercised max. 2×1.5 h/day, T_3 was even reduced, irrespective of whether the rats were sacrificed immediately after the exercise routine or after a rest period. Thus, the mechanisms leading to a higher proportion of V-1 have to overcome the reduced T_3 level. Only when rats were exercised 3×2 h/day, T_3 was increased. The higher thyroid activity was, however, not reflected in a more pronounced redistribution.

One characteristic feature of swimming exercise in the rat is the strong activation of the peripheral adrenergic system. This was reflected in the greatly increased adrenal mass, a marked increase in total norepinephrine and epinephrine content of adrenals and of norepinephrine in ventricles (not shown). The intermittent high peripheral sympathetic drive could thus be considered the correlate for the increased proportion of V-1.

In contrast to swimming exercise, where T_3 and the proportion of V-1 were altered in an opposite way, a change in the feeding schedule involving periods of fasting affected T_3 in a manner which could account at least partially for the observed redistribution of the myosin isoenzyme population. Intermittent feeding involving access to food for one day and food deprivation for either one or two days markedly increased the proportion of V-3 in Wistar rats and SHR [20] (Table 2). A comparable increase has been observed in Wistar rats following a procedure called semistarvation [5]. "Schedule-induced stress" resulting from feeding of food pellets (35 mg every 80 s for 8 h/day) [6] increased the proportion of V-3 to a similar extent as in rats fed on the schedule involving food deprivation for one day (Table 2). Noteworthy is that the effect of an altered feeding schedule has been observed also in SHR, whereby the influence of factors linked with ventricular mass increase and those with intermittent feeding was roughly additive. Although in a group of intermittently fed rats serum T_3 was significantly reduced, additional mechanisms have to be assumed in order to explain the extent of redistribution.

Catecholamine content of ventricles or adrenals was not affected in a way that would indicate a major contribution of the adrenergic system to the observed changes in the myosin isoenzyme population. It should, however, be noted that the number of beta-adrenergic receptors was greatly reduced in ventricles of intermittently fed rats, indicating diminished influence of the adrenergic system (not shown).

Because the effects of swimming exercise and intermittent feeding probably involve different trigger reactions, it was of interest to see how the myocyte would react when both experimental routines were imposed on the same rat. In Wistar/WU rats which were fed intermittently and subjected to swimming exercise, the resulting myosin isoenzyme population was approximately intermediate between the populations which were characteristic for the separate loads. However, in swim-exercised SHR, the effect of the intermittent feeding schedule on the myosin isoenyzme population was much smaller and not statistically significant.

Conclusion

The present data demonstrate that the myocyte reacts to a great number of functional loads with a redistribution of the myosin isoenzyme population, affecting myofibrillar ATPase activity and thus necessarily myocardial energetics. As regards the trigger reactions involved in such remodelling of the myocyte in the intact organism, our knowledge is still very limited. It is well documented that the level of thyroid hormones, particularly of T_3, is decisive and determines the starting point for any further load. The extent to which a redistribution can occur depends critically on the serum T_3 concentration. For example, keeping rats in a room of 34 °C for a prolonged time, results in a reduced blood T_3 and a higher proportion of V-3 [9]. Under those conditions, the effect of pressure load would be small because the limiting state of homogeneous V-3 would be reached much earlier than in a rat kept at a lower temperature. The concentration of T_3 is also decisive with respect to its effect in preventing or overcoming a change in the myosin isoenzyme population observed in the euthyroid rat. For example, chemical denervation using 6-hydroxydopamine did not result in a higher proportion of V-3 when the rats were additionally treated with a pharmacological dose of T_4 [26]. Thyroid hormones have, therefore, to be considered as ranking at the top of a hierarchy involving different regulatory mechanisms. However, although T_3 is important for the myosin isoenzyme population, other mechanisms have to be assumed in order to account for the redistribution of the myosin isoenzyme population in most functional states. Pressure load and intense exercise in the form of swimming are examples where the observed redistribution is most probably not mediated by circulating T_3. As regards the trigger in the pressure loaded heart, the available data suggest a reaction secondary to hypertrophy of the myocytes. One likely candidate could be seen in hypertrophy-induced dilution of sympathetic nerve fibres. In the ventricles of the swim-exercised rat, the higher proportion of V-1 could be assigned to the markedly enhanced adrenergic drive occuring during the exercise routine. Further progress in this field necessitates a broad pharmacological approach, which allows for the possibility that a given redistribution might not arise from a direct action of the drug at the organ level but be secondary to altered circulating thyroid hormones.

Changes in the myosin isoenzyme population are not isolated events and should be considered as an example of a physiological process occuring at a cellular and molecular level which enables the myocyte to cope with a variety of loads. Evidence for functionally comparable changes has been demonstrated for the sarcoplasmic reticulum. For a number of functional loads, an increase in the proportion of V-3 was accompanied by a reduced rate of Ca^{2+}-uptake affecting in a concerted manner both speed of shortening and rate of Ca^{2+} sequestration [28] and thus most probably also rate of muscle relaxation. Further investigation is required to unravel the nature of the trigger reactions operating during remodelling of the myocyte at a molecular and cellular level.

Acknowledgements. This study was supported by the Deutsche Forschungsgemeinschaft. The expert work of Mr. L. Schwarz is gratefully acknowledged.

References

1. Alpert NR, Mulieri LA (1986) Intrinsic determinants of myocardial energetics in normal and hypertrophied hearts. In: Rupp H (ed) Regulation of heart function – basic concepts and clinical applications. Thieme, Stuttgart New York, pp 292–304
2. Chizzonite RA, Zak R (1984) Regulation of myosin isoenzyme composition in fetal and neonatal rat ventricle by endogenous thyroid hormones. J Biol Chem 259:12628–12632
3. Dillmann WH (1985) Methyl palmoxirate increases Ca^{2+}-myosin ATPase activity and changes myosin isoenzyme distribution in the diabetic rat heart. Am J Physiol 248:E602–E606
4. Dillmann WH, Barrieux A, Reese GS (1984) Effect of diabetes and hypothyroidism on the predominance of cardiac myosin heavy chains synthesized in vivo or in a cell-free system. J Biol Chem 259:2035–2038
5. Dillmann WH, Berry S, Alexander NM (1983) A physiological dose of triiodothyronine normalizes cardiac myosin adenosine triphosphatase activity and changes myosin isoenzyme distribution in semistarved rats. Endocrinology 112:2081–2087
6. Falk JL, Tang M, Forman S (1977) Schedule-induced chronic hypertension. Psychosom Med 39:252–263
7. Franks K, Cooke R, Stull JT (1984) Myosin phosphorylation decreases the ATPase activity of cardiac myofibrils. J Mol Cell Cardiol 16:597–604
8. Heyliger CE, Pierce GN, Singal PK, Beamish RE, Dhalla NS (1982) Cardiac alpha- and beta-adrenergic receptor alterations in diabetic cardiomyopathy. Basic Res Cardiol 77:610–618
9. Horowitz M, Peyser YM, Muhlrad A (1986) Alterations in cardiac myosin isoenzymes distribution as an adaptation to chronic environmental heat stress in the rat. J Mol Cell Cardiol 18:511–515
10. Jacob R, Vogt M, Rupp H (1987) Physiological and pathological hypertrophy. In: Dhalla NS, Singal PK, Beamish RE (eds) Pathophysiology of heart disease. Martinus Nijhoff Publishing Boston (in press)
11. Kissling G, Rupp H, Malloy L, Jacob R (1982) Alterations in cardiac oxygen consumption under chronic pressure overload. Significance of the isoenzyme pattern of myosin. Basic Res Cardiol 77:255–269
12. Lompré AM, Nadal-Ginard B, Mahdavi V (1984) Expression of the cardiac ventricular alpha- and beta-myosin heavy chain genes is developmentally and hormonally regulated. J Biol Chem 259:6437–6443
13. Mahdavi V, Strehler EE, Periasamy M, Wieczorek DF, Izumo S, Nadal-Ginard B (1986) Sarcomeric myosin heavy chain gene family: organization and pattern of expression. Med Sci Sports Exerc 18:299–308
14. Martin AF, Robinson DC, Dowell RT (1985) Isomyosin and thyroid hormone levels in pressure overload weanling and adult rat hearts. Am J Physiol 248:H305–H310
15. Morano I, Hofmann F, Zimmer M, Rüegg JC (1985) The influence of P-light chain phosphorylation by myosin light chain kinase on the calcium sensitivity of chemically skinned heart fibres. FEBS Lett 189:221–224
16. Pagani ED, Solaro RJ (1983) Swimming exercise, thyroid state and the distribution of myosin isoenzymes in rat heart. Am J Physiol 245:713–720
17. Pauletto P, Dalla Libera L, Vescovo G, Scannapieco G, Angelini A, Pessina AC, Dal Palu C (1985) Propranolol-induced changes in ventricular isomyosin composition in the rat. Am Heart J 109:1269–1273
18. Rupp H (1982) Polymorphic myosin as the common determinant of myofibrillar ATPase in different haemodynamic and thyroid states. Basic Res Cardiol 77:34–46
19. Rupp H (1983) The determinants of the calcium-dependent activation of myofibrils from rat heart as judged by myofibrillar adenosine triphosphatase and superprecipitation of natural actomyosin. Mol Physiol 3:249–263
20. Rupp H (1985) Association of ventricular myosin heavy chains in functional states which lead to isoenzyme populations encompassing the whole range of possible distribution. Basic Res Cardiol 80:608–616
21. Rupp H, Bukhari AR, Jacob R (1983) Regulation of cardiac myosin isoenzymes – The interrelationship with catecholamine metabolism. J Mol Cell Cardiol [Suppl 1] 15:317

22. Rupp H, Felbier H-R, Bukhari AR, Jacob R (1984) Modulation of myosin isoenzyme populations and activities of monoamine oxidase and phenylethanolamine-N-methyltransferase in pressure loaded and normal rat heart by swimming exercise and stress arising from electrostimulation in pairs. Can J Physiol Pharmacol 62:1209–1218
23. Rupp H, Jacob R (1982) Response of blood pressure and cardiac myosin polymorphism to swimming training in the spontaneously hypertensive rat. Can J Physiol Pharmacol 60:1098–1103
24. Rupp H, Jacob R (1983) The interrelationship between normalization of pressure load of heart and hypertrophy and myosin isoenzyme population in the SHR. J Mol Cell Cardiol [Suppl 2] 15:63
25. Rupp H, Jacob R (1986) Correlation between total catecholamine content and redistribution of myosin isoenzymes in pressure loaded ventricular myocardium of the spontaneously hypertensive rat. Basic Res Cardiol [Suppl 1] 81:147–155
26. Rupp H, Jacob R (1986) Myocardial transitions between fast- and slow-type muscle as monitored by the population of myosin isoenzymes. In: Rupp H (ed) Regulation of heart function – basic concepts and clinical applications. Thieme, Stuttgart New York, pp 271–291
27. Rupp H, Popova N, Jacob R (1983) In: Jacob R, Gülch R, Kissling G (eds) Cardiac adaptation to hemodynamic overload, training and stress. Steinkopff Verlag, Darmstadt, pp 46–52
28. Rupp H, Wahl R, Jacob R (1987) Remodelling of the myocyte at a molecular level – Relationship between myosin isoenzyme population and sarcoplasmic reticulum. In: Dhalla NS, Pierce GN, Beamish RE (eds) Heart function and metabolism. Martinus Nijhoff Publishing, Boston (in press)
29. Samuel J-L, Rappaport L, Mercadier J-J, Lompré A-M, Sartore S, Triban C, Schiaffino S, Schwartz K (1983) Distribution of myosin isozymes within single cardiac cells. An immunohistochemical study. Circ Res 52:200–209
30. Samuel J-L, Rappaport L, Syrovy I, Wisnewsky C, Marotte F, Whalen RG, Schwartz K (1986) Differential effect of thyroxine on atrial and ventricular isomyosins in rats. Am J Physiol 250:H333–H341
31. Schaible TF, Malhotra A, Ciambrone GJ, Scheuer J (1986) Chronic swimming reverses cardiac dysfunction and myosin abnormalities in hypertensive rats. J Appl Physiol 60:1435–1441
32. Sheer D, Morkin E (1984) Myosin isoenzyme expression in rat ventricle: effects of thyroid hormone analogs, catecholamines, glucocorticoids and high carbohydrate diet. J Pharmacol Exp Ther 229:872–879
33. Sreter FA, Faris R, Balogh I, Somogyi E, Sotonyi P (1982) Changes in myosin isozyme distribution induced by low doses of isoproterenol. Arch Int Pharmacodyn Ther 260:159–164
34. Sweeney HL, Stull JT (1986) Phosphorylation of myosin in permeabilized mammalian cardiac and skeletal muscle cells. Am J Physiol 250:C657–C660
35. Tawada Y, Ohara H, Ooi T, Tawada K (1975) Non-polymerizable tropomyosin and control of the superprecipitation of actomyosin. J Biochem 78:65–72
36. Wiegand V, Henniges H, Oberschmidt R, Kreuzer H (1985) Influence of the thyroid state on myocardial myosin in the adult pig heart. Basic Res Cardiol 80:12–17

Authors' address:

Heinz Rupp, Physiologisches Institut II, Gmelinstr. 5, 7400 Tübingen, F.R.G.

Ca-independent regulation of cardiac myosin*

S. Winegrad, G. McClellan, A. Weisberg, L. E. Lin, S. Weindling and R. Horowits

Department of Physiology, School of Medicine, University of Pennsylvania, Philadelphia, U.S.A.

Summary

Calcium-independent regulation of the contractile proteins of cardiac muscle has been studied using hyperpermeable cells from rat ventricles and sections of quickly-frozen rat hearts. These preparations have been used to study maximum Ca-activated force, myosin ATPase activity and the maximum velocity of unloaded shortening. Beta adrenergic activity increases the amount of force and the ATPase activity in accordance with the concentration of the V_1 isozyme of myosin. V_3 activity is decreased at the same time. In tissues containing only V_1, there is no change in maximum velocity in response to beta adrenergic stimulation. These results indicate that beta adrenergic stimulation recruits V_1 force generators and probably regulates the transition between a Ca unresponsive and a Ca responsive force generator. This type of regulation provides the cell with the ability to operate along many different force-velocity relations.

Introduction

Contraction in striated muscle is produced by a rise in the cytosolic concentration of calcium ions and binding of calcium to the troponin complex located in the thin contractile filament. This binding causes a movement of tropomyosin from a position in the thin filament that blocks the interaction between actin and myosin, and thereby allows the force generating reaction to occur. Relaxation follows the lowering of the concentration of calcium ions, release of calcium from troponin, and restoration of tropomyosin to its blocking position. In light of this transition between the relaxed state with the calcium-control site on troponin unsaturated and the contracted state with the calcium control site saturated, it has been customary to think of the cardiac contractile system as having just two physiological states: relaxed and contracted. There is now strong evidence that the contractile system of the hearts of small mammals can exist in three different physiological states: 1) relaxed and Ca unresponsive; 2) relaxed and Ca responsive; and 3) contracted due to calcium activation.

Methods

The existence of a relaxed Ca unresponsive state has been demonstrated by a combination of techniques that permit assessment of the properties of the contractile proteins without interference from other steps in the contractile process with the capacity for modulating the contractile properties of the proteins. The first method involves the use of hyperpermeable cardiac cells, in which the surface membrane has been made highly permeable to

* Supported by grants HL 16010, HL 33294, HL 15835 from the National Institutes of Health

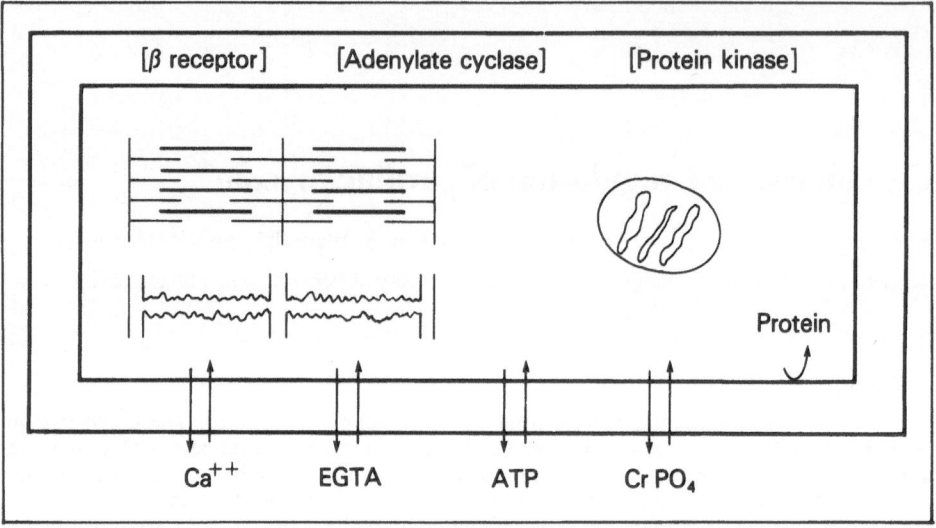

Fig. 1. A schematic drawing of a hyperpermeable fibre depicting properties of the cell that are pertinent to the study of contractile proteins.

ions and small molecules without disturbing the overall integrity of the sarcolemma [9] (Fig. 1). The sarcolemma remains a diffusion barrier for larger molecules and consequently cytoplasmic proteins are retained by the cell. The cell also retains functional beta adrenergic, alpha adrenergic and muscarinic receptors [5], adenylate cyclase and phosphodiesterases [9, β]. The hyperpermeability is produced by exposing the cells to a solution containing 10 mM EGTA at low temperature. Since the membrane of hyperpermeable cells is permeable to Ca ions and EGTA, the intracellular free Ca concentration can be buffered and a specified pCa rigidly maintained. As a consequence, calcium-independent changes in the amount of force generated by the myofibrils can be detected. Thin sections from hearts that have been frozen quickly to preserve the in vivo state of the cells have been used to study the enzymatic properties of myosin. Both Ca and actin activated ATPase activities have been measured as indicators of the state of myosin and its ability to interact with actin in the force producing reaction [13, 14]. Since the sections are about $^1/_3$ of the thickness of a single cell, serial sections include parts of the same cell and provide the opportunity for comparing the responses of the same cell to two different treatments.

Maximum Ca activated force

Activation of the beta adrenergic system at any step between binding of the agonist to the receptor and elevation of the concentration of cAMP increases the maximum Ca activated force in hyperpermeable ventricle cells from the hearts of young rats by an average of about 150% (Fig. 2). In order to demonstrate this phenomenon, it is necessary to include both a phosphodiesterase inhibitor to prevent the breakdown of cAMP and a low concentration of a non-ionic detergent to facilitate the release of the factor responsible for the increase in force from its position of attachment to an intracellular membrane.

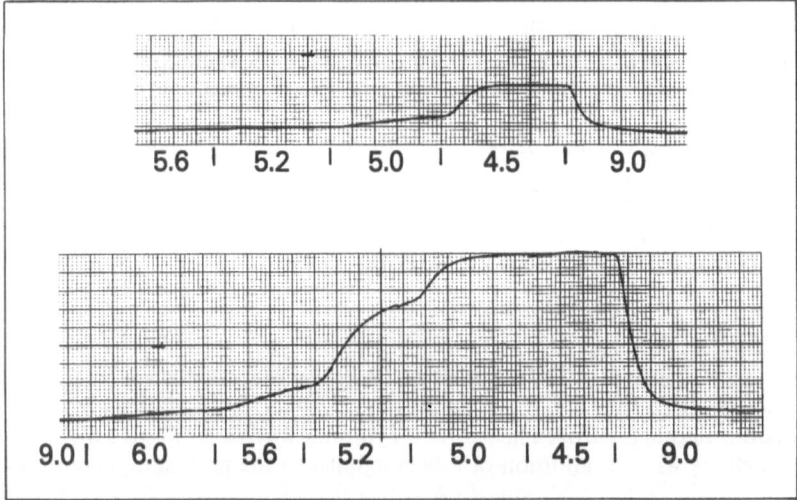

Fig. 2. The force generated by a bundle of hyperpermeable cells. The numbers below indicate values for pCa. In the interval between the upper and lower recordings, the cells were exposed to a solution containing 10^{-6} M cAMP, 5 mM theophylline and 1% Triton X-100.

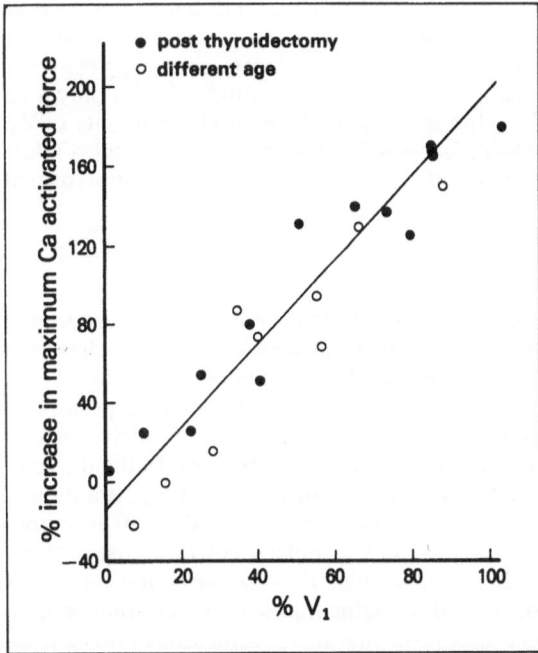

Fig. 3. A plot of the relation between the increase in maximum Ca activated force produced by cAMP or beta adrenergic agonists and the relative amount of V_1 in the cells.

Hearts of young rats contain three isozymes of myosin, two homodimers and the heterodimer, that are formed from two different heavy chains. The alpha heavy chain produces a myosin that has about 3 times the ATPase activity but the same force generating capability as myosin formed with the beta heavy chain. The two homodimers predominate and are called V_1 and V_3 according to whether the heavy chains are respectively alpha or beta [4]. The degree to which activation of the beta adrenergic system increases maximum Ca activated force depends on the percentage of the myosin that is in the form of V_1 (Fig. 3). From this correlation it would appear that the beta adrenergic dependent system that increases force in a Ca independent manner can distinguish between the different isozymes of myosin and act selectively on the fast isozyme V_1.

Myosin ATPase activity

The calcium and actin activated ATPase activities of myosin are also increased by beta adrenergic stimulation. This has been shown in the intact animal, in the isolated perfused heart, and in sections of quickly frozen hearts [14]. Tyramine injection into rats increases myosin ATPase activity as does addition of a beta agonist to the perfusion medium of an isolated heart. Similarly, the addition of cAMP to the medium bathing sections of frozen hearts increases ATPase activity. As in the response of maximum Ca activated force to beta adrenergic stimulation, the increase in ATPase activity is related to the percentage of V_1 myosin present. Another important property of the system can be demonstrated because of the ability to separate the effect of cAMP on V_1 from the effect on V_3. The ATPase activity of V_3 is irreversibly inhibited by exposure to an alkaline pH that has no significant effect on V_1. As a result, it is possible to measure both total ATPase activity due to the sum of V_1 and V_3 (and the small amount of V_2) and the ATPase activity exclusively of V_1 in the same cell. From the two measurements, neglecting the small contribution of V_2, the ATPase activities of V_1 and V_3 can be estimated. Not only does cAMP increase the ATPase activity of V_1, but it also inhibits the ATPase activity of V_3. The consequence of stimulation with cAMP is not only an increase in the total ATPase activity but also an increase in the expression of activity of the V_1 isozyme of myosin at the expense of the V_3 isozyme.

Maximum velocity of shortening

The third parameter of the contractile system that has been examined for its response to beta adrenergic stimulation is the maximum velocity of shortening (V_{max}). This has been done by the slack length technique, in which the time for the unloaded cells to take up a precisely produced amount of slack length is measured [3]. Bundles of hyperpermeable cells from rat ventricles have been used for these experiments. The response depends on the relative amount of the different isozymes of myosin that are present. In the presence of entirely V_1, beta adrenergic activation does not significantly change V_{max} even though it increases maximum Ca activated force. A mixture of isozymes in cells from euthyroid rats produces a different result. There is an increase in V_{max} but a smaller increase in maximum Ca activated force. In cells from other rats, in which the isozyme content is almost entirely or entirely V_3, the V_{max} is about 30% of the value found in the cells that contain only V_1 and beta adrenergic stimulation causes no change in V_{max} and little or no increase in maximum Ca activated force.

The response of shortening velocity to beta adrenergic stimulation is expecially interesting because of what it indicates about the mechanism of the beta adrenergically dependent regulatory system. Since V_{max} is determined by the properties of the individual force generator [1] the failure of beta adrenergic stimulation to alter V_{max}, even though it increased ATPase activity and maximum Ca activated force, indicates that there is probably a transition between a Ca responsive and a Ca unresponsive state rather than a transition between two Ca responsive states that differ in ATPase activity and force generating capacity. The second important conclusion is that the relative amount of V_1 which is Ca responsive can be increased by beta adrenergic stimulation. This follows from the fact that before beta adrenergic stimulation of cells with a mixture of V_1 and V_3, V_{max} is between values for pure V_1 and pure V_3, but after stimulation it has become equal or nearly equal to the value for pure V_1. It has already been shown that the value for V_{max} in rat cells depends on the relative amounts of V_1 and V_3 and the relation between V_{max} and the percentage of V_1 is a continuous function [2, 10, 11]. Some type of mechanical coupling among myosin cross-bridges must exist. Therefore the increase in V_{max} of cells with a mixture of myosin isozymes after beta adrenergic stimulation almost certainly indicates an increase in the relative amount of V_1 that is Ca responsive.

Functional implications of myosin regulation

These results indicate that in the presence of a mixture of myosin isozymes, beta adrenergic stimulation can increase not only the total number of force generators that are Ca responsive and therefore regulates force, but also the ratio of V_1 to V_3 force generators that are Ca responsive and therefore regulates V_{max}. Since both maximum force and maximum velocity of shortening can be altered, each cell is capable of functioning along many different force-velocity curves depending upon the regulatory state of the myosin isozymes (Fig. 4). This capacity provides the heart with a functional versatility similar to that of skeletal muscle. The latter does it by selecting not which isozymes in each cell are Ca

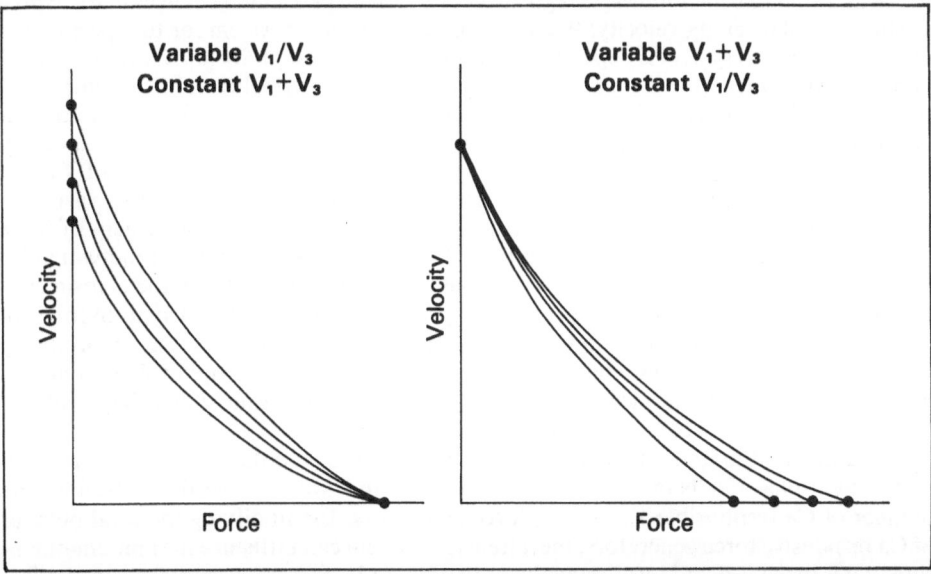

Fig. 4. The effect of variation in either the total number of V_1 and V_3 force generators or their ratio, on the shape of the force-velocity curves of the cells. Change in total number changes maximum force, while change in ratio changes maximum velocity.

responsive but by selecting which cells are stimulated by the nervous system (most skeletal muscles contain cells with different kinds of myosin). It is interesting to note that during most of the active, vigorous part of the life span of the rat, its heart contains substantial amounts of both myosin isozymes, facilitating the regulation of V_{max} [7]. It is only during the first few months and the late portion of the rat's life that the heart is composed primarily or exclusively of one myosin isozyme. During the late period in the animal's life or in the presence of hypothyroidism, when only one isozyme of myosin, V_3, exists in the heart, the regulatory system is not active. Indeed, activity of a system that inhibited V_3 when only V_3 was present would be incompatible with life.

Mechanism of myosin regulation

The factor responsible for the change in the state of V_1 myosin from Ca unresponsive to Ca responsive may be a protein that is bound to an intracellular membrane [6]. Perfusion of the coronary circulation with a cold solution containing 10 mM EGTA and no ATP extracts the regulatory activity from the heart. The activity can be detected by exposing a skinned (or perforated) fibre [12] to the crude extract and then treating the skinned fibre briefly with a low concentration of a non-ionic detergent. The requirement for exposing the skinned fibres to detergent is eliminated by treating the crude extract with detergent and then removing the detergent with adsorbing beads. The combination of dialysis, ultrafiltration and gel filtration on a Sephedex column produces a fraction that contains all of the activity and only a single band on polyacrylamide gels after electrophoresis. The band is due to a 21,000 dalton protein. It is not yet clear whether the same factor that increases the number of V_1 force generators which are Ca responsive is also responsible for the decrease in the number of V_3 force generators that are Ca responsive.

In addition to beta adrenergic regulation of V_1 that can result in changes in force production and shortening velocity, there appears to be another system for the specific regulation of V_3. This system is dependent on alpha adrenergic activity [8]. Treatment of hyperpermeable fibres with the alpha adrenergic agonist, phenylephrine, in concentrations that have little beta activity and in the presence of a beta blocker such as pindolol that has no direct effect on the regulatory systems, produces an increase in maximum Ca activated force. The amplitude of the increase is directly related to the amount of V_3 that is present. The same relationship exists regardless of whether the amount of V_3 varies because of age or thyroid function. There is no indication that the enhancement of force by V_3 force generators is accompanied by a change in the performance of individual V_3 force generators. Regulation of V_3 by this system also disappears when the myosin content of the myocardial cells becomes exclusively V_3, as in fully developed hypothyroidism. The effect of alpha adrenergic stimulation on the contractile proteins is not reproduced by one putative second messenger of the alpha system, inositol triphosphate, but it is mimicked by phorbol ester, which is believed to simulate the action of diacylglycerol, another putative alpha second messenger, on protein kinase C.

In summary, the ventricular cells of rat and rabbit hearts contain systems that are regulated by alpha and beta adrenergic stimulation and are capable of controlling the number of Ca responsible V_1 and V_3 force generators. By modifying the total number of Ca responsive force generators, the adrenergic system can influence the amount of tension that is produced and by altering the ratio of V_1 to V_3 force generators, it can regulate the kinetics of the contraction. The regulatory system controlled by the beta adrenergic system may consist of a protein that is bound, through an anchor peptide, to an intracellular membrane. Beta adrenergic stimulation causes its release not due to phosphorylation of the active protein but presumably through phosphorylation of the anchor peptide. Modification of the V_1 force generator appears to occur as a result of the binding by myosin of the active factor. Preliminary data indicate that the response of these systems to alpha or beta stimulation is sensitive to both the metabolic and the haemodynamic state of the myocardial cells. In this way, the performance of the myocardial cells may be influenced by the response of the regulatory system to the combination of extracellular and intracellular factors.

References

1. Barany M (1967) ATPase activity of myosin correlated with speed of muscle shortening. J Gen Physiol [Suppl] 50:197–218
2. Ebrecht B, Rupp H, Jacob R (1982) Alterations of mechanical parameters in chemically skinned preparations of rat myocardium as a function of isozyme pattern of myosin. Basic Res Cardiol 77:220–234
3. Edman P (1979) The velocity of unloaded shortening and its relation to sarcomere length and isometric force in vertebrate muscle fibres. J Physiol 291:143–159
4. Hoh J, McGrath P, Hale P (1978) Electrophoretic analysis of multiple forms of rat cardiac myosin: Effects of hypophysectomy and thyroxine replacement. J Mol Cell Cardiol 10:1052–1076
5. Horowits R, Winegrad S (1983) Cholinergic regulation of calcium sensitivity in cardiac muscle. J Mol Cell Cardiol 15:277–280
6. Lin LE, Winegrad S (1986) Isolation of a factor that modulates contractility of rat heart. Biophys J 49:450a
7. Lompre A, Mercadier J, Wisnensky C, Bouveret P, Pantaloni C, D'Albis, Schwartz K (1984) Species and age dependent changes in the relative amounts of cardiac myosin isozymes in mammals. Dev Biol 84:286–290
8. McClellan G, Tucker M, Winegrad S (1987) Relationship between adrenergic stimulation and myosin isozyme on the force production of hyperpermeable rat cardiac cells. J Gen Physiol (in press)
9a. McClellan G, Winegrad S (1978) The regulation of calcium sensitivity of the contractile system in mammalian cardiac muscle. J Gen Physiol 72:283–295
9b. McClellan G, Winegrad S (1980) Cycle nucleotide regulation of the contractile proteins in mammalian cardiac muscle. J Gen Physiol 75:283–295
10. Pagani E, Julian F (1984) Rabbit papillary muscle myosin isozymes and the velocity of shortening. Circ Res 54:586–594
11. Schwartz K, Lecarpentier YC, Martin JL, Lompre A, Mercadier J, Swynghedauw B (1981) Myosin isozymic distribution correlates with speed of myocardial contraction. J Mol Cell Cardiol 13:1071–1075
12. Weisberg A, McClellan G, Tucker M, Lin LE, Winegrad S (1983) Regulation of calcium sensitivity in perforated mammalian cardiac cells. J Gen Physiol 81:195–211
13. Weisberg A, Winegrad S, Tucker M, McClellan G (1982) Histochemical detection of specific isozymes in rat ventricular cells. Circ Res 51:802–809
14. Winegrad S, Weisberg A, Lin LE, McClellan G (1986) Adrenergic regulation of myosin adenosin triphosphetase activity. Circ Res 58:83–95

Author's address:

S. Winegrad, Department of Physiology, School of Medicine, University of Pennsylvania, Philadelphia, PA 19104-6085, U.S.A.

Implications of myocardial transformation for cardiac energetics*

G. Kissling, H. Rupp, and R. Jacob

Physiologisches Institut II, Universität Tübingen, F.R.G.

Summary

The influence of isoenzyme pattern of myosin on cardiac energetics was investigated in a modified in situ heart-lung preparation in the rat. Chronic pressure load (spontaneous hypertension, aortic stenosis, Goldblatt hypertension), intermittent feeding, and swim-training elicited redistribution in the concentration of α chains of myosin ranging from 18 to 94%. The influence of isoenzyme pattern of myosin on cardiac energetics could be quantitatively assessed by extrapolation of the regression line of oxygen and substrate consumption related to tension time index. Fast myocardium with 100% α chains had an ATP and oxygen consumption which exceeded that of slow myocardium with 0% α chains by about 60%. This corresponds well to the state of activity of myofibrillar ATPase of fast myocardium which also exceeds that of slow myocardium by about 50%. Furthermore it could be shown that acute increase in the ATPase activity depends on the isoenzyme pattern of myosin. Under the influence of catecholamines the oxygen consumption related to tension time index increased by 30–40% in fast myocardium, whereby in a myocardium with 40% α chains no increase in oxygen consumption per unit tension time index was observed, when catecholamines were applied.

Introduction

The heart has the potential to adapt itself to chronic pressure and volume overload, so as to cope economically with increased load. Adaptational processes occur at the level of the whole organ, as well as at the cellular and even molecular level. For example, chronic pressure overload does not only lead to hypertrophy of the individual myocardial cell with increase in concentration of the contractile proteins but also to concentric hypertrophy of the whole ventricle which – according to the La Place law – makes possible a more favourable conversion of developed tension into pressure. At the same time changes in electromechanical coupling and isoenzyme pattern of myosin also contribute to a more economical functioning of the heart.

In recent years we have focused on the influence of the isoenzyme pattern of myosin on cardiac energetics [5, 9, 10, 11, 13]. The main results of these investigations are summarized in this paper.

The investigations were done on the rat heart since, in this species, three types of isoenzymes of myosin (VM-1, VM-2, and VM-3) with different ATPase activity can be distinguished by electrophoresis [2, 14, 15]. The percentage proportion of the various isoenzymes changes with age as well as under haemodynamic and hormonal influences

---

* Supported by the Deutsche Forschungsgemeinschaft.

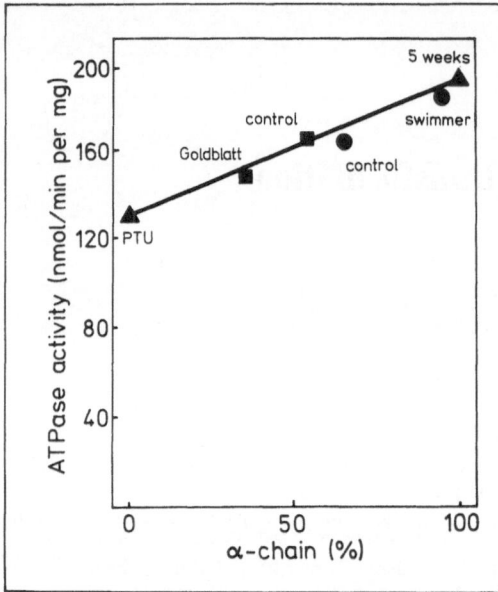

Fig. 1. Relationship between myofibrillar ATPase activity at saturating Ca^{++} concentrations and the relative α chain content of myosin. The differences in isoenzyme pattern were induced by changes in the thyroid state (PTU) as well as by alterations in haemodynamic load (Goldblatt hypertension and swimming-training). (From Rupp [15] with permission of Steinkopff Verlag.)

[2, 5, 10, 14]. The isoenzymes differ in the structure of their heavy chains of myosin [3]. Both VM-1 and VM-3 are homodimers with 2 α and 2 β chains respectively. VM-2 is a heterodimer, having 1 α and 1 β chain. Given the relative proportion of all 3 isoenzymes, the fraction of α and β chains can then be calculated.

The relationship between myofibrillar ATPase activity and the isoenzyme pattern of myosin is shown in Fig. 1. Chronic pressure overload leads to a transformation towards a slower myocardium with reduced concentration of α chains and with reduced ATPase activity. Swimming training causes a transformation towards a faster myocardium with increased concentration of α chains and with increased ATPase activity. In the myocardium of young rats which have almost 100% α chains, the ATPase activity is higher by about 50% than in the myocardium of thyreostatically-treated rats which have hardly any α chains.

The aim of our studies was to find out how far these changes in isoenzyme pattern affect myocardial energetics. For this purpose, left ventricular mechanics, oxygen consumption and substrate uptake were measured in a modified in situ heart-lung preparation using Wistar rats of different ages, rats with chronic pressure overload due to spontaneous hypertension, aortic stenosis or renal hypertension, and swimming-trained rats. The concentration of α chains of myosin in these animals was in the range of 18 to 94%.

Methods

The modified in situ heart-lung preparation has been described in detail in previous publications [8, 9, 10, 11, 13]. Thus we only give a brief description here. Thoracotomy was performed under urethane anaesthesia (1.2–1.5 g/kg body wt.). After removing the pericardium, an electromagnetic flow probe was placed around the pulmonary artery trunk. The left ventricle was pierced with a cannula for pressure measurement and withdrawal of arterial blood samples. Subsequently the superior and inferior venae cavae and the ascending aorta were ligated. The right ventricle – in some experiments the vena cava inferior – was pierced for pressure measurement and withdrawal of venous blood samples. The ligatures cut off the entire systemic circulation with the exception of coronary circulation, whereas pulmonary circulation remained intact.

The following haemodynamic parameters were registered in all animals: 1) pressure amplitude, end diastolic pressure and dP/dt in the left ventricle, 2) pulmonary flow, which under steady-state conditions is identical to cardiac output and coronary perfusion. In some experiments also the pressure in the aortic root, as well as central venous pressure or right ventricular pressure were measured.

At the end of each experiment the end diastolic pressure-volume relationships were recorded as relaxation curves [7, 12]. On the basis of this end diastolic pressure-volume relationships for each beat a corresponding inner ventricular volume could be ascertained.

The oxygen consumption of the preparation was determined by two alternative methods. In most experiments [8, 9, 10, 11], the oxygen consumption has been derived from the gradient of the arterio-venous oxygen concentration and from the cardiac output measured with the flow probe. This experimental procedure allows only a few measurements per animal since the small blood volume in the preparation limits the number of blood samples. In our most recent studies [13] therefore, we determined oxygen consumption on the basis of the minute respiratory volume and oxygen concentration gradient between inspired and expired air.

Besides the oxygen consumption, this preparation also allows measurement of the substrate consumption of the heart [9]. The blood volume in the model was determined using the dye-dilution method. A blood sample was taken prior to ligation and the blood or plasma concentration of glucose, lactate, triglycerides and free glycerine were measured. 30 minutes after ligation the substrate concentrations were determined again. The consumption of each substrate could be calculated from the concentration difference between the initial and final blood samples and the blood volume.

The oxygen consumption and substrate uptake was related to tension time index, peak tension and maximum rate of tension development. These tension parameters were calculated from measured pressure and volume values, assuming a thick walled spherical model [13, 16]. Finally the isoenzyme pattern of myosin was determined for each individual heart using polyacrylamide gel electrophoresis [2, 14].

Statistical analysis of the data was performed using Student's t-test. Differences were considered to be significant when $p < 0.05$. When deviation is given, it represents standard error ($\bar{x} \pm s\bar{x}$).

Myosin isoenzyme pattern and myocardial substrate uptake

Firstly, we would like to summarize the results of our substrate measurements. The consumption of glucose, lactate, and fatty acids was determined in Goldblatt rats and age-matched controls, in rats with aortic stenosis and age-matched controls, and in intermittently fed rats and age-matched swimming-trained rats. In all experiments, the consumption of glucose and lactate was greater the higher the initial available amounts. In contrast, fat was burnt independently of the initial amount. The corresponding oxygen- and ATP equivalent was calculated for every substrate. It could be shown that with increasing total ATP consumption more ATP was derived from the individual substrates. However, at low levels of ATP consumption only carbohydrates were utilized and almost no fat was burnt. Only with higher energy requirement was fat utilized in increasing amounts. This indicates that under our experimental conditions, i.e. under a high afterload, the heart burns preferably carbohydrates and only utilizes fat at higher energy demands. In our experiments approximately 80% of the energy requirement was supplied by carbohydrates and only 20% by fat. In agreement with our results, Keul et al. [6] measured a 20.8% portion of fat in the energy supply of the heart in working human athletes.

The ATP consumption of the heart depends mainly on the mechanical load. In each case, the calculated ATP consumption per gram left ventricular weight and beat related to tension time index, was less in the animals with slower myocardium than in the corresponding controls with faster myocardium. Unfortunately we have determined the isoenzyme pattern of myosin of only the intermittently fed animals and the age-matched swimming-trained rats.

Figure 2 shows the ATP consumption related either to tension time index (a) or to peak tension (b) against the isoenzyme pattern of myosin. Linear extrapolation shows

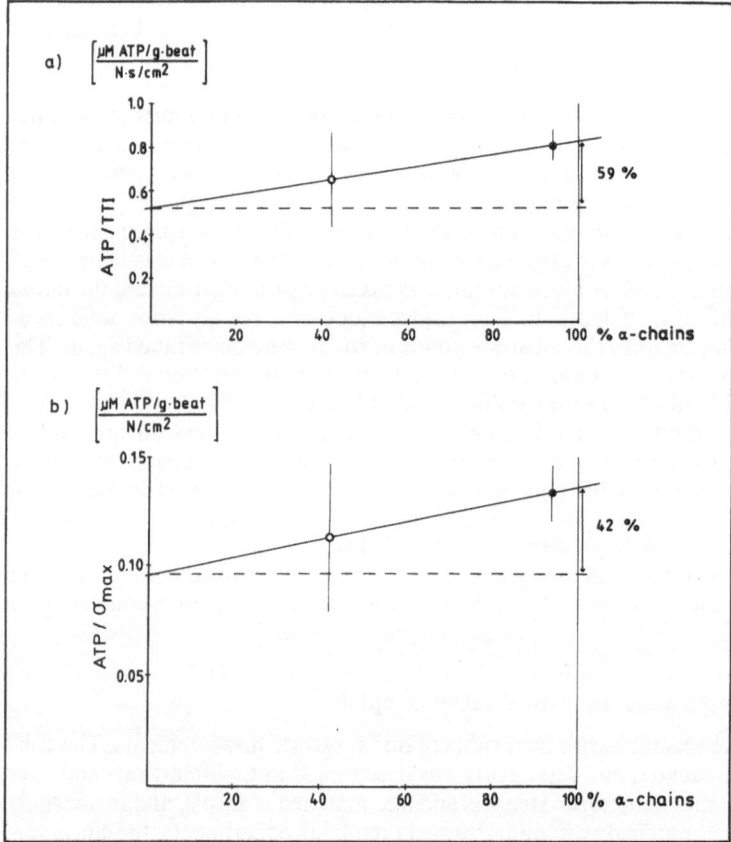

Fig. 2. Influence of the isoenzyme pattern of myosin on the myocardial substrate uptake; intermittently fed rats with 42.5% α chains (○); age-matched swimming trained rats with 94% α chains (●). a) Intermittently·fed rats' ATP consumption per gram left ventricular weight and beat, related to TTI, measured by 0.651 μM ATP/g·beat, and in the swimming-trained animals by 0.806 μM ATP/g·beat. b) ATP consumption related to peak tension was 0.113 μM ATP/g·beat in the intermittently fed rats, and 0.134 μM ATP/g·beat in the swimming-trained rats.

that a fast myocardium with 100% α chains requires 59% more ATP than a slow myocardium with 0% α chains, when the ATP consumption is related to tension time index (Fig. 2a). The ATP consumption related to peak tension increases by 42% from 0 to 100% α chains (Fig. 2b). Both values lie in the same range as the change in ATPase activity shown in Fig. 1.

In the experiments described so far, we assumed that all substrates utilized by the heart originate from the blood. To prove this, in 10 experiments, substrate uptake and oxygen consumption were measured simultaneously. Despite some deviation in the individual experiment, a systematic deviation between the oxygen consumption calculated from substrate uptake and the oxygen consumption actually measured could be excluded. On the basis of these experiments it can be assumed that in our heart-lung preparation most of the substrate utilized by the heart does, in fact, originate from the blood.

Myosin isoenzyme pattern and myocardial oxygen consumption

In this section the results of our oxygen measurements will be summarized. Figure 3 shows the results of a representative experiment in which the oxygen consumption was determined from respiratory minute volume and the oxygen concentration gradient between inspired and expired air. The oxygen consumption per gram ventricular weight and

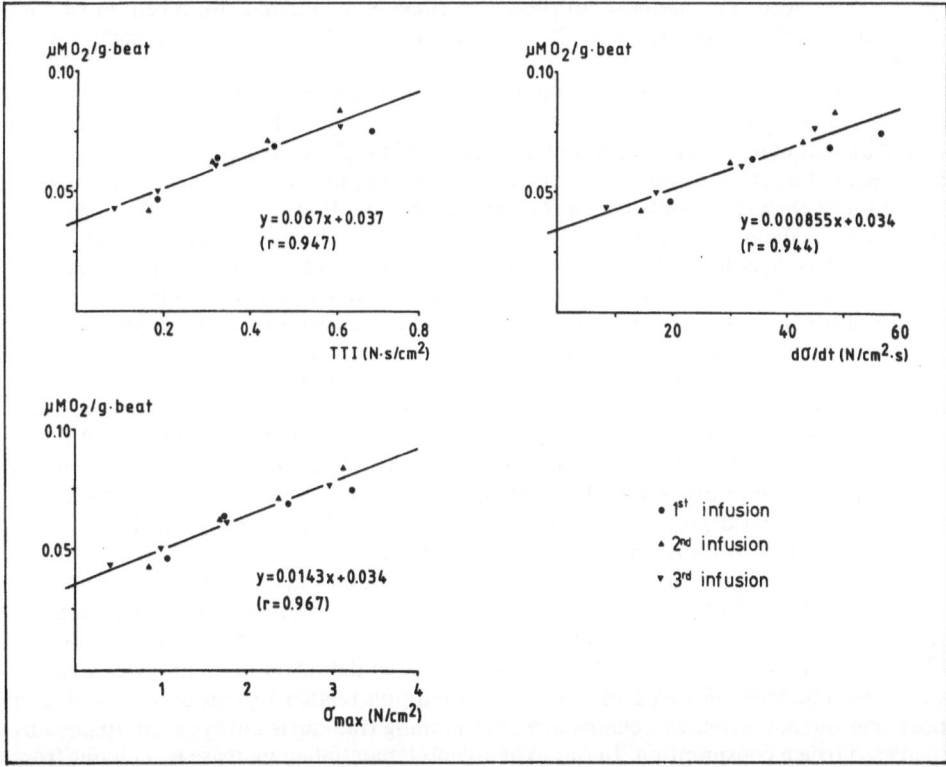

Fig. 3. The oxygen consumption per g left ventricular weight and beat measured at various end-diastolic volumes as a function of TTI, $d\sigma/dt$ and σ_{max}, respectively. (From Kissling and Rupp [13] with permission of Steinkopff Verlag.)

beat is plotted as a function of tension time index, peak tension and maximum rate of tension development. The mechanical activity of the heart was varied by slow blood infusions or blood letting. There is a stringent and linear correlation for all three mechanical parameters. The point of interception of the regression line on the ordinate corresponds to the oxygen consumption, independent of the mechanical activity of the left ventricle. This value includes the resting oxygen consumption of the left ventricle as well as the total oxygen consumption of the other heart compartments and of the lungs. All three relationships give basically the same value. The slope of the regression line gives a measure of the oxygen consumption related to the respective mechanical parameters of the left ventricle. Furthermore, from Fig. 3 it can be seen that the preparation used remains energetically and mechanically stable over a long period of time. The measurements were done during three consecutive infusions, each lasting 30 minutes. Even the values measured after 90 minutes lie on the same regression line as those taken during the initial infusion.

The mechanical parameter which is crucial for oxygen consumption is tension time index. The linear relationship between oxygen consumption and peak tension or the maximum rate of tension development is due to the fact that these parameters also have a linear correlation with tension time index [13].

In those experiments where the oxygen consumption was measured via blood parameters, the amount of oxygen consumed independently of left ventricular activity could not be clearly defined. Furthermore, in these experiments the animals were ventilated with a gas mixture of 95% oxygen and 5% CO_2, and in some the β adrenergic receptors were blocked.

Figure 4 shows the oxygen consumption measured under different experimental conditions and by different methods, related to tension time index (Fig. 4a), and peak tension (Fig. 4b), and plotted as a function of α chains. At first glance there does not seem to be an unequivocal relationship between oxygen consumption and isoenzyme pattern of myosin. For the experiments where the oxygen consumption was measured via respiration there are definitely smaller values than in those experiments where the oxygen consumption was determined from blood parameters. However, when the values that were measured under identical experimental conditions are paired (see dashed lines in Fig. 4a and b), it becomes obvious that in each case the animals with faster myocardium have a higher oxygen consumption related to either tension time index or to peak tension. Furthermore, substantially lower oxygen consumption related to tension time index or peak tension is found in Goldblatt rats as compared to age-matched controls, although the change in concentration of α chains is minimal. As a rule, both the myofibrillar ATPase activity and the maximum unloaded shortening velocity depend to the same extent on isoenzyme pattern of myosin [4]. Interestingly, the maximum shortening velocity is also reduced to a greater degree in the later stages of Goldblatt hypertension which can not be attributed to the myosin isoenzyme pattern alone [1].

The differences in absolute oxygen consumption found due to diverse methods can be eliminated by plotting the changes in oxygen consumption against the changes in isoenzyme pattern (Figs. 4c and 4d). Chronic pressure load arising from aortic stenosis, Goldblatt hypertension, or spontaneous hypertension definitely leads to fewer α chains with a simultaneous decrease in oxygen consumption related to tension time index or peak tension. An increases α chains due to swimming training is always accompanied by increased oxgen consumption. In our experimental material an increase in α chains from 0 to 100% results in an average increase in oxygen consumption per unit tension time index of about 70% (Fig. 4c), and of about 38% when related to peak tension (Fig. 4d). When the values obtained from measurements on Goldblatt rats are excluded from the

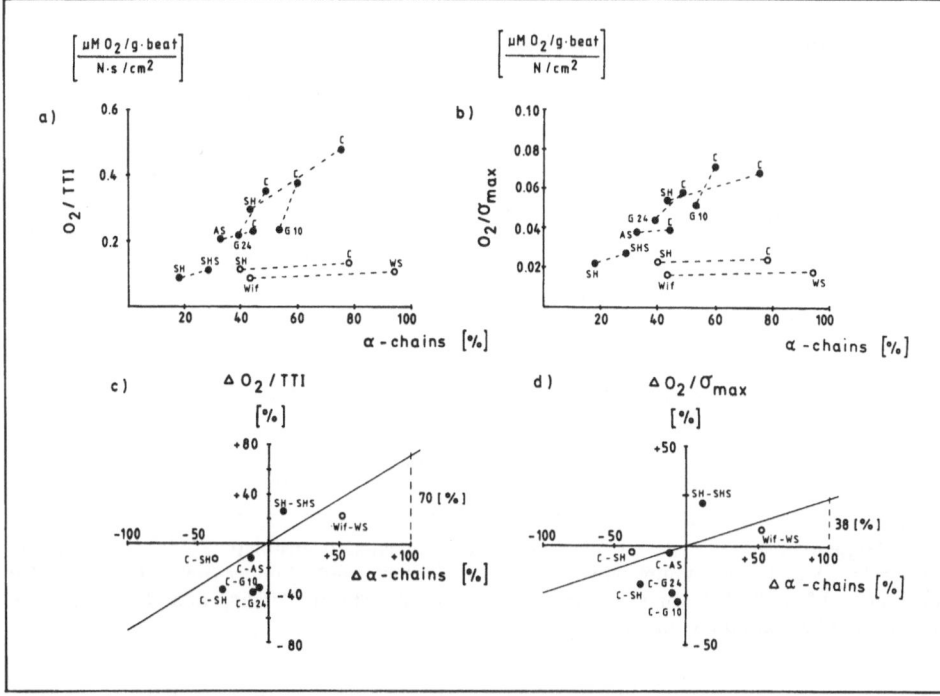

Fig. 4. Myocardial oxygen consumption related to tension time index (a) and to peak tension (b), as a function of α chains. Oxygen consumption determined from blood parameters (●); and via respiration (○). Spontaneous hypertension (SH); aortic stenosis (AS); different stages of Goldblatt hypertension (G); age-matched controls (C); swimming-trained spontaneously hypertensive rats (SHS); intermittently fed rats (W_{if}); age-matched, swimming-trained rats (W_s). Fig. 4c) and d) changes in oxygen consumption against changes in isoenzyme pattern.

correlations, the increase in oxygen consumption related to tension time index is 59%, and related to peak tension 32%. Thus these values again lie in the same range as the changes in ATPase activity.

The influence of myosin isoenzyme pattern on acute enhancement of myofibrillar ATPase activity

In the course of our experiments, we became interested in finding out whether alterations in isoenzyme pattern of myosin also have an effect on the acute enhancement of ATPase activity. On the basis of the close correlation between ATPase activity and oxygen consumption we attempted to draw conclusions regarding acute enhancement of ATPase activity from measurements of oxygen consumption under the influence of catecholamines.

Following administration of catecholamines, the heart develops higher tension within a shorter period of time compared to control conditions. At the same time, there is an increase in the maximum rate of tension development and relaxation, whereas tension

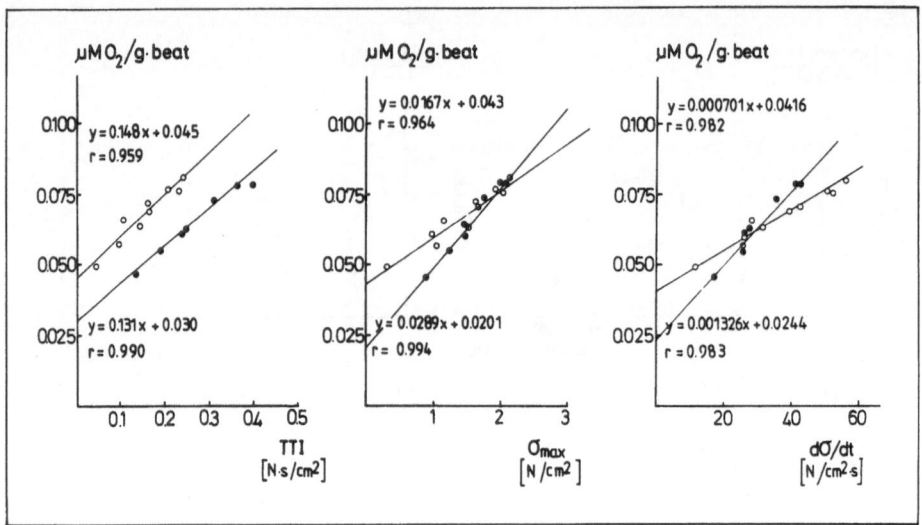

Fig. 5. Oxygen consumption per g left ventricle and beat, related to tension time index (TTI), to peak tension (σ_{max}), and to maximum rate of tension development ($d\sigma/dt$) under control conditions (●) and under 0.5 mg/kg b.w. orciprenaline (○), respectively. (From Kissling and Rupp [13] with permission of Steinkopff Verlag.)

time index remains constant. The quantitative degree of these changes, however, differs in fast and slow myocardium. Following application of catecholamines, at a given heart rate and peak tension the duration of contraction and time to peak tension are significantly longer in slow myocardium than in fast myocardium. Concurrently, maximum rate of tension development and relaxation are significantly lower in slow myocardium. This indicates that in slow myocardium catecholamines have similar effects on the mechanical parameters as increased Ca^{++} concentration.

Figure 5 shows the effect of 0.5 mg/kg body wt. orciprenaline on myocardial oxygen consumption. The results are derived from a representative experiment on a control animal with fast myocardium. The point of interception of the regression line with the ordinate corresponds to the oxygen consumption independent of left ventricular activity. The slope of the regression line corresponds to the left ventricular oxygen consumption related to the respective mechanical parameter. The oxygen consumption independent of left ventricular activity increases under the effect of catecholamines which can be seen in all three diagrams. Whereas the slope of the regression line of oxygen consumption related to tension time index, shows a moderate yet significant increase, that of oxygen consumption related to peak tension and maximum rate of tension development decreases significantly.

Similar measurements on rats with slow myocardium do not show any increase in oxygen consumption related to tension time index following application of catecholamines. Figure 6 summarizes the results of 2 series of experiments done on rats with slow and fast myocardium, respectively. The oxygen consumption related to tension time index before and after application of catecholamines is shown. Whereas in the fast myocardium, the oxygen consumption related to tension time index increases significantly under the influence of catecholamines, the oxygen consumption remains constant in the slow myocardium. Thus we could prove indirectly that the acute increase in ATPase activity is

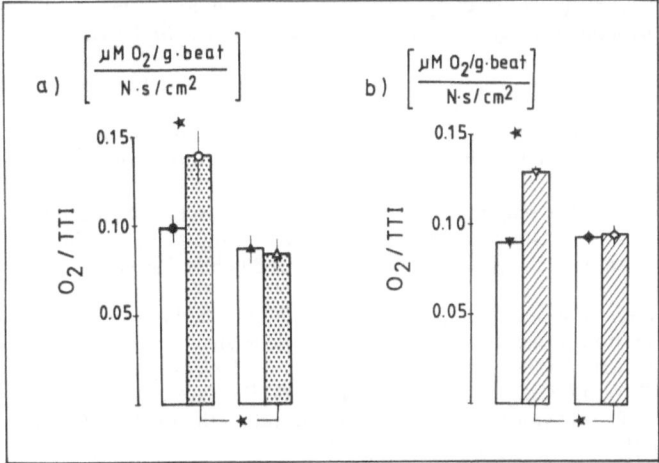

Fig. 6. Influence of catecholamines on oxygen consumption related to tension time index. a) Effect of 1 mg/kg isoproterenol on swimming-trained rats (● ○) with 94% α chains and on intermittently fed rats (▲ △) with 42.5% α chains. b) Effect of 0.5 mg/kg orciprenaline on 12 week-old Wistar rats (▼ ▽) with 78% α chains and on 21 week-old spontaneously hypertensive rats (◆ ◇) with 40.3% α chains. ⋆ p < 0.0005.

influenced by the isoenzyme pattern of myosin. Also Winegrad [17] found a difference in catecholamine sensitivity of fast and slow myocardium.

Conclusions

To summarize our results we would like to draw the following conclusions:

1. Redistribution of isoenzyme pattern of myosin causes a change in myofibrillar ATPase activity.
2. Changes in myofibrillar ATPase activity, measured in vitro, influence myocardial energetics measured in vivo. Following a transformation towards a slow myocardium, a decrease in oxygen and substrate consumption is demonstrable even in the heart in situ. Conversely, transformation to a fast myocardium leads to an increase in the energy turnover of the heart.
3. The degree to which the oxygen consumption of the myocardium increases under the influence of catecholamines correlates with isoenzyme pattern of myosin. We conclude from this that the acute increase in myofibrillar ATPase activity is also influenced by the isoenzyme pattern.
4. The redistribution of the isoenzyme pattern of myosin during chronic pressure or volume overload should be regarded as an adaptational process at the molecular level. It contributes to the heart's ability to cope with the increased load in the most economical way.

References

1. Ebrecht G, Rupp H, Jacob R (1982) Alterations of mechanical parameters in chemically skinned preparations of rat myocardium as a function of isoenzyme pattern of myosin. Basic Res Cardiol 77:220–234

2. Hoh JFY, McGrath PA, Hale PT (1978) Electrophoretic analysis of multiple forms of rat cardiac myosin: Effects of hypophysectomy and thyroxine replacement. J Mol Cell Cardiol 10:1053–1076
3. Hoh JFY, Yeoh GPS, Thomas MAW, Higginbottom L (1979) Structural differences in the heavy chains of rat ventricular myosin isoenzymes. FEBS Lett 97:330–334
4. Jacob R, Rupp H, Ebrecht G, Gülch RW, Kissling G (1981) Chronic reactions of myocardium at the myofibrillar level. The functional significance of the isoenzyme pattern of myosin. Proc Jap Soc Cardiac Metabolism 4:1–28
5. Jacob R, Ebrecht G, Holubarsch Ch, Rupp H, Kissling G (1983) Mechanics and energetics in cardiac hypertrophy as related to the isoenzyme pattern of myosin. In: Alpert NR (ed) Perspectives in cardiovascular research, Vol 7: Myocardial hypertrophy and failure. Raven Press, New York, pp 553–569
6. Keul J, Doll E, Steim H, Fleer U, Reindell H (1965) Über den Stoffwechsel des menschlichen Herzens. III Der oxydative Stoffwechsel des menschlichen Herzens unter verschiedenen Arbeitsbedingungen. Pflügers Arch 282:43–53
7. Kissling G, Gassenmaier T, Wendt-Gallitelli MF, Jacob R (1977) Pressure-volume relations, elastic modulus and contractile behaviour of the left ventricle of rats with Goldblatt II hypertension. Pflügers Arch 369:213–221
8. Kissling G, Ziegler Ch (1978) A new in situ heart preparation for measurement of oxygen consumption under isovolumic conditions. Pflügers Arch 373:R8
9. Kissling G (1980) Oxygen consumption and substrate uptake of the hypertrophied rat heart in situ. Basic Res Cardiol 75:185–192
10. Kissling G, Rupp H, Malloy L, Jacob R (1982) Alterations in cardiac oxygen consumption under chronic pressure overload. Significance of the isoenzyme pattern of myosin. Basic Res Cardiol 77:255–269
11. Kissling G, Malloy L, Rupp H (1983) Energetics of the rat heart in chronic pressure overload. In: Jacob R, Gülch RW, Kissling G (eds) Cardiac adaptation to hemodynamic overload, training and stress. Steinkopff Verlag, Darmstadt, pp 167–173
12. Kissling G, Takeda N, Vogt M (1985) Left ventricular end-systolic pressure-volume relationships as a measure of ventricular performance. Basic Res Cardiol 80:594–607
13. Kissling G, Rupp H (1986) The influence of myosin isoenzyme pattern on increase in myocardial oxygen consumption induced by catecholamines. Basic Res Cardiol [Suppl 1] 81:103–115
14. Rupp H (1981) The adaptive changes in the isoenzyme pattern of myosin from hypertrophied rat myocardium as a result of pressure overload and physical training. Basic Res Cardiol 76:79–88
15. Rupp H (1982) Polymorphic myosin as the common determinant of myofibrillar ATPase in different haemodynamic and thyroid states. Basic Res Cardiol 77:34–46
16. Sandler H, Dodge HT (1963) Left ventricular tension and stress in man. Circ Res 13:91–104
17. Winegrad S, McClellan G, Tucker M, Lin L (1983) Cyclic AMP regulation of myosin isoenzymes in mammalian cardiac muscle. J Gen Physiol 81:749–765

Authors' address:

Prof. Dr. G. Kissling, Physiologisches Institut II, Universität Tübingen, Gmelinstr. 5, D-7400 Tübingen, F.R.G.

Myocardial energetics and diastolic dimensions of the heart in experimental hypertension

P. Friberg

Department of Physiology, University of Göteborg, Göteborg, Sweden

Summary

The present study examined changes in left ventricular design and function in spontaneously hypertensive rats (SHR) and normotensive Wistar Kyoto rats (WKY), and in SHR exposed to voluntary physical exercise in running wheels (R-SHR) and their respective sedentary controls (C-SHR). End-diastolic volumes were obtained in vitro by determining the pressure-volume relationships of isolated hearts arrested in diastole. Cardiac function and myocardial oxygen consumption were also assessed in vitro by means of an antegrade working heart perfusion technique.

Compared with WKY and C-SHR respectively, ordinary SHR and R-SHR had increased end-diastolic volumes, whereas the ratios between wall thickness and internal radius were relatively unchanged. Maximal cardiac performance was elevated in the structurally enlarged SHR heart compared with WKY, whereas it remained unchanged after chronic physical exercise. Long-term voluntary running in SHR caused an elevation of cardiac output due to an increased stroke volume, while arterial pressure was unaltered.

The stimulus for the cardiac redesign to a structurally enlarged heart can probably best be explained by a chronic elevation in cardiac filling. Hence, enlarged left ventricles can then produce higher stroke volumes for given degrees of myocardial fibre shortenings. Thus, despite structurally enlarged left ventricles in SHR and in R-SHR and also increased arterial pressure in SHR compared with WKY (thereby elevating systolic wall stress), cardiac function was maintained and even augmented, which was not associated with an increase in total myocardial oxygen consumption.

Introduction

Sustained alterations of pressure and/or volume load on the heart markedly influence structural and functional characteristics of the myocardium and of the coronary vasculature. In pregnancy, in both women and animals, which represents a sustained natural volume overload, the internal radius of the LV increased (eccentric hypertrophy) and myocardial performance was maintained [1, 2, 3]. Likewise increased myocardial performance has also been found in experimental hyperthyroidism [4, 5], a condition in which an increased LV internal radius was also observed.

The spontaneously hypertensive rat (SHR) has been shown to develop LV hypertrophy in response to the gradual elevation of mean arterial pressure. Some authors have proposed that LV hypertrophy in SHR may be a degenerative phenomenon [6], while other groups have suggested that LV hypertrophy preserves or even improves cardiac function, i.e. it represents a physiological adaptation to the raised arterial pressure [7, 8, 9, 10]. Most of these studies examined maximal LV performance at a high aortic pressure, which meant that a high coronary flow and thereby a good nutritional supply to the heart prevailed. However, when cardiac function was analysed in SHR and in renal hypertensive rats at a fairly low aortic pressure, i.e. at a low coronary perfusion pressure, it was found to be markedly reduced compared with their respective normotensive controls [10, 11, 12].

The influence of physical exercise on blood pressure and other cardiovascular parameters has been extensively studied and conflicting results prevail [13]. Regarding the effects of exercise in SHR, there are reports of both unaltered [14, 15] and lowered blood pressure [16, 17]. In all earlier experimental studies on exercise and cardiovascular dynamics, the animals were forced to exercise. Therefore, in the present study we used a system with voluntary running in wheels [18].

The present study was undertaken to further elucidate the relationship between maximal cardiac performance, coronary flow, myocardial oxygen consumption, cardiac geometry and wall stress patterns in SHR and Wistar Kyoto normotensive rats (WKY) and also to show how these parameters, as well as central haemodynamics, were influenced by voluntary physical exercise. Parts of the results have been previously reported [10, 19].

Methods

Animals

The experiments were performed on rats obtained from Möllegaards Breeding centre Ltd, Denmark. The study consisted of the following groups:
1. 16-week-old SHR and WKY weighing 332 ± 3 g and 342 ± 4 g, respectively, were used for antegrade heart perfusions. Aged-matched SHR and WHY weighing 291 ± 4 g and 299 ± 2 g, repectively, were used for retrograde perfusions.
2. Physical exercise (voluntarily running) was introduced to SHR at the age of 9 weeks. Six to 7 weeks later cardiac function was examined in one group of running SHR (body weight 311 ± 13 g) and non-running SHR (body weight 328 ± 5 g); in another group of exercised- and non-exercised SHR, weighing 323 ± 8 g and 352 ± 10 g respectively, cardiac dimensions were investigated.
3. Coronary flow, coronary vascular resistance and myocardial oxygen consumption were also analysed in 45-week-old SHR and WKY, weighing 402 ± 8 g and 420 ± 5 g, respectively.

Systolic blood pressure (SBP) and heart rate were measured with a tailcuff technique and mean arterial pressure was measured by means of tail artery cannulation. Cardiac output was determined by a cardiogreen dye dilution technique in microscale [20]. The physical exercise procedure consisted of spontaneous running in wheels and was completely voluntary with no superimposed psychogenic stress-element (cf. ref. 18). After a control period of 1 week, the 9-week-old SHR were then running for 6 weeks.

Antegrade heart perfusion technique

This technique has recently been described in detail [10, 19]. The heart was excised and mounted on an aortic cannula and somewhat later the left atrium was also cannulated.

The perfusion system consisted of a 130 cm long oxygenating chamber with a micropore filter at the bottom, from which a peristaltic pump delivered perfusate to the heart via two atrial bubble traps, while the rest of the fluid was pumped to the top of the chamber. Left atrial pressure could be varied by changing the height of a water column above heart level. A Starling resistor was used in order to change the resistance against which the heart pumped. By adjusting this resistor, aortic diastolic pressure could be maintained at any desired level.

Aortic output and coronary flow, i.e., cardiac output, were determined by weighing. Coronary flow was determined by a drop counter before entering the weighing system. The oxygen content of the venous coronary effluent was also measured by a Clark oxygen electrode; thereby, myocardial oxygen consumption (MVO_2) could be calculated.

During the experiment, the hearts were paced at either 300 or 350 beats/min. The contracting isolated ventricles were exposed to mainly three different levels of standardized aortic diastolic pressure levels: 40, 70 and 100 mm Hg respectively. At each level of these pressure loads, left atrial pressure was varied between 2–4 and 15 mm Hg while recordings of cardiac output, coronary flow and oxygen tension of the venous coronary effluent were performed.

Retrograde heart perfusion

This procedure has been described thoroughly by Friberg et al. [3]. The heart was mounted on a aortic cannula through which the coronaries were perfused with oxygenated Tyrode solution. The perfusates used (with and without calcium) were kept in separate water-jacketed glass containers, placed above the heart so as to give a perfusion pressure of about 40 mm Hg.

After starting the retrograde perfusion, contractions reappeared within a few seconds. When all blood had been washed out, the perfusion medium was changed to the calcium-free one, whereupon the heart stopped in diastole. A triple-lumen cannula was now inserted into the non-beating LV via the left atrium and ligated. This catheter consisted of three 0.5 mm wide steel tubes covered by an outer steel tube. Through one of the tubes, the LV could be expanded in a stepwise fashion by repeated injections (42 µl) of Tyrode solution containing Evans blue. Left ventricular pressure was recorded via another tube. The third tube was used for wash-out, whereby the LV could be passively emptied for determination of its initial volume.

Experimental protocol

To provide nutrition for the heart during insertion of the triple-lumen catheter, the coronaries were perfused throughout with oxygenated calcium-free Tyrode and also subsequently to assure complete diastolic arrest. The perfusion was then stopped and a new ligature tied around the aortic root to avoid leakage from the ventricle into the coronaries during the stepwise expansion, for estimation of the pressure-volume (P/V) curves.

From the achieved "basal" pressure level the LV was now rapidly expanded by stepwise injections of 80 µM Evans blue, up to an intraventricular pressure of 15–20 mm Hg. Then the wash-out tube was opened, the ventricle passively emptied and the intraventricular pressure again adjusted to the same "basal" pressure as before. A syringe filled with 10 ml 37 °C calcium-free Tyrode was then connected to the wash-out tube to flush out the remaining Evans blue solution, the concentration of which was later measured spectrophotometrically (i.e., the ventricular content at "basal" pressure).

The aortic root ligature was then removed, whereupon perfusion with the calcium-containing Tyrode solution was started. Soon afterwards, the heart started to beat regularly again, showing that no rigor mortis and consequent abnormal ventricular stiffness had occurred.

Calculations and statistics

Assuming a spheric luminal shape LV diastolic external (r_e) and internal (r_i) radii, wall thickness (w) and the ratio between wall thickness and internal radius (w/r_i) were then calculated from the end-diastolic volume (EDV) at an end-diastolic pressure (EDP) of 7.5 mm Hg and from LV muscle weight, which approximately equals the myocardial volume.

Comparisons were performed using Student's t-test for unpaired data and a value of $p < 0.05$ was considered statistically significant. Values are presented as mean ± S.E.

Results

Cardiac performance in SHR and WKY

Mean arterial pressure was elevated in SHR versus WKY (176 ± 3 mm Hg and 123 ± 3 mm Hg, $p < 0.001$). Relative LV weight was also significantly increased in SHR, being 2.56 ± 0.01 mg/g compared with the normotensive WKY (2.16 ± 0.01 mg/g, $p < 0.001$).

During cardiac pacing the hearts were examined at different aortic pressures. Peak cardiac work, a determinant of maximal LV function, was calculated as aortic pressure minus left atrial pressure times peak stroke volume and heart rate. Peak cardiac work in SHR was markedly elevated compared with WKY (3580 ± 114 versus 2810 ± 125 cm³ × kPa × min⁻¹ × kg⁻¹, $p < 0.001$). This indicates that when SHR is challenged at a high pressure load, the beneficial effect of the hypertrophied LV will improve cardiac performance.

Coronary flow and myocardial oxygen utilization in SHR and WKY

All coronary flow measurements were performed during maximal vasodilatation. The coronary flow levels were linearly related to increasing levels of aortic diastolic pressure

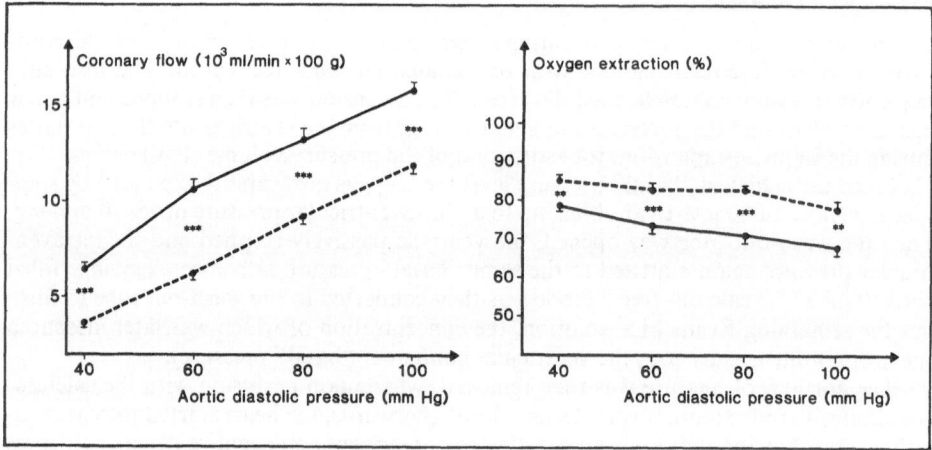

Fig. 1. A. Coronary flow expressed per gram dry heart weight at different aortic diastolic pressures in SHR (dotted line) and WKY (solid line). B. Myocardial oxygen extraction at standardized pressure loads in SHR and WKY. Mean ± S.E. **$p < 0.01$, ***$p < 0.001$.

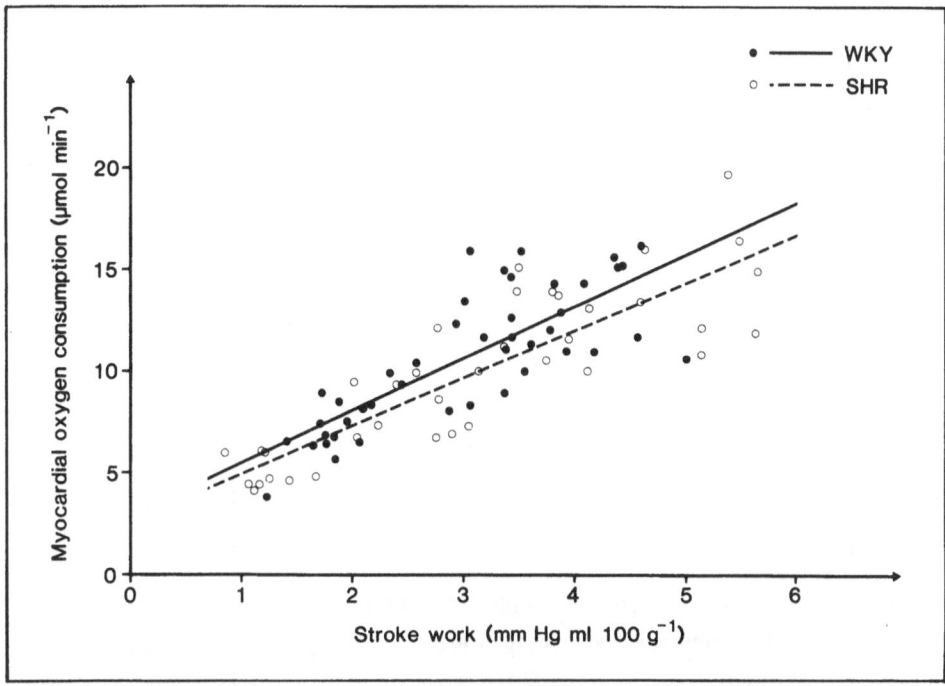

Fig. 2. Linear regression between total myocardial oxygen consumption and stroke work for both SHR and WKY. A significant positive correlation was found for both SHR (r=0.86, p<0.001) and WKY (r=0.80, p<0.001). No statistically significant difference was found between the groups.

levels. As shown in Fig. 1 A, coronary flow was significantly lower in SHR than in WKY at any given aortic diastolic pressure. Minimal coronary vascular resistance was, at all these coronary perfusion levels, consistently elevated in SHR compared with WKY, being about 1.00 mm Hg × min × g/ml for the former and around 0.70 mm Hg × min × g/ml for the latter (p<0.001). This indicates a considerable degree of coronary structural vascular changes present in SHR. Figure 1 B shows the pattern for myocardial oxygen extraction, which was found to be significantly elevated in SHR compared with WKY, at all aortic diastolic pressures examined. In other words, SHR extracted more oxygen, probably to compensate for the reduced coronary flow. The resultant total MVO_2, expressed as µmol/min related to increasing levels of stroke work was relatively unaltered in SHR compared with WKY (Fig. 2). However, both groups showed a significant positive correlation between MVO_2 and stroke work. Since in this study both cardiac work and cardiac oxygen consumption were known, it was possible to calculate cardiac efficiency. The efficiency of the non-hypertrophied isolated WKY hearts was 14.8±2.7% and for the hypertrophied SHR hearts 15.9±0.9% (N.S.).

Cardiac dimensions in SHR and WKY

In this separate set of experiments, LV weight to body weight ratio was, in SHR, increased by 16% (2.20±0.02 versus 2.55±0.05 mg/g, p<0.001), whereas mean arterial pressure was about the same as reported above. As can be seen in Fig. 3 there is a marked,

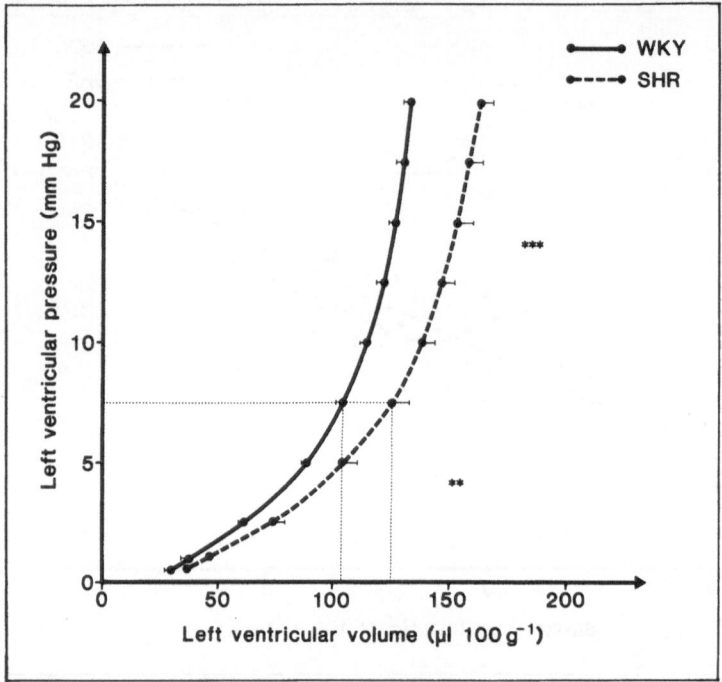

Fig. 3. Diastolic pressure-volume relationships in arrested isolated hearts from SHR and WKY. Mean ± S.E. **p < 0.01, ***p < 0.001.

almost parallel, right-ward shift of the mean diastolic P/V relationship of SHR. Thus, for a given EDP, there is an increased EDV in SHR. Both curves are normalized for 100 g body weight. This difference indicates a true structural out-growth of the LV lumen in SHR. It is not only the internal radius which will be increased in SHR, but the LV will also be expanded externally, reflected as an increased external radius (Fig. 4). Because of these parallel changes, there will only be a slight increase in LV wall thickness in SHR compared with WKY and hence an unchanged ratio between wall thickness and internal radius (Fig. 4). The same type of diastolic dimensional changes in female SHR and WKY have been previously reported [3].

Central haemodynamics and cardiac dimensions in physically exercised SHR

Generally, after six weeks of wheel-running, the rats reached a peak value of running distance, being around 6–7 km/24 hours. Over the first four weeks of exercising, there was a progressive decrease in SBP in SHR compared with non-exercising SHR, but over the last 2–3 weeks SBP again increased for some reason. Heart rate was somewhat lower in exercised animals throughout the whole running period compared with controls. When conscious haemodynamics were examined after 6 weeks of chronic physical exercise there were no significant differences regarding mean arterial pressure, heart rate and total peripheral resistance between R-SHR and C-SHR (Table 1). However, Table 1 also shows

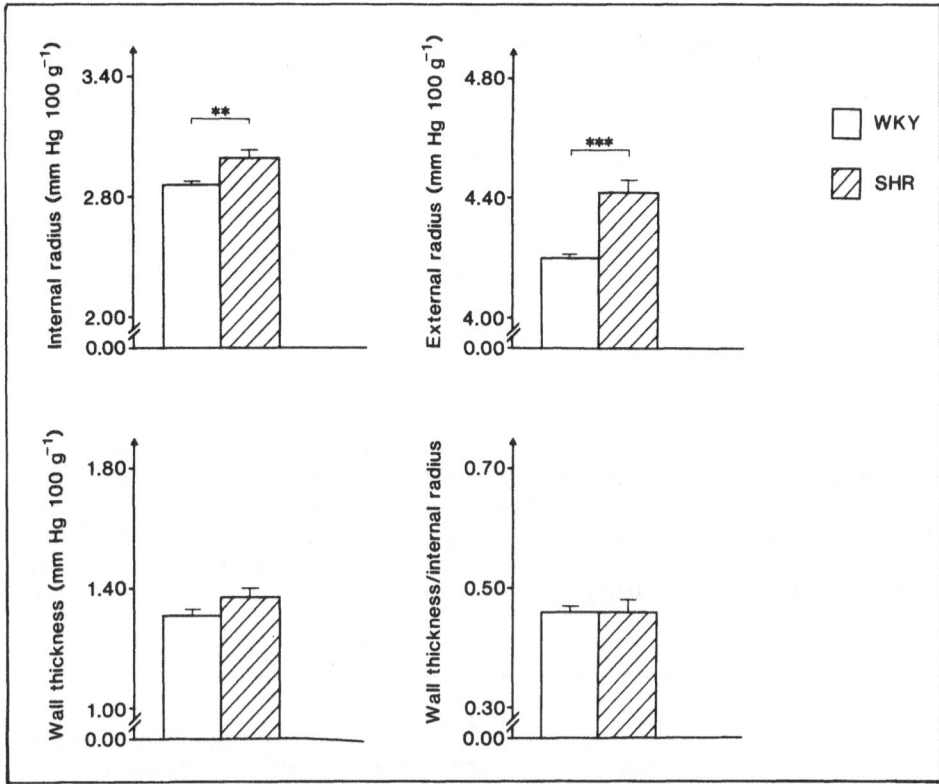

Fig. 4. Diastolic dimensions, i.e. internal and external radii, wall thickness and wall thickness: internal radius ratio at an end-diastolic pressure of 7.5 mm Hg in SHR and WKY. Mean±S.E. **p<0.01, ***p<0.001.

Table 1. Conscious central haemodynamics after 6 weeks of voluntary running in wheels in SHR (R-SHR) and in sedentary control SHR (C-SHR)

	Mean arterial pressure (mm Hg)	Heart rate (beats/min)	Cardiac output (ml/min × 100 g bw)	Stroke volume (µl/100 g bw)	Total peripheral resistance (mm Hg × min × 100 g bw/ml)
R-SHR	173±4	377±8	31.7±0.8[a]	86.4±3.6[a]	5.5±0.6
C-SHR	164±6	391±6	27.9±1.2	74.3±3.6	5.7±0.2

[a] p<0.05

that there was an increase of resting cardiac output in exercised SHR, which was mainly due to a significant elevation of stroke volume. The situation with chronic isotonic exercise will probably create a chronic increase in cardiac filling, resulting in an augmented SV, whereas the anti-hypertensive effect of long-term physical exercise is rather moderate in SHR.

The influence on cardiac dimensions and structure was also investigated in response to spontaneous running. Relative LV weight was increased, though not significantly, in exercised SHR versus control SHR. The mean relationship between left ventricular pressure and volume in diastole was significantly shifted to the right in the group of running SHR compared with WKY, i.e., for a given EDP, EDV was elevated. The internal radius was increased by 5% in R-SHR (from 2.96 ± 0.02 mm/100 g to 3.11 ± 0.01 mm/100 g, $p < 0.05$), whereas wall thickness remained unchanged (1.45 ± 0.02 mm/100 g in C-SHR and 1.43 ± 0.02 mm/100 g in R-SHR). Hence, there was a slight, insignificant decrease of wall thickness to internal radius ratio, being 0.50 ± 0.02 and 0.47 ± 0.02 for controls and running SHR, respectively. Thus, a structural outgrowth of the SHR left ventricle was evident after about 6 weeks of voluntary running, which may be suggested to be a consequence of sustained elevation in cardiac filling.

Cardiac energetics in physically exercised SHR

Another series of exercising and control SHR, handled in exactly the same way as the previously described animals, demonstrated roughly the same pattern regarding SBP and heart rate. However, relative LV weight was significantly increased by exercising, being 2.70 ± 0.06 mg/g in C-SHR and 3.12 ± 0.07 mg/g in R-SHR ($p < 0.001$). In addition, relative right ventricular weight was also increased in the running animals (0.62 ± 0.02 mg/g versus 0.70 ± 0.02 mg/g, $p < 0.01$).

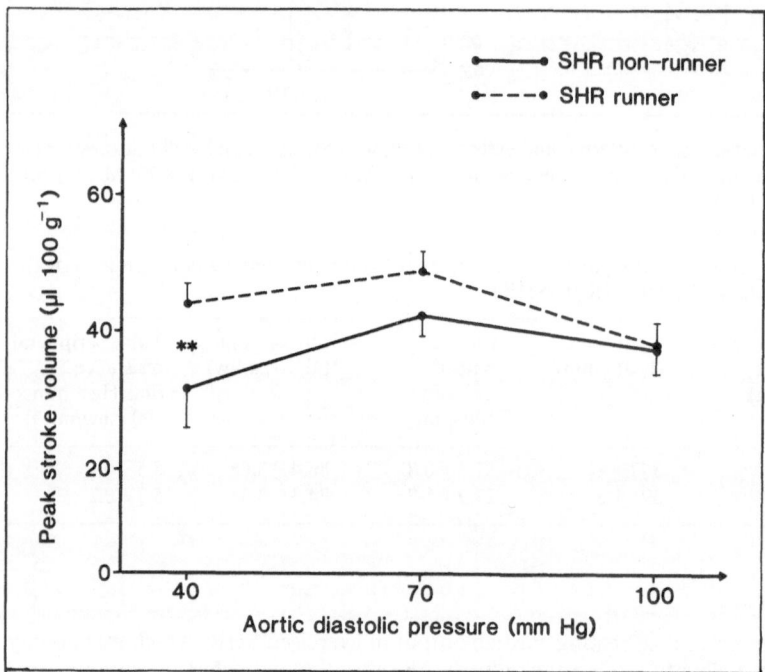

Fig. 5. Peak stroke volume obtained at 3 standardized aortic diastolic pressure levels in SHR exposed to voluntary running and in sedentary SHR (non-runner). Note the marked increase in peak stroke volume at the lowest pressure level in running SHR. Mean $\pm$ S.E. *$p < 0.01$.

Figure 5 shows that peak stroke volume at an aortic diastolic pressure of 40 mm Hg was markedly increased in running SHR compared with non-running SHR. At higher levels of afterloads there were no differences between the groups regarding peak stroke volume, that is, maximal cardiac performance was unaltered by exercise. Coronary flow, oxygen tension in venous coronary effluent, myocardial oxygen consumption and cardiac efficiency were not significantly different between the two groups. Thus, by chronic isotonic exercise there was a clear elevation of cardiac performance compared with controls, when the hearts were perfused at a subnormal perfusion pressure. This finding might be explained by a better nutritional supply due to an increase in capillary surface area; thereby shortening the distances between capillaries and muscle fibres.

Discussion

The present series of experiments were performed on isolated hearts, eliminating all extrinsic and hormonal influences. The experimental models used therefore revealed changes that were solely due to intrinsic cardiac properties.

Challenged at high aortic pressure levels, hearts from 4 month old SHR had a significantly increased performance compared with WKY, a finding consistent with earlier studies reporting maintained [8] and elevated [7, 9] LV function. However, at low aortic pressures (around 40–50 mm Hg) LV performance of 4-month-old SHR were clearly reduced. This latter phenomenon has also been shown in hypertrophied LV from renal hypertensive rats [11, 12, 21].

With increasing levels of aortic pressure, coronary flow expressed as ml/min × g dry weight, increased in both SHR and WKY (Fig. 1 A). This rise was fairly linear, indicating maximal coronary vasodilatation. Figure 1 further shows that coronary flow was consistently reduced in SHR compared with WKY at all perfusion pressures. These findings, of an increased coronary vascular resistance in SHR, are in close agreement with those previously reported by Noresson et al. [22]. Additionally, increased vascular resistance has been found in other vascular circuits in SHR, such as in the skeletal muscle and in the kidneys [23, 24]. This indicates a slow but marked development of structural vascular changes, which, for a given perfusion pressure, will restrict flow.

The importance of the perfusion pressure for cardiac performance was reflected in the finding of a significant positive correlation coefficient between a low aortic pressure and maximal stroke volume in 12-month-old SHR and WKY, whereas no such significant correlation existed at a high aortic pressure [25]. Another indication of the adverse effects of hypertensive structural vascular changes with regard to coronary blood supply was recently described for the hearts of renal hypertensive rats [12]. In that study, the importance of a proper coronary flow through a high resistance vascular bed was stressed in order to maintain a normal cardiac performance. Thus, subnormal perfusion pressure levels in a structurally autoregulated hypertensive coronary vascular bed, may well compromise cardiac pumping performance. This could emanate from a situation where coronary flow is not enough to secure a proper nutritional supply to the myocardium, i.e., when demand exceeds supply.

As mentioned above, coronary flow was reduced in SHR compared with WKY. To balance this reduced flow capacity, oxygen extraction was increased in SHR (Fig. 1 B right part). Similar results have been reported by others [26], showing increased oxygen extraction in hearts from renal hypertensive rats. Hence, due to the increased oxygen extraction, SHR can probably maintain a reasonable tissue oxygen tension despite the reduced coronary flow.

The advantage of the antegrade working heart technique, described in the present study, is that most parameters determining cardiac performance and MVO_2 can be selectively controlled. For example, the hearts can be paced to the same rate and both left atrial and aortic pressures can be changed independently of each other. Therefore, normotensive and hypertensive hearts can be examined during identical pressure conditions. In addition, the oxygen content in the venous coronary effluent could be continuously measured, allowing calculations of MVO_2. MVO_2 is mainly dependent on myocardial wall stress, heart rate and inotropic state [27]. Changes in wall stress are, in turn, dependent on alterations in ventricular volume, pressure and/or wall thickness. Changes in pressure are generally regarded as influencing MVO_2 to a greater extent than do changes in ventricular volume [27].

When compared at equal arterial pressure levels, as in the present study, total MVO_2 was about the same in the SHR group as in the WKY group (Fig. 2). Thus, SHR can produce an augmented cardiac performance without necessarily elevating total MVO_2. Moreover, regarding cardiac efficiency, our data does not indicate any difference between SHR and WKY.

A close relationship between total MVO_2 and increasing levels of stroke work was found both in SHR and WKY, as is shown in Fig. 2. The present results are in contrast with those presented by Strauer [28], who reported a poor correlation between MVO_2 and ventricular function. He instead showed a linear relationship between total MVO_2 and increasing degree of ventricular hypertrophy in primary hypertension. Furthermore, the same study demonstrated a marked increase in coronary perfusion pressure and vascular resistance. Despite these changes, a net increase in coronary flow was observed, which is somewhat surprising since the structural autoregulation will have a flow restricting effect. These differences in results between the two studies could very well be due to the lack of control of the various cardiovascular parameters when performing analysis in vivo.

The present study also explored LV geometry in SHR and WKY and in exercised SHR compared with sedentary SHR, by examining diastolic P/V curves and from these calculating w, r_i and w/r_i ratio. With respect to early established SHR primary hypertension (in 16-week-old rats), the P/V relationship was considerably shifted to the right of that of WKY (Fig. 3). This indicates a greater EDV for a given EDP, i.e. an eccentric type of LV hypertrophy. However, a similar, even more pronounced rightward shift of the P/V curve has been obtained in 20-week-old female SHR compared with WKY [3].

An eccentric type of cardiac adaptation seems to be present in the early phases of hypertension in SHR. This may be the result of a sustained elevation of cardiac filling, consequent to sympathetic overactivity that also includes the capacitance vessels. Young borderline hypertensive SHR also have a sympathetically mediated increase in cardiac output compared with normotensive rats, during both rest and mental arousal [29]. Furthermore, since SHR are physically more active than WKY [30], cardiac filling may be more enhanced in young SHR than in WKY. In borderline hypertensive patients, increased diastolic diameter in association with slight elevation of cardiac function have been observed [31, 32]. This also points to an eccentric hypertrophy due to sustained increase in cardiac filling. Thus, a situation with a structurally enlarged LV will be more favourable, since myocardial fibre shortening could be less than in a smaller sized ventricle, in order to produce the same stroke volume. As hypertension advances, cardiac adaptation will be determined more by the pressure load and a concentric hypertrophy becomes superimposed on top of the eccentric one, resulting in a greatly increased LV wall thickness.

Six weeks of spontaneous exercise in SHR created a slight blood pressure reduction, but at the end, mean arterial pressure did not differ significantly between exercised and non-exercised SHR (Table 1). There are contrasting results in the literature regarding the antihypertensive effect of exercise in SHR. Some groups have shown a reduction in arterial pressure [16, 17], whereas other groups have not been able to show any reduction in blood pressure [14, 15]. The type of training program, duration and intensity, are probably important factors when considering the differences in obtained results. In our laboratory it was recently shown that SHR, which began voluntary running at 5 weeks of age and continued for about 10 weeks, did not show any preventative effect on hypertension development (Friberg et al., unpublished observations). One explanation may be that the strong genetic predisposition in SHR will confound any blood pressure lowering effect achieved by exercise.

Usually, chronic exercise is associated with the development of LV hypertrophy, as also demonstrated in the present study. In addition, we also observed a development of right ventricular hypertrophy, which indicates hypertrophy ensuing from a volume overload, since the pressure in the lung circuit is expected to be normal. It is then reasonable to assume that the LV was also exposed to a volume overload.

Voluntary running in SHR resulted in a significant rightward shift of the P/V curve compared with non-running SHR. This is consistent with findings reported earlier by Pfeffer et al. [14] who with another approach showed increases in LV luminal size both in SHR and WKY in response to chronic swimming exercise. In man, Sugishita et al. [33] found ventricular hypertrophy and increased diastolic diameter in professional runners, i.e., resulting in an eccentric form of LV hypertrophy. In runners, no structural vascular changes would be expected and the LV was suggested to be in an advantageous situation, since the greater EDV would mean that a higher stroke volume can be expelled for the same degree of myocardial fibre shortening compared with, for example a smaller sized heart, which has to produce a greater systolic fibre shortening in order to deliver the same stroke volume.

Generally, a structural increase in r_i may imply considerable advantages, as long as myocardial contractility remains normal and w/r_i is well adapted to the prevailing mean arterial pressure level. In fact, the present study showed an increase in cardiac output in exercised SHR compared with non-trained SHR (Table 1), which was entirely due to an elevation of stroke volume. Likewise, when female SHR were chronically given an antihypertensive calcium antagonist (felodipine) an augmented stroke volume was observed [34]. As stroke volume was more enhanced after chronic (12 weeks) than after acute administration of felodipine, a structural adaptation of the LV may have contributed. Thus, in the chronic felodipine treated SHR a structural luminal increase (increased r_i) of the LV was seen [34]. This reconstruction was probably a consequence of enhanced cardiac filling due to the hyperkinetic circulation created by the chronic vasodilatation caused by felodipine.

The question then arises of how LV luminal widening will affect cardiac function and oxygen consumption. Since systolic wall stress emanates from systolic pressure and wall thickness to internal radius ratio, it will be close to the ventricular afterload imposed on the LV wall. Consequently, EDV of the LV and systolic wall stress are important determinants of cardiac function.

As already mentioned, maximal cardiac function was elevated in 4 month old SHR and in somewhat older SHR; total MVO_2 was unchanged for a given level of external work, as compared with WKY. Thus, despite a 20% elevation of EDV in SHR, maximal LV performance was enhanced and MVO_2 unaltered. The finding of an enhanced cardiac function is also supported by other groups [31, 32] who have reported increased LV diastolic dimensions and elevations of systolic wall stress in combination with an augmented cardiac function in early established human primary hypertension. Exercised

SHR, which also showed an increase in internal radius compared with non-exercised SHR, demonstrated a maximal LV performance and MVO_2 close to sedentary SHR, whereas myocardial performance was markedly increased in the exercised group at a subnormal perfusion pressure (Fig. 5). Furthermore, pregnancy (a natural volume overload) has also been shown to induce eccentric hypertrophy both in women [1] and in animals [2, 3], and with maintained cardiac performance. The eccentric form of structural cardiac adaptation therefore by no means implies an impairment of LV performance, provided that myocardial contractility is maintained. The results from the present study further demonstrated that in these hypertrophied hearts, increased systolic wall stress and external work were not associated with any augmentation of total oxygen consumption.

In conclusion, the present experiments showed that maximal cardiac performance was elevated in the structurally enlarged eccentric hypertrophied SHR heart. Likewise, in response to sustained voluntary running, SHR showed further development of eccentric hypertrophy with a relatively unchanged wall thickness to internal radius ratio. Furthermore, spontaneous exercise in SHR caused an increase in cardiac output, entirely due to elevation of stroke volume. The stimulus for the cardiac redesign to a structurally enlarged heart can probably best be explained by a chronic elevation of cardiac filling. Thus, these enlarged left ventricles can produce higher stroke volumes for given myocardial fibre shortenings. Coronary flow was markedly decreased in SHR compared with WKY, due to enhanced coronary vascular resistance, which is caused by structural vascular changes. To balance the reduced flow, myocardial oxygen extraction was increased in SHR. The resulting total myocardial oxygen consumption for given levels of stroke work, was fairly equal in SHR compared with WKY. Furthermore, despite structurally enlarged left ventricles in SHR and exercised SHR, and also increased arterial pressure in SHR compared with WKY (thereby elevating systolic wall stress), cardiac function was maintained and even elevated. These events were not associated with an increase in oxygen consumption.

Acknowledgements. The technical assistance of Gunnel Andersson and Ulla Axelsson is gratefully acknowledged. This study was supported by grants from the Medical Research Council (no 00016) and from the medical faculty, University of Göteborg.

References

1. Katz R, Karliner JS, Resnik R (1978) Effects of a natural volume overload state (pregnancy) on left ventricular performance in normal human subjects. Circulation 58(3):434–441
2. Morton M, Tsang H, Hohimer R, Ross R, Thornburg K, Faber J, Metcalfe J (1984) Left ventricular size, output, and structure during guinea pig pregnancy. Am J Physiol 226:R40–R48
3. Friberg P, Folkow B, Nordlander M (1985) Structural adaptation of the rat left ventricle in response to changes in pressure and volume loads. Acta Physiol Scand 125:67–79
4. Goldman S, Olajos M, Friedman H, Roeske WR, Morkin E (1982) Left ventricular performance in conscious thyrotoxic calves. Am J Physiol 242:H113–H121
5. Friberg P, Wåhlander H, Nordlander M (1986) Dimensional and functional behavior of the rat heart exposed to various haemodynamic alterations. J Hypertension 4:S127–S129
6. Saragoca MA, Tarazi RC (1981) Left ventricular hypertrophy in rats with renovascular hypertension. Alterations in cardiac function and adrenergic responses. Hypertension [Suppl II] 3:II-171–II-196
7. Noresson E, Ricksten S-E, Hallbäck-Nordlander M, Thoren P (1979) Performance of the hypertrophied left ventricle in spotaneously hypertensive rat. Effects of changes in preload and afterload. Acta Physiol Scand 107:1–8

8. Pfeffer JM, Pfeffer MA, Fishbein MC, Frohlich ED (1979) Cardiac function and morphology with aging in the spontaneously hypertensive rat. Am J Physiol 237(4):H461–H468

9. Lundin S, Friberg P, Hallbäck-Nordlander M (1982) Left ventricular hypertrophy improves cardiac performance in spontaneously hypertensive rats. Acta Physiol Scand 114:321–328

10. Friberg P, Nordlander M, Lundin S, Folkow B (1985) Effects of ageing on cardiac performance and coronary flow in spontaneously hypertensive and normotensive rats. Acta Physiol Scand 125:1–11

11. Friberg P, Nordborg C (1986) Functional, morphological and metabolic characteristics of isolated hearts from normotensive and spontaneously hypertensive rats before, during and after renal hypertension. Acta Physiol Scand 126:161–171

12. Shimamatsu K, Fouad-Tarazi F (1986) Basal inotropic state in rats with renal hypertension: influence of coronary flow and perfusion pressure. Cardiovasc Res 20:269–274

13. Scheuer J, Tipton CM (1977) Cardiovascular adaptations to physical training. Ann Rev Physiol 39:221–251

14. Pfeffer MA, Ferrell BA, Pfeffer JM, Weiss AK, Fishbein MC, Frohlich ED (1978) Ventricular morphology and pumping ability of exercised spontaneously hypertensive rats. Am J Physiol 235(2):H193–H199

15. Weiss L (1978) Adaptive cardiovascular changes to physical training in spontaneously hypertensive and normotensive rats. Cardiovasc Res 12(6):329–333

16. Tipton CM, Matthes RD, Callahan A, Tcheng T, Lais LT (1977) The role of chronic exercise on resting blood pressure of normotensive and hypertensive rats. Med Sci Sports 9(3):168–177

17. Evenwel R, Struyker-Boudier H (1979) Effect of physical training on the development of hypertension in the spontaneously hypertensive rat. Pflügers Arch 381:19–24

18. Shyu BC, Andersson SA, Thoren P (1984) Spontaneous running in wheels. A microprocessor assisted method for measuring physiological parameters during exercise in rodents. Acta Physiol Scand 121:103–109

19. Friberg P, Nordlander M (1986) Influence of long-term antihypertensive therapy on cardiac function, coronary flow and myocardial oxygen consumption in spontaneously hypertensive rats. J Hypertension 4:165–173

20. Stage L (1978) Rapid determination of cardiac output in small animals from dye dilution measurements. Acta Physiol Scand 102:43A

21. Friberg P (1985) Structural and functional adaptation in the rat myocardium and coronary vascular bed caused by changes in pressure and volume load. Acta Physiol Scand [Suppl 540] 124:1–47

22. Noresson E, Hallbäck M, Hjalmarsson Å (1977) Structural "resetting" of the coronary vascular bed in spontaneously hypertensive rats. Acta Physiol Scand 101:363–365

23. Folkow B, Hallbäck M, Lundgren Y, Weiss L (1970) Structurally based increase of flow resistance in spontaneously hypertensive rats. Acta Physiol Scand 79:373–378

24. Göthberg G, Folkow B (1983) Age-dependent alterations in the structurally determined vascular resistance, pre- to postglomerular resistance ratio and glomerular filtration capacity in kidneys, as studied in aging normotensive rats and spontaneously hypertensive rats. Acta Physiol Scand 177:547–555

25. Nordlander M, Wåhlander H, Friberg P (1986) Myocardial and vascular structural adaptation to chronic pressure overload. J Cardiovasc Pharmacol (in press)

26. Alfaro A, Schaible TF, Malhotra A, Yipintsoi T, Scheuer J (1983) Impaired coronary flow and ventricular function in hearts of hypertensive rats. Cardiovasc Res 17:553–561

27. Braunwald E (1971) Control of myocardial oxygen consumption. Physiologic and clinical considerations. Am J Cardiol 27:416–432

28. Strauer BE (1984) The coronary circulation in hypertensive heart disease. Hypertension [Suppl III] 6:74–80

29. Lundin SA, Hallbäck-Nordlander M (1980) Background of hyperkinetic circulatory state in young spontaneously hypertensive rats. Cardiovasc Res 14:561–567

30. Knardahl S, Sagvolden T (1979) Open-field behavior of spontaneously hypertensive rats. Behavioral and Neural Biology 27:187–200

31. Wikstrand J (1984) Left ventricular function in early primary hypertension. Functional consequences of cardiovascular structural changes. Hypertension [Suppl III] 6:108–116
32. Lutas EM, Deveraux RB, Reis G, Alderman MH, Pickering TG, Borer JS, Laragh JH (1985) Increased cardiac performance in mild essential hypertension. Left ventricular mechanics. Hypertension 7:979–988
33. Sugishita Y, Susumu K, Matsuda M, Yamaguchi T, Ito I (1983) Myocardial mechanics of athletic hearts in comparison with diseased hearts. Am Heart J 105:273–280
34. Friberg P, Folkow B, Nordlander M (1986) Cardiac dimensions in spontaneously hypertensive rats following different modes of blood pressure reduction by antihypertensive treatment. J Hypertension 4:85–92

Author's address:

Dr. Peter Friberg, Department of Physiology, University of Göteborg, P.O. Box 33031, S-400 33 Göteborg, Sweden

Myocardial contractility and left ventricular myosin isoenzyme pattern in cardiac hypertrophy due to chronic volume overload*

N. Takeda, T. Ohkubo, T. Hatanaka, A. Takeda, I. Nakamura, and M. Nagano

Department of Internal Medicine, Aoto Hospital, Jikei University, Tokyo, Japan

Summary

Chronic volume overloaded cardiac hypertrophy was induced by abdominal arteriovenous shunt in 10-week-old male Wistar rats. Ten weeks after the operation, myocardial mechanical examination was performed with isolated left ventricular papillary muscles. Left ventricular myosin isoenzyme pattern was also examined by pyrophosphate gel electrophoresis. In addition, morphological study by electron microscope was performed. In arteriovenous shunt rats, isometric developed tension (T) and its first derivative (dT/dt) were decreased, resting tension (RT) was increased and time to peak tension (TPT) was prolonged as compared to sham-operated control rats (T: 2.4 ± 0.5 vs 2.8 ± 0.8 g/mm^2, NS. dT/dt: 23.9 ± 4.8 vs 31.9 ± 7.0 g/mm$^2 \cdot$s, $p < 0.05$. RT: 1.2 ± 0.2 vs 0.9 ± 0.1 g/mm^2, $p < 0.05$. TPT: 151.7 ± 20.2 vs 128.3 ± 10.3 msec, $p < 0.05$). Myocardial mechanical responses to isoproterenol (10^{-7} mol/l) and dibutyryl cyclic AMP (10^{-5} mol/l) were both depressed in hypertrophied myocardium, although not significant statistically. Left ventricular myosin isoenzyme pattern was shifted towards VM-3 by chronic volume overload. Electron microscope study revealed increased collagen content in hypertrophied myocardium.

Introduction

In the pressure overloaded rat's cardiac hypertrophy, myocardial contractility, especially unloaded maximum shortening velocity, is depressed with the shift in ventricular myosin isoenzyme pattern towards VM-3, the isoenzyme possessing the lowest electrophoretic mobility and ATPase activity. This alteration in myosin isoenzyme pattern can be considered as an adaptive alteration of the pressure overloaded myocardium [6, 7]. In this study we examined alterations in myocardial contractility and ventricular myosin isoenzyme pattern using hypertrophied myocardium by another method, i.e. volume overload. We also examined myocardial catecholamine responsiveness and myocardial morphological changes by electron microscope.

Materials and methods

Chronic volume overloaded cardiac hypertrophy was induced by abdominal arteriovenous shunt using 10-week-old male Wistar rats. Arteriovenous shunt was produced in

---

* Supported by Tanaka Memorial Medical Research Fund.

rats under a microscope according to Flaim et al. [2] and Mercadier et al. [8]. Rats were anaesthetized with pentobarbital (40 mg/kg, i.p.) and an abdominal midline incision was made. The abdominal aorta and vena cava were exposed and separated from posterior tissue between the origin of the renal arteries and the aortic bifurcation. This part of the aorta and vena cava (approximately 10 mm in length) was isolated by two small vascular clamps and twisted so that the vena cava was located above the aorta. The vena cava was incised longitudinally, approximately 1 mm in length, with a dissecting blade. After the blood in this segment of vena cava was driven out, a small part of the median wall was protruded using a microsurgical suture (9-0), and resected with microscissors. At this moment an opening of approximately 1 mm in diameter was made. Then the inside of the two vessels was washed with physiological saline solution and the incision opposing the opening of the median common wall was closed with 3-4 stitches using a microsurgical suture (9-0). Vascular clamps were removed and the abdominal incision was closed. Control rats underwent sham-operation without arteriovenous shunt.

Ten weeks after the operation, myocardial mechanical and biochemical changes were examined. Mechanical study was performed using isolated left ventricular papillary muscles; left ventricular free walls were used for the determination of myosin isoenzyme pattern. Papillary muscles were stimulated at 32 °C with a frequency of 0.2 Hz and a voltage 30% above threshold and were perfused with Tyrode solution containing 1.1 mmol/l Ca^{2+}. After the steady state was obtained at L_{max}, developed and resting tension, dT/dt, time to peak tension, and total contraction time were recorded. The responses of mechanical parameters to isoproterenol (10^{-7} mol/l) were estimated also at L_{max}, following the interposition of Tyrode solution for 15–20 min and 10^{-5} mol/l dibutyryl cyclic AMP was administered. The response of each parameter was obtained by comparing two paired values. One value was that of steady state prior to isoproterenol or dibutyryl cyclic AMP administration and the other was that of maximum value after isoproterenol or dibutyryl cyclic AMP administration, respectively.

Polyacrylamide gel electrophoresis in the presence of pyrophosphate was carried out as described elsewhere [1, 4, 9]. The gel contained 3.8% acrylamide and 0.12%, N,N′-methylene-bis-acrylamide. Electrophoresis buffer was 20 mmol/l $Na_4P_2O_7$ (pH 8.8) in the presence of 10% glycerol. Native myosin from the left ventricle was extracted with a solution consisting of 100 mmol/l $Na_4P_2O_7$ (pH 8.8), 5 mmol/l 1,4-di-thioerythritol, 5 mmol/l EGTA, 5 µg/ml leupeptin. Electrophoresis was carried out for 30 h at 2 °C and a voltage gradient of 13.3 V/cm. At the end of mechanical examination, stretched papillary muscles at L_{max} were fixed in 2% glutaraldehyde, 0.1 mol/l PB solution and myocardial ultrastructural changes were examined by electron microscope.

Stastitical comparisons were carried out using Student's t-test.

Results

As is shown in Table 1, body weight and both ventricular weights were heavier, and ventricular weight: body weight ratio was greater in arteriovenous shunt rats than in controls. There were no significant differences in papillary muscles between the two groups, although papillary muscles of arteriovenous shunt rats were slightly bigger than those of controls.

In arteriovenous shunt rats, isometric developed tension (T) was decreased (not significant); dT/dt was decreased, resting tension (RT) was increased and time to peak tension (TPT) was prolonged (Table 2). Myocardial mechanical responses to isoproterenol and dibutyryl cyclic AMP were both depressed in hypertrophied myocardium, although

Table 1. Body weight, heart weight and papillary muscle size

	BW (g)	LVW (mg)	RVW (mg)	$\dfrac{LV+RV}{BW} \times 1000$	Papillary muscle	
					L (mm)	CSA (mm^2)
Control (n=6)	568.3±25.6	1039.5± 59.3	254.8±22.4	2.28±0.08	5.8±0.3	0.88±0.22
AV shunt (n=6)	631.7±46.2	1487.8±165.3	562.3±88.9	3.26±0.49	6.2±0.3	0.93±0.20
	P<0.02	P<0.001	P<0.001	P<0.001	ns	ns

BW: body weight, LVW: left ventricular weight, RVW: right ventriculat weight, L: length, CSA: cross sectional area, Values are means±SD, ns: not significant.

Table 2. Myocardial mechanics

	AT (g/mm^2)	RT (g/mm^2)	dT/dt (g/mm$^2 \cdot$ s)	TPT (ms)	TCT (ms)
Control (n=6)	2.8±0.8	0.9±0.1	31.9±7.0	128.3±10.3	454.2±45.7
AV shunt (n=6)	2.4±0.5	1.2±0.2	23.9±4.8	151.7±20.2	516.7±67.1
	ns	P<0.05	P<0.05	P<0.05	ns

AT: active tension, RT: resting tension, TPT: time to peak tension, TCT: total contraction time, Values are means±SD, ns: not significant.

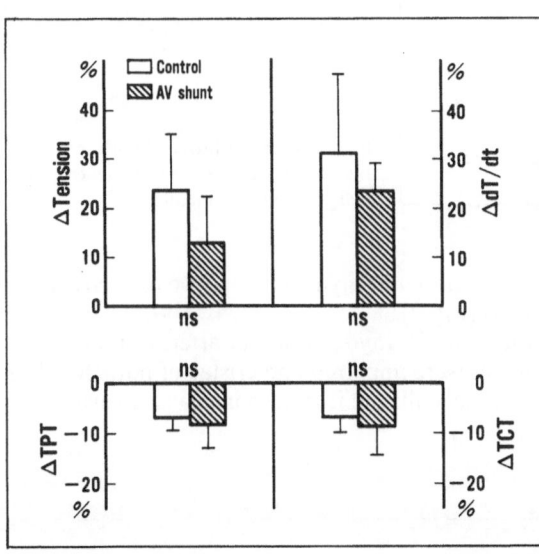

Fig. 1. Comparisons between Δtension, ΔdT/dt, Δtime to peak tension (ΔTPT) and Δtotal contraction time (ΔTCT) due to isoproterenol (10^{-7} mol/l) between arteriovenous shunt rats and sham-operated control rats. Vertical lines indicate SD; ns: not significant.

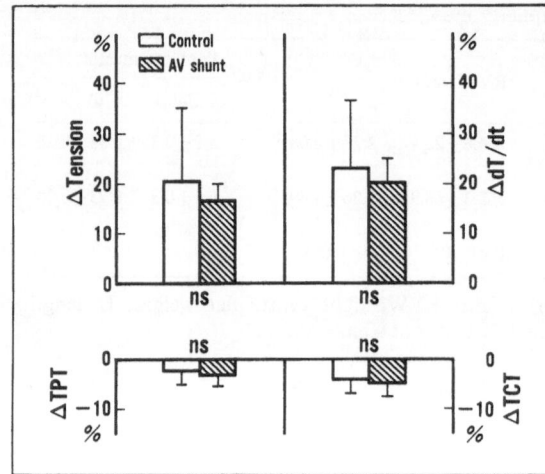

Fig. 2. Comparisons between the same parameters as Fig. 1, but due to dibutyryl cyclic AMP (10^{-5} mol/l) between arteriovenous shunt rats and sham-operated control rats. Vertical lines indicate SD; ns: not significant.

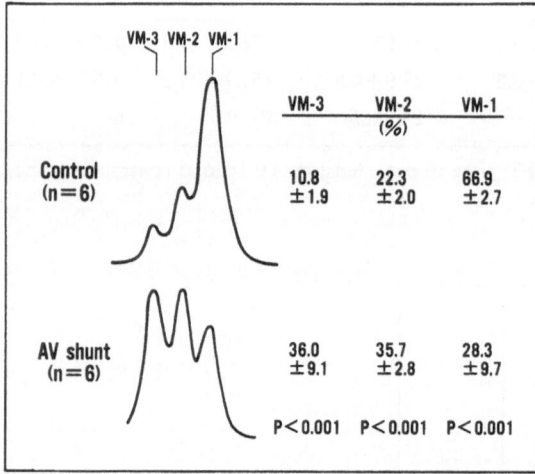

Fig. 3. Representative absorbancy profiles of pyrophosphate gel electrophoresis. Values are means ± SD.

not statistically significant (Figs. 1, 2). Left ventricular myosin isoenzyme pattern shifted towards VM-3 by chronic volume overload (Fig. 3).

Electron microscope study revealed that, in the myocardium of arteriovenous shunt rats, there were some areas where myofibrils were impaired and cristae of mitochondria were disarrayed (Fig. 4). Myocardial interstitial collagen fibre also increased in arteriovenous shunt rats.

Fig. 4. Comparisons between morphological changes in arteriovenous shunt rats (A) and control rats (B) by electron microscope, showing intercalated disc (in), myofibril (mf), mitochondria (mt), collagen fibre (co).

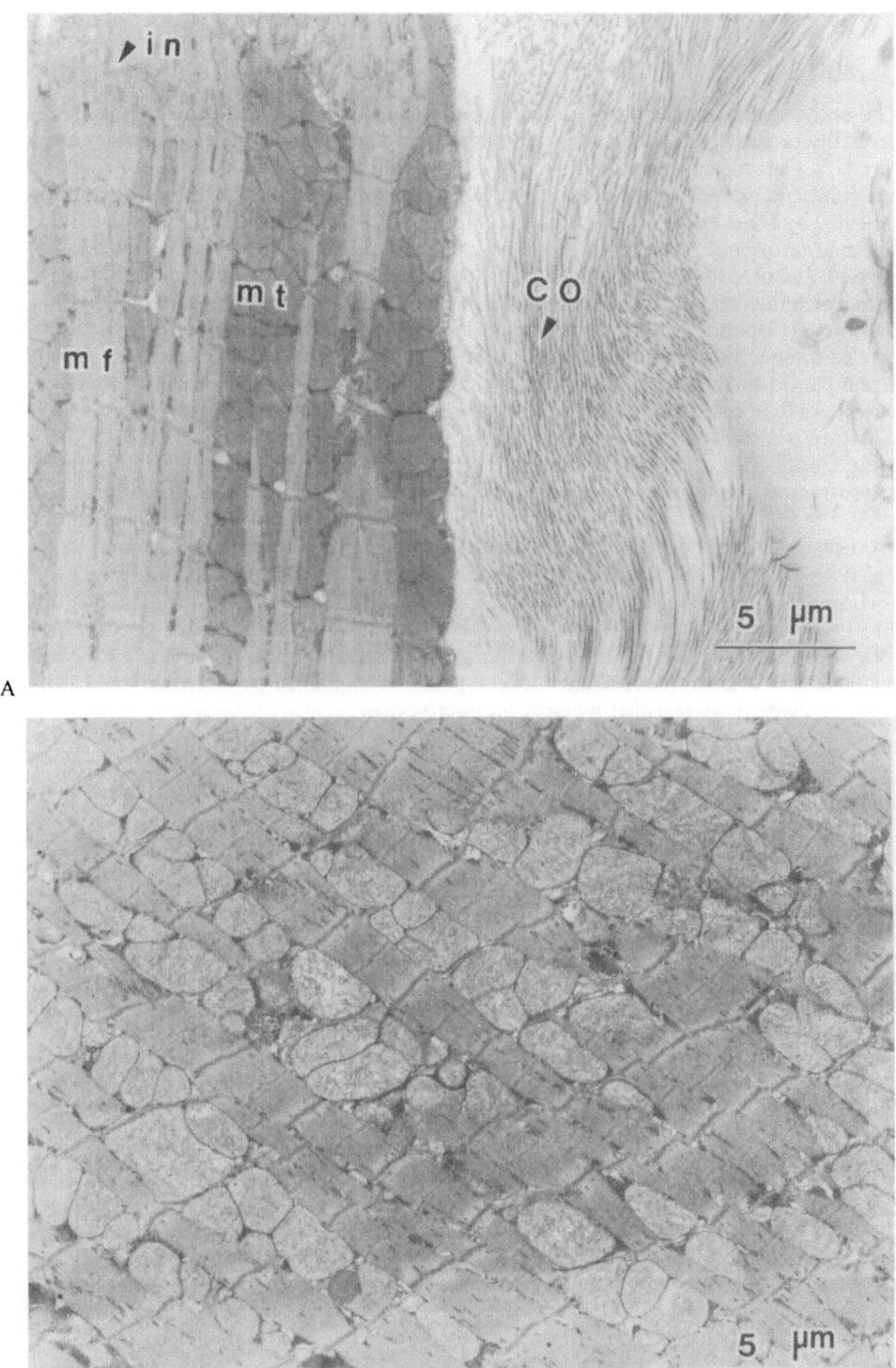

Discussion

The body weight of arteriovenous shunt rats was heavier than that of controls. This is probably because circulatory congestion and fluid retention, as reported in other literature [12]. Chronic cardiac volume overload by arteriovenous shunt produced right and left ventricular hypertrophy. The degree of hypertrophy was a little different from that reported by other authors [3]. We assume that it must depend on the size of the fistula.

In arteriovenous shunt rats, developed tension was slightly depressed, dT/dt was decreased, and time to peak tension and total contraction time were both prolonged as compared with that of controls. These results mean that chronic volume overload can affect tension development, which might partly be explained by impaired myofibrils, as shown by electron microscope, and a change in the myocardium to one which contracts slowly, accompanied by the shift of myosin isoenzyme towards VM-3, the isoenzyme possessing lowest ATPase activity. Myocardial responses to isoproterenol and to dibutyryl cyclic AMP were both depressed in arteriovenous shunt rats, although not statistically significant. Considering that elevated plasma epinephrine and norepinephrine levels in chronic arteriovenous shunt rats were reported [2], myocardial adrenoceptors might have been altered in the hypertrophied myocardium in the present study. It has been indicated in previous studies that post-receptor processes may also play a role in myocardial mechanical responsiveness to catecholamines in swim-trained rats [10, 11]. As dibutyryl cyclic AMP is thought to pass the myocardial surface membrane and exert its positive inotropic effect without direct stimulation of β-receptors [5], there is a possibility that post-receptor processes also play a role in the myocardium of our present animal model. Increased resting tension in arteriovenous shunt rats can partly be explained by increased myocardial collagen content, as revealed by electron microscope.

References

1. d'Albis A, Pantaloni C, Becher J-J (1979) An electrophoretic study of native myosin isoenzymes and of their subunits content. Eur J Biochem 99:261–272
2. Flaim SF, Minteer WJ, Nelles SH, Clark DP (1979) Chronic arteriovenous shunt: evaluation of a model for heart failure in rat. Am J Physiol 236:H698–H704
3. Flaim SF, Minteer WJ (1980) Ventricular volume overload alters cardiac output distribution in rats during exercise. J Appl Physiol 49:482–490
4. Hoh JFY, McGrath PA, Hale PT (1978) Electrophoretic analysis of multiple forms of rat cardiac myosin: effects of hypophysectomy and thyroxine replacement. J Mol Cell Cardiol 10:1053–1076
5. Imai S, Otorii T, Takeda K, Katano Y, Horii D (1974) Effects of cyclic AMP and dibutyryl cyclic AMP on the heart and coronary circulation. Jpn J Pharmacol 24:499–510
6. Jacob R, Rupp H, Ebrecht G, Gülch RW, Kissling G (1981) Chronic reaction of myocardium at the myofibrillar level. In: Nagano M, Seki I (eds) Cardiac structure and metabolism, vol 4. Roppo (Tokyo), pp 1–28
7. Jacob R, Kissling G, Ebrecht G, Holubarsch C, Medugorac I, Rupp H (1983) Adaptive and pathological alterations in experimental cardiac hypertrophy. In: Chazov E, Saks V, Rona C (eds) Advances in myocardiology, vol 4. Plenum Publishing Corporation, New York, pp 55–77
8. Mercadier JJ, Lompre AM, Wisnewsky C, Samuel JL, Bercovici J, Swynghedauw B, Schwartz K (1981) Myosin isoenzymic changes in several models of rat cardiac hypertrophy. Circ Res 49:525–532
9. Rupp H, Jacob R (1982) Response of blood pressure and cardiac myosin polymorphysm to swimming training in the spontaneously hypertensive rat. Can J Physiol Pharmacol 60:1098–1103

10. Takeda N, Dominiak P, Türck D, Rupp H, Jacob R (1985) The influence of endurance training on mechanical catecholamine responsiveness β-adrenoceptor density and myosin isoenzyme pattern of rat ventricular myocardium. Basic Res Cardiol 80:88–99
11. Takeda N, Dominiak P, Türck D, Rupp H, Jacob R (1985) Myocardial catecholamine responsiveness of spontaneously hypertensive rats as influenced by swimming training. Basic Res Cardiol 80:384–391
12. Taylor RR, Covell JW, Ross J Jr (1968) Left ventricular function in experimental aorto-caval fistula with circulatory congestion and fluid retension. J Clin Invest 47:1333–1342

Authors' address:

Dr. N. Takeda, Department of Internal Medicine, Aoto Hospital, Jikei University, Aoto 6-41-2, Katsushika-ku, Tokyo 125, Japan

Decreased L-carnitine transport
in mechanically overloaded rat hearts

Z. El Alaoui-Talibi and J. Moravec

I.N.S.E.R.M. U2, Hôpital Léon Bernard, Limeil-Brévannes Cédex, France

Summary

The transport of L[^{14}C] carnitine was studied in rat hearts with a three-month-old aorto-caval fistula. Tissue TG content was determined in order to assess the state of FFA utilization. The hearts were perfused with a bicarbonate buffer containing 11 mM glucose and variable concentrations (10–200 µM) of L[^{14}C] carnitine. In some experiments, the active component of carnitine transport was suppressed by the adjunction of 0.05 mM mersalyl acid. The subtraction of passive from total transport allowed us to reconstruct the saturation curves of the net active transport of L-carnitine. Our results suggest that at physiological carnitine concentration (50 µM) the uptake of L-carnitine is significantly depressed in mechanically overloaded hearts. These changes are not related to alterations of coronary perfusion, since coronary flow rates (ml/min/g dry wt) are quite comparable in both groups tested. According to the Lineweaver-Burk analysis of the kinetics of saturable transport, the affinity of the membrane carrier for L-carnitine is considerably diminished in the overloaded hearts (K_m[carnitine] 125 instead of 83 µM). The alterations of the kinetics of carnitine transport do not seems to be related to the decrease of the transmembrane gradient of sodium: the intracellular sodium content of the hypertrophied, but non-failing, hearts is quite similar to that of control hearts. In addition, carnitine deficiency does not lead to TG accumulation, at least under in situ conditions.

Introduction

According to metabolic studies performed in situ and on hearts perfused in vitro [3, 9, 17], the FFA oxidation provides up to 70% of the metabolic energy of the normal myocardium. However, for the FFA utilization to proceed without difficulty a functionally unlimited acyl-carnitine translocase is required [7, 11] in order to transfer cytosolic long chain acyl moieties into the mitochondria and at the same time recycle free carnitine back to the cytosol [8]. One of the conditions which may alter the fatty acid utilization in the heart is therefore the lack of L-carnitine in the tissue [13, 18, 26].

In this respect is seems interesting that, according to several authors, a pronounced depletion of tissue L-carnitine regularly occurs in mechanically overloaded hearts of different species [6, 22, 27]. In some cases, the carnitine deficiency has been associated with altered free fatty acid utilization by isolated mitochondria [19] and by tissue homogenates from mechanically overloaded hearts [27]. Therefore, it has been suggested that low tissue carnitine may contribute to the onset of metabolic disorders responsible for the development of the contractile failure as observed in situ [20, 27].

However, direct evidence that carnitine deficiency of mechanically overloaded hearts is indead associated with a decreased FFA oxidation and TG accumulation in the myocardium is lacking. Another point which remains unsolved is the mechanism leading to decreased carnitine content of mechanically overloaded hearts. The purpose of this work was to throw more light on these two questions using the same experimental model of sustained volume overload as used in our earlier experiments [6, 15, 16]. According to these data [6], three months after the opening of the aorto-caval communication, the tissue carnitine content is about 30% lower in hearts of rats with haemodynamically compensated cardiac hypertrophy. Such a depletion of tissue carnitine may result from a dysfunction in the saturable carrier-mediated transport of L-carnitine [23] which seems to be driven by the energy derived from its co-transport with sodium [25]. We therefore studied the state of L-carnitine transport in the mechanically overloaded hearts using the experimental approach developed by Vary and Neely for normal and diabetic hearts [23, 24]. Tissue TG and sodium content were also determined in order to assess the state of lipid metabolism and intracellular sodium concentrations, respectively, in the mechanically overloaded hearts.

Methods

A chronic volume overload was induced in two-month-old female rats of Wistar strain, by surgically created aorto-caval fistula [16]. The sham-operated animals from the same litters were used as controls. All animals were sacrificed three months after surgery. The hearts of rats which did not present any clinical signs of congestive heart failure were used either for biochemical determination of tissue TG and sodium content or for the perfusion experiments in the presence of L-[^{14}C] carnitine [23].

The animals used for the determinations of tissue TG and tissue sodium content were anaesthetized with penthotal (40 mg/kg). The hearts were arrested by a slow injection of 1 ml of buffered 2% xylocaine into the vena cava and thoroughly washed with intracoronary perfusion of 200 ml cold potassium phosphate buffer (pH 7,4), delivered under pressure via the abdominal aorta. They were then excised and briefly rinsed in the same solution. After freeze drying, the left ventricular samples were divided into two aliquots for TG or sodium determination. Tissue glycerides were determined enzymatically according to Altmann et al. [1] using chloroform/methanol extraction. The samples for sodium determination were deproteinized in 3 ml of 0.6 N perchloric acid, and sodium concentrations in the neutralized supernagents of tissue extracts determined by flame photometry in the presence of 10 mM $LiCl_2$.

The hearts of two other series of control and mechanically overloaded rats were perfused at 70 Torr according to Langendorff. The perfusate consisted of a bicarbonate buffer supplemented with 11 mM glucose and 10 IU of insulin. L-[^{14}C] carnitine concentration (New England Nuclear) varied from 10 to 200 µM (spec-activity of about 450 cpm/nmol).

After the initial 10 min period, necessary to remove any blood and stabilize the heart rate at about 240 bpm, 200 ml of buffer containing 50 µM L-[^{14}C] carnitine (specific activity of about 450 cpm/nmd) was recirculated for variable periods (10, 20, 30 or 45 min). At the end of this period, the hearts were switched to a perfusate (Krebs-Henseleit buffer) which did not contain any carnitine and left for 6 min. This proved necessary in order to eliminated the radioactive label from the extracellular spaces [23]. At the end of the perfusion, the hearts were rapidly frozen by cool Wollenberger clamps and stored in liquid nitrogen. The tissue was then lyophilized and aliquots of about 100 mg dry wt. introduced into the polycarbonate capsules. The samples were mineralized in a catalytic oven (Intertechnique IN 4101) and [^{14}C]O_2 quantitatively trapped in the scintillation liquid containing phenylethylamine. The amount of L-carnitine taken by the hearts was expressed in nmol/g dry wt. after having determined the specific activity of L-carnitine in the perfusate (cpm/nmol). In some experiments, 0.05 mM salyrganic acid was added to the perfusate containing the L-carnitine, in order to dissociate the passive component from total carnitine transport to the heart [23]. The analysis of saturation curves thus obtained allowed us to compare the activity of carrier-mediated transport in control and mechanically overloaded hearts. The V_{max} and apparent K_m of the saturable carnitine transport were determined using the Lineweaver-Burk method [25].

Results

When the control hearts were perfused with physiological concentration of L-carnitine (50 µM, 450 cpm/nmol), their radioactivity progressively increased. The same linear re-

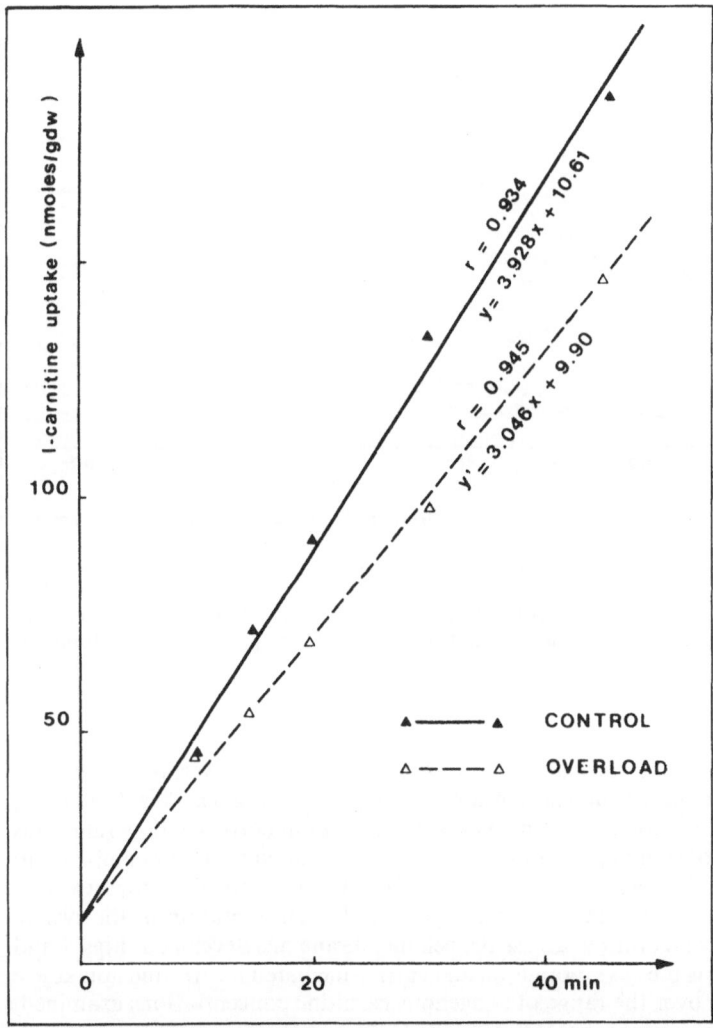

Fig. 1. L-[^{14}C] carnitine uptake as function of time in control and overloaded rat hearts. Each point represent the mean of 5–10 observations.

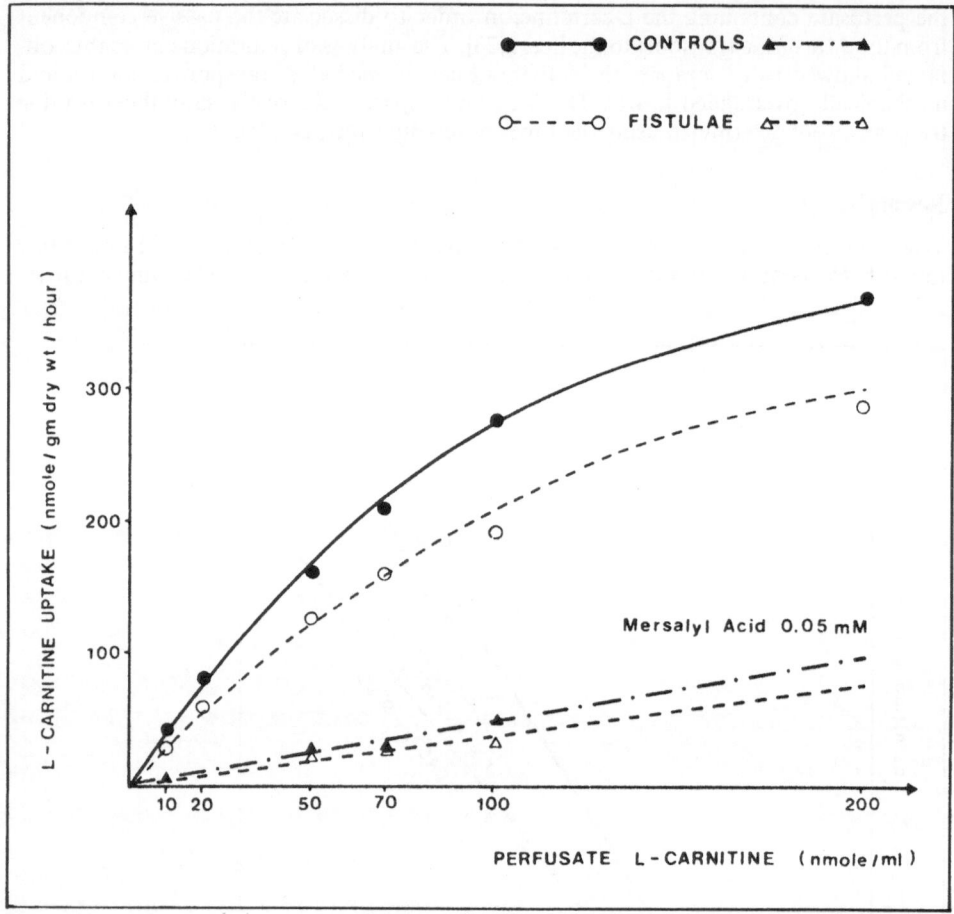

Fig. 2. Uptake of L-[^{14}C] carnitine by control and overloaded hearts as a function of different carnitine concentration perfusates. Mean values (n = 5–10) of both total and passive carnitine transport are indicated.

lationship also occurred in the mechanically overloaded hearts. However, is this case, the slope of L-carnitine accumulation was considerably depressed (Fig. 1). This decrease of L-carnitine transport does not seem to be related to changes in coronary flow rate. This latter parameter was not significantly modified in our overloaded hearts as compared to control-ones (64.1 ± 6.3 instead of 64.9 ± 7.1 ml/g of dry wt per min). Therefore it would seem that during the constitution of cardiac hypertrophy, an alteration in the system transporting exogenous L-carnitine across the cell membrane has developed. Figs. 2 and 3 show clearly that this is the case: both total and carrier-mediated L-carnitine uptake are significantly depressed over the range of exogenous carnitine concentrations examined. On the other hand, the passive (concentration gradient-mediated) L-carnitine transport, as observed in the presence of 0.05 mM salyrganic acid, remains unaffected in the mechanically overloaded hearts. According to the results of the Lineweaver-Burk analysis of

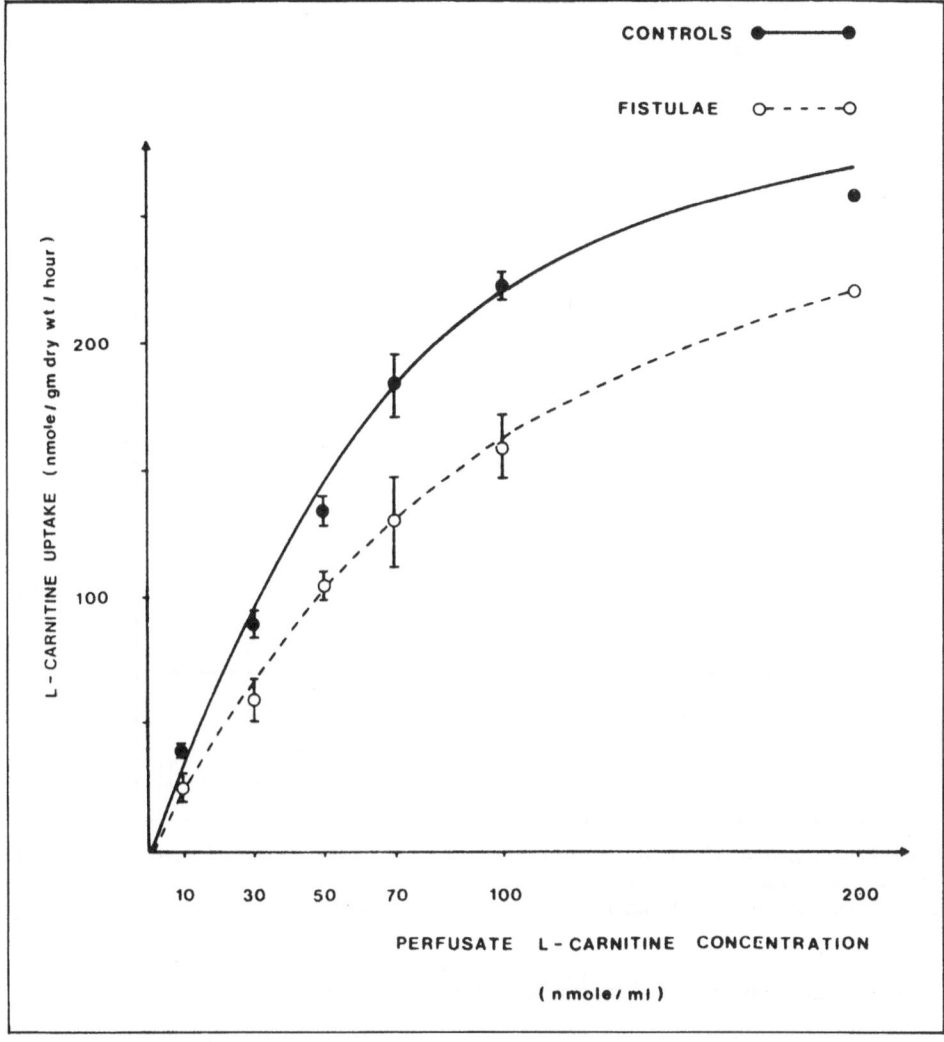

Fig. 3. Saturations curves of the carrier mediated transport as observed in control and mechanically overloaded rat hearts (obtained by substraction of values given in Fig. 2).

the respective saturation curves (Fig. 4), a decreased affinity of carnitine carrier for L-carnitine (higher apparent K_m[carnitine], V_{max} unchanged) may explain the above decrease of L-carnitine transport and possibly the low tissue levels of L-carnitine [6]. The present observations seems to strengthen the contention expressed in this latter study that decreased L-carnitine content of mechanically overloaded hearts does not result from either decreased disposability of this compound (the plasma levels of L-carnitine having been unchanged in rats with three-month-old aorto-caval fistula [6]), nor from its accelerated destruction in this tissue.

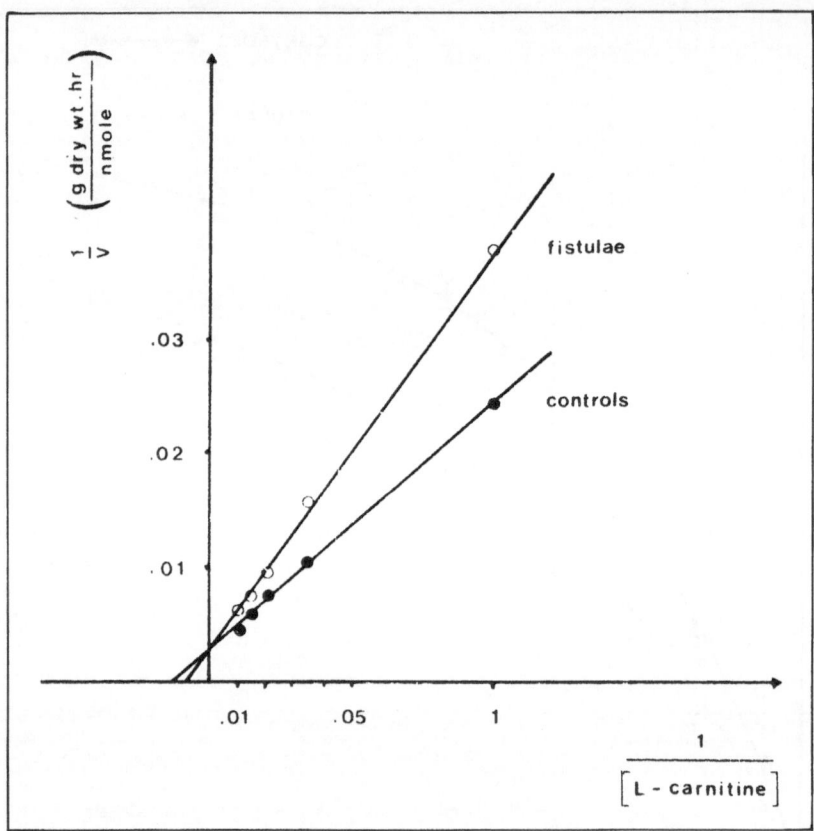

Fig. 4. Kinetic analysis of saturable carnitine transport; the reciprocal plots of Lineweaver and Burk. Note increased apparent K_m[carnitine] in volume overloaded hearts (125 instead of 83 µM in controls). No change of V_{max} (380 nmol/g of dry wt. per h) could be detected.

Table 1. Tissue triglycerides and sodium content in control and mechanically overloaded rat hearts

	Tissue content of		Estimated intracellular
	TG (µmol/g of dry wt.)	Intracellular Na (mmol/g of dry wt.)	Na concentration (mM)
Controls (n = 10)	13.21 ± 1.00	25.18 ± 0.5	11.44 ± 0.10
Fistulae (n = 10)	11.80 ± 0.30	25.98 ± 1.6	11.81 ± 0.32
	NS	NS	NS

m ± SE$_M$, tissue intracellular water content estimated as 44% (Vary and Neely 1982).

Another factor which can be ruled out is the altered transmembrane gradient of sodium. According to our data given in Table 1, the intracellular sodium content, as well as the estimated intracellular sodium concentration, of the non-failing mechanically overloaded hearts are quite the same as those of control hearts (Table 1). This seems to exclude the possibility of altered energy supply for L-carnitine/sodium co-transport [25]. Another result given in Table 1 refers to our TG determinations in the homogenates of control and mechanically overloaded hearts. These data suggest that, despite of low levels of tissue carnitine, the mechanically overloaded hearts do not present any TG accumulation, at least in situ (Table 1).

Discussion

Low tissue levels of L-carnitine were previously described in different types of mechanical overload [6, 20, 22, 27] as well as in Syrian hamsters with hypertrophic cardiomyopathy [12, 26]. In both of these conditions a defective carnitine transport has been predicted, but a direct proof is still missing. According to our data, carnitine deficiency of mechanically overloaded hearts may result from an alteration of carrier-mediated transport of L-carnitine [5]. This latter process has been recently characterized in normal myocardium by Bahl [2] and Neely [23] and it is believed to be necessary for the intracellular concentrations of L-carnitine to be maintained about 50 fold higher than that of circulating blood.

Our kinetic analysis of the saturable carrier-mediated carnitine transport, as observed on isolated hearts perfused with variable concentrations of exogenous carnitine, does suggest that the affinity of carnitine carrier [5, 23] for L-carnitine decreases during the development of cardiac hypertrophy (K_m[carnitine] of 125 instead of 85 μM, V_{max} unchanged). These changes do not seem to be related either to the presence of any competitive inhibitor (we controlled the composition of our perfusion medium fairly well) nor to the accumulation of free L-carnitine in the cytosol. According to our previous data [6] the degree of L-carnitine esterification is not affected in the normoxic mechanically overloaded hearts. At the same time, total carnitine (90 per cent of which is cytosolic) [18] decreased by about 33 per cent. Therefore a decrease rather than an accumulation of free L-carnitine in the cytosol of the overloaded hearts can be expected. Another parameter which should be considered in relation to altered carnitine transport is the intracellular sodium content of the ventricular myocardium. According to recent studies, the energy necessary for the up hill carnitine transport to the myocardium is supplied by sodium co-transport driven by the transmembrane concentration gradient of this ion [25]. According to our data, no modification in sodium content in the ventricular myocardium of mechanically overloaded hearts can be detected. The estimated intracellular sodium concentrations are also very similar in control and overloaded hearts. It would seem that either a defect of the sarcolemmal carnitine-binding protein [28] or alterations in membrane fluidity, secondary to changes in membrane phospholipides [21] should be considered in order to explain the decrease of carnitine transport which we observed in the mechanically overloaded hearts.

As to the impact of decreased carnitine concentration on the energy production of mechanically overloaded hearts, it has often been suggested analogously to the observations made on isolated cardiac mitochondria [12, 19], that it may compromise the FFA oxidation and lead to TG accumulation in the tissue. This has been considered to occur in the mechanically overloaded hearts of different species [22, 27] and in Syrian hamsters with congenital cardiomyopathy [12]. In both of these conditions a dysfunction of carnitine acyl transferase has been predicted, since tissue carnitine levels were 40–50 per cent lower than those of control hearts. However low tissue carnitine can sometimes be accompanied by a decrease of total tissue CoA [26] so that the mass action ratio of carnitine acyl transferase:

$$\frac{[\text{long chain acyl-CoA}] + [\text{free carnitine}]}{[\text{long chain acyl carnitine}] + [\text{free CoA}]}$$

need not be markedly different between the control and diseased hearts [14, 26]. In other words, the carnitine acyl transferase complex need not become rate-limiting for FFA oxidation. In our previous work on the hearts of rats with chronic aorto-caval fistula we noted a slight decrease in total tissue CoA [15] as well as a significant decrease in tissue carnitine [6]. It has been suggested that these two changes, when they occur simultaneously, may avoid a major TG accumulation and protect the mitochondrial function of carnitine deficient hearts [26]. However, direct measurements of $[C^{14}]O_2$ production by hearts perfused with $[C^{14}]$-labelled fatty acids will be necessary in order to ascertain that the lack of TG accumulation, which we observed in this study, does not reflect merely a decreased availability of blood FFA resulting from the chronic anorexia of rats with aorto-caval fistula [16].

Acknowledgements: Mrs F. Simonet of the INSTN Saclay is thanked for having given us the opportunity to use her catalytic oven.

References

1. Altmann A, Bach A, Metais P (1967) Dosage des TG; détermination enzymatique du glycérol. In: Biochimie des Lipides, Simep Editions, Lyon, pp 157–161
2. Bahl J, Navim T, Manian A, Bressler R (1981) Carnitine transport in isolated adult rat heart myocytes. Circ Res 48:378–385
3. Bing RJ (1965) Cardiac metabolism. Physiol Rev 45:171–213
4. Bishop SP, Altschuld RA (1971) Evidence for increased glycolytic metabolism in cardiac hypertrophy and congestive heart failure. In: Alpert NR (ed) Cardiac hypertrophy. Academic Press, New York, pp 567–570
5. Bohmer T, Mølstad P (1980) Carnitine transport across the plasma cell membrane. In: Frenkel RA, McGarry JD (eds) Carnitine biosynthesis, metabolism and functions. Academic Press, New York, pp 73–89
6. Bowe C, Nzonzi J, Corsin A, Moravec J, Feuvray D (1984) Lipid intermediates in chronically volume-overloaded rat hearts. Pflügers Arch 102:317–320
7. Bremer J (1962) Carnitine in intermediary metabolism. The metabolism of fatty acid esters of carnitine by mitochondria. J Biol Chem 237:3628–3636
8. Bremer J (1983) Carnitine metabolism and function. Physiol Rev 63:1420–1480
9. Crass MF, McCaskill ES, Shipp JC (1969) Effect of pressure development on glucose and palmitate metabolism in perfused heart. Am J Physiol 216:1569–1576
10. Folch J, Lees M, Sloanstanley GM (1961) In: Glick D (ed) Methods in biochemical analysis, vol II. Interscience Publisher, New York, pp 106–112
11. Fritz IB, Yke KTN (1962) Long chain carnitine acyl transferase and the role of acyl carnitine derivatives in the catalytic increase of fatty acid oxidation induced by carnitine. J Lipid Res 4:279–288
12. Hoppel CL, Tandler B, Parland W, Turkaly JS, Albers LD (1982) Hamster cardiomyopathy: a defect in oxidative phosphorylation in cardiac mitochondria. J Biol Chem 257:1540–1548
13. Leidke AJ, Nellis SH, Whitesell LF, Mahar CO (1982) Metabolic and mechanical effects of L- and D-carnitine in working swine hearts. Am J Physiol 243:H691–H697
14. Lopaschuk GD, Hansen CA, Neely JR (1986) Fatty acid metabolism in hearts containing elevated levels of CoA. Am J Physiol 250:H351–H359

15. Moravec J (1980) Possible relationship between tissue levels of long chain acyl-CoA and the ability of the overloaded hearts to oxidize an excess of reduced pyridine nucleotides. FEBS Lett 113:134–137
16. Moravec J, Moravec M, Hatt PY (1981) Rate of pyridine nucleotide oxidation and cytochrome oxidase interaction with intracellular oxygen in hearts of rats with compensated volume overload. Pflügers Arch 392:106–114
17. Neely JR, Morgan HE (1974) Relationship between carbohydrate and lipid metabolism and energy balance of the heart. Annu Rev Physiol 36:413–459
18. Neely JR, Robishaw J, Vary T (1982) Control of myocardial levels of CoA and carnitine. J Mol Cell Cardiol [Suppl 3] 4:30–42
19. Ramsay RR, Tubbs PK (1975) The mechanism of FFA uptake by heart mitochondria. FEBBS Lett 54:21–25
20. Reibel DK, Uboh CE, Kent RL (1983) Altered CoA and carnitine metabolism in pressure overloaded hypertrophied hearts. Am J Physiol 244:H839–H843
21. Reibel DK, O'Rurke B, Foster KA, Hutchinson H, Uboh CE, Kent RL (1986) Altered phospholipid metabolism in pressure-overloaded hypertrophied hearts. Am J Physiol 250:H1–H6
22. Revis NW, Cameron AJV (1979) Metabolism of lipids in experimentally hypertrophic hearts of rabbits. Metabolism 28:601–613
23. Vary T, Neely JR (1982) Characterization of carnitine transport in isolated perfused adult rat hearts. Am J Physiol 242:H585–H592
24. Vary T, Neely JR (1982) A mechanism for reduced myocardial carnitine levels in diabetic animals. Am J Physiol 243:H154–H158
25. Vary T, Neely JR (1983) Sodium dependence of carnitine transport in isolated perfused rat hearts. Am J Physiol 244:H247–H252
26. Whitmer JT (1986) Energy metabolism and mechanical function in perfused hearts of Syrian hamsters with dilated on hypertrophic cardiomyopathy. J Mol Cell Cardiol 18:307–317
27. Wittels B, Spann JF (1968) Defective lipid metabolism in the failing heart. J Clin Invest 47:1787–1794
28. York CM, Cantrell CR, Borum PR (1983) Cardiac carnitine deficiency and altered carnitine transport in cardiomyopathic hamsters. Arch Biochem Biophys 221:526–533

Authors' address:

J. Moravec, M.D., Dr. Sc., I.N.S.E.R.M. U2 Hôpital Léon Bernard, 94456 Limeil-Brévannes Cédex, France, 2. El Alaoui Talibi is a fellow of the Marocain Dept of Education and Scientific Research.

New aspects of excitation-contraction coupling in cardiac muscle:
Two types of Ca^{++} entry promotion with and without involvement of cyclic AMP and Mg^{++} ions

A. Fleckenstein, F. Späh, W. L. Wagner, and M. Frey

Physiological Institute, University of Freiburg, Study Group for Calcium Antagonism

I. Slow-channel mediated Ca^{++} influx and cardiac inotropism

Overwhelming evidence has accumulated during the last two decades indicating that the most important regulatory system for cardiac contractile force is the "slow" trans-sarcolemmal Ca^{++} channel. This carrier mechanism controls Ca^{++} entry through the excited myocardial fibre membranes quantitatively, and thereby promotes or inhibits the chain reaction which finally leads – via splitting of ATP by the Ca^{++}-activated myofibrillar ATPase – to the appearance of contractile energy. It is certainly a highlight in experimental cardiovascular research that drugs have been found which act as keys for this Ca^{++}-dependent physiological regulatory system. More precisely, with the help of specific Ca^{++}-antagonists discovered in our laboratory, transmembrane Ca^{++} influx can be inhibited and contractile energy expenditure reduced to any desired extent [3–6]. Conversely, positive inotropic effects are initiated by many drugs or procedures which promote transmembrane Ca^{++} entry. For instance, a most simple way of augmenting contractile force consists of an increase in extracellular Ca^{++} concentration that favors Ca^{++} uptake from the environment.

a. Promotion of Ca^{++} influx through formation of cyclic AMP

It is widely accepted that, under physiological conditions, i.e. at a constant extracellular Ca^{++} level, the enhancement of slow-channel-mediated Ca^{++} influx results from facilitation of transmembrane inward Ca^{++} movements. In all probability, the physiological "opener" of the slow Ca^{++} channels is cyclic AMP (cAMP). This adenine nucleotide is produced (see Fig. 1) following:

  1. Stimulation of adrenergic β-receptors [10, 15, 16, 19, 27, 29, 31–33] or histaminergic H_2-receptors [7, 9, 10, 13, 14, 18],
  2. Receptor-independent activation of adenylate cyclase with forskolin [11, 17, 22, 24, 26], or
  3. Administration of dibutyryl cAMP [1, 9, 15, 16, 19, 23].

   Then, in the further sequence of events, a membrane located phosphokinase seems to use the cAMP for phosphorylation of sarcolemma membrane proteins [34]. The appearance of new additional phosphate groups in the superficial layers of the sarcolemma membrane and in the slow channels seems to increase the Ca^{++} binding capacity and

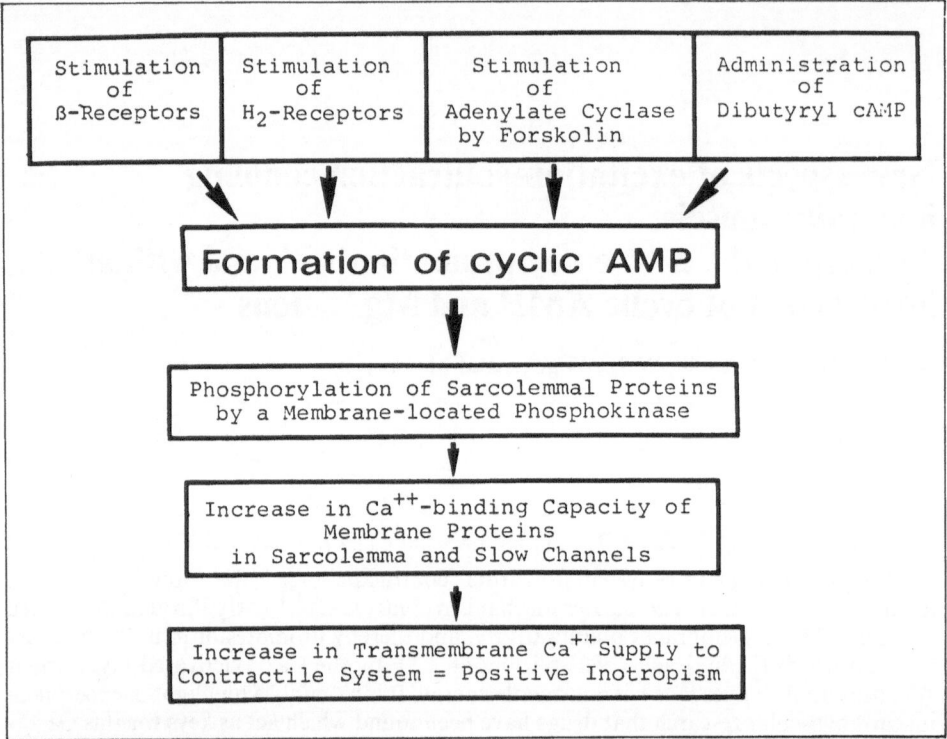

Fig. 1. Supposed key-role of cyclic AMP in the promotion of transsarcolemmal Ca^{++} influx.

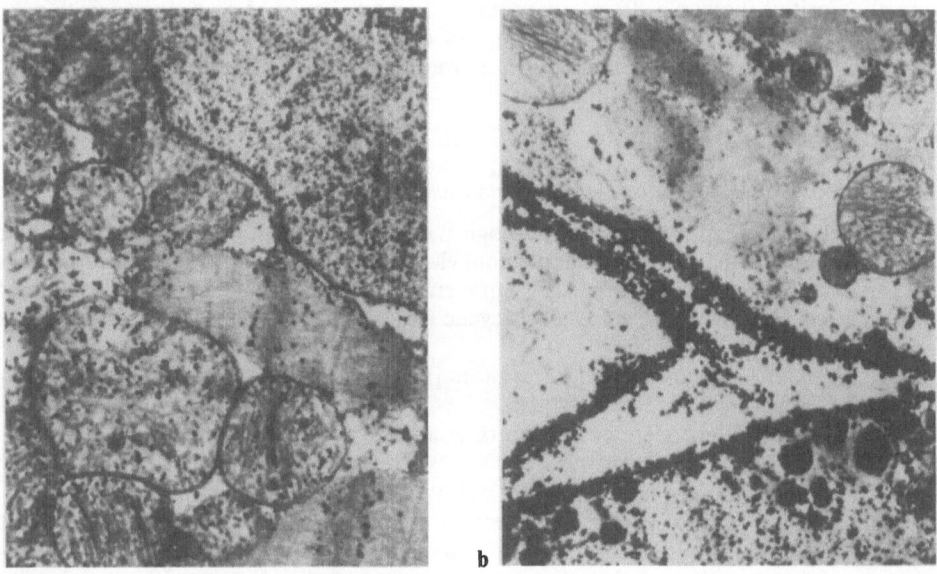

a   b

Fig. 2

transmembrane Ca^{++} conductivity. Thereby, the supply of activator Ca^{++} from extracellular sources (and the additional Ca^{++}-triggered Ca^{++} release from intracellular stores) is potentiated, so that positive inotropic effects are elicited. For instance, the electron micrograph of Fig. 2 shows directly the strong accumulation of Ca^{++} ions in the sarcolemma membrane of a rat ventricular fibre following subcutaneous injection of the adrenergic β-receptor stimulant isoproterenol.

b. Promotion of Ca^{++} influx without formation of cyclic AMP

However, there is still another potentiation of transmembrane inward Ca^{++} fluxes which does *not* involve the usual mediation by cAMP [8]. The fact that stimulation of Ca^{++} influx by Bay K 8644 and CGP 28392 takes place in a cAMP-independent way, was reported by Kokubun and Reuter, who studied these drug effects on cultured cardiac cells from neonatal rats [8]. Wagner and Fleckenstein [30] also stressed this peculiarity of a cAMP-independent Ca^{++} entry promotion. On uterine smooth muscle too, Bay K 8644

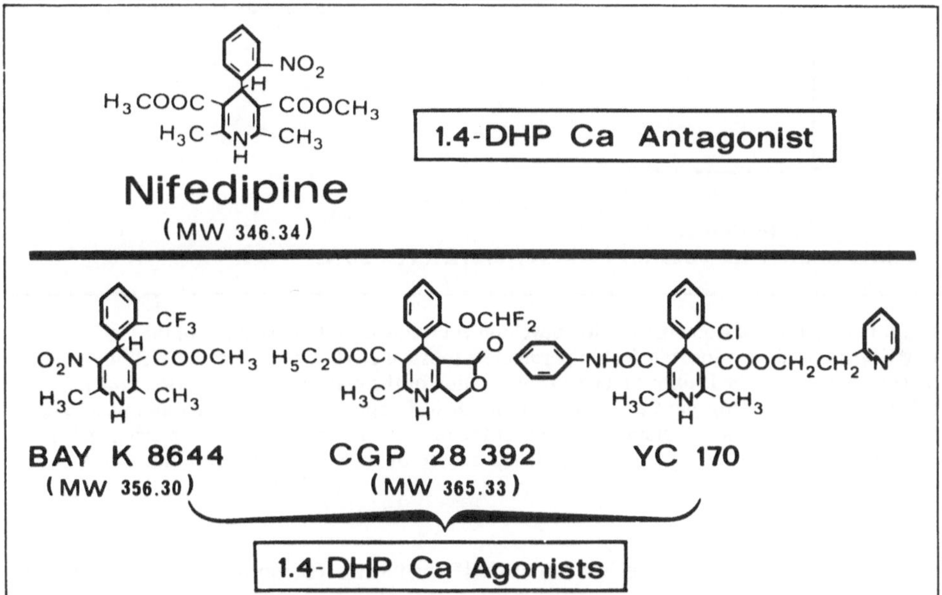

Fig. 3. Positive inotropism by new „calcium agonists" without involvement of cyclic AMP. Fundamental mechanism: facilitation of Ca^{++} entry by operating at the same sarcolemmal receptor groups that otherwise would produce slow channel blockade under the influence of Ca^{++}-antagonistic 1,4-dihydropyridine derivatives (nifedipine etc.).

Fig. 2. Electron-optical visualization of Ca^{++} with the potassium pyroantimonate technique (a) in a normal myocardial fibre of a control rat, and (b) in a myocardial fibre from a rat treated with the β-stimulant isoproterenol (30 mg/kg, injected subcutaneously). This dose of isoproterenol always leads to a tremendous accumulation of Ca^{++} in the sarcolemma membrane, followed by intrusion of Ca^{++} into the cytoplasm. Precipitated calcium appears in the form of electron-dense black spots.

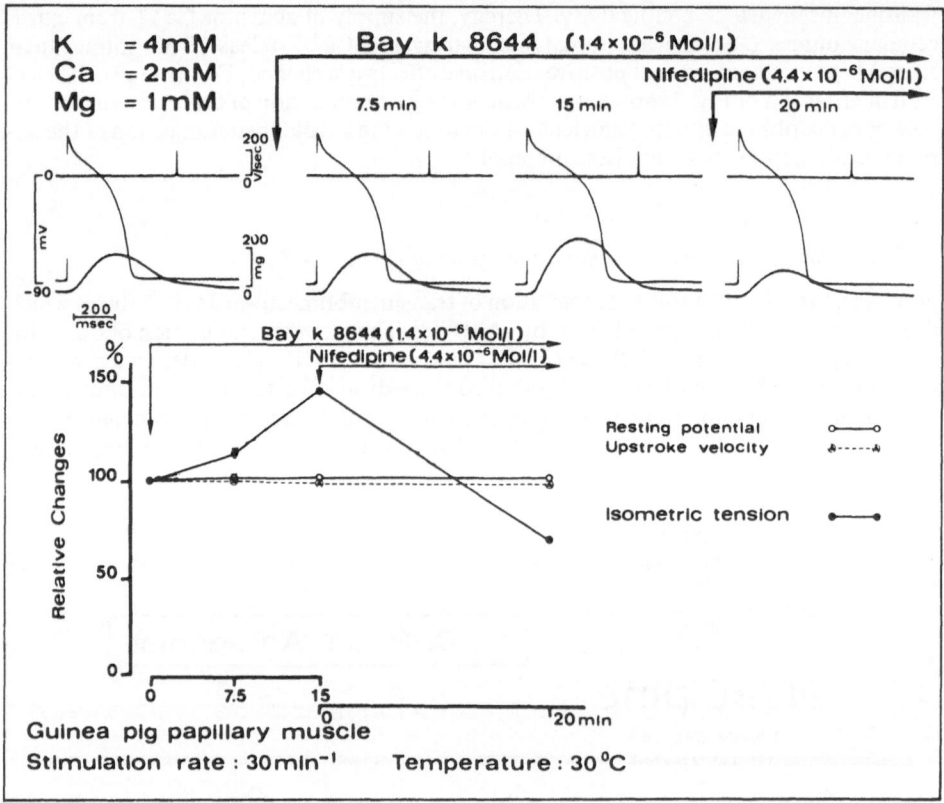

Fig. 4. Potentiation of isometric tension development following administration of the Ca^{++} agonist Bay K 8644, whereas most bioelectric parameters (upstroke velocity, overshoot, resting potential) remain constant. There is only a modest prolongation of the plateau phase of action potential. The enhancement of electrogenic Ca^{++} entry by Bay K 8644 appears to counteract and postpone the repolarizing outward K^+ current at the end of the plateau. The bioelectric and mechanical effects of the Ca^{++} agonist Bay K 8644 are readily neutralized by the Ca^{++} antagonist nifedipine.

was lacking any promoter effect on cAMP formation (personal communication by Dr. H. Metzger, Pharmacological Research Laboratories, Hoechst AG, Frankfurt).

This cAMP-independent Ca^{++} transport mechanism is activated by three new 1,4-dihydropyridine derivatives, which were found in the research laboratories of three different drug companies (Bayer AG, Ciba-Geigy Inc., Yamanuchi Inc.). The three compounds act as "calcium agonists" (see Fig. 3). They operate at the same receptor groups as the original Ca^{++}-antagonistic substance nifedipine does, but instead of blocking Ca^{++} entry in the usual manner, the new Ca^{++} agonists intensify Ca^{++} inflow and contraction (Fig. 4). The most potent Ca^{++} agonist is Bay K 8644. Its fundamental action, i.e. potentiation of contractile force by enhancement of Ca^{++} entry, was first described by Schramm et al. [20, 21]. As it could be expected, all Ca^{++} entry promoters, whether they are acting through formation of cAMP or in a cAMP-independent way, can be neutralized by suitable Ca^{++} antagonists (Fig. 4).

II. Reciprocal suppression of electrogenic Mg^{++} effects bei cAMP-producing Ca^{++} entry promoters

a. Observations of Ca^{++}- or Mg^{++}-dependent action potentials of partially depolarized myocardial fibres

The physiology and pharmacology of the slow Ca^{++} channels is currently a favourite research topic. But one should realize in this context that observations which focus exclu-

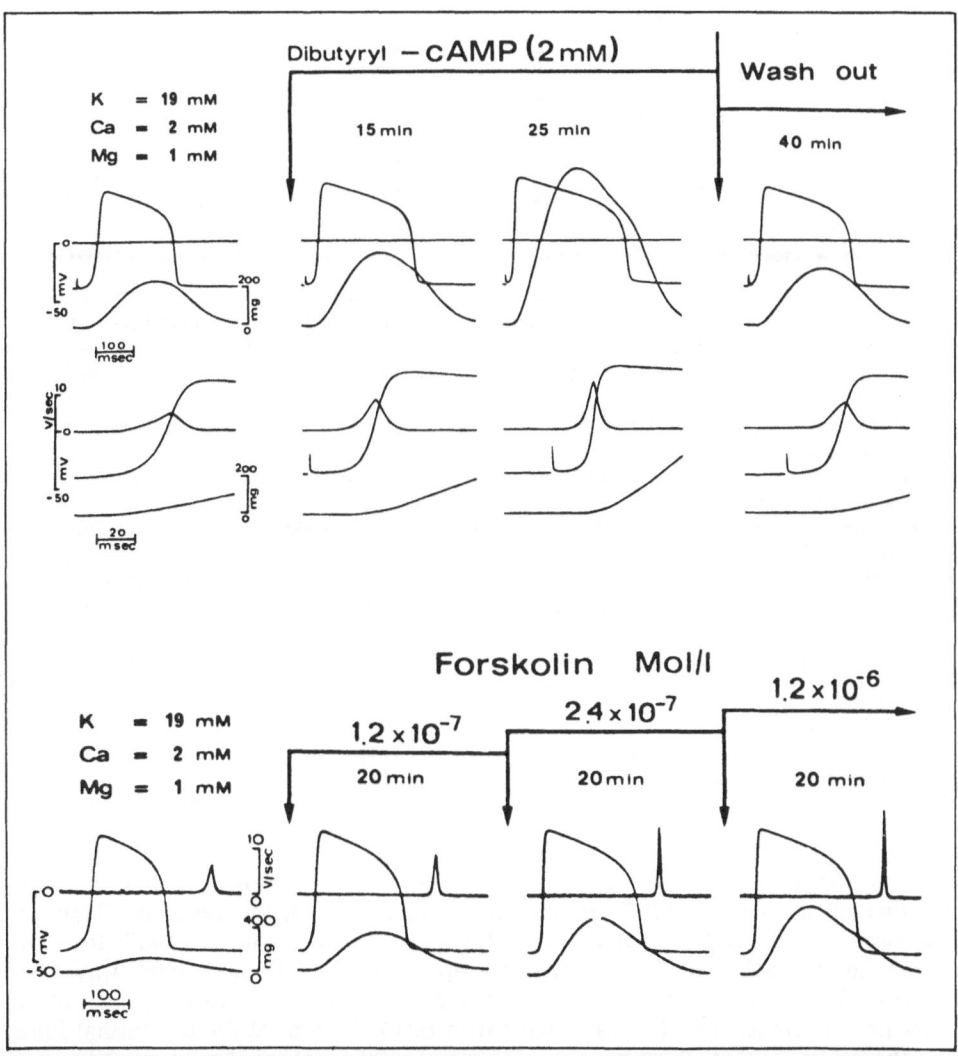

Fig. 5. Parallel potentiation of the bioelectric parameters (upstroke velocity, height and duration) of Ca^{++}-carried action potentials and of contractile force of partially depolarized guinea-pig papillary muscles under the influence of dibutyryl cAMP (upper panel) and forskolin (lower panel). Upstroke velocity (dV/dt max) of the Ca^{++}-carried action potentials (indicating the intensity of Ca^{++} influx) rose from 5–7 V/s to 12–20 V/s, i.e. by 2–3 times. Isometric peak tension increased simultaneously to a similar extent.

Table 1. Characteristics of action potentials of partially depolarized guinea-pig papillary muscles mediated by Ca or Mg ions[a]

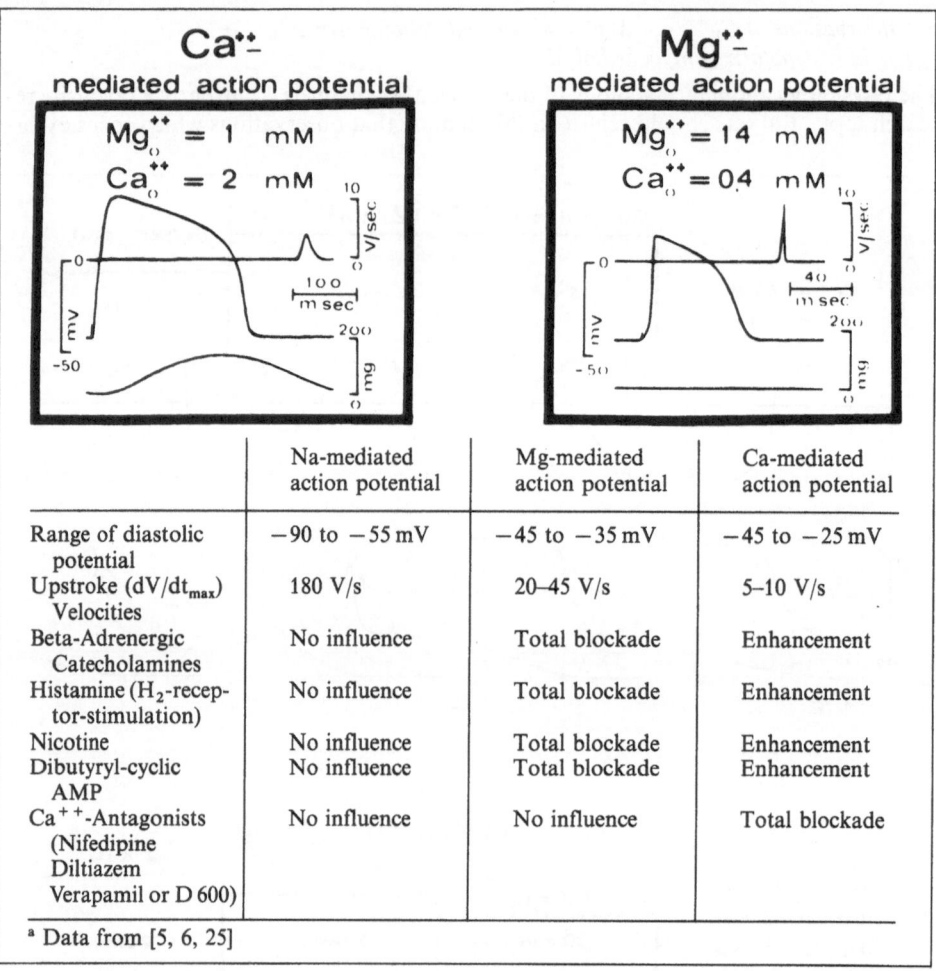

	Na-mediated action potential	Mg-mediated action potential	Ca-mediated action potential
Range of diastolic potential	−90 to −55 mV	−45 to −35 mV	−45 to −25 mV
Upstroke (dV/dt_{max}) Velocities	180 V/s	20–45 V/s	5–10 V/s
Beta-Adrenergic Catecholamines	No influence	Total blockade	Enhancement
Histamine (H_2-receptor-stimulation)	No influence	Total blockade	Enhancement
Nicotine	No influence	Total blockade	Enhancement
Dibutyryl-cyclic AMP	No influence	Total blockade	Enhancement
Ca^{++}-Antagonists (Nifedipine Diltiazem Verapamil or D 600)	No influence	No influence	Total blockade

[a] Data from [5, 6, 25]

sively on the transmembrane Ca^{++} movements are rather incomplete, as long as simultaneous interactions with Mg^{++} ions are not included in such considerations. Thus we emphasized years ago [5, 6, 25], that all substances which promote contractile force by mediation of cAMP exert their positive-inotropic actions strictly in a double way, that is to say: *firstly*, they potentiate trans-sarcolemmal Ca^{++} entry, but *secondly*, they exert a reciprocal suppression of electrogenic effects of Mg^{++} ions at the myocardial fibre membrane. We have obtained evidence of this reciprocal membrane action exerted by the cAMP-producing Ca^{++} entry promoters, in experiments on *partially depolarized* myocardium of guinea-pigs (isolated papillary muscles). It is well known that the fast TTX-sensitive Na^+ transport system in the myocardial fibre membranes becomes inactivated if the resting potential falls from its normal height of about −85 mV to a lower level of about −45 mV. This critical fall can easily be induced by application of a K^+-rich

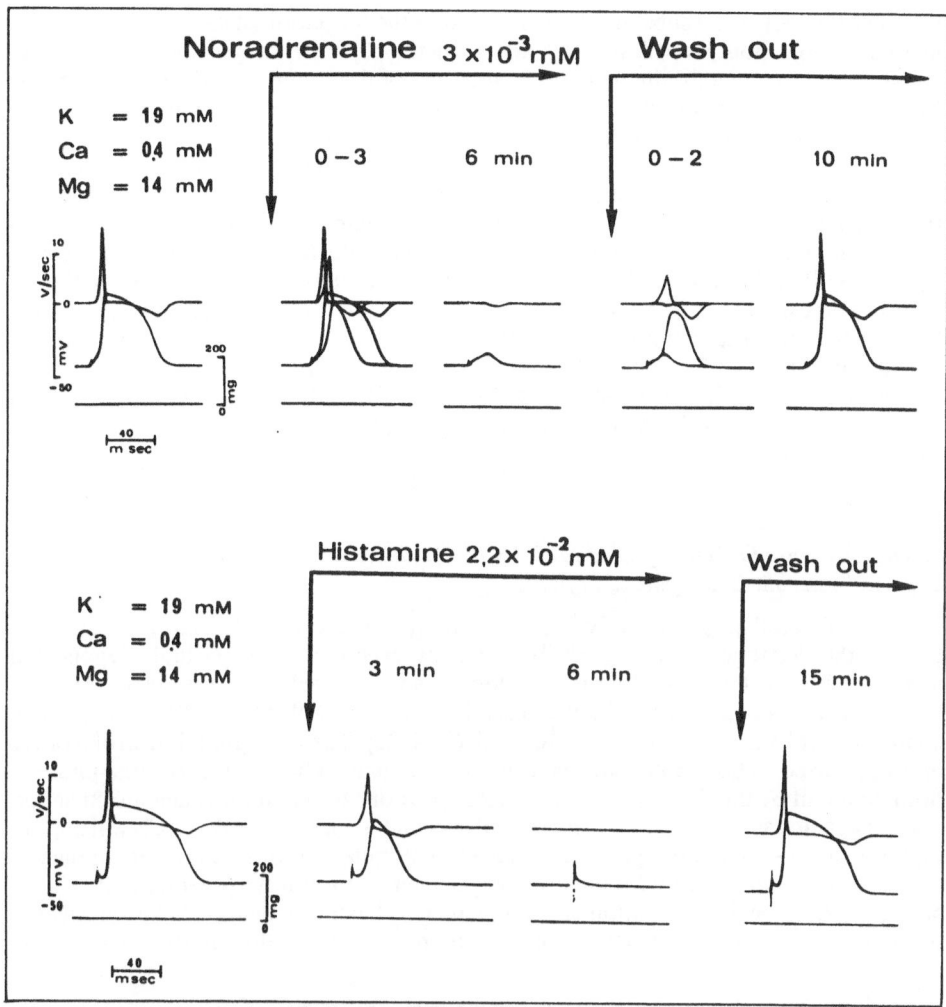

Fig. 6. Suppression of Mg^{++}-mediated action potentials by β-receptor stimulation with noradrenaline (upper experiment) or by H_2-receptor stimulation with histamine (lower experiment). If the stimulants are washed out or neutralized by β- or H_2-receptor blockade, the Mg^{++}-mediated action potentials revive. Decline and subsequent recovery of upstroke velocity (dV/dt max) of the Mg^{++}-mediated action potentials is shown in the upper tracing of each experiment. The studies were carried out on partially depolarized guinea-pig papillary muscles in a special Tyrode solution with a high K^+ (19 mM), high Mg^{++} (14 mM), and a low Ca^{++} (0,4 mM) concentration. Stimulation rate: 30/min; temperature: 30 °C.

(19 mM K^+) Ringer or Tyrode solution. Nevertheless, the partially depolarized myocardial fibres are still able to respond to electrical stimuli with the generation of action potentials under two conditions [3–6, 12, 25, 26, 28].

Firstly, as long as the depolarizing Ringer or Tyrode solutions contain a sufficient concentration of Ca^{++} (2 mM), Ca^{++} ions are able to replace the Na^+ ions as electric charge carriers. Such Ca^{++}-carried action potentials (and contractile force) are stimulated enormously by all substances which enhance the formation of cAMP. For instance, in Fig. 5 (top), dibutyryl cAMP was added to the K^+-rich Ringer solution and consequently led to a massive increase in the intensity of Ca^{++} influx and contractile force. Moreover, β-receptor stimulation, H_2-receptor stimulation or forskolin produced identical effects. Figure 5 (bottom) shows the action of forskolin [24, 25].

A second possibility, however, is the maintenance of the electrogenesis of action potentials in partially depolarized myocardium by means of an elevated Mg^{++} content of the environmental solution, if not enough Ca^{++} is available (see Table 1). But in contrast with the stimulation of Ca^{++} influx, the electrogenic effects of Mg^{++} ions are reciprocally [5, 6, 25, 26] suppressed by all Ca^{++} entry promoters which act by mediation of cAMP. This can be seen from our experiments with noradrenaline, histamine or forskolin (see Fig. 6). Our observations illustrate that cAMP does not produce a simple increase in transmembrane Ca^{++} supply, but in addition always damps the influence of Mg^{++} ions which would otherwise restrict Ca^{++}-dependent contractile activation as natural Ca^{++} antagonists. In other words, by counteracting Mg^{++}, cAMP maximizes the positive inotropic influence of Ca^{++}.

b. Observations of mixed biphasic Mg^{++}/Ca^{++}-induced action potentials of partially depolarized myocardial fibres

There is still another, particularly elegant technique to assess the differential drug influence on the electrogenic Ca^{++} and Mg^{++} effects, even in a single partially depolarized myocardial fibre. The procedure is as follows: As we showed several years ago (Fig. 7), biphasic action potentials can be produced in a K^+-rich (19 mM K^+) medium, which contains 2 mM Ca^{++} and 14 mM Mg^{++} [4–6, 24–26]. The rapid initial upstroke phase of these mixed action potentials is Mg^{++}-dependent, whereas the second phase is brought about by the slow inward Ca^{++} current. If one differentiates such mixed action potentials, a first sharp Mg^{++} peak (No. 1) and a second blunt Ca^{++} peak (No. 2) will appear. Figure 8 shows – as an example – that β-receptor stimulation with adrenaline (in the upper part), and isoproterenol (in the lower experiment) enormously potentiate the Ca^{++} peak (No. 2) in the differentiation curve, together with contractile force. Conversely, the sharp Mg^{++} peak No. 1 becomes more and more flat and finally disappears when the

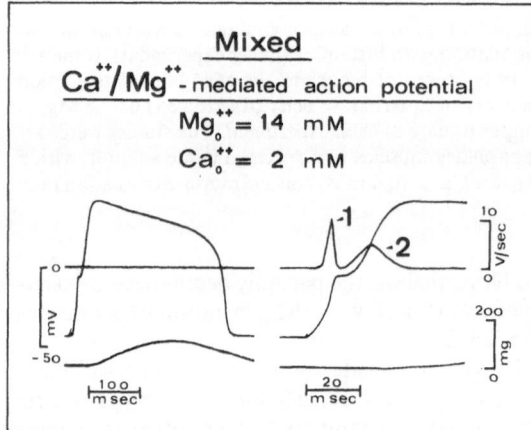

Fig. 7. Partially depolarized guinea-pig papillary muscle fibres in a K^+-rich (19 mM K^+) Tyrode solution respond, in the presence of 2 mM $CaCl_2$ and 14 mM $MgCl_2$, to electric stimulation of elevated strength with „mixed" biphasic action potentials. A notch in the upstroke phase signifies the point of transition from the first (rapid) Mg^{++}-dependent part to the second (slow) Ca^{++}-dependent part. If the bioelectric membrane activity is registered with a higher speed, and, in addition, differentiated, two distinct peaks of upstroke velocity (dV/dt max) become apparent, i.e. a first sharp Mg^{++} peak No. 1, followed by a second blunt Ca^{++} peak No. 2.

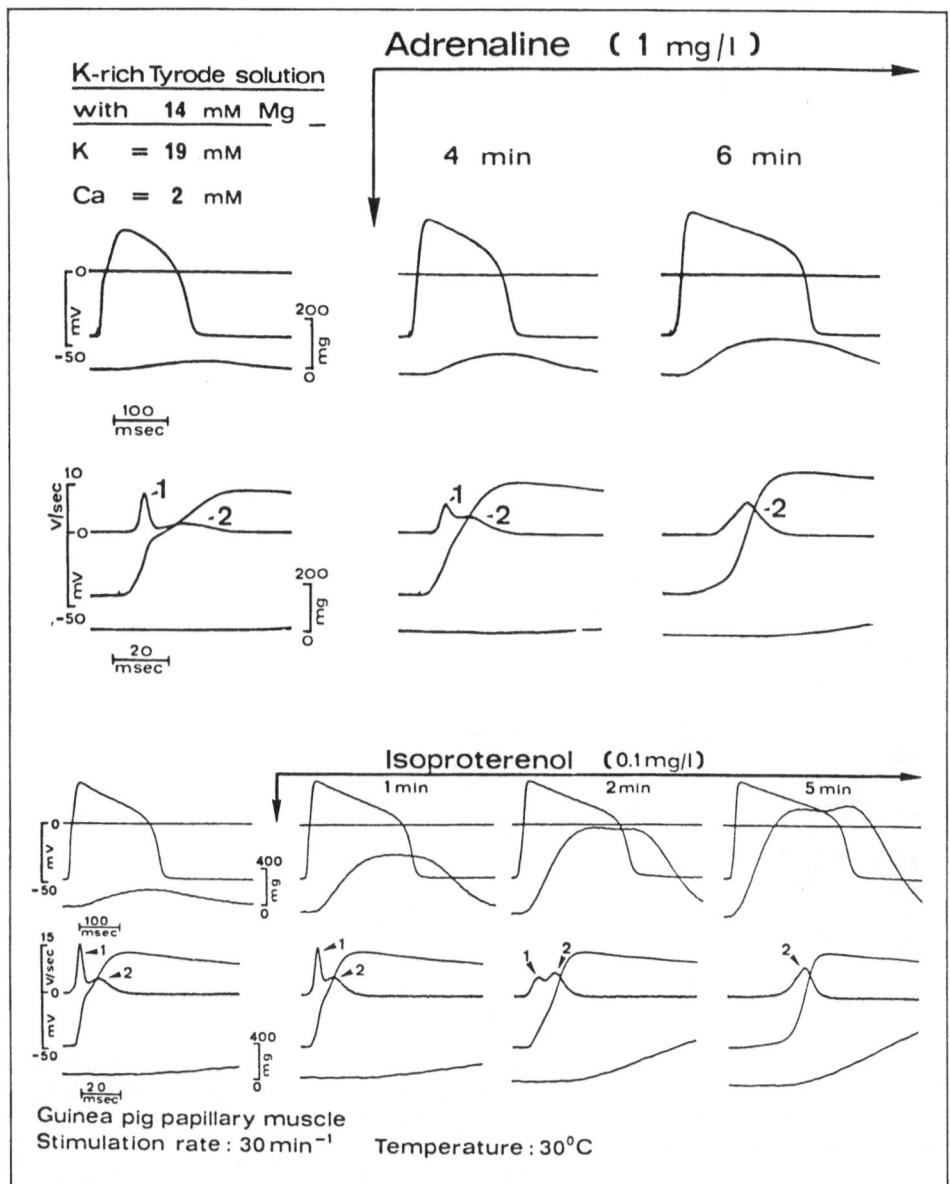

Fig. 8. Reciprocal effects of adrenergic β-receptor stimulation with adrenaline or isoproterenol on the Mg^{++} and Ca^{++}-mediated bioelectric membrane activities: suppression of the Mg^{++}-dependent peak No. 1 of upstroke velocity (dV/dt max) of mixed Mg^{++}/Ca^{++}-mediated action potentials simultaneously with a maximum reinforcement of the Ca^{++}-dependent peak No. 2. Isometric tension grows with decrease of the Mg^{++} peak, and with potentiation of the Ca^{++} peak which reflects the raise in slow-channel-mediated Ca^{++} influx.

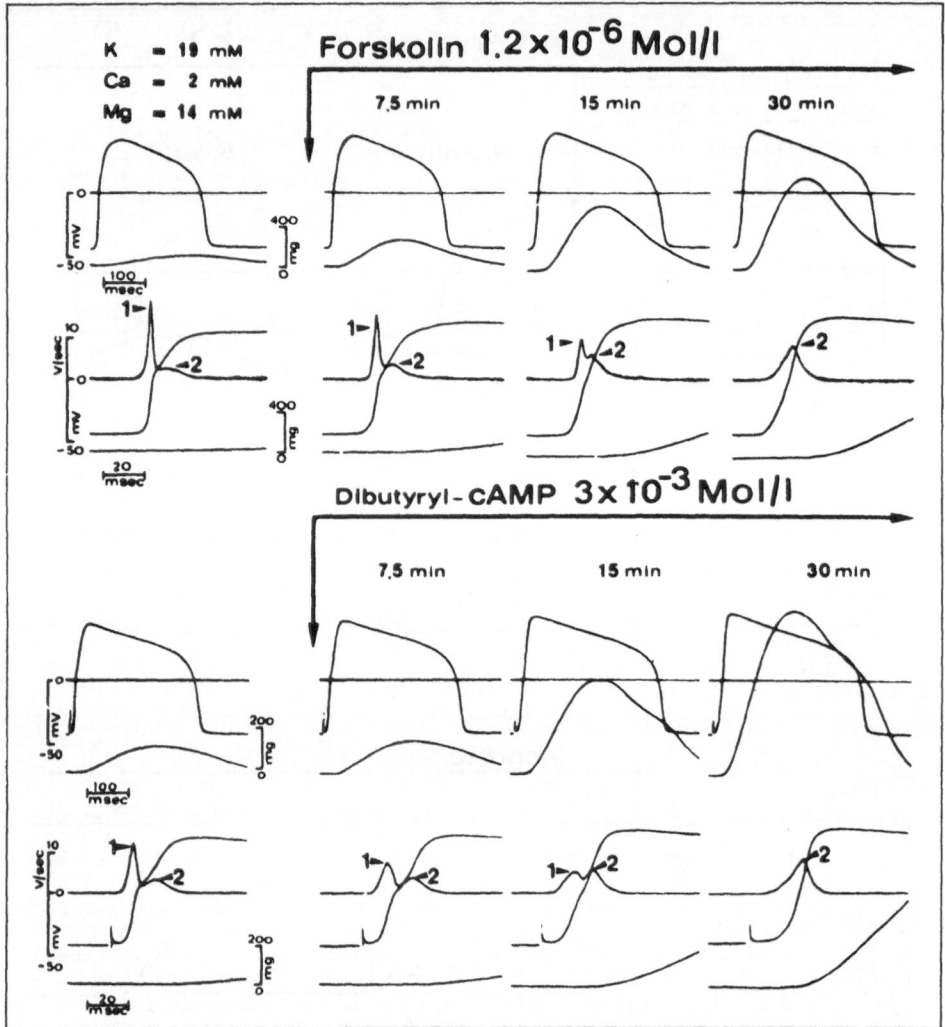

Fig. 9. Reciprocal influence on the Mg^{++}- and Ca^{++}-dependent peaks of mixed action potentials following intermediate formation of cyclic AMP by activation of adenylate cyclase with forskolin (upper) or by administration of dibutyryl cyclic AMP as precursor (lower experiment). Same methods as in the Figs. 7 and 8.

Ca^{++} peak No. 2 reaches its maximum. This reciprocal suppression of the Mg^{++} peak No. 1, coupled with a simultaneous large increase of the Ca^{++} peak No. 2, is indeed, the most spectacular effect which is uniformly exerted by all cAMP producing drugs, whether this happens after β-receptor or H_2-receptor stimulation or in consequence of administration of forskolin or dibutyryl cAMP (Fig. 9). Noradrenaline releasing agents such as tyramine or nicotine act in a similar way (see Fig. 10).

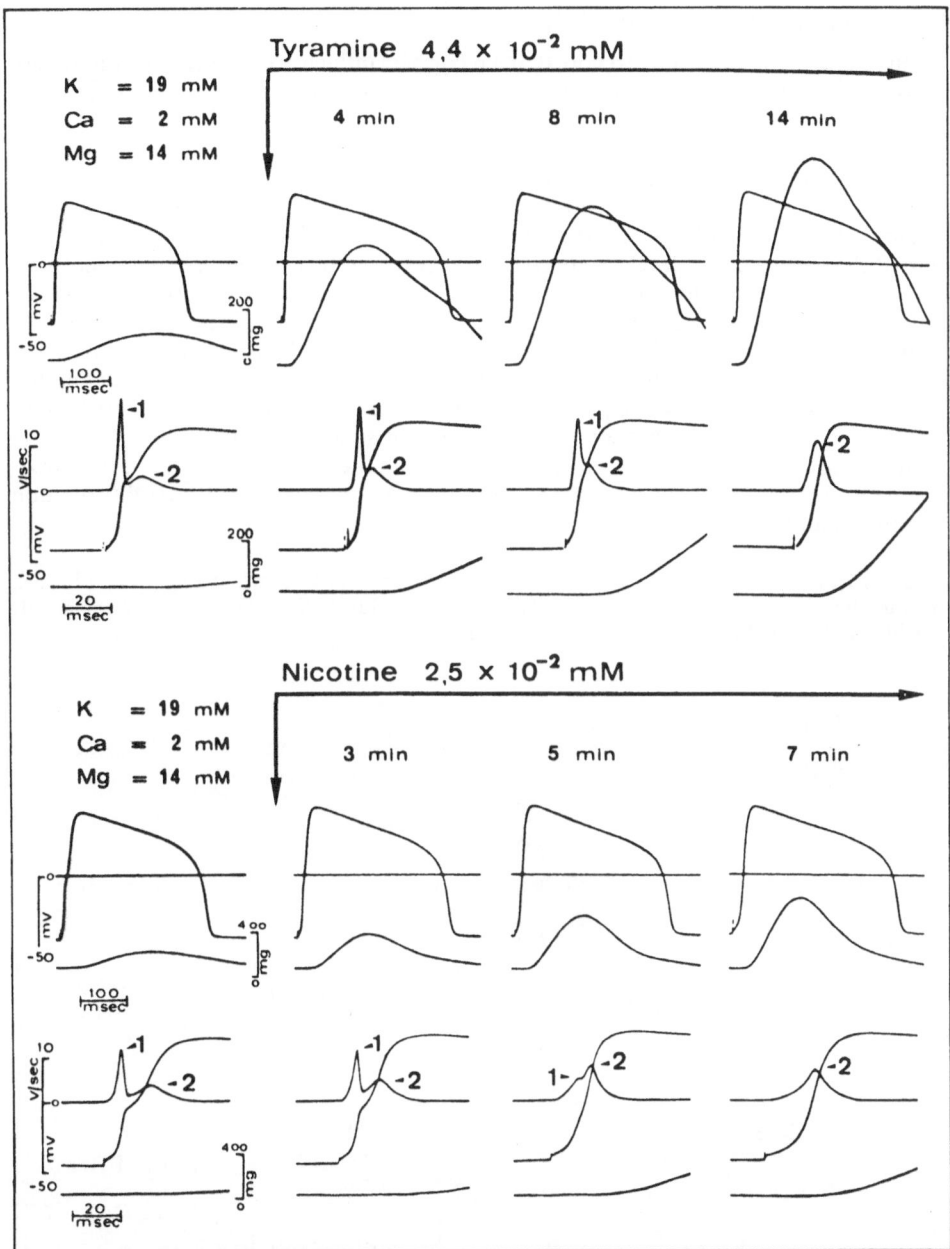

Fig. 10. Reciprocal influence of the catecholamine releasers tyramine and nicotine on the electrogenic Mg^{++} and Ca^{++} effects on partially depolarized guinea-pig papillary muscles. Compensatory suppression of the Mg^{++}-dependent peak No. 1 of upstroke velocity (dV/dt max) in mixed Mg^{++}/Ca^{++}-mediated action potentials, simultaneously with extreme potentiation of the Ca^{++} peak No. 2. Experimental conditions as in Figs. 7–9.

c. Isotope studies with ^{45}Ca and ^{28}Mg

In this context it is interesting to note that a reciprocal influence of β-receptor stimulation on myocardial Ca^{++} and Mg^{++} uptake has also become evident from our studies, with radioactive $^{45}Ca^{++}$ and $^{28}Mg^{++}$ *. Rats were injected with 10 μCi ^{45}Ca or ^{28}Mg intraperitoneally, and the net uptake in the left and right ventricular myocardium measured with a Tricarb 460 C scintillation counter. Both the β- and γ-spectra of ^{28}Mg were used for quantitation. Figure 11 shows that isoproterenol (30 mg/kg, subcutaneously injected) intensified myocardial Ca^{++} uptake within 6 h by roughly 300% above normal. Elevation of the plasma Ca^{++} level by approx. 50% (resulting from one single previous injection of vitamin D_3 in a dose of 300,000 I.U./kg given intramuscularly 4 days beforehand), potentiated the uptake of labelled Ca^{++} even more, i.e. up to 670% above normal. In contrast, the net uptake of labelled Mg was always inhibited by isoproterenol under the same experimental conditions (mean inhibition between 20% and 40%). In the vitamin D_3-treated rats with an elevated plasma Ca^{++} level, the inhibition of ^{28}Mg incorporation was particularly marked. These observations are consistent, too, with the hypothesis that there might be a competitive removal of Mg^{++} from the superficial layers of the sarcolemma membrane if local Ca^{++} binding increases as consequence of cAMP formation (see Figs. 2 and 3).

---

* Carrier-free radioactive ^{28}Mg was kindly provided by Professor G. Stöcklin and Dr. H. Michael, Institute for Nuclear Chemistry, Nuclear Research Center Jülich, West Germany. We are greatly indebted to these colleagues for their decisive help.

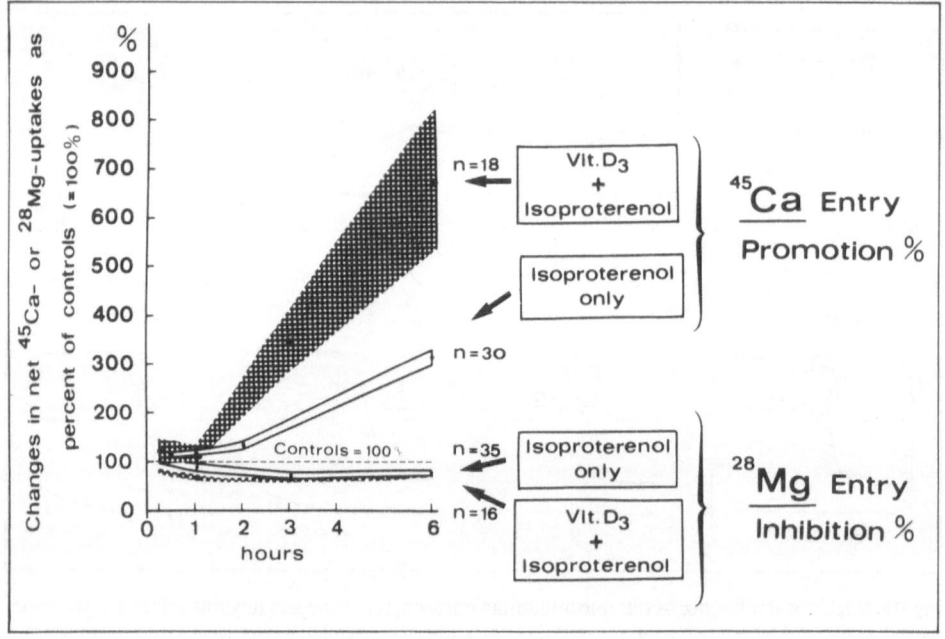

Fig. 11. Large increase in radiocalcium ($^{45}Ca^{++}$) uptake of rat ventricular myocardium upon subcutaneous administration of 30 mg isoproterenol/kg body weight. Potentiation of the isoproterenol effect by a single previous dose of vitamin D_3 (300,000 I.U. intramuscularly). In contrast, the myocardial Mg^{++} uptake ($^{28}Mg^{++}$) is inhibited by isoproterenol by 20%–40% at the same time.

III. Absence of reciprocal Mg^{++}-antagonistic effects
in promotion of Ca^{++} entry with Ca^{++} agonists of the Bay K 8644-Type

In sharp contrast to the cAMP-producing Ca^{++} entry promoters, the new 1,4-dihydro-pyridine Ca^{++} agonists of the Bay K 8644-type, which enhance myocardial Ca^{++} uptake and contractile force without mediation by cAMP, did not exert in our experiments, an inhibitory influence on the electrogenic Mg^{++} effects on partially depolarized myocardium. As shown by Wagner and Fleckenstein [30] with Bay K 8644, this peculiarity can be visualized in two ways:

1. Bay K 8644 is unable to suppress Mg^{++}-mediated action potentials of partially depolarized myocardium in a K^{+}-rich (19 mM K^{+}), Ca^{++}-deficient (0.4 mM Ca^{++}) Tyrode solution with 14 mM Mg^{++}. The absence of any inhibitory action on Mg^{++} becomes particularly obvious if the previous observations with cAMP-producing drugs (Fig. 6) are compared with those made under the influence of Bay K 8644. In fact, Bay K 8644, in a concentration which clearly enhances transmembrane Ca^{++} supply, leaves the Mg^{++}-mediated action potentials entirely unchanged (see Fig. 12, upper panels). Conversely, the Mg^{++}-mediated action potentials rapidly disappear if isoproterenol is administered (lower panels in Fig. 12) as a cAMP-producing drug.

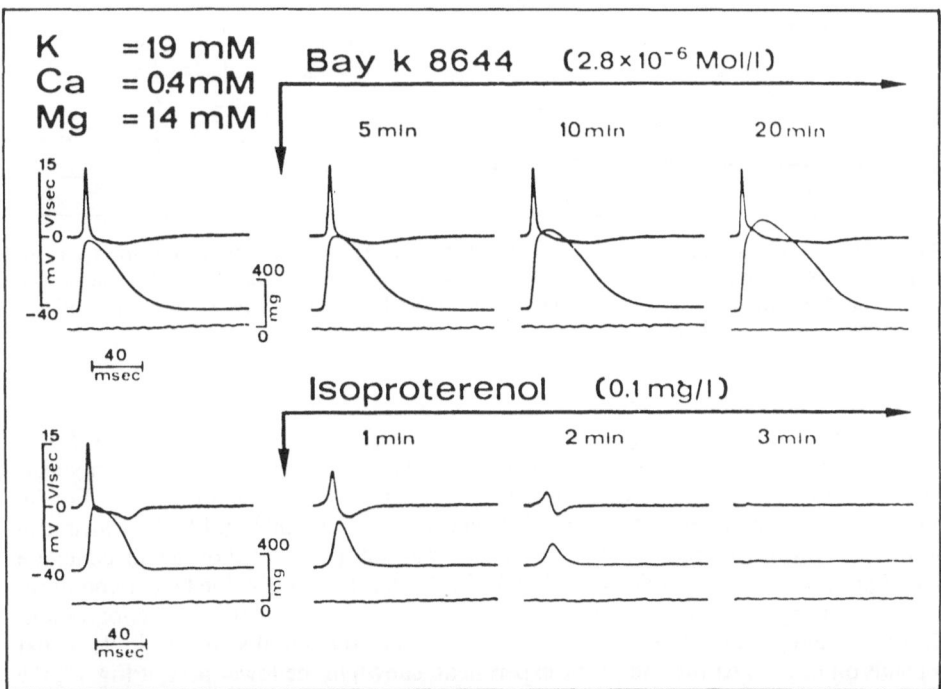

Fig. 12. In contrast with isoproterenol (or other β- or H$_2$-receptor stimulants), the Ca^{++} agonist Bay K 8644 which does not produce cyclic AMP, is unable to suppress Mg^{++}-dependent action potentials of partially depolarized guinea-pig papillary muscles in Tyrode solution with 19 mM K^{+}, 0,4 mM Ca^{++} and 14 mM Mg^{++} (Experimental conditions as in Fig. 6).

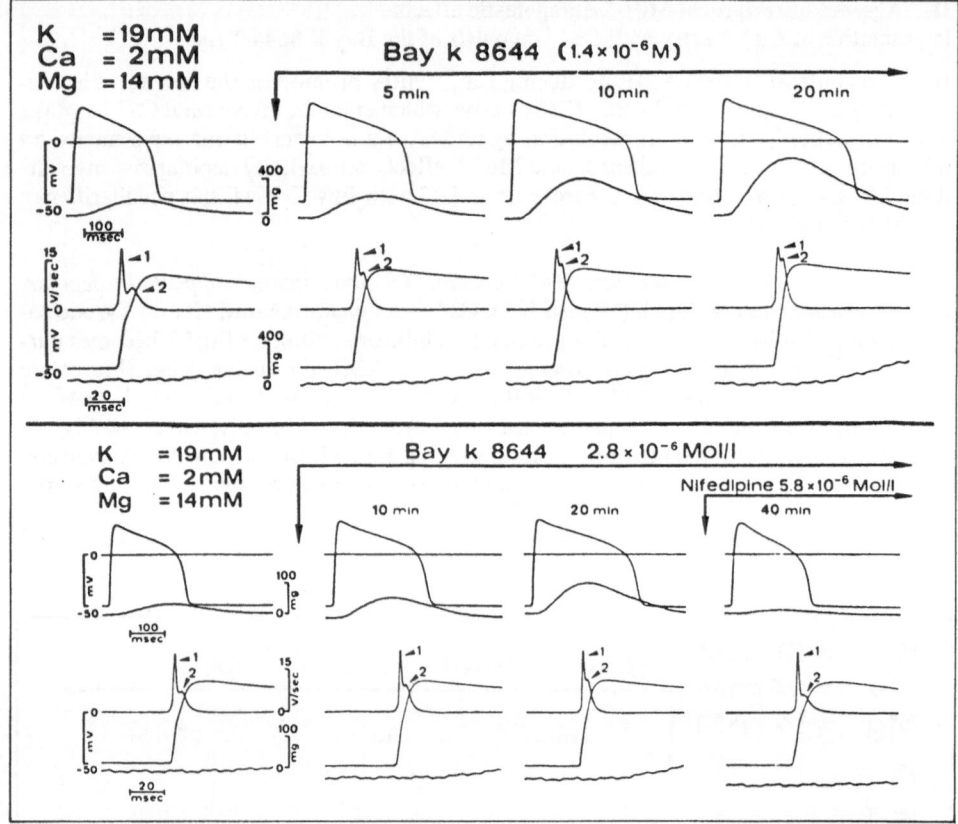

Fig. 13. In contrast with the producers of cyclic AMP (β- or H_2-receptor stimulants, forskolin, dibutyryl cyclic AMP in Figs. 8–10), there is no reciprocal suppression of the Mg^{++}-dependent peak No. 1 in the differentiated Mg^{++}/Ca^{++} action potentials when the Ca^{++} peak No. 2 is potentiated by administration of Bay K 8644.

2. Bay K 8644 is unable to suppress the Mg^{++}-mediated initial component (peak No. 1) of the mixed Mg^{++}/Ca^{++}-mediated action potentials of partially depolarized myocardium in a Tyrode solution with 19 mM K^+, 2 mM Ca^{++}, and 14 mM Mg^{++}. For instance, in two experiments, the results of which are shown in Fig. 13, the transmembrane Ca^{++} influx, represented by the Ca^{++}-dependent peak No. 2 of the mixed action potentials, is strongly potentiated by Bay K 8644. Simultaneously, the height and duration of the mixed action potentials grow together with the isometric mechanograms. However, there is in no case a reciprocal depression of the initial sharp peak No. 1 that depends on Mg^{++}. At the end of the experiment, shown in the lower part of Fig. 13, the highly specific calcium antagonist nifedipine was added. In consequence, the Ca^{++} peak No. 2 was selectively reduced, together with the duration of action potential and contractile force. However, the Mg^{++} peak No. 1 retained its initial magnitude. All these results on partially depolarized myocardium discussed in Sections II and III, indicate that the Ca^{++} influx promotion by mediation of cAMP is rigidly coupled with opposite effects on Mg^{++}, whereas the basic action of the new 1,4-dihydropyridine Ca^{++} agonists (particularly Bay K 8644) consists of a selective *unilateral* facilitation of Ca^{++} entry through the slow channels. As shown by Reuter and others, the new Ca^{++} agonists prolong the open state of the Ca^{++} channels and increase the probability of opening. But this effect is obviously not accompanied by any interference with Mg^{++}.

IV. Concluding remarks

The present studies have further confirmed that there are two classes of positive inotropic Ca^{++} entry promoters in mammalian heart muscle fibres, namely
1. Agents which potentiate the slow transmembrane inward Ca^{++} current, via intermediate formation of cAMP. This mode of action is probably the physiological way in which β-adrenergic and H_2-receptor stimulants (as well as forskolin and dibutyryl cAMP) increase contractile force.
2. Agents which facilitate Ca^{++} entry through the slow channels in a more direct, cAMP-independent manner. This "artificial" type of Ca^{++} entry promotion is represented by the new *"calcium agonists"* Bay K 8644, CGP 28392, and YC 170, the Bayer compound being the strongest.

The best evidence of the fundamental difference in the mode of action of these two classes of Ca^{++} entry promoters could be obtained on partially depolarized mammalian myocardium (guinea-pig papillary muscles).

a. Agents acting through cAMP

All agents which promote contractile force by mediation of cAMP exert their positive inotropic actions strictly in a double way, 1. by potentiation of transsarcolemmal Ca^{++} entry, and 2. by counteracting reciprocally the influence of Mg^{++} ions on the myocardial fibre membrane.

Thus all cAMP-producing Ca^{++} promoters, so far tested, suppress the Mg^{++}-mediated action potentials in partially depolarized, Ca^{++}-deprived myocardium, while simultaneously, the Ca^{++}-carried action potentials are enhanced, together with contractile force. Similarly, in biphasic Mg^{++}/Ca^{++}-carried action potentials of partially depolarized myocardium, the initial Mg^{++}-dependent portion is abolished by the cAMP-producing drugs whereas, conversely, a dramatic potentiation of the subsequent Ca^{++}-dependent portion takes place. Moreover, as visualized directly with electron microscopy and tracer techniques (radioactive ^{45}Ca and ^{28}Mg), the intermediate formation of cAMP, initiated by isoproterenol, leads to a tremendous intensification of superficial Ca^{++} binding and intracellular Ca^{++} accumulation. But simultaneously, the net ^{28}Mg uptake is inhibited. We should like to propose that the formation of cAMP and the phosphorylation of membrane proteins by a local protein kinase in the vicinity of the slow channels or in the slow channels increases the number of fixed negative phosphate groups which accumulate Ca^{++} ions from the environment. However, this high-grade accumulation of Ca^{++} by cAMP-producing agents presumably displaces Mg^{++} ions from the fibre surface and thereby counteracts all Mg^{++}-dependent bioelectric membrane phenomena.

b. Agents acting without involvement of cAMP

The new 1,4-dihydropyridine Ca^{++} agonists (particularly Bay K 8644) seem to potentiate Ca^{++} influx through the slow channels by a direct influence on the gating mechanisms, as for instance by prolongation of the "open state" (see Reuter [19 b]). Our results indicate that the new Ca^{++} agonists of the Bay K 8644-type which do not produce cAMP, only facilitate the trans-sarcolemmal Ca^{++} influx, whereas they are unable to promote sarcolemmal Ca^{++} accumulation. Probably, for this reason, the new Ca^{++} agonists do not exert competitive inhibitory effects on Mg^{++} actions.

If one compares the two types of drug-induced enhancement of transmembrane Ca^{++} supply, the positive inotropism by means of cAMP – as for instance with β-adrenergic catecholamines – is probably the more efficient mode of contractile potentiation, because there is, in this situation, a superposition of two factors: (a) an increase in Ca^{++} supply, and (b) a simultaneous restriction of Mg^{++} uptake which in principle can counteract the function of Ca^{++} ions in excitation-contraction coupling (Fabiato and Fabiato [2]). Thus, through formation of cAMP, the positive inotropism will be maximized.

References

1. Drummond GI, Hemmings SJ (1972) Inotropic and chronotropic effects of dibutyryl cyclic AMP. In: Greengard P, Robison GA (eds) Advances in cyclic nucleotide research. Raven Press, New York, pp 307–316
2. Fabiato A, Fabiato F (1978) Calcium-induced release of calcium from the sarcoplasmic reticulum of skinned cells from adult human, dog, cat, rabbit, rat, and frog hearts and from fetal and newborn rat ventricles. Ann NY Acad Sci 307:491–522
3. Fleckenstein A (1981) Fundamental actions of calcium antagonists on myocardial and cardiac pacemaker cell membranes. In: Weiss GB (ed) New perspectives on calcium antagonists. Clin Physiol Series, Am Physiol Soc, pp 59–81
4. Fleckenstein A (1983) History of calcium antagonists. In: Schwartz A, Taira N (eds) Calcium channel-blocking drugs: A novel intervention for the treatment of cardiac disease. Monograph Nr 95, part II, vol 52, Nr 2. Am Heart Ass Circ Res, pp 3–16
5. Fleckenstein A (1983) Calcium antagonism in heart and smooth muscle – experimental facts and therapeutic prospects. Monograph, John Wiley Publishing Company, New York Chichester Brisbane Toronto Singapore
6. Fleckenstein A, Späh F (1981/82) Excitation-contraction uncoupling in cardiac muscle. In: Yoshida H, Hagihara Y, Ebashi S (eds) Advances in pharmacology and therapeutics II, vol 3, cardio-renal and cell pharmacology, 8th Internat. Congress of pharmacology, Tokyo 1981. Pergamon Press, Oxford New York, pp 97–110
7. Klein I, Levey GS (1971) Activation of myocardial adenyl cyclase by histamine in guinea-pig, cat, and human heart. J Clin Invest 50:1012–1015
8. Kokubun S, Reuter H (1984) Dihydropyridine derivatives prolong the open state of Ca channels in cultured cardiac cells. Proc Natl Acad Sci 81:4824–4827
9. Kukovetz WR, Pöch G (1970) Cardiostimulatory effects of cyclic 3',5'-adenosine monophosphate and its acylated derivatives. Naunyn-Schmiedebergs Arch exper Path Pharmakol 266:236–254
10. Kukovetz WR, Pöch G, Wurm A (1973) Effects of catecholamines, histamine and oxyfedrine on isotonic contraction and cyclic AMP in the guinea-pig heart. Naunyn-Schmiedebergs Arch exper Path Pharmakol 278:403–424
11. Lindner E, Dohadwalla AN, Bhattacharya BK (1978) Positive inotropic and blood pressure lowering activity of a diterpene derivative isolated from Coleus forskohli: forskolin. Arzn-Forsch (Drug Res) 28:84–89
12. Mascher D (1970) Electrical and mechanical responses from ventricular fibres after inactivation of the sodium carrying system. Pflügers Arch 317:359–372

13. McNeill JH, Muschek LD (1972) Histamine effects on cardiac contractility, phosphorylase and adenyl cyclase. J Mol Cell Cardiol 4:611–624
14. McNeill JH, Verma SC (1974) Blockade by burimamide of the effects of histamine and histamine analogs on cardiac contractibility, phosphorylase activation and cyclic adenosine monophosphate. J Pharmacol Exp Ther 188:180–188
15. Meinertz T, Nawrath H, Scholz H (1973) Dibutyryl cyclic AMP and adrenaline increase contractile force and ^{45}Ca uptake in mammalian cardiac muscle. Naunyn-Schmiedebergs Arch exper Path Pharmakol 277:107–112
16. Meinertz T, Nawrath H, Scholz H (1973) Stimulatory effects of DB-c-AMP and adrenaline on myocardial contraction and ^{45}Ca exchange. Experiments at reduced calcium concentration and low frequencies of stimulation. Naunyn-Schmiedebergs Arch exper Path Pharmakol 279:327–338
17. Metzger H, Lindner E (1981) The positive inotropic-acting forskolin, a potent adenylate cyclase activator. Arzn-Forsch (Drug Res) 31:1248–1250
18. Reinhardt D, Schmidt U, Brodde OE, Schümann HJ (1977) H_1- and H_2-receptor mediated responses to histamine on contractility and cyclic AMP of atrial and papillary muscles from guinea-pig hearts. Agents and Actions 7:1-12
19. a. Reuter H (1974) Localization of beta-adrenergic receptors, and effects of noradrenaline and cyclic nucleotides on action potentials, ionic currents and tension in mammalian cardiac muscle. J Physiol (Lond) 242:429–451
19. b. Reuter H (1985) Calcium movements through cardiac cell membranes. Med Res Rev 5:427–440
20. Schramm M, Thomas G, Towart R, Franckowiak G (1983) Activation of calcium channels by novel 1,4-dihydropyridines. A new mechanism for positive inotropics or smooth muscle stimulants. Arzn-Forsch (Drug Res) 33:1268–1272
21. Schramm M, Thomas G, Towart R, Franckowiak G (1983) Novel dihydropyridines with positive inotropic actions through activation of Ca^{2+} channels. Nature 303:535–537
22. Seamon KB, Daly JW (1981) Forskolin: a unique diterpene activator of cyclic AMP-generating systems. J Cyclic Nucleotide Res 7:201–204
23. Skelton CL, Levey GS, Epstein SE (1970) Positive inotropic effects of dibutyryl cyclic adenosine 3′,5′-monophosphate. Circ Res 26:35–43
24. Späh F (1984) Forskolin, a new positive inotropic agent, and its influence on myocardial electrogenic cation movements. J Cardiovasc Pharmacol 6:99–106
25. Späh F, Fleckenstein A (1979) Evidence of a new, preferentially Mg-carrying, transport system besides the fast Na and the slow Ca channels in the excited myocardial sarcolemma membrane. J Mol Cell Cardiol 11:1109–1127
26. Späh F, Fleckenstein A (1984) Influence of forskolin, a direct promoter of the formation of cAMP in heart muscle, on Na-, Ca-, and Mg-dependent myocardial membrane phenomena, vol 400. Frühjahrstagung der Deutsch Physiol Gesell, Dortmund, März 1984. Pflügers Arch Ges Physiol [Suppl] R6
27. Sutherland EW, Robison GA, Butcher RW (1968) Some aspects on the biological role of adenosine 3′,5′-monophosphate (cyclic AMP). Circulation 37:279–306
28. Tritthart H, Volkmann R, Weiss R, Fleckenstein A (1973) Calcium-mediated action potentials in mammalian myocardium: Alteration of membrane response as induced by changes of Ca or by promoters and inhibitors of transmembrane Ca inflow. Naunyn Schmiedebergs Arch Pharmakol 280:239–252
29. Tsien RW, Giles W, Greengard P (1972) Cyclic AMP mediates the effects of adrenaline on cardiac Purkinje fibers. Nature New Biol 240:181–183
30. Wagner WL, Fleckenstein A (1984) Electric and mechanical effects of Bay K 8644, a new positive inotropic dihydropyridine calcium agonist, on mammalian ventricular myocardium, vol 402. 60. Meeting der Deutsch Physiol Gesell (Fall Meeting), Dortmund Okt 1984 . Pflügers Arch [Suppl] R22, 77
31. Watanabe AM, Besch HR Jr (1974) Cyclic adenosine monophosphate modulation of slow calcium influx channels in guinea-pig hearts. Circ Res 35:316–324
32. Watanabe AM, Besch HR Jr (1974) Subcellular myocardial effects of verapamil and D 600: comparison with propranolol. J Pharmacol Exp Ther 191:241–251

33. Watanabe AM, Besch HR Jr (1975) The relationship between adenosine 3',5'-monophosphate levels and systolic transmembrane calcium flux. In: Fleckenstein A, Dhalla NS (eds) Basic functions of cations in myocardial activity. Recent advances in studies on cardiac structure and metabolism, vol 5. University Park Press, Baltimore London Tokyo, pp 95–102
34. Wollenberger A (1975) The role of cyclic AMP in the adrenergic control of the heart. In: Nayler WG (ed) Contraction and relaxation in the myocardium. Academic Press, New York, pp 113–190

Author's address:

Prof. Dr. A. Fleckenstein
Physiologisches Institut der
Universität Freiburg
Hermann-Herder-Str. 7
7800 Freiburg

Energetic aspects of inotropic interventions in rat myocardium

G. Hasenfuss, Ch. Holubarsch, H. Just, E. Blanchard, L. A. Mulieri and N. R. Alpert

Medizinische Klinik III, University of Freiburg, F.R.G. and
Department of Physiology & Biophysics, University of Vermont, Burlington, Vermont, U.S.A.

Summary

Contractile force of the myocardium can be increased by different molecular mechanisms, and therefore different energetic consequences may result. The influence of the inotropic substances isoproterenol and UDCG-115 on myocardial energetics in isometrically contracting left ventricular rat papillary muscles was investigated by means of highly sensitive antimony bismuth thermopiles.

Isoproterenol increased total heat and initial heat by 147% (p < 0.01) and 69% (p < 0.02) when normalized to tension-time integral, respectively. No significant change of both heat terms occurred due to UDCG-115. Initial heat was separated into tension-independent heat ("calcium cycling") and tension-dependent heat ("cross-bridge cycling") by means of a new method using 2,3-butanedione monoxime. Both tension-dependent heat per tension-time integral and tension-independent heat increased significantly, due to isoproterenol, from 4.9 ± 1.17 to 7.6 ± 2.72 μcal/g·cm·s (p < 0.05) and from 0.15 ± 0.06 to 0.22 ± 0.04 mcal/g (p < 0.01). UDCG-115 influenced neither tension-independent heat nor tension-dependent heat per tension-time integral significantly.

Thus, the economy of force development was not significantly altered due to UDCG-115 whereas isoproterenol significantly increased the energy necessary for activation, i.e. calcium cycling, and the energy necessary for force production, i.e. cross-bridge cycling. The basic mechanisms of these energetic changes are discussed.

Introduction

In search of therapeutic alternatives in the treatment of congestive heart failure, several new inotropic drugs are currently under investigation [4, 18]. Although most of these drugs have shown to improve haemodynamics in acute and chronic heart failure, some investigators have demonstrated poor long term haemodynamic efficiency or even a deterioration in heart failure due to inotropic stimulation [7, 13, 19]. Detrimental effects of inotropic drugs on the myocardium may result from increased myocardial oxygen consumption, the amount of which depends on the molecular mechanism by which inotropism is achieved [6, 9]. To evaluate the energetic costs of different inotropic interventions, we measured the heat liberated from rat papillary muscles simultaneously with the development of isometric force and isometric force-time integral [12]. Using this method the energetic changes due to isoproterenol and UDCG-115 (a new noncatecholamine nonglycoside inotropic agent) were investigated.

Methods

Thermopile equipment and definitions of used heat terms

The experiments were performed using antimony bismuth thermopiles with a temperature sensitivity of 1.20 mV/°C. The heat liberated from a papillary muscle is divided into different compartments. When the muscle is stimulated, initial heat is liberated during the isometric contraction phase. Initial heat is composed of heat liberation due to calcium cycling (tension-independent heat), and heat liberation due to cross-bridge cycling (tension-dependent heat). The recovery heat, mainly released after the end of contraction, represents the metabolic costs of resynthesis of high energy compounds. Total heat is the sum of initial heat and recovery heat.

Animals and experimental protocol

The experiments were performed using left ventricular papillary muscles of Wistar Kyoto rats (aged 10–12 weeks). The papillary muscles were removed as previously described [11]. By means of silk loops the muscles were mounted between the force transducer and a fixed glass hook. Thus the muscle was in contact with the central hot junctions of the thermopile [12]. After a two hour equilibration period in Krebs-Ringer solution, l_{max} (muscle length with maximum developed force) was defined evaluated. The muscle was stimulated at 0.2 Hz, 20% above threshold at 21 °C. For heat recordings the chamber was drained, and temperature and tension signals were recorded synchronously. After the

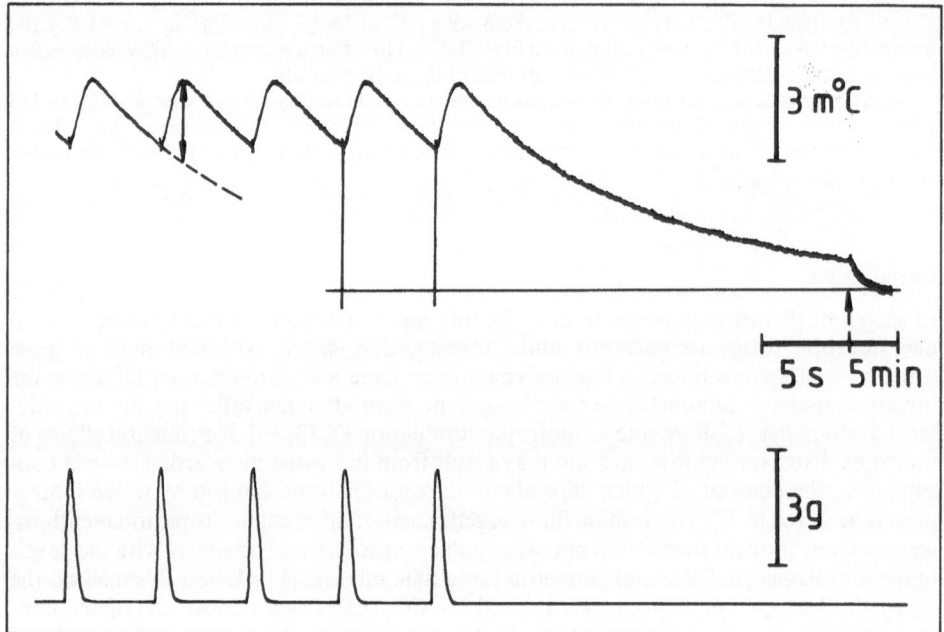

Fig. 1. A typical record of the temperature signal (upper trace) and the isometric tension (lower trace). Note that temperature increases rapidly when the contraction starts, and decreases slowly ("cool-off") when the contraction is finished.

recording of a cool-off curve after activity (Fig. 1), the muscle was physically heated using a high frequency AC generator (10^6 Hz) connected directly to the thermopile. Then, a cool-off curve after physical heating was recorded. From this cool-off curve the cool-off time constant was derived. For differentiation between tension-independent heat and tension-dependent heat, BDM (2,3-butanedione monoxime, 8 mM) was added to the solution [1]. After an equilibration period of 15 minutes, temperature and tension signals were recorded again. Under these conditions, maximum tension decreased by more than 95%. After a washout period of one hour, the developed tension reached control values again.

Pharmacological interventions

After measurements were performed under control conditions, isoproterenol (10^{-6} M) was added, and left for a period of 10–15 minutes to achieve steady state conditions. The measurements were then repeated. For washout, the solution was changed three times and the muscle was left in normal Krebs-Ringer solution for one hour. After this period, control measurements were performed again. UDCG-115 ($200 \cdot 10^{-6}$ M) was then applied, and after 15 min measurements were repeated.

Myothermal analysis

Initial heat was calculated from the distance between the peak of the oscillating temperature signal (arrow in Fig. 1) and the extrapolation of the cool-off curve (obtained after stimulation was stopped) to the previous temperature record (dotted line; Fig. 1). This temperature difference was multiplied by the heat loss coefficient (0.54 mcal/°C·s) and the cool-off time constant. From initial heat and the corresponding tension-time integral under control and BDM conditions, tension-independent heat was calculated as described earlier [10]. Tension-dependent heat was calculated as the difference between initial heat and tension-independent heat. Total heat was evaluated from the area between the oscillating temperature signal and the temperature baseline during one twitch interval (unbroken line, Fig. 1), according to Bungard [3]. Recovery heat was the difference between total heat and initial heart.

Statistical analysis

Values are expressed as mean ± standard deviation. Comparisons between control measurements and measurements after application of isoproterenol or UDCG-115 were made using the paired t-test. A value of $p < 0.05$ was accepted as significant.

Results

Mechanical measurements

Isoproterenol decreased peak twitch tension from 5.56 ± 0.92 to 5.05 ± 0.88 g/mm^2 ($p < 0.02$) and tension-time integral from 3.18 ± 0.74 to 1.53 ± 0.31 g·s/mm^2 ($p < 0.01$) (Fig. 2; Table 1). Peak twitch tension and tension-time integral were 5.86 ± 1.53 g/mm^2 and 2.68 ± 0.45 g·s/mm^2 before, and 5.78 ± 1.55 g/mm^2 and 2.60 ± 0.85 g·s/mm^2 after administration of UDCG-115 (not significant), respectively (Fig. 2, Table 1).

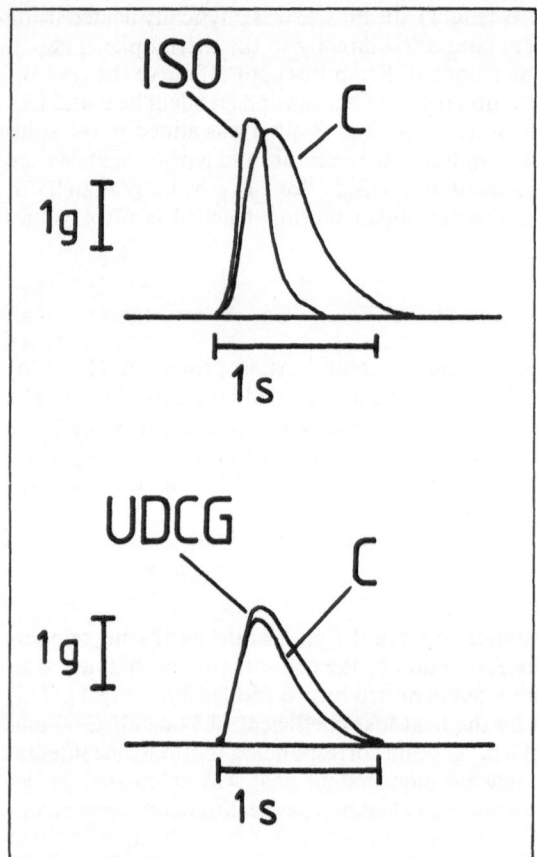

Fig. 2. Isometric twitch contraction before and after administration of isoproterenol (ISO, 10^{-6} M) (upper panel) and UDCG$_{115}$ (UDCG, $200 \cdot 10^{-6}$ M) (lower panel). C=control.

Table 1. Mechanical and myothermal effects of isoproterenol and UDCG-115

	δ (g/mm^2)	$\int \delta \cdot t$ (g·s/mm^2)	IH (mcal/g)	TH (mcal/g)	R/I	TIH (mcal/g)	TDH (mcal/g)
Control (n=5)	5.56±0.92	3.18±0.74	1.66±0.33	3.02±0.28	0.86±0.29	0.15±0.06	1.52±0.29
Isoproterenol	5.05±0.88	1.53±0.31	1.35±0.31	3.59±0.66	1.73±0.51	0.22±0.04	1.12±0.29
(n=5)	p<0.02	p<0.01	p<0.05	p<0.05	p<0.02	p<0.01	p<0.02
Control (n=5)	5.86±1.53	2.68±0.45	1.45±0.36	2.77±0.88	0.89±0.12	0.21±0.04	1.24±0.34
UDCG-115	5.78±1.55	2.60±0.85	1.58±0.50	2.94±0.74	0.87±0.13	0.15±0.07	1.42±0.44
(n=5)	n.s.	n.s.	n.s.	n.s.	n.s.	n.s.	n.s.

δ=peak twitch tension; $\int \delta \cdot t$= tension-time integral; IH=initial heat; TH=total heat; R/I=recovery to initial heat ratio; TIH=tension-independent heat; TDH=tension-dependent heat.

Myothermal measurements

Initial heat decreased by 19% (p<0.05) and tension-dependent heat decreased by 26% (p<0.02) due to isoproterenol (Table 1). Tension-independent heat increased by 47% (p<0.01) and the total heat was increased by 19% (p<0.05). The recovery to initial heat ratio increased from 0.86±0.29 to 1.73±0.51 (p<0.02) due to isoproterenol. None of these heat terms was significantly changed due to UDCG-115.

Relationship between myothermal and mechanical measurements

Initial heat per tension-time integral was 5.39±1.33 µcal/g·cm·s during control conditions and significantly increased by 69% to 9.13±3.15 µcal/g·cm·s due to isoproterenol (Fig. 3). Initial heat per tension-time integral did not significantly change due to UDCG-115 (5.51±1.55 before and 6.46±2.59 µcal/g·cm·s with UDCG-115). Analysis of tension-dependent heat per tension-time integral yielded similar results: a significant increase from 4.90±1.17 to 7.60±2.72 µcal/g·cm·s (55%) due to isoproterenol, and no significant change due to UDCG-115 (4.72±1.46 before and 5.81±2.24 µcal/g·cm·s with UDCG-115 (Fig. 4).

Total heat per tension-time integral was 9.96±2.69 µcal/g·cm·s during control conditions and significantly increased to 24.65±8.55 µcal/g·cm·s due to isoproterenol. Total heat per tension-time integral was 10.51±3.68 µcal/g·cm·s before, and 12.07±4.17 µcal/g·cm·s after application of UDCG-115 (not significant; Fig. 5).

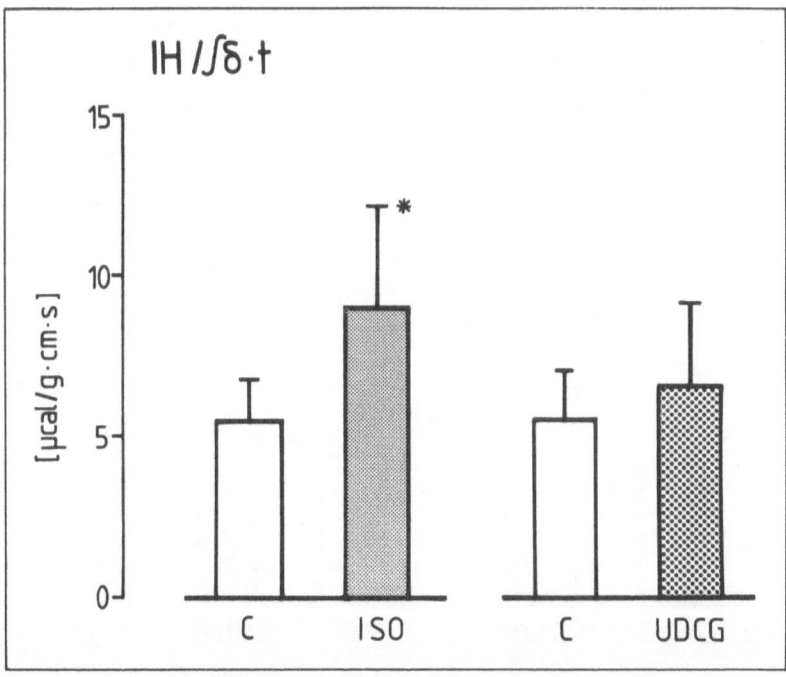

Fig. 3. Influence of isoproterenol (ISO, 10^{-6} M) and UDCG-115 (UDCG, $200 \cdot 10^{-6}$ M) on initial heat per tension-time integral (IH/∫δ·t). C=control. *p<0.05.

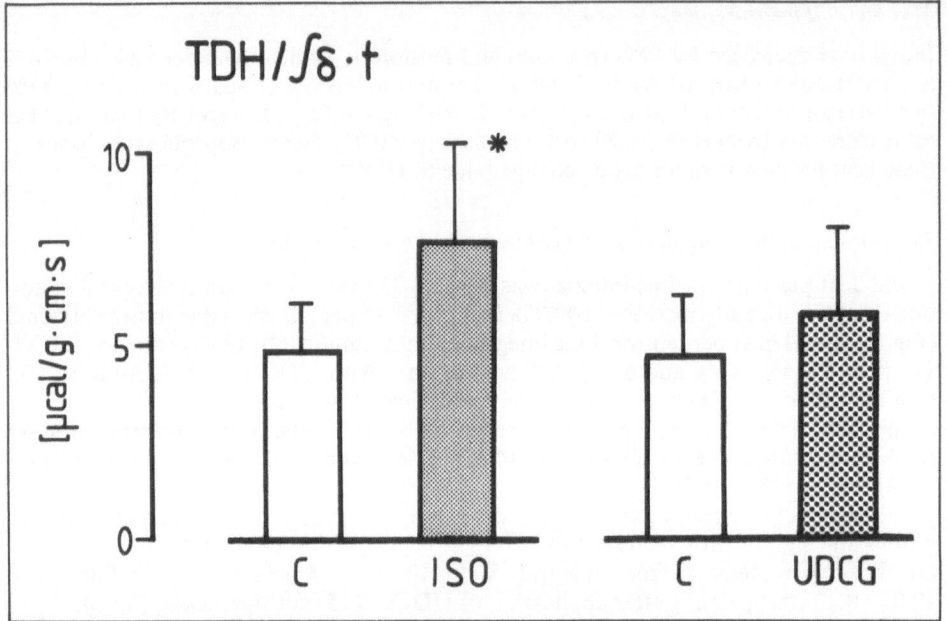

Fig. 4. Influence of isoproterenol (ISO, 10^{-6} M) and UDCG-115 (UDCG, $200 \cdot 10^{-6}$ M) on tension-dependent heat per tension-time integral (TDH/$\int\delta \cdot$ t). C = control. *p < 0.05.

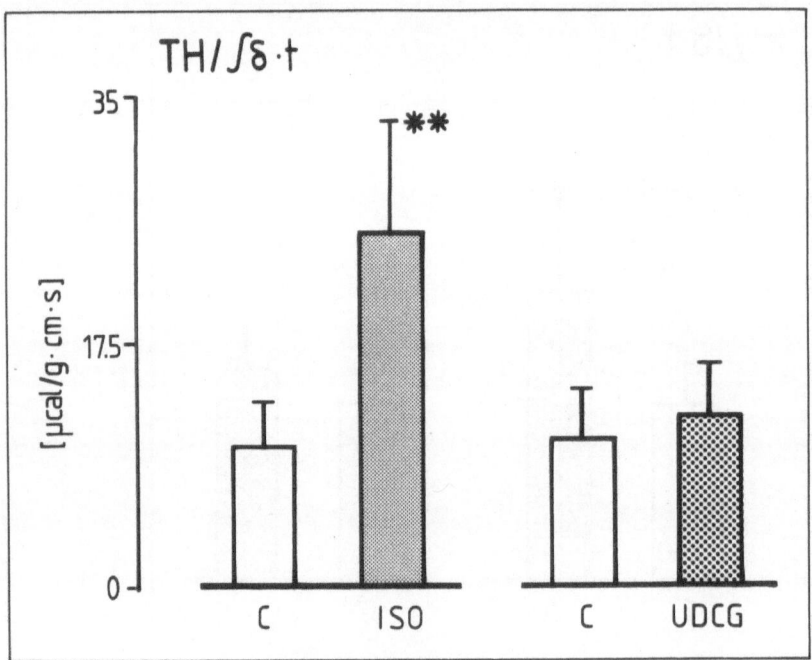

Fig. 5. Influence of isoproterenol (ISO, 10^{-6} M) and UDCG-115 (UDCG, $200 \cdot 10^{-6}$ M) on total heat per tension-time integral (TH/$\int\delta \cdot$ t). C = control. **p < 0.01.

Discussion

Methodological considerations

To evaluate the economy of isometric force development under different inotropic interventions, the liberated heat must be analysed in relation to its mechanical determinants. As discussed in detail elsewhere [9], we believe that the energy consumption during isometric contraction is mainly determined by the tension-time integral. Therefore the ratio between heat and tension-time integral was used as an inverse measure of the economy of myocardial contraction.

In the present experiments, which were performed at a calcium concentration of 2.5 mM, UDCG-115 did not significantly increase developed peak twitch tension. This may be due to the fact that the contractile apparatus of rat myocardium is almost maximally activated at 2.5 mM calcium at 21 °C [5]. The decrease of peak twitch tension due to isoproterenol may reflect either an increased rate of deactivation (calcium re-uptake) or a decreased affinity of troponin-C to calcium ions due to phosphorylation of troponin I [14, 15, 16].

In contrast to isoproterenol, UDCG-115 did not significantly influence mechanical parameters at a calcium concentration of 2.5 mM. However the influence of UDCG-115 on rat myocardium at a concentration used in the present experiments ($200 \cdot 10^{-6}$ M) can be demonstrated when the calcium concentration is reduced to 0.625 mM [9].

Myocardial energetics

The application of isoproterenol resulted in a significant increase in initial heat per tension-time integral. This increase can be analysed in terms of tension-independent heat and tension-dependent heat per tension-time integral. Tension-independent heat reflects the energy consumption for calcium-cycling. Tension-independent heat increased significantly (by 47%) due to isoproterenol. This increase may indicate that the quantity of calcium which cycled during one beat was increased by about 50% due to isoproterenol. Tension-dependent heat per tension-time integral increased by 55%, which reflects the increased energy consumption of the contractile apparatus per unit tension-time integral due to isoproterenol.

As discussed earlier [2], each cross-bridge cycle can be described in terms of on-time (force generating position) and off-time (non-force-generating position). Assuming that a given amount of energy is consumed for one single cross-bridge cycle, tension-dependent heat results from the total number of all cross-bridge cycles during the whole period of one twitch. Furthermore, assuming that each cross-bridge generates a constant amount of force, tension-time integral results from the total number of all cross-bridge cycles during the whole period of one twitch and the on-time of the single cross-bridge. Consequently, changes in the relationship between tension-dependent heat and tension-time integral indicate changes in the on-time. The increased tension-dependent heat per tension-time integral due to isoproterenol may therefore indicate changes in cross-bridge kinetics, namely a decrease in the on-time. Catecholamine induced changes in cross-bridge kinetics are further supported by the findings of Hoh and Rosmanith [8] who demonstrated increased cycling rates of the cross-bridges using a pertubation analysis and by Winegrad et al. [20] who demonstrated increased activity of myosin ATPase due to catecholamines.

Total heat is composed of initial heat and recovery heat. Recovery heat represents the ATP resynthesis and is mainly released after the twitch. Total heat, as well as the recovery to initial heat ratio, increased significantly due to isoproterenol. The cause of increased recovery to initial heat ratio remains speculative. It may be a consequence of extra heat liberation due to aftercontractions which have been observed following the administration of isoproterenol. Alternatively, it may reflect altered economy of ATP resynthesis due to isoproterenol.

In contrast to isoproterenol, UDCG-115 increased neither tension-independent heat nor tension-dependent heat per tension-time integral significantly. In addition, total heat and the recovery to initial heat ratio remained unchanged. These findings are consistent with the data from Rüegg [17] who demonstrated a sensitization of the contractile proteins to calcium ions, due to UDCG-115. Increased sensitivity of the troponin-tropomyosin system to calcium may increase the contractile force without changes in the quantity of calcium delivery and without changes in cross-bridge kinetics.

In conclusion, the influence of inotropic drugs on myocardial energetics depends on the pharmacological mode of action. Isoproterenol increases the calcium delivery and calcium re-uptake by the saroplasmic reticulum and thereby increases the tension-independent heat. Furthermore, it may change cross-bridge kinetics resulting in reduced economy of tension development. In contrast, UDCG-115, supposed to act by increasing the sensitivity of the contractile proteins to calcium, neither increases the tension-independent heat nor the tension-dependent heat per tension-time integral significantly.

References

1. Alpert NR, Blanchard EM, Mulieri LA (1987) The energy requirements of calcium transport in the presence of inotropic agents and for normal and hypertrophied rabbit hearts. This book
2. Alpert NR, Mulieri LA (1984) Hypertrophic adaptation of the heart to stress: A myothermal analysis. In: Zak R (ed) Growth of the heart in health and disease. Raven Press, New York, pp 363–379
3. Bungard L (1934) The relation between total and initial heat in single muscle twitches. J Physiol 82:509–519
4. Colucci WS, Wright RF, Braunwald E (1986) New positive inotropic agents in the treatment of congestive heart failure. N Engl J Med 314:349–358
5. Forrester GV, Mainwood GW (1974) Interval dependent inotropic effects in the rat myocardium and the effect of calcium. Pflügers Arch 352:189–196
6. Hasenfuss G, Holubarsch Ch, Heiss HW, Bonzel T, Meinertz Th, Just H (1986) Influence of enoximone and isoproterenol on myocardial energetics in the human heart. Circulation 74:398
7. Hasenfuss G, Kasper W, Meinertz Th, Busch W, Hofmann Th, Krause Th, Holubarsch Ch, Lehmann M, Just H (1985) Does long term oral levodopa therapy improve cardiac function in congestive heart failure? Circulation 72:303
8. Hoh JFY, Rosmanith HG (1983) Cross-bridge dynamics in rat papillary muscle containing V1 and V3 isomyosins: Effects of adrenaline. J Mol Cell Cardiol 15:65
9. Holubarsch Ch, Hasenfuss G, Blanchard E, Alpert NR, Mulieri LA, Just H (1986) Myothermal economy of rat myocardium, chronic adaptation versus acute inotropism. Basic Res Cardiol 81:95–102
10. Holubarsch Ch, Goulette RP, Mulieri LA, Alpert NR (1983) Heat liberation in experimentally induced tetanic contractions of myocardium from normal and Goldblatt rats. In: Jacob R (ed) Cardiac adaptation of hemodynamic overload, training and stress. Steinkopff Verlag, Darmstadt, pp 158–166
11. Holubarsch Ch, Alpert NR, Goulette R, Mulieri LA (1982) Heat production during hypoxic contracture of rat myocardium. Circ Res 51:777–786
12. Mulieri LA, Luhr G, Trefry J, Alpert NR (1977) Metal-film thermopiles for use with rabbit right ventricular papillary muscles. Am J Physiol 233:C146–C156

13. Packer M, Medina N, Yushak M (1984) Hemodynamic and clinical limitations of long-term inotropic therapy with amrinone in patients with severe chronic heart failure. Circulation 70:1038–
1047
14. Ray K, England P (1976) Phosphorylation of the inhibitory subunit of troponin and its effects
on the calcium dependence of cardiac myofibrillar ATPase. FEBS Lett 70:11–16
15. Robertson S, Johnson D, Holroyde J, Kranias E, Patter J, Solaro J (1982) The effect of TNI
phosphorylation on the state and kinetic Ca^{++} binding of cardiac TNC. J Biol Chem 257:260–
263
16. Rüegg JC, Pfitzer G (1984) Myokardkontraktilität und Phosphorylierung der kontraktilen Pro-
teine. In: Keul J, Dickhuth HH (eds) Herzinsuffizienz. Pathophysiologie, Klinik und Therapie.
Perimed Fachbuch-Verlagsgesellschaft, Erlangen, pp 53–56
17. Rüegg JC (1986) Effects of new inotropic agents on Ca^{++} sensitivity of contractile proteins. Circulation 73:78–84
18. Scholz H (1984) Inotropic drugs and their mechanisms of action. J Am Coll Cardiol 4:389–
397
19. Shah PK, Amin DK, Hulse S, Shellock F, Swan HJC (1985) Inotropic therapy for refractory
congestive heart failure with oral fenoximone (MDL-17,043): Poor long-term results despite
early hemodynamic and clinical improvement. Circulation 71:326–331
20. Winegrad S, Weisberg A, Lin LE, McClellan G (1986) Adrenergic regulation of myosin
adenosine triphosphate activity. Circ Res 58:83–95

Authors' address:

Gerd Hasenfuß, M.D., University of Freiburg, Department of Internal Medicine III (Cardiology),
Hugstetter Str. 55, D-7800 Freiburg, F.R.G.

III. Cardiac energetics in hypoxia and ischaemia

Mechanics of rat myocardium revisited: Investigations of ultra-thin cardiac muscles under high energy demand*

R. W. Gülch and G. Ebrecht

Physiologisches Institut II der Universität Tübingen, F.R.G.

Summary

Disregarding the influence of thickness on elevated strength of isolated preparations inevitably leads to erroneous tension-frequency relations, especially in the range of high frequencies. Thus, much of the confusion in interpreting the atypical negative staircase phenomenon of the rat heart is due to this. In view of the fact that the rat has become the preferred laboratory animal in cardiological research, it was imperative to reinvestigate force-frequency relations using ultra-thin preparations of the rat right ventricle (d < 0.1 mm). Contrary to popular opinion, it could be demonstrated that the rat myocardium shows a positive staircase in the range of physiological heart rates. An increase in tension is still attainable even at frequencies up to 600 min^{-1}. The interval-strength relations exhibit a minimum at frequencies of 60–120 min^{-1}, being shifted to higher frequencies with increasing diameter, vanishing completely for thick preparations (d > 1.0 mm). At high extracellular Ca^{++} concentration the positive staircase even of ultra-thin muscles is flattened. However, it can be reinforced when the strength, and thus the energy expenditure, is reduced by lowering the extension. The same is true for contractions under hypoxia.

From these findings it seems probable that many investigations on isolated heart muscles of the rat, as well as other species, are objectionable when done under high energy demand, as diffusion problems will certainly limit any rise in contractility.

Introduction

Experiments on isolated cardiac muscles must always be performed under physiological conditions that guarantee adequate oxygenation and diffusion of metabolites. Especially, the energy supply and elimination of metabolites in isolated muscles performing under elevated contractile state are quickly exhausted, thus leading to contractile impairment. In view of this, the thickness of isolated heart preparations is of paramount significance. Even under the energy saving contractile conditions of low temperature and low stimulation rate, compared to physiological conditions isolated papillary muscles and trabeculae exhibit an inverse relationship between developed tension and diameter, respectively cross-sectional area (Fig. 1 a), in accordance with the findings of Frezza & Bing [5] in the rat or Koch-Weser [15] in the cat. Disregarding the influence of insufficient diffusion due to thickness on elevated performance of isolated preparations, inevitably leads to an underestimation of positive inotropic alterations. Above all, force-frequency relationships will be falsified in the range of high stimulation rates where the energy turnover is over-proportionally high. Thus, much of the confusion in interpreting the atypical commonly accepted negative staircase phenomenon of the rat heart [2, 4, 8, 10, 18, 20, 22]

---

* Supported by the Deutsche Forschungsgemeinschaft.

is due to this. In view of the fact that the rat has become the preferred laboratory animal in cardiological research we considered it imperative to reinvestigate force-frequency relations with respect to heart muscle dimensions and contractile state.

Methods and materials

The experiments were performed on heart preparations of 15 Wistar rats (mean weight $\pm$ S.D. $= 370 \pm 75$ g). Under deep ether anaesthesia the hearts were rapidly excised, washed out and prepared in a preparation dish superfused with carbogen-equilibrated Tyrode bathing solution. All measurements were synchronically done on isolated right ventricular trabecula and papillary muscle, removed from the same heart. Trabeculae suitable for investigation had to be of the following size: diameter < 0.15 mm, length > 2 mm. In approximately half of the hearts at least one appropriate trabecula was found. The diameters of the papillary muscles were selected so as to allow for the greatest possible range in diameter. Whereas the preparation and handling of papillary muscles of normal size do not present any difficulties [6], isolating ultra-thin trabeculae requires extremely delicate work. In the latter case it is of primary importance to exclude possible risks involved in manual handling by using micromanipulators instead. Both ends of the preparations were pierced in situ with 2 mm long tips of glass microelectrodes which served as suspension in the experimental set-up (Fig. 1 b). After inserting the glass tips into a small piece of plasticine, the preparations were then removed from the ventricular wall. Still fixed to the plasticine, the preparations were transferred in a small transport bath to the actual muscle bath, where they were fixed to the force transducer and the isotonic lever, as shown in Fig. 1 b. The muscle baths were continuously perfused at a temperature of 37 °C with carbogen-equilibrated Tyrode solution of the following composition in mmol/l: NaCl 130, NaHCO$_3$ 20, NaH$_2$PO$_4$ 1.2, KCl 4.1, MgCl$_2$ 1.2, CaCl$_2$ 1.1,

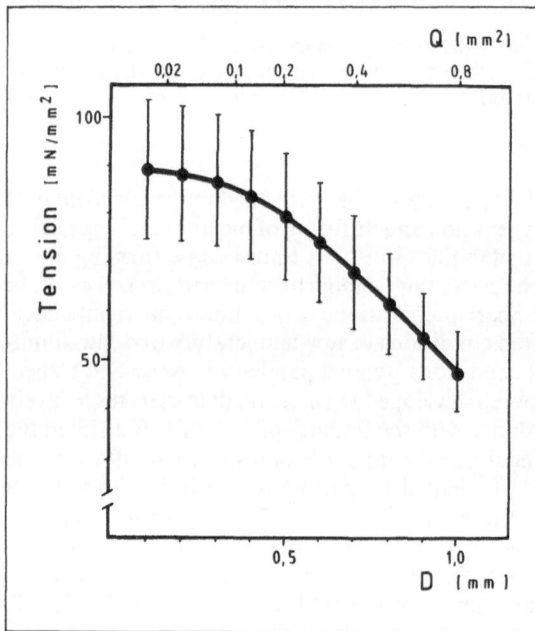

Fig. 1 a. Mean developed isometric tension of 180 trabeculae and papillary muscles of rat heart as a function of diameter D resp. cross-sectional area Q under the following conditions: temperature $= 30$ °C, stimulation rate $= 0.5$ Hz, extracellular Ca^{++} concentration $= 1.1$ mmol/l, muscle length $= L_{max}$. The vertical bars indicate standard deviation (S.D.). Heart preparations with diameters > 0.3 mm exhibit an almost linear inverse relationships between tension and diameter which certainly may be ascribed to an expanding central region of hypoxia with increasing diameter.

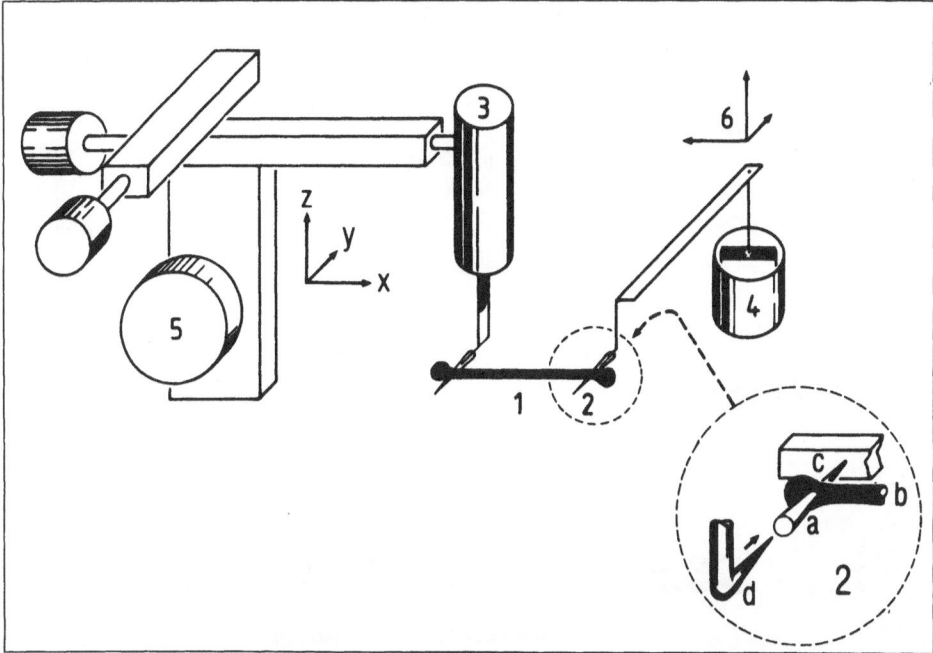

Fig. 1 b. Diagram of the experimental set-up. The heart preparation (1) is attached by means of tips of glass microelectrodes (2) to a piezoresistive force transducer (3) (AE 801 ame, resonance frequency 3 kHz) and to an isotonic electrically controlled lever system (4) consisting of a rebuilt dc microservomotor (M915L, Portescap). The effective mass of the moving part of the lever system being pivoted in jewel bearings with negligible friction is less than 10 mg. Both elements (3) and (4) can be independently moved in all three directions (x/y/z) by means of micromanipulators (5) and (6) to allow for careful attachment of the preparation. This procedure is schematically demonstrated in the inset (2): The tip of microelectrode (a) piercing the preparation (b) penetrates a small piece of plasticine (c) for transport. In the muscle bath small tips of stainless needles (d) which are connected to the force transducer and the isotonic lever system are pushed into the glass tips by means of the micromanipulators. Afterwards the preparations fixed to the set-up are withdrawn from plasticine.

glucose 23 (when not otherwise stated). Hypoxia was attained by aerating the Tyrode solution with a gas mixture consisting of 95% N_2 + 5% CO_2 instead of with carbogen gas (95% O_2 + 5% CO_2).

The heart preparations were stimulated with 1 ms squarewave pulses of an amplitude 10% above threshold via the electric field of two platinum electrodes placed along them. All force-frequency data were recorded under strictly stationary conditions, following variations in stimulation frequency.

Results and discussion

The physiological heart rate of the rat ranges from about 300 to 420 min^{-1}, corresponding to 5–7 Hz [3, 16]. Stimulation of isolated heart muscles of usual diameter with such frequencies leads to a decrease in force, as has been described many times and designated as negative staircase. This phenomenon is peculiar to the rat when compared with hearts

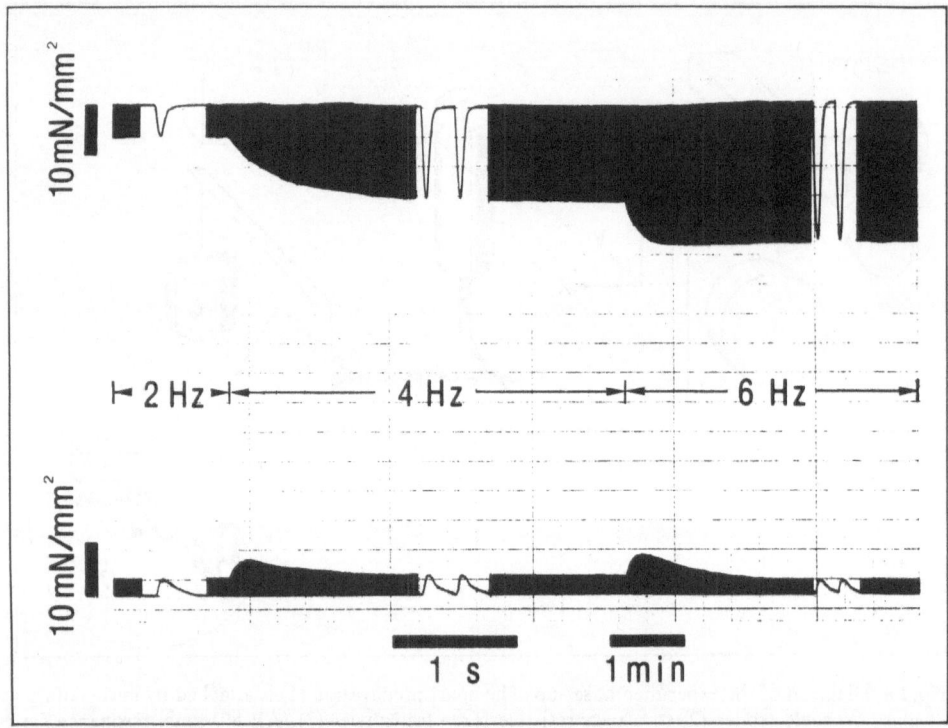

Fig. 2. Simultaneous low-speed recording of isometric mechanograms of an ultra-thin trabecula (D = 0.085 mm) (upper trace) and of a papillary muscle (D = 0.9 mm) (lower trace) at 37 °C and 1.1 mmol/l Ca^{++} concentration under variation of frequency in two steps. The trabecula reaches an augmented steady state tension monoexponentially.

of other warm-blooded animals. On the other hand, investigations on perfused rat hearts [9, 19] seem to indicate that an optimally supplied rat myocardium can show a positive staircase in the range of higher frequencies. We therefore decided to reinvestigate the tension-frequency concept of rat myocardium using ultra-thin heart preparations where almost no diffusion problems should exist. Figure 2 depicts a representative original registration of isometric mechanograms of a relatively thick and of an ultra-thin heart muscle under stepwise variation of the stimulation frequency. Similar changes in amplitude are found when the contractions are registered under isotonic conditions. Following an increase in frequency, the thick preparations show only a transient increase in force, whereas ultra-thin heart muscles exponentially reach steady state values which increase with higher frequencies. Figure 3 shows characteristic tension-frequency relations of heart muscle with varying diameters, which were all registered in steady-state, following successive increments in frequency. At frequencies <1 Hz, a negative staircase is common to all relations. Minimum developed force was registered at frequencies between 1 and 2 Hz for ultra-thin preparations and at higher frequencies for thicker muscle preparations up to a diameter of 1 mm. The minimum vanished completely in muscles with diameters >1 mm.

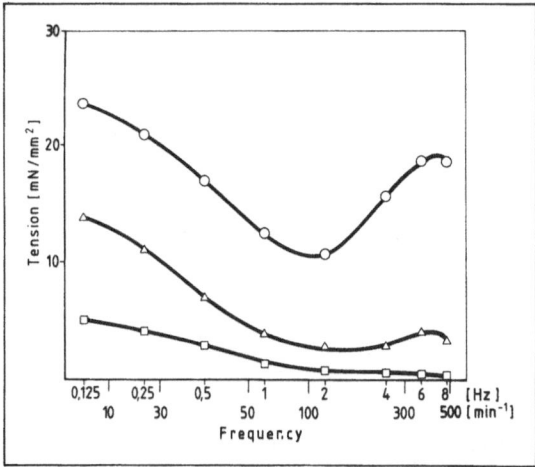

Fig. 3a. Tension-frequency relation of three papillary muscles with different diameters (–o– D = 0.25 mm, –△– D = 0.85 mm, –□– D = 1.3 mm). In accordance to Fig. 1a the tension level is inversely related to the diameter.

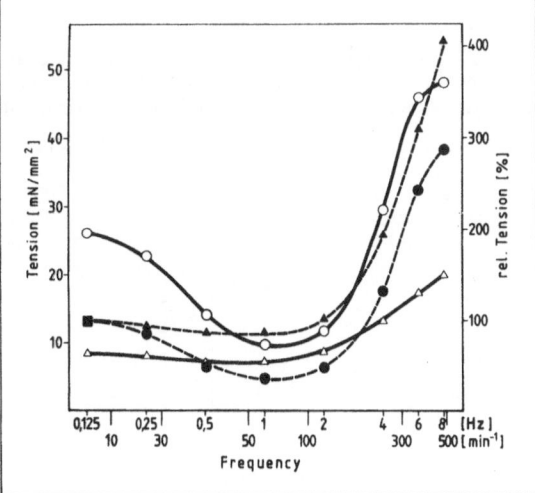

Fig. 3b. Two different types of tension-frequency relations of ultra-thin heart preparations. Type A: –△– heart preparations of a mean diameter ± S.D. = 0.05 ± 0.02 (n = 6) which probably belong to the conducting system of heart. This fact may explain the relatively low tension development. Type B: –o– trabeculae of mean diameter ± S.D. = 0.11 ± 0.03 mm (n = 6). The dashed curves are based on the relative tension values related to the tension at 0.125 Hz.

Two types of tension-frequency relations were found in ultra-thin preparations. Curves exhibiting a distinct minimum of force were found in ultra-thin preparations and to a lesser degree in thicker preparations of the myocardium. In contrast, extremely thin preparations (diameter 70 μm) revealed tension-frequency relations which were altogether shifted to a lower level of tension and ran parallel to the abscissa at lower frequencies. Presumably, the preparations belong to the conducting system of the heart. From Fig. 3 it can be primarily concluded that the thinner the preparations, the more accentuated the increase in tension with increasing stimulation frequencies in the range of physiological heart rates. The degree of steepness of the staircase can be characterized by the mean slope $\Delta T/\Delta f$ in the range of 4–6 Hz, as defined in the inset to Fig. 4a. This index, when plotted as a function of diameter or cross-sectional area, yields a clearly inverse relationship (Fig. 4a). For diameters of >1 mm, $\Delta T/\Delta f$ becomes negative, corresponding

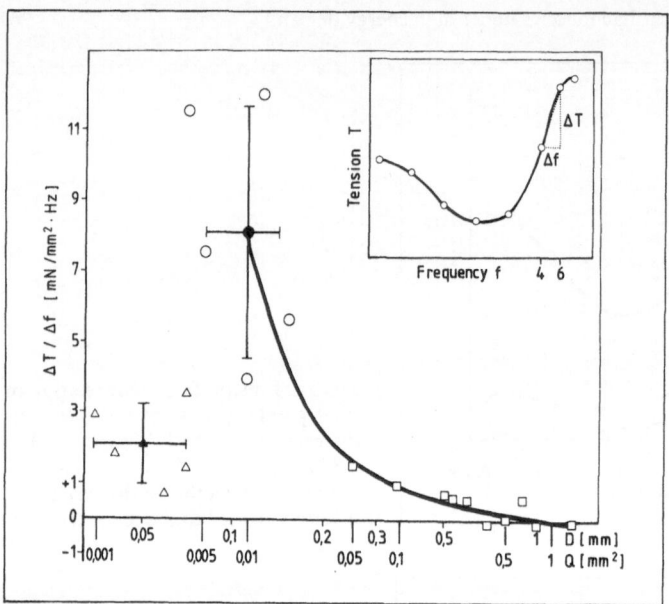

Fig. 4a. The mean slope of the tension-frequency relations $\Delta T/\Delta f$ at 300 min^{-1} as a measure of the positive staircase phenomenon correlated with the diameter resp. cross-sectional area. $\Delta T/\Delta f$ is defined in the inset. For the two groups of Fig. 3 b the statistical mean is expressed by the filled symbols (vertical and horizontal bars = ±S.D.

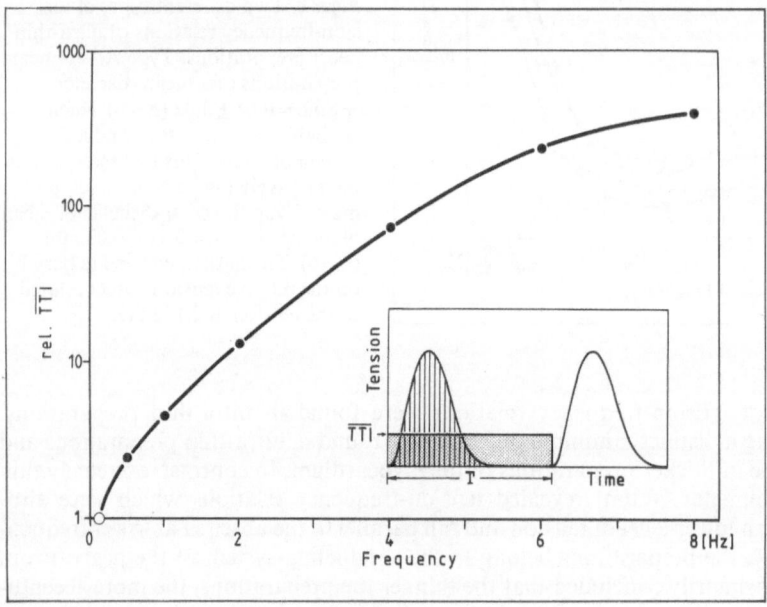

Fig. 4b. Mean tension-time integral *TTI* (=shaded rectangular area in the inset) which is calculated by the product TTI × frequency (TTI=hatched area underneath the mechanogram) as a function of frequency. *TTI* is supposed to correlate well with the O_2-consumption. The 400-fold increase in *TTI* under 8 Hz compared to the value of 0.125 Hz would mean a 400-fold increase in O_2 consumption to a first approximation which had to be met solely by O_2 diffusion from the outer surface of the isolated heart muscles.

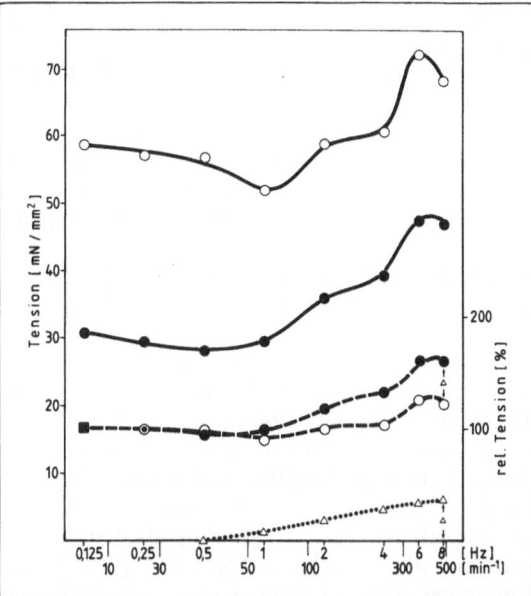

Fig. 5 a. Influence of high Ca^{++} concentration (4.4 mmol/l) on the tension-frequency relation of type B (Fig. 3 b) (–o–). Repeating the measurements under low muscle extension (0.9 L$_{max}$) and thus under lower O$_2$ consumption leads to a relation (–•–) exhibiting a more pronounced positive staircase. This fact is more clearly demonstrated by the relative tension-frequency relations (dashed curves) resp. the difference curve (dotted curve).

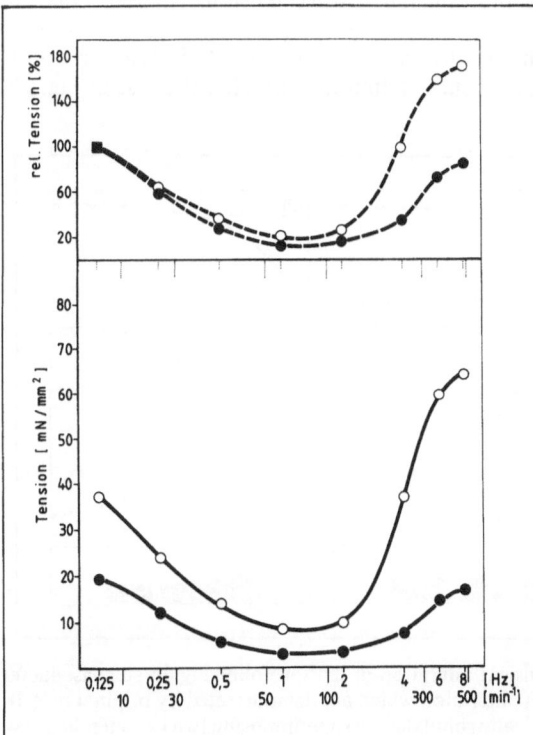

Fig. 5 b. Influence of hypoxia on the tension-frequency relation of type B (Fig. 3 b). In addition to the overall negative inotropic effect of O$_2$ shortage the positive staircase is attenuated, being more clearly demonstrated by the relative tension-frequency relations in the upper diagram (hypoxia = –•–).

to a negative staircase. The extremely thin preparations of the conducting system ($D = 29$–70 µm) do not, however, obey this relation.

It can be postulated that at high stimulation rates, even very thin heart muscles rapidly approach the limits of their diffusion capacity within in vitro systems. Tension time index, TTI, which can be calculated as time integral underneath the isometric mechanogram (see inset to Fig. 4b) can be used to estimate O_2 consumption under isometric contractions. The time average TTI can be obtained by dividing TTI by the cycle length of stimulation, T, or by multiplying it with the frequency. In the rat it correlates very well with the O_2 consumption [14]. By plotting TTI as a function of stimulation frequency, it becomes evident that O_2 consumption, which must be met solely by diffusion from the surrounding solution, is, at 8 Hz, approximately 400 fold the amount obtained at 0.1 Hz.

We therefore suspected that even in very thin preparations, there must be some impairment of positive staircase at high frequencies, under elevated contractile state, as can be attained by augmenting the extracellular Ca^{++} concentration to 4.4 mmol/l. Figure 5 verifies this assumption: at very high stimulation frequencies the positive staircase turns negative, as has been described for thicker preparations along the entire range of frequencies [4]. The less accentuated force minimum in the range of 1–2 Hz (in contrast to the one in Fig. 3b) must be due to changes in electromechanical coupling elicited by high Ca^{++} concentrations. Reducing the force to values similar to those in Fig. 3b by destretching the preparations causes an amplification of the positive staircase, on the one hand. On the other hand the decline in force at frequencies > 6 Hz is attenuated. This can most clearly be demonstrated by plotting the relative tensions (dashed curve in Fig. 5a).

Supposing that the steepness of the positive staircase is strongly influenced by O_2 supply, then under hypoxia the slope $\Delta T / \Delta f$ should diminish, which is the case in Fig. 5b.

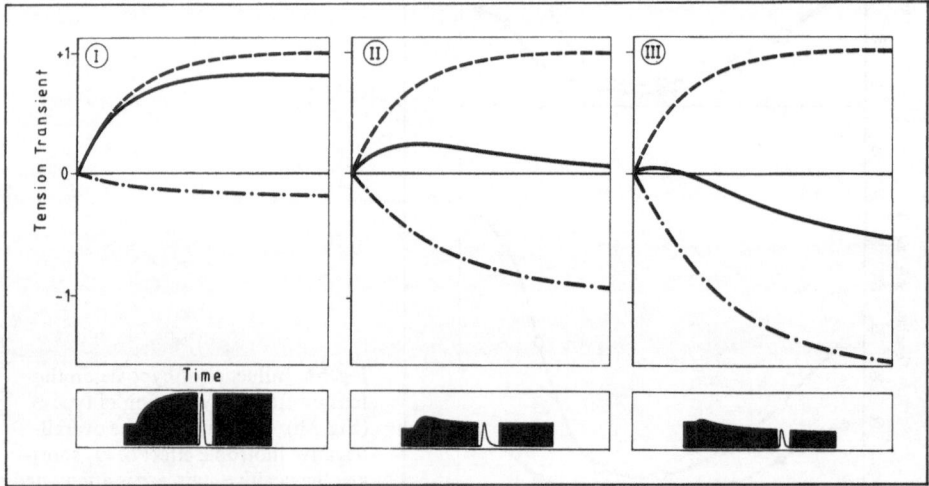

Fig. 6. Mathematical simulation of the transient alteration in tension following an stepwise increment in frequency. All three cases of tension alteration which are demonstrated by original recordings in the lower diagrams can be mathematically simulated by superimposing two exponential functions. Recording I is typical for ultra-thin preparations ($D < 0.1$ mm), II for preparations with $D < 1$ mm, III for very thick preparations ($D > 1$ mm).

The higher the frequencies and thus O_2 requirements the greater is the depression of tension-frequency relation under hypoxia (Tyrode solution aerated with a gas mixture of 95% N_2 + 5% CO_2). This again is best depicted by plotting relative tension.

According to Henry [9], the glucose content of the bathing solution influences the course of the tension-frequency curves. He described the transition from a positive staircase under high glucose content (30 mmol/l) to a negative one under lower glucose concentration (5 mmol/l). This phenomenon could not be verified by the present study where ultra-thin preparations were used. Under 5 mmol/l glucose a positive staircase, which was only slightly attenuated, could still be obtained at frequencies > 2 Hz.

The present findings show clearly that contractile response of isolated rat heart muscles to stepwise increments in frequencies within the range of physiological heart rates, greatly depends on diffusion conditions, respectively, diameter of the preparations. Thus, the tension-frequency relations are inevitably falsified more or less, depending on the cross-sectional area of the preparations. Thereby, the reactions can be divided into three categories as Fig. 6 demonstrates: Type I, peculiar to the ultra-thin heart muscles is characterized by an almost monoexponential increase in tension to a higher steady state value following stepwise increments in frequency. In thicker preparations (diameter > 0.1 mm), the force reaches different maxima depending on the diameters (Type II and III), to finally approach lower steady state values asymptotically. In very thick preparations of Type III (diameter > 1 mm) this steady state value is even lower than the starting level, which corresponds to a negative staircase. All three types of tension transients can be simulated by superimposing two counteracting exponential effects, described as follows:

$$f = f_p - f_n$$

whereby

$$f_p = a_p \left[1 - \exp(-k_p t)\right]$$
$$f_n = a_n \left[1 - \exp(-k_n t)\right]$$

with

$$k_p > k_n.$$

f_p is designated as positive inotropic effect of increased frequencies, as can be registered under optimum diffusion conditions, e.g. in situ. It is independent on diameter. Conversely, negative inotropic effect f_n is a function of diameter whereby it suffices to vary a_n alone in order to simulate the curves I–III. Thick preparations have a greater a_n compared to a_p in contrast to ultra-thin preparations with $a_p \gg a_n$. Both effects f_p and f_n are not identical to PIEA (positive inotropic effects of activation) and NIEA (negative inotropic effects of activation) which were introduced by Koch-Weser & Blinks [16], since the authors themselves recognized these diameter-nondependent effects as being entirely characteristic of the myocardial cells. In our opinion, the diameter-dependent negative inotropic influence is definitely due to the limitation of diffusion, the latter being rapidly exhausted under high stimulation frequencies. Insufficient diffusion could influence the process of contraction in two ways by creating: (1) a lack in the supply of some requisite metabolites from the surrounding bathing solution, (2) diminished washout of certain metabolites which inhibit contraction.

The first hypothesis addresses mainly the substrates O_2 and glucose both of which, when reduced, cause a reduction in intracellular PCr/ADP ratio and a subsequent loss of contractile function [1, 11]. This hypothesis is hardly compatible with the facts depicted in Fig. 2, since, beginning from an initially high frequency level, every increment in frequency effects an at least transient increase in force, even with thick preparations. Given severe shortage of PCr respectively ATP, it would be difficult to resolve the fact that an abrupt increase in frequency, which from the very onset naturally causes higher ATP-consumption per unit time, is accompanied by an at least transient increase in force development. The second hypothesis, in our opinion, is more plausible; possibly, accumulation of lactate and H^+ feedback to inhibit contraction. Recent works by Jacobus et al. [12]; Steenebergen et al. [24] as well as Roos & Boron [23] have demonstrated the close coupling between intracellular acidosis and contractility. In thicker preparations and especially under high stimulation rates, intracellular acidosis arising from anaerobic glycolysis are unavoidable. Insufficient diffusion of lactate or H^+, which would cause growing acidification and a subsequent depression of contractility, inevitably explains the impairment of the positive staircase at higher stimulation frequencies in the case of preparations of medium size on the one hand, and the negative staircase in very thick preparations on the other.

In our opinion, the ideal form of tension-frequency relation for the rat myocardium is fairly accurately reproduced by the curve (–o–) in Fig. 3b or in Fig. 2 [7]. By way of intimation a similar course is indicated in the work of Kruta & Stejskalová [17], using preparations of rat auricles, or in the study on isolated perfused rat by Henry [9]. The diminishing steepness of the staircase at stimulation frequencies > 6 Hz is certainly a result of the insufficient diffusion in preparations with diameter of approx. 0.1 mm, since in extremely thin preparations, e.g. of the conducting system type, where $D = 30 \mu m$, it remains unaltered up to 10 Hz. The negative staircase in the region of low stimulation frequencies, as well as the accentuated minimum of force between 1 and 2 Hz, seems to be peculiar to the rat myocardium. Qualitatively, the course of the curve can be fairly well explained by two counteracting processes which influence the contraction-effective intracellular Ca^{++} concentration in dependence on frequency. On the one hand, in the rat myocardium, the duration of the action potential decreases exponentially with stimulation frequency, as has been described by Payet et al. [21]. Thus, given unaltered Ca^{++} loading of SR, a decrease in Ca^{++} release must, in parallel, result under any frequency enhancement. With increasing rates of the action potential on the other hand, in time average the Ca^{++} loading of intracellular Ca^{++} stores will become more effective since, in the myocardial cell of the rat, each action potential is also coupled with transmembrane Ca^{++} currents [13]. Assuming further that both effects are superimposing additively or multiplicatively, the transition from the negative course, over the intermediate minimum, to the positive segment of the tension-frequency relation can be conclusively explained.

Conclusions

1. In ultra-thin isolated heart preparations of the rat, a considerably steep positive staircase is found in the range of physiological heart rates (300–420 min^{-1}).
2. In many investigations so far published on the rat which showed a negative staircase, the heart preparations were too thick not to hinder diffusion.
3. It must always be considered that contractions of isolated heart preparations under elevated contractile state are more or less impaired by some negative inotropic effect, depending on the cross-sectional area.
4. The transient alteration in tension after an increment in frequency can be interpreted as the sum of two counteracting inotropic effects where the negative one is assumed to be a function of muscle thickness.

5. The negative inotropic effect is probably due to insufficient diffusion of some anaerobic metabolite like lactate or H^+ rather than lack of PCr and ATP, since following increments in frequency there is at least a transient increase in tension, even in thick preparations.
6. These effects must always be considered when interpreting measurements under high energy demand on isolated heart muscles irrespective of animal species.

References

1. Brooks WM, Haseler LJ, Clarke K, Willis RJ (1986) Relation between the phosphocreatine to ATP ratio determined by ^{31}P nuclear magnetic resonance spectroscopy and left ventricular function in underperfused guinea-pig heart. J Mol Cell Cardiol 18:149–155
2. Buckley NM, Penefsky ZJ, Litwak RS (1972) Comparative force-frequency relationships in human and other mammalian ventricular myocardium. Pflügers Arch 332:259–270
3. DiCara LV, Miller NE (1969) Heart rate learning in noncurarized state, transfer to the curarized state, and subsequent retraining in the noncurarized state. Physiol Behavior 4:621–624
4. Forester GV, Mainwood GW (1974) Interval dependent inotropic effects in the rat myocardium and the effect of calcium. Pflügers Arch 352:189–196
5. Frezza WA, Bing OHL (1976) PO_2-modulated performance of cardiac muscle. Am J Physiol 231:1620–1624
6. Gülch RW, Jacob R (1975) Length-tension diagram and force-velocity relations of mammalian cardiac muscle under steady-state conditions. Pflügers Arch 355:331–346
7. Gülch RW (1986) The concept of "end-systolic" pressure-volume and length-tension relations of the heart from a muscle physiologist's point of view. Basic Res Cardiol [Suppl 1] 81:51–57
8. Henderson AH, Brutsaert DL, Parmley WW, Sonnenblick EH (1969) Myocardial mechanics in papillary muscles of the rat and cat. Am J Physiol 217:1273–1279
9. Henry PD (1975) Positive staircase effect in the rat heart. Am J Physiol 228:360–364
10. Hoffman BF, Kelly JJ Jr (1959) Effect of rate and rhythm on contraction of rat papillary muscle. Am J Physiol 197:1199–1204
11. Jacobus WE, Taylor GJ, Hollis DP, Nunnally RL (1977) Phosphorus nuclear magnetic resonance of perfused working rat hearts. Nature 265:756–758
12. Jacobus WE, Pores IH, Lucas SK, Weisfeldt ML, Flaherty JT (1982) Intracellular acidosis and contractility in the normal and ischemic heart as examined by ^{31}P NMR. J Mol Cell Cardiol [Suppl 3] 14:13–20
13. Josephson IR, Sanchez-Chapula J, Brown AM (1984) Early outward current in rat single ventricular cells. Circ Res 54:157–162
14. Kissling G, Rupp H (1986) The influence of myosin isoenzyme pattern on increase in myocardial oxygen consumption induced by catecholamines. Basic Res Cardiol [Suppl 1] 81:103–115
15. Koch-Weser J (1963) Effect of rate changes on strength and time course of contraction of papillary muscle. Am J Physiol 204:451–457
16. Koch-Weser J, Blinks JR (1963) The influence of the interval between beats on myocardial contractility. Pharmacol Rev 15:601–652
17. Kruta V, Stejskalová J (1960) Allure de la contractilité et fréquence optimale du myocarde auriculaire chez quelques mammifères. Arch Intern Physiol 68:152–164
18. McDowall RJS, Munro AF, Zayat AF (1955) Sodium and cardiac muscle. J Physiol 130:615–624
19. Meijler FL (1962) Staircase, rest contractions, and potentiation in the isolated rat heart. Am J Physiol 202:636–640
20. Nilius B, Boldt W, Fechner G (1976) Auswirkungen der Hypertrophie auf das Potentiationsverhalten isolierter Ventrikelstreifen der Ratte. Acta Biol Med Germ 35:1657–1664
21. Payet MD, Schanne OF, Ruiz-Ceretti E (1981) Frequency dependence of the ionic currents determining the action potential repolarization in rat ventricular muscle. J Mol Cell Cardiol 13:207–215

22. Penefsky ZJ, Buckley NM, Litwak RS (1972) Effect of temperature and calcium on force-frequency relationships in mammalian ventricular myocardium. Pflügers Arch 332:271–282
23. Roos A, Boron WF (1981) Intracellular pH. Physiol Rev 61:296–434
24. Steenbergen C, Delleuw G, Rich T, Williamson JR (1977) Effects of acidosis and ischemia on contractility and intracellular pH of rat heart. Circ Res 41:849–858

Authors' address:

Prof. Dr. R. W. Gülch, Institute of Physiology II, University of Tübingen, Gmelinstraße 5, D-7400 Tübingen (F.R.G.)

Changes in myocardial distensibility in rat papillary muscle: Fibrosis, KCl contracture, hypoxic contracture, oxygen and glucose deficiency contracture, and experimental tetanus

Ch. Holubarsch

Physiologisches Institut, Universität Tübingen, F.R.G.
Institute of Physiology and Biophysics, University of Vermont, Burlington, Vermont, U.S.A., and
Medizinische Klinik III, Universität Freiburg, F.R.G.

Summary

We describe diastolic properties of the myocardium in terms of stress-strain relations. In a mathematical analysis, the equation $\sigma = \alpha(e^{\beta \cdot \varepsilon} - 1)$ of the stress-strain curve can be changed by an increase in the exponent β or the multiplicative constant α. It can be experimentally shown that in hypertrophied myocardium of rats with essential hypertension and renal hypertension (SHR and Goldblatt rats, respectively), the steepening of the stress-strain curve is associated with an increase in the exponent β (stiffness constant) and an enhancement of collagen content.

On the other hand, in acute hypoxic myocardium, the slope of the stress-strain curve is increased without a significant change of the exponent β. Our results from heat measurements and quick-release experiments indicate that the hypoxic contracture (H) and the oxygen and glucose deficiency contracture (HG) are in contrast to the results of depolarization contracture (KCl) and experimental tetanus (T). In H and HG, the cross-bridge cycling rate was found to be slowed by a factor of 2 000 compared to KCl and T. This means that ATP demand for force development and maintenance is 2 000 times less in H and HG than in KCl and T. We will further discuss the meaning and implications of these experimental findings.

Introduction

Changes in diastolic properties of the left and right ventricular wall may occur for many different reasons: on the one hand, pericardial diseases – constrictive pericarditis, carcinosis of the pericardium, pericardial effusion – influence the distensibility of the ventricles; on the other hand, the filling of one ventricle interacts with that of the neighbour ventricle. Additionally, the distensibility of the ventricular wall may be diminished by an increase of ventricular muscle mass in the case of concentric hypertrophy. In excentric cardiac hypertrophy the configuration of the ventricle plays an important role.

Beside these extracardial and geometrical causes for inhibited distensibility of the ventricle, intrinsic changes of the myocardial tissue per se may be responsible for altered elastic properties: content of collagen, fibre orientation, contracture, oedema, and insufficient relaxation.

The separation between geometry-dependent and tissue-dependent impairments of ventricular distensibility is difficult, because (1) only a short part of the diastolic pressure-volume relationship is available in vivo, and (2) the transformation of pressure-volume data into normalized stress-strain data, which indicate the elastic properties of tissue, is based on several assumptions [15, 18]. The present study demonstrates how some basic questions about diastolic myocardial properties can be resolved by means of papillary muscle experiments in which myocardial force and length can be easily controlled. Changes of myocardial distensibility were investigated for different types of cardiac hypertrophy as well as for hypoxic contracture and contracture due to oxygen and substrate deficiency. In contrast to myocardial fibrosis, myocardial contracture is an active process which is necessarily energy-dependent. In order to study of whether hypoxic contracture and contracture due to oxygen and glucose deficiency are low or high energy consumption processes, we performed heat measurements and quick release experiments on papillary muscles. These studies were performed because hypoxic contracture might be only a symptom of low ATP concentration at the contractile proteins (rigor-like cross-bridges) or rather a calcium-activated high energy consuming process (fast cycling cross-bridges) accelerating cell death.

Methods

Theoretical considerations and definitions

Myocardial distensibility is described by stress-strain relations, when stress is defined as force per area and strain is the relative length change of the muscle [$\varepsilon = (l - l_o / l_o)$] according to Lagrange.

The stress-strain relationship of biological tissue is an exponential function $\sigma = \alpha(e^{\beta \cdot \varepsilon} - 1)$, when β is the exponential coefficient or so called stiffness constant.

The tangent modulus is the slope at any point of the curve.

Figure 1 demonstrates two theoretical possibilities: from a mathematical point of view, stress-strain relations may become steeper in two different ways under pathological conditions.

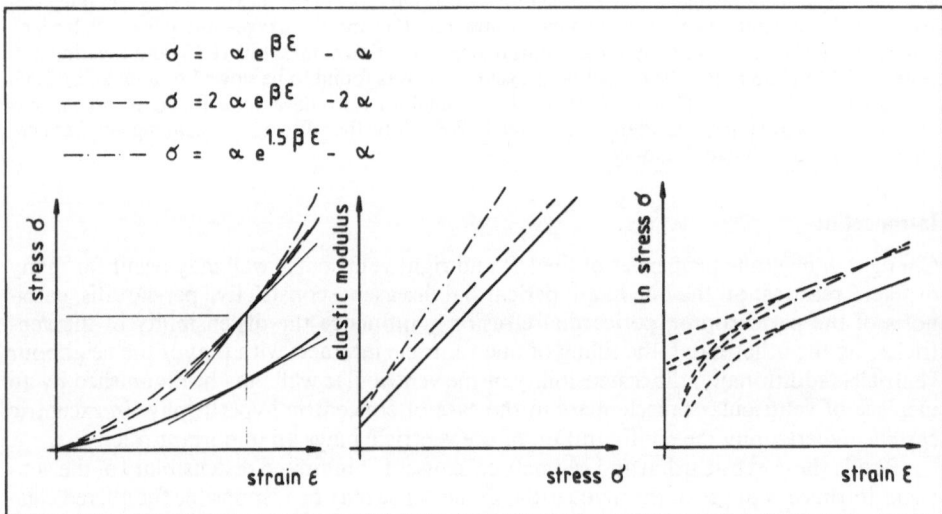

Fig. 1. Three different stress-strain equations are shown. The control (——) stress-strain relation can be changed either by the multiplicative constant α (—-—), or by the exponent β (—·—·—). For further explanation see Methods.

In the first case, the stiffness constant β is increased by 50%, in the second case, the constant α is doubled. At the intercept of both curves, the curves have the same stress and strain values, but different slopes, i.e., different values of the tangent modulus. If the tangent modulus is plotted as a function of stress, the differences are obvious either in the intercept (increase in α) or in the slope (increase in β) (Fig. 1, middle diagram). The same characteristics of stress-strain relationships can be demonstrated by the plot of "ln stress versus strain" (Fig. 1, right diagram). However, this relationship is not linear at low strain values. Two conclusions can be drawn from these considerations.

1) Stress-strain relationships may change due to two different mathematical possibilities.

2) A shift of the stress-strain relations towards higher stress-values is not necessarily combined with an increase in the coefficient β, i.e., stiffness constant.

Experimental procedure

All investigations were performed in left ventricular papillary muscles of rats.

Chronic experiments

Renal hypertension was induced by narrowing the left renal artery by means of a silver clip (diameter 0.2 mm). The investigations on papillary muscles were performed four and eight weeks after operation. The systolic blood pressure was 180 mm Hg in Goldblatt rats and 95 mm Hg in the control animals, on average. The left ventricular weight to body weight ratio increased by 50–80%.

As a second model of cardiac hypertrophy, spontaneously hypertensive rats (SHR) were chosen.

At 40 weeks-of-age, the systolic blood pressure was 150 mm Hg. At 80 weeks-of-age, it was 180 mm Hg on average. LV mass (relative to body weight) increased by 44% in the early stage, and by 90% in the late stage of cardiac hypertrophy.

In the thyrotoxic model of cardiac hypertrophy, rats were treated with L-thyroxine (0.1 mg/100 g body wt. per day, intraperitoneal injections over a period of two weeks). Placebos were injected in control rats. The increase in LV-mass (relative to body weight) was 76% on average.

Resting tension curves were obtained by stretching and releasing the muscle preparation by 0.1 mm every 30 seconds. l_o is the muscle length with diastolic stress zero, l_{max} is the muscle length with systolic stress maximum [10].

Hypoxic contracture, and contracture due to oxygen and glucose deficiency

Papillary muscles of the rat left ventricle were stimulated at a frequency of 1 Hz and contracted isometrically. Oxygen was withdrawn abruptly so that $_pO_2$ fell from about 600 mm Hg to less than 10 mm Hg. The glucose concentration was 5 mmol/l in hypoxic contracture.

In case of oxygen and glucose deficiency experiments, the papillary muscle was kept in a solution with a $_pO_2$ of about 600 mm Hg containing 21 mmol/l glucose. When the $_pO_2$ was lowered to less than 10 mm Hg, the glucose concentration was decreased to zero.

After ten minutes, in the case of hypoxia, and after 30 minutes in the case of oxygen and glucose deficiency, length-tension curves were obtained as described above. In experiments with high potassium chloride contracture, the sodium was exchanged with potassium isotonically. Exact details of myothermal measurements using antimony-bismuth thermopiles and of quick release experiments are given elsewhere [11, 12].

Results

Figure 2 shows the models and stage of cardiac hypertrophy investigated. Statistically significant decreases in myocardial distensibility were found in the Goldblatt model at the 4 week as well as at the 8 week stages and in the 80 week stage of the SHR. In all these cases, the relationship between tangent modulus and stress was steeper than in the control preparations. No significant change of stress-strain relations and tangent modulus-stress relations were found for the 40 week stage of SHR and the thyrotoxic model. Furthermore, we analysed the collagen content of the hypertrophied myocardium which was sig-

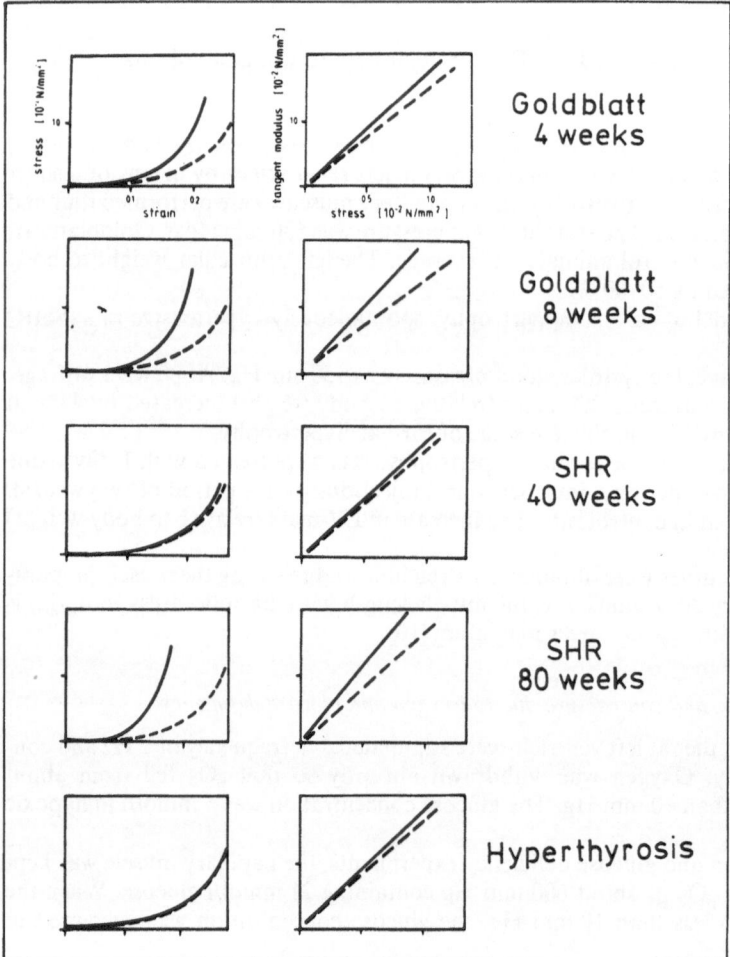

Fig. 2. Stress versus strain relations and tangent modulus versus stress relations in three different models of hypertrophy: Goldblatt, SHR and thyrotoxicosis. The hypertrophied myocardium (——) is compared to myocardium of age-matched controls (– – –).

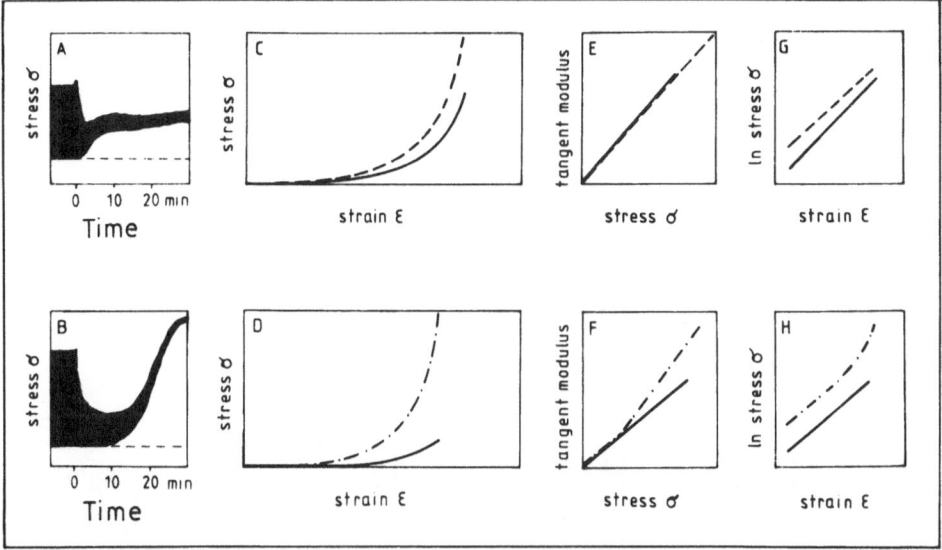

Fig. 3. A and B = time course of systolic and diastolic tension in hypoxia, and oxygen and glucose deficiency, respectively. A plateau of diastolic tension is reached after 10 min in A and after 30 min in B. C and D = respective stress-strain relations. The function of tangent modulus versus stress (E) is not changed in hypoxic contracture. In oxygen and glucose deficiency contracture, the function of tangent modulus versus stress is only changed at higher stress values (F). The same alterations are obvious when *ln* stress is plotted as a function of strain (G and H).

nificantly increased in the pressure-overload hypertrophy but unchanged in the thyro-toxic model [9, 17].

Figure 3 demonstrates the time course of hypoxic contracture (A) and oxygen and glucose deficiency contracture (B); Figs. C and D demonstrate the respective stress-strain relationships. In hypoxic contracture, the tangent modulus versus stress relation, and the *ln* stress versus strain relation is not changed with respect to the slope (Fig. 3 F and G). In glucose and oxygen deficiency contracture, the slope is only increased at higher stress values (Fig. 3 F and H).

In another study, the energetic consequences of hypoxic contracture were compared with those of KCl contracture. From KCl contracture it is known that calcium activates the acto-myosin system due to depolarization. In this study (Fig. 4), we measured the resting heat (R) and calculated the contracture heat (C) from the absolute value of contracture tension multiplied by the ratio of total heat and the tension-time integral of a twitch (cross-hatched columns). The sum of resting heat (R) and contracture heat (C) is then the RPC heat (resting plus contracture heat). In the case of KCl contracture, there is no difference between the measured RPC heat (white column) and the predicted RPC heat (cross-hatched column). However, in hypoxic contracture (H) and glucose and oxygen contrac-ture (HG), the calculated RPC heat is much bigger than the actually measured RPC heat. The measured RPC heat represents only 50% of resting heat which accounts for the rest-ing heat rate under hypoxic conditions. Therefore, no or very little heat is liberated in H and HG contracture which is due to contracture tension.

In a further study, quick release experiments were performed (Fig. 5). In hypoxic (H) and oxygen and glucose deficiency contracture (HG), the abrupt fall in tension is not fol-

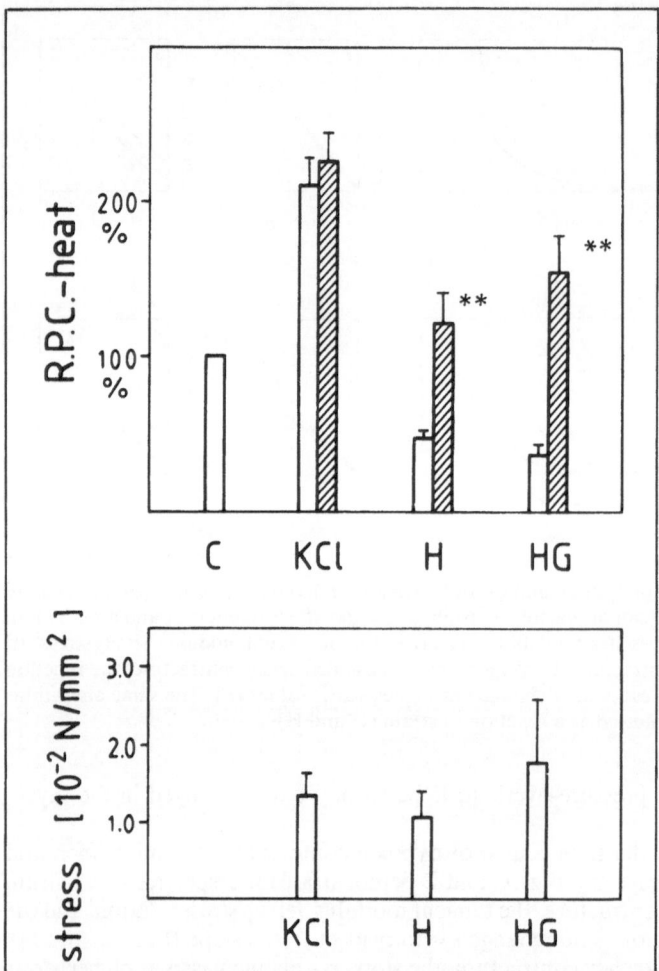

Fig. 4. Heat measurements in left ventricular rat papillary muscle preparations: C = Resting heat rate under control conditions (100%). RPC is the resting plus contracture heat liberation in KCl (potassium chloride), H (hypoxic) and HG contracture (hypoxia and glucose deficiency). The white columns represent the actually measured heat values, whereas the cross-hatched columns are predicted values. The heat liberation was predicted assuming that contracture is developed and maintained by the same calcium-induced cross-bridge cycling as in twitches. Therefore, the contracture heat was calculated from the contracture tension multiplied by the ratio of heat versus developed tension-time integral of a twitch. Measured resting heat and calculated contracture heat were added to predict RPC heat. In KCl-contracture the predicted values correspond to the measured values. In H and HG contracture, the measured heat values represent only resting heat rate under hypoxic conditions. No heat is observed which corresponds to the contracture tension seen in the lower diagram. **p < 0.01.

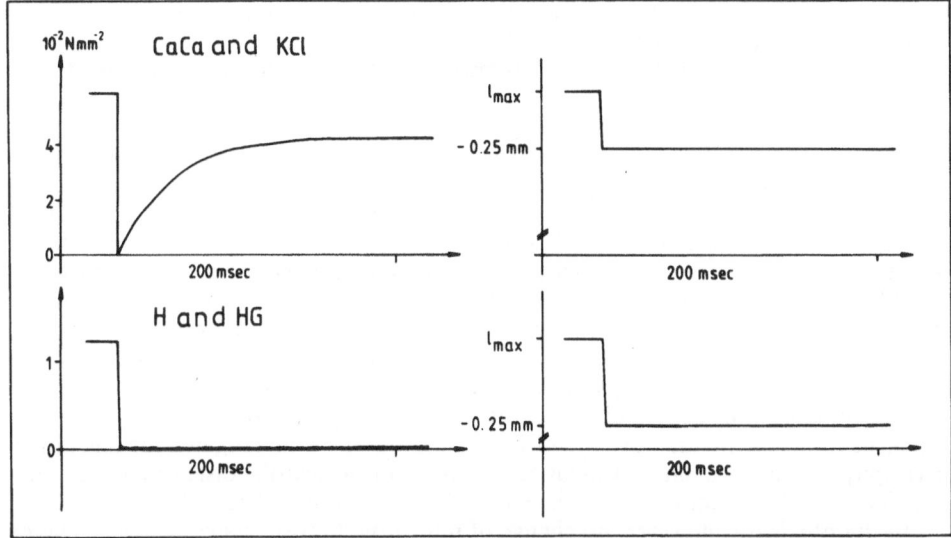

Fig. 5. Quick release experiments in myocardial tetanus obtained in high calcium (11 mM) and caffeine (7.5 mM) solutions and in KCl contracture (upper traces). After an abrupt length change by 0.25 mm (upper right diagram), there is a correspondingly abrupt drop in tension followed by a fast tension recovery phase (upper left).

In hypoxic contracture (H) or oxygen and glucose deficiency contracture (HG), there is no fast recovery phase after the initial tension drop (lower trace). The length changes correspond to 5% of muscle length.

lowed by a quick recovery of tension, which is observed for KCl contracture and experimental tetanus (CaCa). However, records with a slow time course show some very slow recovery of tension in H and HG contracture (no figure). This increase in tension following the abrupt release is about 2000 times slower than the quick tension recovery following the abrupt release in KCl contracture and in CaCa tetanus.

Discussion

Figure 1 showed that the slope of stress-strain relations may become steeper by changes of either the exponential coefficient β (type I) or the multiplicative constant α (type II). In the present study the question was investigated whether certain pathophysiological changes of myocardial distensibility can be attributed to type I or II.

Chronic experiments

Changes in myocardial distensibility were found in the Goldblatt rat myocardium (4- and 8-week stage) as well as in the 80-week stage of the SHR rats. In all of these animal groups, the exponential coefficient β (stiffness constant) was increased. In the 40-week stage of SHR and in the thyrotoxicosis model, no changes in myocardial distensibility were observed. To explain the decreased myocardial distensibility in the Goldblatt and SHR myocardium, three hypotheses have been proposed:

(1) By an increased diastolic cytosolic calcium concentration, actin-myosin cross-bridges may remain activated in diastole [14].

(2) The passive parallel elastic structures within the myocytes may alter, due to the hypertrophy process [5, 13].

(3) The collagen content of the interstitial space may be increased [4].

In a separate study [9], we were able to demonstrate the diastolic stress-strain relations of the hypertrophied Goldblatt myocardium to be steeper than those of the normal myocardium even in calcium-free EGTA solutions, because in such a solution a possible diastolic coupling of calcium-activated cross-bridges would have disappeared. Therefore, hypothesis [1] can be rejected – at least for the Goldblatt model of cardiac hypertrophy.

If the parallel-elastic structures within the myocytes characterized the diastolic stress-strain behaviour of the normal and hypertrophied myocardium, we would also expect a change in myocardial distensibility in the thyrotoxicosis model of hypertrophy. In this model, indeed, no change of myocardial distensibility is observed.

Our findings of decreased diastolic distensibility in hypertrophied myocardium can rather be explained by the existence of fibrosis. In all of the investigated models of hypertrophy in which we did find significant changes of myocardial distensibility, significant fibrosis was present when evaluated chemically [9] or histochemically [17].

In thyrotoxicosis, however, no change of interstitial fibrosis tissue was seen. Therefore, we postulate that an increase in the collagen content is the cause for the altered stiffness constant (coefficient β) of pressure overloaded hypertrophied myocardium.

Acute experiments

After ten minutes of hypoxia, the resting tension was doubled whereas no change in the stiffness constant was obvious. After thirty minutes of oxygen and glucose deficiency, we found no change in the stiffness constant at low stress values and a significant change of the stiffness constant at high stress values. The explanation of this phenomenon comes from morphological observations [1, 10]. Whereas in the hypoxic contracture the sarcomere lengths are homogeneously distributed in the whole muscle preparation, there exist two populations of sarcomeres in the oxygen and glucose deficiency contracture: one population with small sarcomere lengths (1.5 μ) and one population with large lengths (2.5 μ). It seems that there is a balance of force between the sarcomeres in the hypoxic contracture, whereas the "stronger" sarcomeres have stretched the "weaker" sarcomeres, in the case of oxygen and glucose deficiency. This explains why two components occur in the tangent modulus versus stress function in the HG contracture.

What are the basic mechanisms generating contracture tension? Three pathophysiological explanations have been proposed:

(1) Cytosolic calcium accumulation may occur because of an impaired calcium pump. The impairment may be due to ATP deficiency or other intracellular disturbances due to hypoxia [6, 7, 16].

(2) The sensitivity of the contractile proteins versus calcium may be increased, when the sarcoplasmic concentration of Mg^{2+}, ATP and H-ions are disturbed [11].

(3) A quick and pronounced decrease of high energy phosphates may lead to the generation of so-called rigor-complexes [1, 2, 8].

In order to prove these three hypotheses, two different studies with two different techniques were performed.

Heat measurements in papillary muscles demonstrate that neither in hypoxic contracture nor in glucose and oxygen deficiency contracture, the generated contracture tension is accompanied by an essential additional heat output. This is in contrast to the KCl contracture. In this type of contracture, we are able to exactly predict the energy costs, if we assume an active process of force generation which resembles the twitch force generation. In H and HG contracture, however, the predicted heat is much higher than the measured heat, if the same type of force generation is assumed as in a twitch. Because the measured heat in H and HG contracture is about half of the resting heat under control conditions, it can be suggested that this heat represents only, or mostly, resting heat under hypoxic conditions. Therefore, only very little energy consumption can be due to contracture tension development. It must by concluded from this data that the increase in contracture tension (H and HG contracture) costs much less energy compared to twitch tension generation or KCl contracture. Therefore the hypothesis which postulates rigor-like complexes is supported by this study.

Furthermore, quick release experiments were performed. Again, the hypothesis of rigor-like complexes was supported. The very fast tension recovery after the quick release – within 160 ms – which was observed in KCl contracture and CaCa tetanus, was not observed in H or HG contracture. We observed rather a very slow tension recovery which was about 2000 times slower than the fast one in KCl contracture and CaCa tetanus. Again, we conclude a very slow cross-bridge cycling due to H and HG contracture.

However, this conclusion implies very low Mg-ATP concentrations at the contractile proteins according to hypothesis [3].

On the other hand, when other researchers measured ATP concentrations in ischaemic or hypoxic myocardium going into or being in contracture, the ATP concentrations were lowered by a factor of less than 10 rather than 100 [2]. However, these biochemical analytical results represent overall ATP concentrations, without considering intercellular or intracellular compartmentalisation of ATP [1, 12]. However, whether calcium ions are involved in the initiation of contracture or not, the contracture tension is developed without a pronounced ATP consumption. This is of clinical relevance, because rigor-like cross-bridges maintain elevated diastolic tension without wasting a lot of extra energy in hypoxia or ischaemia, preventing early aneurysmal bulging and paradoxial systolic movement.

References

1. Bing OHL, Fishbein MC (1979) Mechanical and structural correlates of contracture induced by metabolic blockade in cardiac muscle from the rat. Circ Res 45:298–308
2. Bremel RD, Weber A (1972) Cooperation within actin filament in vertebrate skeletal muscle. Nature (New Biol) 238:97–101
3. Diamond G, Forrester JS (1972) Effect of coronary artery disease and acute myocardial infarction on left ventricular compliance in man. Circulation 45:11–19
4. Gay WA, Johnson EA (1967) An anatomical evaluation of the myocardial length-tension diagram. Circ Res 21:33–43
5. Grimm AF, Whitehorn WV (1966) Characteristics of resting tension of myocardium and localization of its elements. Am J Physiol 210:1362–1368
6. Grossman W, Barry WH (1980) Diastolic pressure-volume relations in the diseased heart. Fed Proc 39:148–155
7. Harris P (1975) A theory concerning the course of events in angina and myocardial infarction. Eur J Cardiol 3:157–163
8. Hearse DJ, Garlick PB, Humphrey SM (1977) Ischemic contracture of the myocardium: Mechanism and prevention. Am J Cardiol 39:986–993
9. Holubarsch Ch, Holubarsch T, Jacob R, Medugorac I, Thiedemann KU (1983) Passive elastic properties of myocardium in different models and stages of hypertrophy: A study comparing mechanical, chemical, and morphometric parameters. Perspectives in Cardiovasc Res 7:323–336

10. Holubarsch Ch, Jacob R (1981) Diastolic tension of rat cardiac muscle during deficiency of oxygen and glucose. Stress-strain relationships and reversibility. Basic Res Cardiol 76:690–703
11. Holubarsch Ch, Alpert NR, Goulette R, Mulieri LA (1982) Heat production during hypoxic contracture of rat myocardium. Circ Res 51:777–786
12. Holubarsch Ch (1983) Force generation in experimental tetanus, KCl contracture, and oxygen and glucose deficiency contracture in mammalian myocardium. Pflügers Arch 396:277–284
13. Maruyama KR, Natori R, Nonomura Y (1976) New elastic protein from muscle. Nature 262:58–60
14. Meerson FZ (1976) Insufficiency of hypertrophied heart. Basic Res Cardiol 71:343–354
15. Mirsky I, Parmley WW (1973) Assessment of passive elastic stiffness for isolated heart muscle and the intact heart. Circ Res 33:233–243
16. Nayler WG, Poole-Wilson PA, Williams A (1979) Hypoxia and calcium. J Mol Cell Cardiol 11:683–706
17. Thiedemann KU, Holubarsch Ch, Medugorac I, Jacob R (1983) Myocardial stiffness and connective tissue in pressure induced cardiac hypertrophy. A combined study of morphologic, morphometric, mechanical and biochemical parameters. Basic Res Cardiol 78:140–155
18. Wong AYK, Rautaharju PH (1968) Stress distribution within the left ventricular wall approximated as a thin ellipsoidal shell. Am Heart J 75:649–662

Authors' address:

PD Dr. Ch. Holubarsch, Internal Medicine, Dept. of Cardiology, Hugstetter Str. 55, 7800 Freiburg, F.R.G.

Ultrastructural observations on the effects of different substrates on ischaemic contracture in global subtotal ischaemia in the rat heart

E. van der Merwe, I. S. Harper, P. Owen*, A. Lochner**, S. Wynchank and L. H. Opie*

MRC Research Institute for Medical Biophysics and
** MRC Molecular and Cellular Cardiology Research Unit, Tygerberg and
* MRC Ischaemic Heart Disease Research Unit, University of Cape Town, Republic of South Africa

Summary

Ultrastructural morphology was examined in myocardia perfused with different substrates, following the hypothesis that ultrastructural changes or differences would indicate the mechanism of early ischaemic contracture. Substrates used were glucose, acetate and no substrate (substrate-free). Isolated rat hearts were subjected to global low flow ischaemia followed by perfusion fixation at either no contracture, early contracture (5%) or full contracture (100%). Total tissue adenosine triphosphate (ATP) levels were analysed from myocardia of each group. Glucose perfused myocardia did not develop contracture for at least 30 minutes and maintained ultrastructural and ATP normality. Substrate-free and acetate perfused myocardia developed contracture after 10–12 minutes of ischaemia at which time myofibrillar morphology was altered within small sparse foci of developing mild oedema with contracture and hyperextension of myofibrils. ATP levels were normal. At full ischaemic contracture, both substrate-free and acetate perfused myocardia showed severe mitochondrial damage, oedema, myofibrillar contracture and hyperextension with accompanying distortion and swelling of the sarcoplasmic reticulum.

Introduction

Ischaemic contracture occurs when there is a decline in the cellular ATP levels with a resulting loss of normal calcium homeostasis [5], but thus far it has not been possible to distinguish between decline in ATP and calcium gain as a causal mechanism. Substrate supply has been shown to influence total tissue high energy phosphate levels and to alter the rate of development of ischaemic contracture [3]. In this study we explore the hypothesis that the substrate supply to the subtotal ischaemic rat heart could influence the myocardial ultrastructure, thereby indicating possible mechanics of early myocardial contracture.

Methods

Long-Evans rats (weighing 250–300 g) were anaesthetized and their hearts rapidly excised and plunged into cold saline. Each isolated heart was attached to a cannula and perfused retrogradely on a Langendorff system. Hearts were maintained at 37 °C in a water-jacketed chamber and perfused at 100 cm H_2O for a 15 minute equilibrium period, after which global low flow ischaemia was induced by reducing the coronary flow from approximately 10 ml per minute to 0.5 ml per minute [2]. The hearts were perfused with Krebs-Henseleit bicarbonate buffer containing either 11 mM glucose or 5 mM acetate, or no exogenous substrate. Contractile force was measured by means of a Grass-force displacement transducer attached to the apex of the heart. Hearts with no contracture, 5% contracture and 100% contracture were either freeze-clamped for ATP analysis or perfusion-fixed with glutaraldehyde for electron microscopy [4]. Samples for microscopy were taken from the left ventricular subendocardium and were further fixed by immersion, dehydrated in ethanol and embedded in Spurr's Epoxy resin. Semi-thin sections (5 μm) for light microscopy were used for orientating the myofibrils on the longitudinal axis and for evaluating the extent of ischaemic damage. Thin sections (60 nm) were stained with uranyl acetate and lead citrate and examined in a Philips 420 transmission electron microscope at 80 kV. Micrographs were randomly selected for analysis.

Results

Development of contracture. Glucose perfused hearts did not develop ischaemic contracture during low flow ischaemia within the experimental period. In substrate-free and acetate perfused hearts there was evidence of ischaemic contracture after 10–12 minutes of ischaemia, culminating in full contracture after approximately 15 minutes.

Microscopy. Glucose perfused myocardium had well preserved ultrastructure illustrating normal mitochondria with densely packed cristae and matrix, abundant glycogen, and relaxed myofibrils. Substrate-free and acetate perfused myocardia, sampled prior to con-

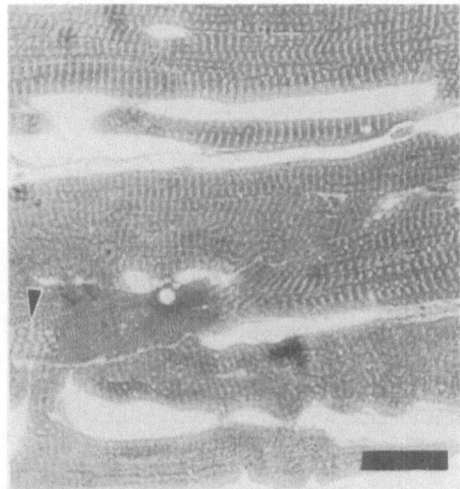 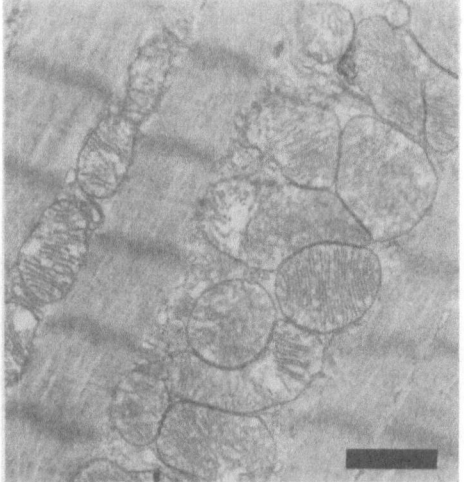

Fig. 1. Light micrograph of acetate perfused myocardium at 5% contracture showing a single myocyte in severe contracture and myofibrillar hyperextension of the adjacent myocytes (arrow). Bar = 200 μm.

Fig. 2. Electron micrograph of acetate perfused myocardium at 100% contracture showing myofibrillar contracture, swollen mitochondria with cleared matrix and intermyofibrillar oedema. Bar = 1 μm.

tracture, showed similar morphology to glucose perfused myocardia. At 5% contracture both substrate-free and acetate perfused myocardia displayed normal ultrastructure when examined by random sampling for electron microscopy. However, light microscopy revealed small foci of disturbed myocytes with mild oedema, moderate contracture and hyperextension of the myofibrils representing less than 5% of the myocardium (Fig. 1). Full contracture in both groups on light microscopic examination showed extensive distortion of cellular architecture, moderate to severe oedema and myofibrillar contracture with accompanying hyperextension. Hyperextension of myofibrils was predominant in substrate-free perfused hearts. Ultrastructure showed moderate intermyofibrillar oedema, slight to moderate swelling of mitochondria with cleared matrix density and separation of cristae, reduced glycogen and severe myofibrillar contracture (Fig. 2).

ATP levels. Total tissue ATP in all groups taken at 10–12 minutes of ischaemia showed no significant differences between substrate-free, acetate and glucose perfused hearts. Values were 3.34 ± 0.08, 3.15 ± 0.12 and 3.23 ± 0.25 µmol/g wet wt. respectively.

Discussion

Previous studies [2, 3] have used the influence of substrate supply to investigate factors affecting the development of contracture in low flow ischaemia. The present study, using the same model, shows that total tissue ATP levels do not correlate with development of ischaemic contracture in the subtotal ischaemic myocardium. During early contracture the morphological changes occur as a series of focal events with the number of cells so small as to have a negligible effect on the total tissue ATP level. In addition, in some myocytes, contracture precedes mitochondrial damage, on observation also reported for isolated myocardial strips subjected to hypoxia and glycolytic blockade [1]. These results, in revealing such focal irregularities, are consistent with the concept that the development of myocardial contracture is not necessarily an homogeneous event but may proceed from focal sequelae.

References

1. Bing OHL, Fishbein MC (1979) Mechanical and structural correlates of contracture induced by metabolic blockade in cardiac muscle from the rat. Circ Res 45:299–308
2. Bricknell OL, Daries PS, Opie LH (1981) A relationship between adenosine triphosphate, glycolysis and ischaemic contracture in the isolated rat heart. J Mol Cell Cardiol 13:941–945
3. Bricknell OL, Opie LH (1978) Effect of substrates on tissue metabolic changes in isolated rat heart during underperfusion and on release of lactate dehydrogenase and arrhythmias during reperfusion. Circ Res 43:102–115
4. Edoute Y, van der Merwe E, Sanan D, Kotze JCN, Steinmann C, Lochner A (1983) Normothermic ischaemic cardiac arrest of the isolated working rat heart: effects of time and reperfusion on myocardial ultrastructure, mitochondrial oxidative function, and mechanical recovery. Circ Res 53:663–677
5. Hearse DJ, Garlick PB, Humphrey SM (1977) Ischemic contracture of the myocardium: mechanisms and protection. Am J Cardiol 39:986–993

Authors' address:

E. van der Merwe, MRC Research Institute for Medical Biophysics, P.O. Box 70, Tygerberg, South Africa

Diastolic relaxation abnormalities during ischaemia and their association with high energy phosphate depletion, intracellular pH and myocardial blood flow

W. Grossman, S. I. Momomura, and J. S. Ingwall

Charles A. Dana Research Institute and Harvard-Thorndike Laboratory
of the Beth Israel Hospital and Department of Medicine (Cardiovascular Division),
Beth Israel Hospital, Brigham and Women's Hospital and Harvard Medical School,
Boston, Massachusetts

Summary

Diastolic relaxation of cardiac muscle is an energy requiring process dependent upon restoration of a very low cytosolic calcium concentration. During demand ischaemia, induced by an increase in myocardial oxygen demand in the setting of restricted coronary blood flow, myocardial relaxation is impaired with a resultant decrease in left ventricular chamber distensibility. Metabolic studies suggest that high energy phosphate depletion alone cannot account for these relaxation abnormalities. Changes in myocardial pH and coronary blood flow modulate the impaired relaxation. The most likely hypothesis to explain impaired relaxation during demand ischaemia is that slow calcium uptake and incomplete calcium sequestration are characteristic of the physiology of angina pectoris and demand ischaemia, underlying the increased diastolic stiffness seen in these conditions.

Introduction

Diastolic relaxation is a complex, energy requiring process which is dependent upon restoration of an extremely low cytosolic concentration of calcium ions $[Ca^{++}]_i$, following excitation-contraction coupling. Ordinarily, extracellular calcium concentration is approximately 10^{-3} m, whereas $[Ca^{++}]_i$ in the quiescent myocardial cell is 10^{-7} m. Thus, the normal myocardial cell is capable of maintaining a 10,000-fold concentration gradient for calcium ions. This is accomplished by means of potent calcium ion pumps located in the membrane of the sarcoplasmic reticulum and possibly the sarcolemma as well. During phase II of the action potential, influx of ionic calcium through slow channels serves as a trigger for the release of calcium stores within sarcoplasmic reticulum. This calcium-triggered calcium release results in an abrupt rise in $[Ca^{++}]_i$, which can be recorded by the photoluminescent protein aequorin as the calcium transient [11, 12].

As might be expected, considerable amounts of energy are needed to maintain this normal 10,000-fold calcium ion gradient. There is currently some controversy as to the source of this energy, but evidence suggests that ATP produced primarily from glycolysis serves a special role in providing energy for the calcium-ATPase pumps located on sarcoplasmic reticular membranes [1, 5, 17]. Thus, it seems possible that impaired relaxation might result from depression in high energy phosphate production in some circumstances, but not in others. Specifically, a decrease in high energy phosphate production by oxidative phosphorylation might have less of an effect on diastolic relaxation than an equivalent depression of high energy phosphate production by glycolytic flux.

Another important consideration to the study of myocardial relaxation concerns intracellular pH. The relationship between myofilament force generation and intracellular calcium concentration is sigmoid in shape. However, this relationship can be shifted by changes in pH, such that a decrease in pH (increase in hydrogen ion concentration) results in a decrease in myofilament force generation for any given calcium ion concentration. Recent evidence suggests that this is a result of a change in the configuration of troponin C, induced by hydrogen ions [3].

A variety of *mechanical factors* influence left ventricular diastolic distensibility, which may be defined as the diastolic pressure needed to distend the left ventricular cavity to a given volume. Extrinsic compression of the left ventricle (by pericardium or right ventricle) can decrease its distensibility. Engorgement of the coronary vasculature with blood may decrease distensibility of the left ventricular chamber due to what has been termed the erectile effect [16]. With myocardial ischaemia, changes in coronary vascular volume may be dramatic, and can easily contribute to changes in left ventricular diastolic distensibility.

Clinical studies have shown that left ventricular diastolic pressure rises during angina pectoris, and diastolic pressure is increased at any chamber volume [2, 4, 6–8]. This decrease in left ventricular diastolic distensibility is associated with slowing of myocardial relaxation [4, 8]. To investigate potential mechanisms for these alterations in diastolic distensibility and myocardial relaxation, we have utilized an animal model of transient demand ischaemia, designed to duplicate the physiology of angina pectoris.

Methods

Pacing induced ischaemia

An angina physiology model, developed in our laboratory [9, 10, 13–15], was used. Mongrel drops weighing 17–35 kg were anaesthetized with intravenous alpha-chloralose (100 mg/kg body wt.), after premedication with a subcutaneous injection of ketamine (10 mg/kg body wt.) As in our previous studies, propranolol (0.5 mg/kg body wt.) was injected to prevent ventricular fibrillation during pacing and to suppress the heightened sympathetic tone associated with the anaesthetized state. Respiration was maintained with room air by a Harvard pump via an endotracheal tube. A left thoracotomy was performed, usually at the fifth intercostal space, and the pericardium was opened widely to make a pericardial cradle. Both proximal left circumflex and left anterior descending arteries were freed from adipose tissue, and electromagnetic flow probes (Biotronex BL-5030, 5025) were placed around the arteries. A high fidelity micromanometer catheter (Millar instruments PC-480) was inserted into the left ventricle via the right carotid artery. To assess regional wall motion, a pair of ultrasonic crystals was implanted subendocardially and parallel to the short axis of the left ventricle in an area perfused by the left anterior descending artery and another pair in an area perfused by the circumflex coronary artery (Fig. 1). Crystals were placed in the inner third of the myocardium because of the relatively homogeneous fibre orientation at this distance from the endocardial surface. The position of these crystals was confirmed at the end of each experiment. Pacing electrodes were sutured on the left atrial appendage.

Blood flow in the left anterior descending and the left circumflex coronary arteries was reduced using small metal clips while phasic flow was monitored using the electro-

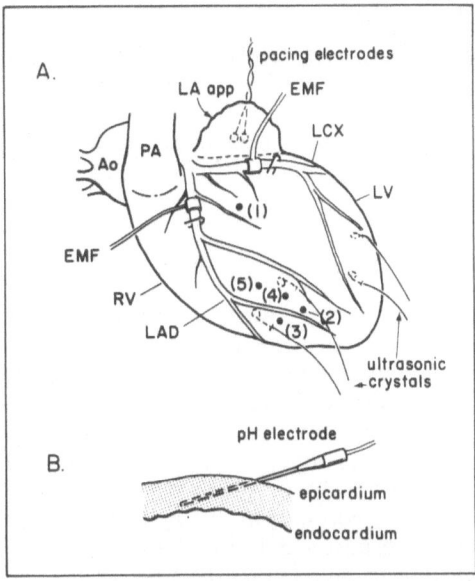

Fig. 1. In the presence of proximal stenoses on LAD and LCX coronary arteries, biopsy locations, in the left ventricular wall, included: (1) nonischaemic area, (2) ischaemic area, before pacing tachycardia, and (3) ischaemic area, immediately after pacing tachycardia. After removal of stenoses, control before coronary (LAD) occlusion (4), and at 3 minutes of coronary occlusion (5). EMF = electromagnetic flow probe, Ao = aorta, PA = pulmonary artery, La app = left atrial appendage, RV = right ventricle, LV = left ventricle. B. Polymer membrane coated pH electrode was implanted in LAD area with the tip placed in the inner third of the left ventricular free wall. Reproduced from ref. 10, with permission.

magnetic flow meter. The gap diameter of the clips was adjusted using a commercially available gap gauge so that the peak diastolic deflection of phasic coronary flow was reduced by 50%, which corresponds to more than a 90% reduction in diameter [13, 15]. If a stenosis of excessive severity was created, obvious systolic bulge in segmental wall motion occurred and the stenosis was immediately reduced in severity until segmental function was restored, as assessed visually and by the amplitude of segmental shortening measured by ultrasonic crystals. Thus, critical stenoses were created on both proximal left anterior descending (LAD) and left circumflex (LCX) coronary arteries (Fig. 1). After recording left ventricular pressure, dP/dt and myocardial segment lengths (Honeywell/ Electronics-for-Medicine Research Recorder), rapid atrial pacing (1.9 times resting heart rate) was performed for three minutes. Haemodynamic parameters were recorded continuously during and after pacing tachycardia.

Brief coronary occlusion

After the pacing tachycardia protocol, stenoses clips were removed and haemodynamic parameters were monitored for approximately thirty minutes to allow for complete recovery. The left anterior descending coronary artery was then occluded completely for three minutes, at the point where the stenosis clip was placed, then the occlusion was released and haemodynamic recovery during reperfusion was observed.

Pacing tachycardia without coronary stenosis

In seven dogs, control experiments were carried out with pacing tachycardia in the absence of coronary stenoses. The heart was paced at 1.7 times resting heart rate for three minutes. Haemodynamic parameters and ultrasonic crystal measurements before, during and immediately after pacing tachycardia were obtained.

Coronary occlusion without preceding pacing tachycardia

To determine whether the first episode of ischaemia (pacing tachycardia in the presence of critical coronary stenoses) affected the results observed during the second episode of ischaemia (brief coronary occlusion), the proximal left anterior descending artery was occluded for three minutes without preceding pacing induced ischaemia, in seven dogs, and haemodynamic changes were compared with those in experiments where coronary occlusion followed preceding pacing induced ischaemia.

Biopsy technique

A high speed air-turbine biopsy drill connected via a vacuum line to a bottle filled with liquid nitrogen was constructed [10] and used to take 2 mm diameter transmural myocardial biopsies. In 14 dogs, biopsies were taken before pacing tachycardia from a normally perfused area and an area perfused by the left anterior descending artery distal to the stenosis. Immediately following pacing tachycardia, after observing 5 to 10 beats and recording haemodynamic parameters, another biopsy was obtained from the left anterior descending coronary artery region. In the coronary occlusion protocol, a myocardial sample was obtained from the left anterior descending coronary artery region before and after three minutes of left anterior coronary artery occlusion. Before each biopsy, the epicardium was stained with Gentian violet or methylene blue so that the epicardial end of the tubular biopsy specimen could be distinguished from the endocardial end. Myocardial transmural biopsy samples were small (<2 mm in diameter) and bleeding was easily stopped by gently pressing a finger over the biopsy hole, without impairment of regional wall motion. Since ischaemia is usually more profound in subendocardium than in subepicardium, each frozen sample was divided, using tweezers, into subendocardial and subepicardial halves in a plastic saucer filled with liquid nitrogen, and then stored at $-70\,°C$. Location and sequence of biopsy samples is indicated in Fig. 1.

Myocardial pH

In ten dogs, a hydrogen ion selective polymer membrane pH electrode (LifeSpan 100 TM pH monitor, Biochem International, WI) was implanted into the subendocardium in the left anterior descending artery region (Fig. 1) and myocardial tissue pH was measured continuously. The pH sensing element consisted of a silver wire chlorided at the tip (0.4 mm in diameter), coated with a gel electrolyte layer and encapsulated by an ion specific polymer. A built-in silver to silver chloride reference is contained in the sensor. The electromotive force of the sensors is a function of the hydrogen ion concentration of the media being measured. The sensor yields a linear end response over pH range of 4.0 to 9.0. Time response of this pH sensor is rapid, so that a 90% step change is achieved in less than 2 s. The pH sensor was calibrated using buffers with pH 6.8 and 7.4 before insertion into myocardium, and again at the end of each experiment. Respiration was controlled so that arterial blood pH was 7.40 ± 0.04 (Radiometer blood gas system PMH72, Radiometer, Copenhagen).

Microsphere protocol

To study regional myocardial blood flow in the pacing induced ischaemia and coronary occlusion models, radioactive microspheres were used in 13 dogs. Microspheres (diameter $=15\,\mu$, New England Nuclear, MA) labelled with 5 species of radioactive material

(Co 57, Sn 113, Ru 103, Nb 95 and Sc 46) were suspended in 0.01% tween-80 and 5% Dextran solution and agitated in an ultrasonic bath for 5 minutes before injection. 30–50 µCi (1.5–2.5×10^5 microspheres) of each different isotope-labelled microsphere preparation were injected into the left atrium via a short (10–15 cm) 6F catheter: (a) before creation of coronary stenoses (b) with stenoses; (c) during the latter half of three minutes pacing tachycardia plus stenoses and (d) during coronary occlusion. A blood sample was collected for 90 seconds via a catheter placed in the aorta, using a withdrawal pump with a rate of 10–15 ml/min, for each injection of microspheres. At the end of each experiment, methylene blue dye was injected at the level of the stenoses first into the left anterior descending and then the left circumflex coronary arteries to confirm areas perfused by coronary arteries distal to the level at which stenoses were created. Then, the heart was excised, and muscle samples were obtained from the following areas: A) the left anterior descending artery territory distal to the stenosis; B) the left circumflex artery territory distal to the stenosis; C) normally perfusd (non-ischaemic) area. Usually, myocardium in border zones between these three areas was discarded. Six small muscle pieces weighing 1.0–1.5 g were obtained from the centre of each ischaemic area, and each piece was divided into subendocardial, mid- and subepicardial layers for separate analyses.

Radioactivity in the samples and blood was counted at the specific window of each radio-nuclide, using a Packard 5130 gamma scintillation well counter. Myocardial blood flow was determined using Tektronix 4052 computer system, as follows: The counts of radioactivity were corrected for background and for concomitant activity, contributed by the associated radio-nuclides. Blood flow of each myocardial sample is given by the following assumptions; $F_M = F_B \times C_M/C_B \times 1/W$ where F_M = myocardial blood flow (ml/g/min), F_B = flow rate of withdrawing pump (ml/min), C_M = corrected counts in myocardial sample at specific window (counts/min), C_B = counts in reference blood sample withdrawn, and W = weight of the sample in grams.

Measurement of high energy phosphate content

Myocardial tissue concentrations of the primary nucleotides and nucleosides in the heart were measured using high pressure liquid chromatography on Waters Model 440 and 450 chromatographs equipped with manual and automatic sample injectors and Waters Model M730 and Hewlett Packard Model 338A integrators. Frozen portions of the biopsies were homogenized in 0.6 N perchloric acid. The homogenate was neutralized with saturated K_2PO_4 and precipitated salts and proteins removed by centrifugation. Aliquots of a supernatant were applied to a Partisil SAX ion exchange column, 4.6 mm × 25 cm (Whatman). Nucleosides and nucleotides were eluted at room temperature isocratically in $0.16M$ KH_2PO_4 with $0.1M$ KCl at pH 6.5 and flow rate of 1.4 ml/min. Resulting chromatograms were analysed at 254 nm for ATP content and amounts calculated using external standards. Creatine phosphate was analysed using a fluorometric coupled enzyme method. The protein concentration of the perchloric acid extract was determined, and the concentration of each metabolite was expressed as nmol/mg of non-collagen cardiac protein.

Data analysis

Left ventricular pressure and myocardial segment lengths were digitized at 5 ms intervals and pressure – segment length loops were constructed using a Tektronix 4052 graphic computer system. Time constant T of left ventricular isovolumic relaxation was calculated from both the logarithm of pressure and the first derivative of pressure. These time constants were termed T_L and T_D respectively. Haemodynamic data before and after pacing, and before and after coronary occlusion were compared using Student's t-test. Metabolite contents, myocardial pH and myocardial blood flow were compared using analysis of variance. Differences were considered to be statistically significant at <0.05. All values were expressed as the mean ± standard error of the mean.

Results

In dogs with coronary stenoses, pacing tachycardia resulted in an upward shift in the pressure – segment length for ischaemic regions. Pacing tachycardia in dogs without coronary stenoses was not associated with any post-pacing change in diastolic pressure – segment length relations.

Haemodynamics in association with different types of ischaemia showed different patterns. Left ventricular pressure was $(135\pm4)/(8\pm1)$ mm Hg prior to pacing tachycardia in dogs with coronary stenoses. During pacing tachycardia, pressure changed to $(101\pm4)/(15\pm1)$ mm Hg, and in the immediate post-pacing period, pressure was $(126\pm5)/(17\pm2)$ mm Hg. Thus, left ventricular end-diastolic pressure increased substantially with demand ischaemia, and this change was most obvious following the cessation of pacing. In the control state, before coronary occlusion, left ventricular pressure was $(127\pm5)/(7\pm1)$ mm Hg, and at 3 minutes of coronary occlusion at the same heart rate this changed to $(115\pm6)/(11\pm1)$ mm Hg. This increase in end-diastolic pressure was exactly matched by an appropriate increase in end-diastolic segment length, so that the change in diastolic pressure could not be attributed to alteration in left ventricular myocardial stiffness. Time constants of relaxation were calculated by both logarithmic and derivative methods. Using the derivative method, which does not assume an asymptote of zero, the time constant was 42 ± 2 ms in the control state, and increased to 68 ± 5 ms during demand ischaemia; that is, immediately following discontinuation of 3 minutes of pacing tachycardia. A repeat control prior to coronary occlusion showed a return of the time constant to 38 ± 2 ms, and this increased slightly (but significantly) to 49 ± 3 ms after 3 minutes of coronary occlusion.

As seen in Fig. 2, the end-diastolic pressure – end-diastolic segment length point – moved upward following pacing-induced ischaemia in dogs with coronary stenoses. In contrast, the end-diastolic pressure – segment length point – moved rightward during coronary occlusion. Thus, demand ischaemia, which closely mimics the physiology of angina pectoris, is associated with a decrease in left ventricular diastolic distensibility; primary ischaemia (also called supply-side ischaemia) is associated with systolic failure and a rightward shift of the diastolic pressure – segment relation.

In association with these changes in function, pH declined precipitously with coronary occlusion, but only moderately with demand ischaemia in dogs with coronary stenoses. The decline in intramyocardial pH was 236% greater at 3 minutes with coronary occlusion ischaemia, as compared to the demand ischaemia associated with pacing tachycardia in dogs with coronary stenoses. Changes in pH were rapidly reversible. After termination of pacing tachycardia, intramyocardial pH returned to control values within 3–5 minutes. Similarly, following release of coronary occlusion, intramyocardial pH returned to control levels within 3–5 minutes.

High energy phosphate changes are shown in Table 1. As can be seen, there was approximately an 11% decline in subendocardial ATP in the left anterior descending distribution following pacing tachycardia in dogs with coronary stenoses. This compares with a 12% decline in subendocardial ATP at 3 minutes of coronary occlusion in the same dogs. Thus, similar depletions in ATP were seen with both types of ischaemia, even

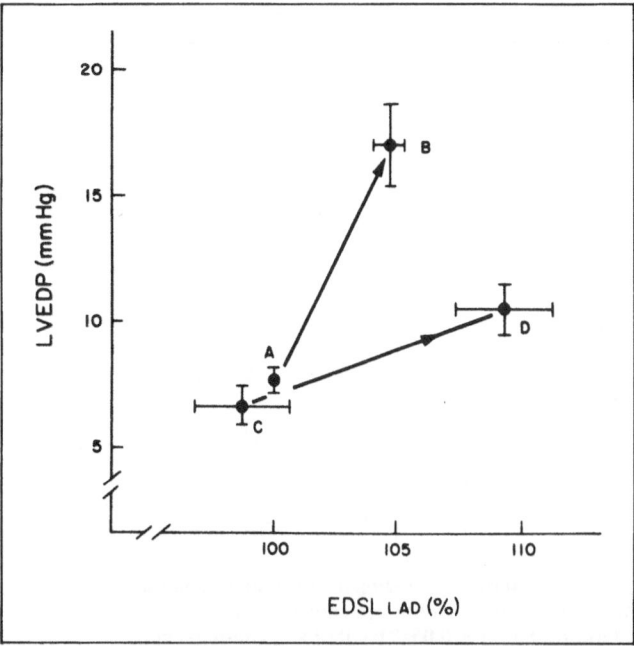

Fig. 2. Left ventricular end-diastolic pressure (LVEDP) to end-diastolic segment length (EDSL) points in pre-pacing with coronary stenoses (A), immediately after 3 minutes of pacing tachycardia (B), control before coronary occlusion (C), and at 3 minutes of coronary occlusion (D). Segment length was measured in LAD region. EDSL in pre-pacing was defined as 100%. LVEDP and its increase were significantly higher in pacing-induced ischaemia (A→B), and LVEDSL and its increase were greater in coronary occlusion (C → D). Reproduced from ref. 10, with permission.

though diastolic abnormalities were substantially different with the two types of isch-aemia. Creatine phosphate depletion was much more marked with coronary occlusion than with pacing tachycardia in dogs with coronary stenoses. As might be expected, cor-onary occlusion, which represents a complete deprivation of oxygen delivery, should re-sult in a major fall in high energy phosphate produced by mitochondrial respiration. High energy phosphate produced by aerobic metabolism generates creatine phosphate, which moves by the creatine phosphate shuttle to cytosolic sites where ATP is generated for lo-cal utilization. The marked decrease in creatine phosphate may also reflect the more sub-stantial accumulations of hydrogen ions in dogs with coronary occlusion, since the cre-atine kinase reaction is highly pH sensitive.

Myocardial blood flow changes are shown in Table 2. As can be seen, the placement of coronary stenoses produced no significant change in subendocardial or subepicardial blood flow. Thus, absolute and relative (endo/epi) blood flows were preserved, although there was a tendency for the endo/epi ratio to decrease. With pacing tachycardia, endo-cardial blood flow distal to a coronary stenosis decreased substantially. Of interest, epi-cardial blood flow actually increased, although not as much as it did in non-ischaemic myocardium. Thus, the endo/epi ratio fell substantially. As might be expected with cor-onary occlusion flow decreased markedly in all layers of myocardium, and the endo/epi ratio also diminished.

Table 1. Nucleotide and nucleoside content in biopsy samples (nmols/mg protein)

	CrP	ATP
Non-ischaemic area (n=14)		
pre-pacing		
epi	60.2±3.1	35.0±1.2
endo	55.3±2.5	34.5±1.3
LAD area		
pre-pacing (n=14)		
epi	56.3±2.4	33.3±1.5
endo	53.8±2.1	31.3±1.4
immediately post-pacing (n=14)		
epi	50.0±3.1	30.8±0.8
endo	39.6±2.5[b,h,e]	27.9±1.0[b,g,e]
control before occlusion (n=7)		
epi	57.3±2.3	33.9±0.8
endo	51.1±2.3	33.2±0.5
occlusion 3' (n=7)		
epi	18.3±2.9[d]	32.1±1.5
endo	7.8±1.4[d,f]	29.2±2.6[c]

CrP = creatine phosphate; LAD = the left anterior descending artery; (Adenosine was not detected in the biopsies analyzed and thus no data are shown for adenosine.) epi = subepicardium; endo = subendocardium; [a] $p<0.05$, [b] $p<0.01$ vs. Non-ischaemia area; [c] $p<0.05$; [d] $p<0.01$ vs. control before occlusion; [e] $p<0.05$; [f] $p<0.01$ vs. epi; [g] $p<0.05$; [h] $p<0.01$ vs. pre-pacing.
Note: These data are reproduced from reference 10.

Table 2. Regional myocardial blood flow (ml/min/g) in pacing-induced ischaemia and coronary (left anterior descending) occlusion

	Control (n=7)	Stenoses (n=7)	Pacing tachycardia (n=7	LAD occlusion (n=6)
LAD area				
epi	0.87±0.06	1.05±0.10	1.22±0.15[b]	0.26±0.12[d,b,c]
mid	0.97±0.06	1.08±0.09	0.73±0.07[a]	0.11±0.04[b,c]
endo	1.01±0.05	1.02±0.07	0.47±0.05[b,c,e]	0.10±0.03[d,b,c]
endo/epi	1.20±0.06	1.04±0.04	0.40±0.05[b,c]	0.50±0.11[b,c]
Non-ischaemic area				
epi	0.93±0.11	1.16±0.14	1.46±0.12[b]	1.09±0.13
mid	1.04±0.09	1.28±0.15	1.31±0.11	1.31±0.16
endo	0.95±0.07	1.17±0.14	0.99±0.11[e]	1.11±0.16
endo/epi	1.09±0.06	0.94±0.04	0.71±0.10[b]	1.02±0.07[d]

epi = subepicardial blood flow; mid = blood flow in mid layer; endo = subendocardial blood flow; endo/epi = transmural blood flow ratio values are expressed as mean±SE; [a] $p<0.05$ compared to "Control"; [b] $p<0.01$ compared to "Control"; [c] $p<0.01$ compared to "Stenoses"; [d] $p<0.01$ compared to "Pacing tachycardia"; [e] $p<0.01$ compared to "epi".
Note: These data are reproduced from reference 10.

Discussion

This study again demonstrates the previously reported finding that myocardium subjected to demand ischaemia exhibits slowed relaxation and decreased diastolic distensibility. The impaired relaxation and altered distensibility may well be related, and we have previously suggested that demand ischaemia is associated with both slowing of myocardial relaxation and incompletion of myocardial relaxation (Fig. 3). This may well be associated with impairment in the rate and ultimate extent of calcium sequestration in sarcoplasmic reticulum. Specific studies to assess intracellular calcium are not possible in this intact heart model of demand ischaemia. Indirect evidence concerning the metabolic mechanisms responsible for altered relaxation and distensibility can be gleaned from an examination of the high energy phosphate and pH data. Since essentially equal decreases in cytosolic ATP resulted from both demand ischaemia and supply ischaemia (coronary occlusion), it seems reasonable to conclude that a decline in total intracellular ATP is not responsible for the impaired relaxation and altered distensibility. This is further supported by the fact that declines in ATP in both types of ischaemia were quite modest. A decrease in ATP sufficient to produce contracture/rigor would be considerably larger. Decreases in creatine phosphate were more marked with supply ischaemia, even though abnormalities in relaxation and distensibility were seen to a much lesser degree (if at all) with supply ischaemia. Thus, it also seems unlikely that deficiency in creatine phosphate is responsible for the abnormal relaxation and distensibility seen primarily with demand ischaemia. It is possible (in fact it may well be likely) that changes in high energy phosphate pools within the myocardium may be of critical importance in explaining the abnormalities seen in this study. Detection of changes in specific subcellular pools would not be possible using the methodology that we employed. Indeed it is extremely difficult to differentiate ATP produced from glycolytic metabolism, from that resulting from aerobic metabolism in the intact heart. Specific experiments using different metabolic inhibitors and/or stimulators of specific metabolism may be helpful in differentiating the role of high energy phosphate produced by glycolysis from that produced by aerobic metabolism. There is evidence [5, 17] that glycolytic ATP plays a particular role in maintaining diastolic relaxation and preventing contracture. Whether this can be generalized to the types of transient abnormalities in relaxation and distensibility seen in association with the physiology of angina pectoris remains uncertain.

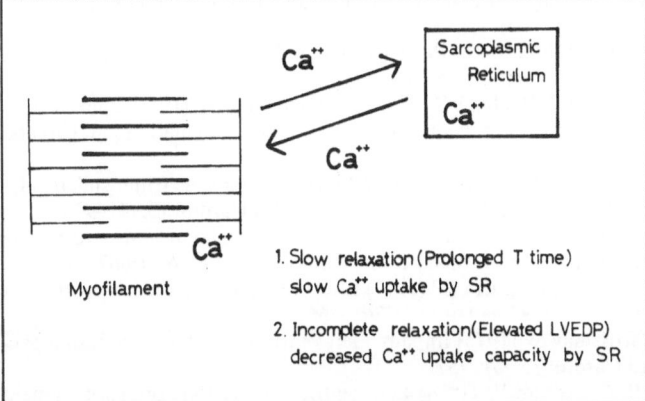

Fig. 3. Proposed mechanism of altered diastolic relaxation and distensibility in demand ischaemia.

Changes in intracellular pH may also be important in explaining the differences in diastolic distensibility seen in these two models of ischaemia. Hydrogen ion shifts the force – $[Ca^{++}]_i$ curve in such a way that a lower force is generated by myofilaments for any given $[Ca^{++}]_i$. Thus, any impaired relaxation and increased diastolic stiffness (increased myocardial force/segment length) that might have been seen during supply ischaemia could well have been offset by the substantial hydrogen ion accumulation in this model. The pH decline was 236% greater during supply ischaemia than during demand ischaemia. There was a decrease in pH with demand ischaemia as well, and this may have tended to blunt the diastolic relaxation abnormalities actually seen. It will be of interest to determine whether artificial maintenance of a more normal pH (buffers, hyperventilation) will tend to exacerbate the physiological abnormalities of diastole seen during demand ischaemia.

Finally, it is of interest to note that myocardial blood flow was actually well preserved during demand ischaemia in subepicardial and mid-wall layers of the myocardium. There was, however, a relative ischaemia developing in subendocardium. This may reflect increased diastolic pressure with compression of subendocardial vessels. Alternatively, slow relaxation of subendocardial muscle may restrict coronary blood flow through this layer, and thus be responsible for the diminished blood flow. This "chicken and egg" problem cannot be solved by further extrapolation from these experiments.

In summary, impaired myocardial relaxation and diminished diastolic distensibility are seen primarily with demand ischaemia (physiology of angina pectoris) and only to a minor degree with supply ischaemia (coronary occlusion). Changes in high energy phosphate failed to explain these differences in relaxation. Changes in myocardial pH and coronary blood flow modulate these changes and probably contribute to the physiology observed. The most likely hypothesis to explain the observations of this and other studies is that slow calcium uptake and incomplete calcium sequestration are characteristic of the physiology of angina pectoris and demand ischaemia, and underlie the impaired relaxation and increased diastolic stiffness seen in this condition.

References

1. Apstein CS, Deckelbaum L, Hagopian L, Hood WB Jr (1978) Acute cardiac ischemia and reperfusion: contractility, relaxation, and glycolysis. Am J Physiol 235:H637–H648
2. Aroesty JM, McKay RG, Heller GV, Royal HD, Als AV, Grossman W (1985) Simultaneous assessment of left ventricular systolic and diastolic dysfunction during pacing-induced ischaemia. Circulation 71:889–900
3. Blanchard EM, Solaro RJ (1984) Inhibition of the activation and troponin calcium binding of dog cardiac myofibrils by acidic pH. Circ Res 55:382–391
4. Bourdillon PD, Lorell BH, Mirsky I, Paulus WJ, Wynne J, Grossman W (1983) Increased regional myocardial stiffness of the left ventricle during pacing-induced angina in man. Circulation 67:316–635
5. Bricknell OL, Daries PS, Opie LH (1981) A relationship between adenosine triphosphate, glycolysis and ischaemic contracture in the isolted rat heart. J Mol Cell Cardiol 13:941–945
6. Lorell BH, Turi Z, Grossman W (1981) Modification of left ventricular response to pacing tachycardia by nifedipine in patients with coronary artery disease. Am J Med 71:667–675
7. Mann T, Brodie BR, Grossman W, McLaurin LP (1977) Effect of angina on the left ventricular diastolic pressure-volume relationship. Circulation 55:761–766
8. McLaurin LP, Rolett EL, Grossman W (1973) Impaired left ventricular relaxation during pacing induced ischaemia. Am J Cardiol 32:751–757
9. Momomura SI, Bradley AB, Grossman W (1984) Left ventricular diastolic pressure-segment length relations and end-diastolic distensibility in dogs with coronary stenoses: an angina physiology model. Circ Res 55:203–214

10. Momomura SI, Ingwall J, Parker JA, Sahagian P, Ferguson JJ, Grossman W (1985) The relationship of high energy phosphates, tissue pH and regional blood flow to diastolic distensibility in the ischaemic dog myocardium. Circ Res 57:822–835
11. Morgan JP, Wier WG, Hess P, Blinks JR (1983) Influence of Ca^{++}-channel blocking agents on calcium transients and tension development in isolated mammalian heart muscle. Circ Res [Suppl 1] 52:47–52
12. Morgan JP, DeFeo TT, Morgan KG (1984) A chemical procedure for loading the calcium indicator aequorin into mammalian working myocardium. Pflugers Arch 400:338–340
13. Paulus WJ, Serizawa T, Grossman W (1982) Altered left ventricular diastolic properties during pacing-induced ischaemia in dogs with coronary stenosis. Potentiation by caffeine. Circ Res 50:218–227
14. Paulus WJ, Grossman W, Serizawa T, Bourdillon PD, Pasipoularides A, Mirsky I (1985) Different effects of two types of ischaemia on regional left ventricular systolic and diastolic function. Am J Physiol 248:H719–H728
15. Serizawa T, Carabello BA, Grossman W (1980) Effect of pacing-induced ischemia on left ventricular diastolic pressure-volume relations in dogs with coronary stenoses. Circ Res 46:430–439
16. Vogel WM, Apstein CS, Briggs LC, Gaasch WH, Ahn J (1982) Acute alterations in left ventricular diastolic chamber stiffness. Role of the "erectile" effect of coronary arterial pressure and flow in normal and damaged hearts. Circ Res 52:465–478
17. Weiss J, Hiltbrand B (1985) Functional compartmentation of glycolytic versus oxidative metabolism in isolated rabbit heart. J Clin Invest 75:436–447

Author's address:

William Grossman, M.D., Chief, Cardiovascular Division, Beth Israel Hospital, 330 Brookline Avenue, Boston, MA 02215

Inotropic changes in ischaemic and non-ischaemic myocardium and arrhythmias within the first 120 minutes of coronary occlusion in pigs*

Hj. Hirche, M. Hoeher, and J. H. Risse

Institut für Angewandte Physiologie, Universität Köln, F.R.G.

Summary

In 16 anaesthetized open-chest pigs occlusion of the distal third of the LAD was performed. Local myocardial contractility within and outside the ischaemic area, measured using electromagnetic and ultrasonic probes, arrhythmias and plasma catecholamine concentrations were monitored during 180 min before and 120 min after occlusion of the distal LAD. 5 pigs developed ventricular fibrillation (VF) during phase 1 a of arrhythmias (1–6 min post occlusion) and a further 5 pigs during phase 1 b (10–30 min post occlusion). Segment lengthening within the ischaemic area, measured at the beginning of ventricular ejection, started within a few seconds of ischaemia and reached 11–13.5% of end-diastolic length prior to occlusion after 2 min of ischaemia. Between 2 min and 120 min after the onset of occlusion, no further segment lengthening was observed. There were no significant differences in segment lengthening between VF and non-VF pigs. However, within the VF group, mechanical alterations 2 min after the onset of ischaemia were more marked in pigs developing VF within phase 1 a than those with VF in phase 1 b. VES increased markedly after 50 min of occlusion and seemed to occur independently of further mechanical alterations, VF within the first 30 min of ischaemia often occurred without preceding VES.

Introduction

Myocardial contractility has been shown to change within a few seconds after the onset of ischaemia [14]. During regional ischaemia it has been reported that, coincident with the deterioration in contractility in the ischaemic area, shortening in the non-ischaemic area increases [11]. Fibre stretching within the ischaemic myocardium and at the borderline between ischaemic and non-ischaemic tissue may be involved in the genesis of ischaemia-induced arrhythmias [7], in addition to other known factors such as ion shifts and catecholamine release [5]. Clinical data suggest a link between myocardial contractility and cardiac arrhythmias, since the risk of sudden cardiac death due to ventricular arrhythmias is increased in patients suffering from decreased left ventricular function after myocardial infarction [3].

The purpose of this study was to investigate the time course of changes in myocardial contractility and segment shortening in both the ischaemic and the non-ischaemic area during the first two hours of regional ischaemia. Secondly we wanted to investigate the relationship between changes in myocardial function and the type and incidence of postischaemic arrhythmias.

* Supported by Deutsche Forschungsgemeinschaft (Hi 137/8-1).

Methods

Instrumentation

We studied 16 young domestic pigs (9 female, 7 male) weighting 43 ± 2 kg ($\bar{x} \pm$ SEM). Anaesthesia was induced by intramuscular injection of ketamine (25 mg/kg body wt.) and droperidol (1 mg/kg body wt.), followed by intravenous injection of pentobarbital sodium (15 mg/kg), fentanyl dihydrogencitrate (0.2 mg) and pancuronium bromide (4 mg). After intubation, anaesthesia was maintained by ventilation using a $NO_2 : O_2$ mixture (2:1) and continuous intravenous infusion of droperidol (0.25 mg/h), fentanyl (0.02 mg/h), and pancuronium (0.2 mg/h). After mid-sternal thoracotomy the heart was exposed and supported in a pericardial cradle. A 7F two-sensor micro-tip catheter (Millar, PC-771) was introduced via the left carotid artery into the left ventricle for measurement of left ventricular and aortic pressure, as well as LV dp/dt by means of an analog differentiator (Gould-Brush 13/4214/01). An 8F angiographic catheter was placed in the coronary sinus via the right external jugular vein, for blood sampling. The left vena azygos was occluded since it drains into the coronary sinus in pigs. A loose ligature was placed around the LAD about $^2/_3$ of the distance from its origin.

Local myocardial contractility was measured within the ischaemic area, parallel ($long_i$) and perpendicular ($circ_i$) to the LAD as well as transmurally ($trans_i$) using an electromagnetic distance measuring system, recently developed in our institute [8]. This system uses one common transmitter probe and three receiver probes, all 1.5–2.0 mm in diameter. The probes were positioned subepicardially 12–14 mm ($long_i$, $circ_i$) and subendocardially 7–10 mm ($trans_i$) from the transmitter probe. Segmental myocardial contractility within the non-ischaemic myocardium was measured using ultrasonic crystals (Parks Electronics) positioned subendocardially on the anterior left ventricular wall perpendicular to the LAD ($circ_{ni}$).

Data recording

Segmental lengths, left ventricular pressure (LVP), LV dp/dt, aortic pressure (AP) and surface ECG were continuously recorded on an 8-channel recorder (Gould-Brush 480) and, with the exception of AP, on a 7-channel FM tape recorder (Bell-Howell VR 3000). Segment lengths were analysed from pressure-length diagrams printed using an XY-recorder (Rhode & Schwarz ZSKT 301.9010.02). Arterial and coronary sinus blood samples were taken for measurement of noradrenaline and adrenaline concentration, as well as for intermittent monitoring of pH, O_2 saturation and potassium concentration.

Plasma catecholamine analysis

Plasma catecholamines were analysed by HPLC (Waters, M6000A) with electrochemical detection, using DHBA (dihydroxybenzylamine) as internal standard. HPLC data were as follows: flow, 0.8 ml/min; pressure, 2,000 PSI; column, 5 μm C 18 resolve; injection volume, 25 μl. The eluent consisted of: 50 mM sodium acetate, 20 mM citric acid, 0.5 mM sodium octane sulphonate, 1 mM di-n-butylamine, 0.1 mM disodium edetate in water:methanol (950:50). Electrochemical detector: Waters M 460 or BAS LC17 4B, electrode voltage 0.5 V, offset 0–10 nA, range 0.2 nA. The Waters Automatic Datamodule was used for registration and calculation of catecholamine concentration (for more detailed description see ref. 6).

Experimental protocol

After complete instrumentation the pigs were studied from 3 hours before (phase I) to 2 hours after (phase II) LAD-occlusion. The observation period pre-occlusion was to exclude time, rather than occlusion dependent effects. During phase I, data were recorded every half hour. During phase II, haemodynamics and ECG were recorded continously. Local contractility data were continuously recorded only within the first 10 minutes of occlusion. Afterwards data were taken at 10, 20, 30, 60, 90, and 120 minutes after onset of occlusion. In animals in which ventricular fibrillation (VF) developed following coronary occlusion, resuscitation was performed by intrathoracic defibrillation and direct cardiac massage. After 2 hours of LAD-occlusion the size of the ischaemic area was measured by intravenous injection of disulphine blue. The pigs were sacrificed and after excision of the heart the wet weight both of the ischaemic tissue, which remained unstained, and the non-ischaemic tissue of both ventricles was measured.

Data analysis

Segmental length data, as well as changes in segmental length, were calculated as a percentage of end-diastolic length at the beginning of the observation period (phase I) and the occlusion period (phase II) respectively. Segmental shortening was calculated as percent change in segmental length from end-diastole to end-systole. During phase I, haemodynamics, segmental lengths and catecholamine concentrations were correlated with time (Pearson's correlation). Data from beginning and end of phase I were compared using Student's t-test for paired samples. During phase II pigs were divided into those that developed ventricular fibrillation and those that did not. Data from both groups were compared using Student's t-test for unpaired data. Data are shown as mean values $\pm$ SEM.

Results

180 min observation period prior to occlusion

Segmental shortening was initially $5.6 \pm 1.3\%$ in $long_i$, $12.1 \pm 1.3\%$ in $circ_i$, $-18.3 \pm 3.2\%$ in $trans_i$ and $12.5 \pm 1.3\%$ in $circ_{ni}$. During the 3 hours pre-occlusion there was a slight decrease of about 2–3.5% in segmental length at the beginning of the ejection phase in $long_i$, $circ_i$ and $circ_{ni}$ and an increase of $7.2 \pm 2.3\%$ in $trans_i$, with a corresponding decrease in segmental shortening of 1–4%. LVP and left ventricular end-diastolic pressure (LVEDP) were unchanged but increases in heart rate from 100 ± 7 to 125 ± 7 bpm and LV dp/dt_{max} from 2285 ± 144 to 2657 ± 263 mm Hg/s were observed. There was a corresponding increase in arterial noradrenaline from 0.15 ± 0.05 to 0.48 ± 0.11 ng/ml and in coronary sinus noradrenaline from 0.19 ± 0.05 to 0.56 ± 0.14 ng/ml whereas adrenaline concentration remained unchanged (Fig. 1). Both arterial and coronary sinus noradrenaline showed a weak positive correlation ($r = 0.37/0.36$, $p = 0.06/0.07$) to duration of the observation period. There were no differences in catecholamine levels between pigs that developed VF during phase II and those that did not.

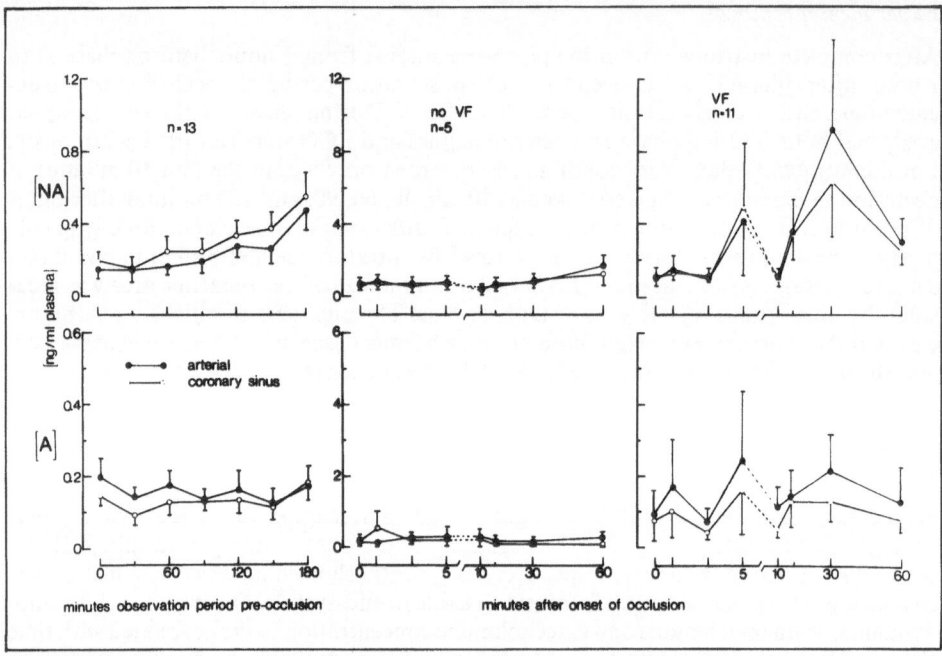

Fig. 1. Arterial and coronary sinus plasma noradrenaline (NA) and adrenaline (A) concentrations during 3 h pre-occlusion and the first 60 min of ischaemia for pigs that developed VF and those that did not ($\bar{x} \pm$ SEM).

Post-occlusion period

Arrhythmias

All pigs had ventricular arrhythmias after occlusion of the distal half of the LAD, but there were marked interindividual differences in the incidence of ventricular extrasystoles (VES) and VF. Table 1 shows the incidence and the time at which VF occurred, listed separately for each pig. In Fig. 2 the incidence of both VES and VF within the same time periods is shown for each animal. In 5/16 pigs (31%) VF did not occur. Five pigs developed VF during phase 1a (1–6 min post occlusion) and a further 5 pigs developed VF for the first time during phase 1 B (10–30 min post occlusion). 3/5 pigs with VF during phase 1a could not be resuscitated, whereas all later occuring VF episodes could be terminated. 3/16 pigs developed VF after the first hour of occlusion but only one for the first time. In contrast to VF, the incidence of VES was relatively low during phase 1a and 1 b, but there was a marked increase after the 50th minute of occlusion (Fig. 3). Within the second hour the number of VES/min remained almost constant. Further analysis of the VES revealed that the VES during 1 a and 1 b were often polymorphic, closely coupled and repetitive, whereas those occuring after the 50th minute showed more regular patterns or accelerated idioventricular rhythm. There was no correlation between incidence of VES and VF. During phase 1 a and 1 b, 8/19 VF episodes (42%) occurred without other preceding arrhythmias (Fig. 2).

Table 1. Time of occurrence of ventricular fibrillation for each pig. In 3 animals with VF within phase 1a reanimation was unsuccessful.

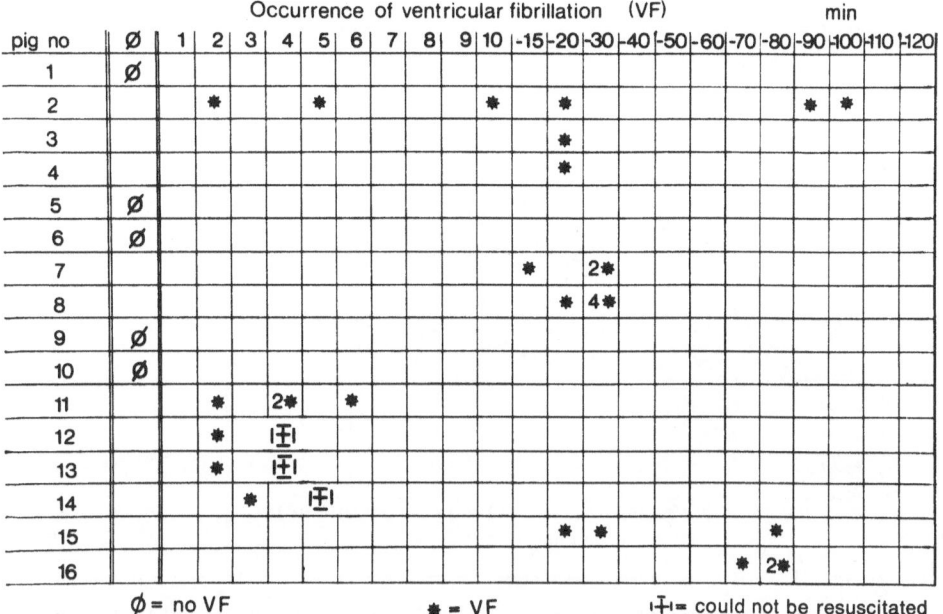

Occurrence of ventricular fibrillation (VF)　　　　　　　min

pig no	∅	1	2	3	4	5	6	7	8	9	10	-15	-20	-30	-40	-50	-60	-70	-80	-90	-100	-110	-120
1	∅																						
2			❋		❋							❋	❋							❋	❋		
3													❋										
4													❋										
5	∅																						
6	∅																						
7												❋	2❋										
8												❋	4❋										
9	∅																						
10	∅																						
11			❋		2❋		❋																
12			❋		⊞																		
13			❋		⊞																		
14					❋		⊞																
15												❋	❋						❋				
16																	❋	2❋					

∅ = no VF　　　　　❋ = VF　　　　　⊞ = could not be resuscitated

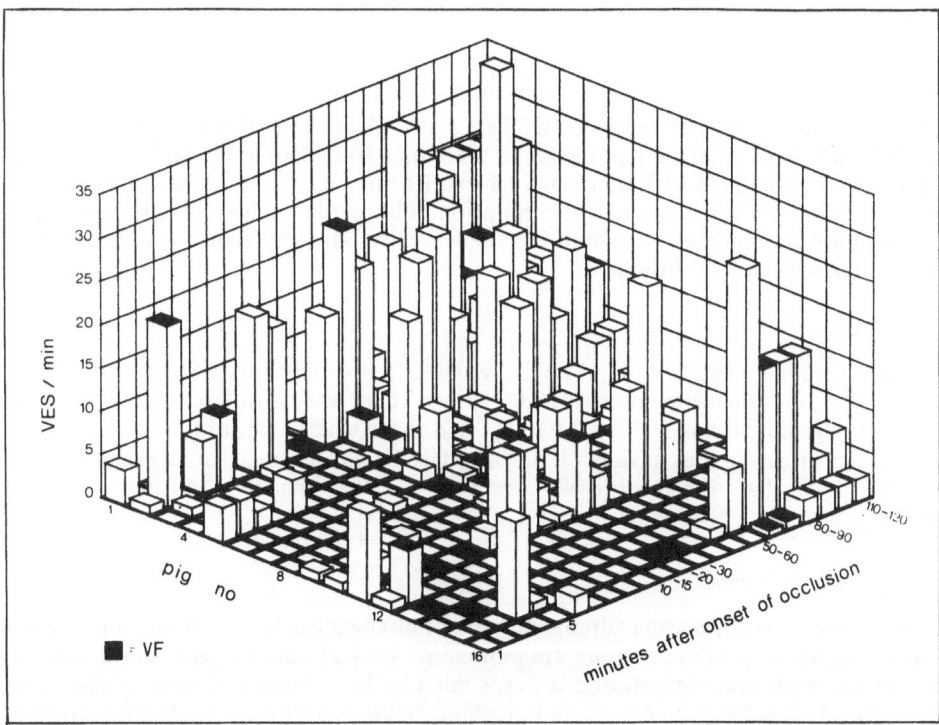

Fig. 2. Incidence of VES (columns) and VF (black fields) during the first 120 min of coronary occlusion in each pig.

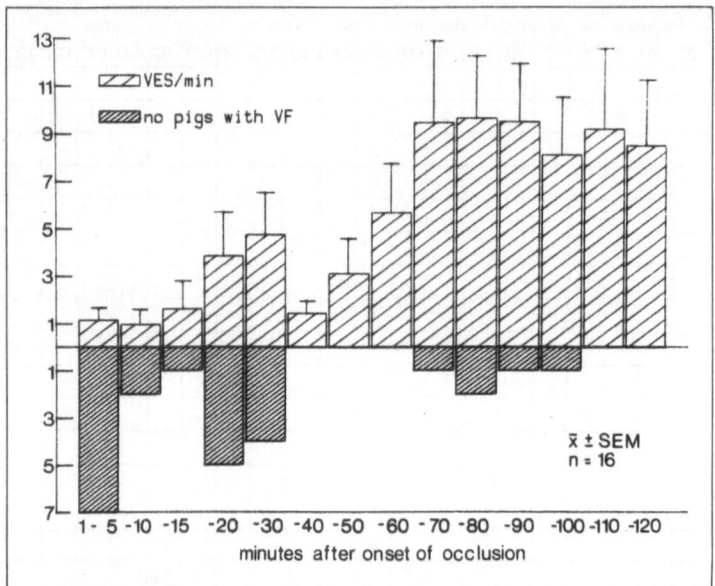

Fig. 3. Incidence of ventricular extrasystoles (VES) and ventricular fibrillation (VF) within the first 120 min of coronary occlusion.

Haemodynamics

Within the first 2 minutes of ischaemia there was a slight but not significant decrease of LVP from 110 ± 6 to 99 ± 5 mm Hg and LV dp/dt_{max} from 2567 ± 245 to 2051 ± 154 mm Hg/s and an increase in LVEDP from 6 ± 1 to 8 ± 1 mm Hg, whereas heart rate remained constant. Except for a temporary deterioration of haemodynamics (probably caused by arrhythmias during phase 1a and 1b) there was no further change during the first 120 min of coronary occlusion.

Infarct size

The ischaemic area was $19.1 \pm 1.1\%$ of the wet weight of both ventricles (ischaemic tissue: 29.1 ± 1.8 g; non-ischaemic tissue: 122.8 ± 5.0 g). There was no difference in infarct size between pigs developing VF $(18.7 \pm 1.3\%)$ and those which did not $(20.0 \pm 1.8\%)$, but within the VF group pigs having VF during phase 1a had a somewhat larger infarct size $(21.4 \pm 1.8\%)$ than those with VF occuring for the first time during phase 1b $(17.0 \pm 1.5\%, p = 0.09)$.

Plasma catecholamines

In Fig. 1 (right panel) plasma adrenaline and noradrenaline concentrations from arterial and coronary sinus blood samples are plotted for VF and non-VF pigs. In the non-VF group, catecholamine concentrations were similar to those during the foregoing observation period. There was no change in adrenaline but a slight time dependent increase in noradrenaline. At the time of occlusion, catecholamine levels were similar in pigs which later developed VF and those which did not, but both adrenaline and noradrenaline in-

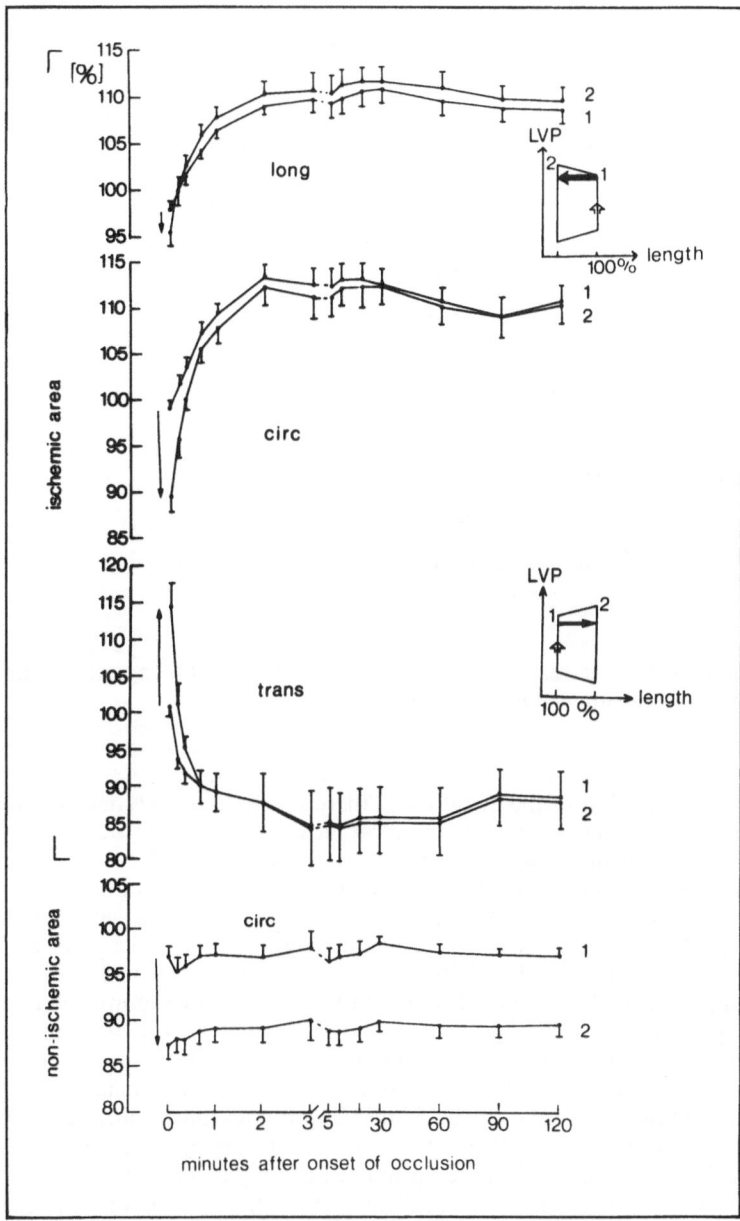

Fig. 4. Time course of changes in segment length at the beginning (1) and end (2) of the ejection phase pre- and post-occlusion within the ischaemic (long$_i$, circ$_i$, trans$_i$) and non-ischaemic (circ$_{ni}$) myocardium ($\bar{x} \pm$ SEM, n = 16). For positioning of the probes, see text.

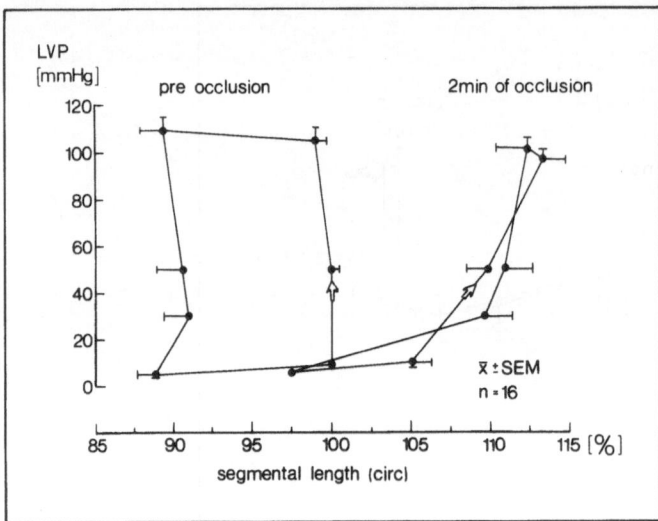

Fig. 5. Pressure-length diagram showing ischaemia-induced changes in segment length (circ$_i$) ($\bar{x} \pm$ SEM, n = 16).

creased dramatically after VF and resuscitation. Peak values of up to 13.9 ng/ml adrenaline and 32.7 ng/ml noradrenaline were observed.

Segmental lengths

Within the first 20 s of occlusion of the distal LAD there was a 4.1 ± 1.3% increase in segment length in the ischaemic area (circ$_i$), measured at the beginning of ventricular ejection (Fig. 4). XY-diagrams at this time revealed bulging of 8.4 ± 1.3% of end-diastolic length at early diastole (circ$_i$). Within the first minute of occlusion active contraction of the ischaemic myocardium was completely extinguished in long$_i$ and almost completely extinguished in circ$_i$. There was passive segment lengthening of 6–7% in the ischaemic area during each cardiac cycle (Fig. 5). In addition, there was an increase in end-diastolic baseline segment length of 5–6.5%. Segment lengths in the ischaemic area showed no further change after 2 min of occlusion (Fig. 4). In the non-ischaemic area, local segment length and segment shortening remained constant, except for a temporary decrease within the first minute of occlusion. Changes in segment lengths were similar in pigs which developed VF and those without VF. Dividing VF-pigs into those which had VF within phase 1 a and those which had VF for the first time during phase 1 b, segment lengthening increased faster in those of phase 1 a than of phase 1 b. Two minutes after the onset of occlusion, epicardial segment length was greater in pigs with VF in phase 1 a (117 ± 1%/119 ± 3%) (beginning or end of ejection phase) than in those developing VF in phase 1 b (112 ± 3%/111 ± 3%) (p = 0.03/0.008).

Discussion

During an experimental period of 2 hours after coronary occlusion it might be important to differentiate between time- and ischaemia-dependent effects. During the 3 h observation period, our data show a slight decrease in segment shortening, together with an increase in heart rate and noradrenaline concentration, with no change in adrenaline concentration. Similar effects were seen within the second hour after LAD occlusion. This phenomenon seems not to be caused simply by a lack of volume or insufficient anaesthesia, since in such cases one would except an increase in adrenaline and more marked changes in LVEDP. The probably time-dependent noradrenaline increase under open-chest conditions was similar in magnitude to catecholamine increases seen during experiments with repeated occlusions over a comparable time period (unpublished data from our laboratory).

Typical mechanical alterations following acute coronary occlusion include paradoxical systolic lengthening or bulging in the ischaemic zone [2, 14] and an increase in segment shortening in the non-ischaemic myocardium [11, 13]. In our experiments the first mechanical alterations were seen as early-diastolic bulging ($long_i$, $circ_i$) occurring within the first 10 s after coronary occlusion. A decrease in the duration of myocardial contraction has been reported as the reason for this phenomenon [1, 13]. This early relaxation has been shown to cause shortening of the ventricular ejection period with subsequent reduction of left ventricular output [12]. In our study, active shortening of ischaemic myocardium ceased within the first minute of occlusion. Changes in segment lengths of ischaemic tissue reached a maximum within 2 min of occlusion. Whereas there was no further systolic shortening in $long_i$, there remained a slow systolic shortening of 1–1.5% in $circ_i$ (Figs. 4 and 5). Similar results using tension-length instead of pressure-length loops were recently shown by Akaishi et al. [2]. They demonstrated that an increased LV preload caused a decrease in systolic bulging and in shortening during isovolumic relaxation (recoil) in the ischaemic myocardium by reaching steeper parts (i.e. lower compliance) of the exponential tension-length curve of the ischaemic myocardium. In this study, coronary occlusion did not result in a marked decrease in global left ventricular function and thus LVEDP did not increase markedly. In contrast to data from other investigators [11, 13] there were no changes in segment length and segment shortening in the non-ischaemic area (Fig. 4). Increases of segment shortening within the non-ischaemic area has been explained on the basis of a model of serially linked non-ischaemic and ischaemic myocardium, resulting in an intraventricular unloading of the non-ischaemic myocardium, and from an increase in preload caused by reduced global ventricular function [10, 13, 15]. For the non-ischaemic area adjacent to the ischaemic zone, a mechanical tethering effect resulting in a decrease in segment shortening has been proposed [4]. In our data both effects may be involved, since $circ_{ni}$ was positioned on the anterior LV wall about 1.5 cm from the ischaemic border.

Ion shifts with subsequent increases in extracellular potassium and maintained depolarisation, as well as catecholamine release, are generally accepted as causes of arrhythmias following acute coronary occlusion [5, 9]. Although there are some experimental data from isolated preparations suggesting that stretching of Purkinje fibers induces depolarisation [7], the role of this factor in the genesis of post-ischaemic arrhythmias still remains unclear. Our data showed that during phase 1 a the incidence of VES was relatively low, although almost all mechanical alterations occurred within this period. Furthermore, there were no differences in passive segment lengthening between pigs developing VF and those that did not. Within the VF group mechanical alterations 2 min after the onset of ischaemia were more marked in pigs developing VF within phase 1 a than in those with VF in phase 1 b. This suggests that passive segment lengthening may be a co-factor in the induction of early VF. VES increased markedly during the second hour of occlusion, at a time when bulging of the ischaemic area showed no further change. These later arrhythmias, which less often result in VF, seem to occur independently of further mechanical alterations in the ischaemic myocardium at this time.

References

1. Akaishi M, Schneider RM, Mercier RJ, Agarwal JB, Helfant RH, Weintraub WS (1986) Analysis of phases of contraction during graded acute myocardial ischemia. Am J Physiol 250:H778–H785
2. Akaishi M, Weintraub WS, Schneider RM, Klein LW, Agarwal JB, Helfant RH (1986) Analysis of systolic bulging. Mechanical characteristics of acutely ischemic myocardium in the conscious dog. Circ Res 58:209–217
3. Bigger JT Jr, Fleiss JL, Kleiger R, Miller JP, Rolnitzky LM and The Multicenter Post-Infarction Research Group (1984) The relationship between ventricular arrhythmias, left ventricular dysfunction and mortality in the two years after myocardial infarction. Circulation 69:250–258
4. Guth BD, White FC, Gallagher KP, Bloor CM (1984) Decreased systolic wall thickening in myocardium adjacent to ischemic zones in conscious swine during brief coronary artery occlusion. Am Heart J 107:458–464
5. Hirche Hj, Franz C, Bös L, Bissig R, Lang R, Schramm M (1980) Myocardial extracellular K^+ and H^+ increase and noradrenaline release as possible cause of early arrhythmias following acute coronary artery occlusion in pigs. J Mol Cell Cardiol 12:579–593
6. Hirche Hj, McDonald FM, Polwin W, Addicks K (1985) Vicious cycle of catecholamines and K^+ in cardiac ischemia. J Cardiovasc Pharmacol [Suppl 5] 7:71–75
7. Kaufmann R, Theophile U (1967) Automatie-fördernde Dehnungseffekte an Purkinje-Fäden, Papillarmuskeln und Vorhofrabekeln von Rhesus-Affen. Pflügers Arch 297:174–189
8. Kebbel U, Hirche Hj (1985) An electromagnetic distance measuring system. Med Prog Technol 10:213–223
9. Kleber AG, Janse MJ, van Capelle FJL, Durrer D (1978) Mechanism and time course of S–T and T–Q segment changes during acute regional myocardial ischemia in the pig heart determined by extracellular and intracellular recordings. Circ Res 42:603–613
10. Lew WYW, Ban-Hayashi E (1985) Mechanisms of improving regional and global ventricular function by preload alterations during acute ischemia in the canine left ventricle. Circulation 72:1125–1134
11. Lew WYW, Chen Z, Guth B, Covell JW (1985) Mechanisms of augmented segment shortening in nonischemic areas during acute ischemia of the canine left ventricle. Circ Res 56:351–358
12. Lewartowski B, Sedek G (1974) Mechanical performance of the left ventricle at early stages of experimental ischaemia. Mechanism of shortening of ejection period. Cardiovasc Res 8:593–601
13. Smalling RW, Ekas RD, Felli PR, Binion L, Desmond J (1986) Reciprocal functional interaction of adjacent myocardial segments during regional ischemia: An intraventricular loading phenomenon affecting apparent regional contractile function in the intact heart. JACC 7:1335–1346
14. Tennant R, Wiggers CJ (1935) The effect of coronary occlusion on myocardial contraction. Am J Physiol 112:351–361
15. Tyberg JV, Parmley WW, Sonnenblick EH (1969) In vitro studies of myocardial asynchrony and regional hypoxia. Circ Res 25:569–579

Authors' address:

Prof. Dr. med. Hj. Hirche, Institut für Angewandte Physiologie,
Universität Köln, Robert-Koch-Str. 39, 5000 Köln 41, West Germany

Phosphocreatine and adeninenucleotides in postasphyxial hearts with normal basal function and normal oxygen demand*

H. M. Hoffmeister, G. Stein, R. Storf, and L. Seipel

Medizinische Klinik III, Universität Tübingen, F.R.G.

Summary

We investigated whether there is a relationship between the prolonged dysfunction after myocardial ischaemia and the postischaemic phosphocreatine overshoot phenomenon. In 16 open-chest rats 3 periods of 4 minutes of oxygen deficiency were performed and basal haemodynamic variables and the myocardial oxygen demand were determined during the recovery period. At the end of the 20 minutes recovery period, left ventricular pressure, dp/dt_{max}, ejection fraction, and myocardial oxygen demand were completely recovered. High energy phosphate levels, however, were still altered. The sum of adeninenucleotides was decreased to $78 \pm 4\%$ of control ($\bar{x} \pm$ SEM, $p < 0.05$). The level of phosphocreatine was markedly elevated to 162 ± 14 ($\bar{x} \pm$ SEM). The persistence of the phosphocreatine overshoot phenomenon, while basal function was already normalized, indicates that a reduced function and thus a reduced energy demand of the contractile apparatus are not the cause of the phosphocreatine overshoot. We found no close relationship between high energy levels and basal function or oxygen demand in myocardium after mild oxygen deficiency.

Introduction

A phosphocreatine overshoot is a common phenomenon in myocardium with reversible postischaemic dysfunction. In reversibly injured cardiac tissue, often an elevation of the phosphocreatine level during reperfusion is observed, which is associated with a reduced content of adeninenucleotides [8, 14]. Irreversibly damaged myocardium (i.e. infarction), on the other hand, is characterized by a reduction of phosphocreatine as well as of adeninenucleotides. The generation of phosphocreatine in reversibly injured myocardium during reperfusion probably indicates properly working oxidative phosphorylation [10, 11]. The overshoot phenomenon, however, which is to date not completely understood [15], might be caused by a reduced energy utilization at the level of the contractile apparatus. To evaluate whether there exists a causal or only a casual relationship between the basal myocardial function and the postischaemic content of phosphocreatine, we measured haemodynamic data and cardiac tissue levels of high energy phosphates in open-chest rats after short periods of oxygen deficiency.

* Supported by the Deutsche Forschungsgemeinschaft (Ho 1003/1-1).

Methods

Male wistar-rats were anaesthetized with 50% urethane (2.5 ml/kg body wt.) intraperitoneally. The chest was opened by a median sternotomy and the pericardium was removed.
Respiration was maintained using a Starling-type respirator. Short fluid-filled metal ca-
nulas connected to Statham pressure transducers were positioned in the proximal left
carotid artery and in the left ventricle (via the apex). Aortic pressure (AoPm), left ventricular pressure (LVSP), and dp/dt were registered on a multi-channel inkjet recorder.
To measure stroke volume (except coronary flow) an electromagnetic flow-probe (2 mm
internal diameter) was fitted around the ascending aorta and the cardiac output [stroke
volume × heart rate (HR)] was determined. Myocardial oxygen demand was calculated
according to Hütter et al. [7] as

$$C_0 + C_1 \times LVSP \times HR + C_2 \times dp/dt_{max} \times HR.$$

At the end of the experiments we ligated the hearts at the base and unloaded the right
ventricle by an incision. We then recorded the pressure-volume relationship of the left
ventricle by filling the previously completely emptied ventricle with a defined volume and
calibrating the re-emptying. From this pressure-volume relationship, the volumes corresponding to end-diastolic pressure were derived and thus the ejection fraction (stroke vol-
ume/end-diastolic volume) could be calculated.

For biochemical analysis at the end of the experiments, myocardial tissue was rapidly
frozen in liquid nitrogen between two precooled blocks of aluminium. After a perchloric
acid extraction and neutralization, the levels of phosphocreatine (PCr) and of adeninenucleotides (AN) were determined using bioluminescence techniques according to Ellis and
Gardner [2]. In controls, the mean content of PCr was 4.3 ± 0.3 µmol/gww and the mean
total sum of adeninenucleotides (AN) was 7.2 ± 0.2 µmol/gww (n = 6, means + SEM).

After a 15 minutes stabilization period the initial haemodynamic data were recorded
(LVSP = 116 ± 3 mm Hg, dp/dt$_{max}$ = 10537 ± 524 mm Hg/s, heart rate (HR) = 275 ± 11/
min, AoPm = 79 ± 2 mm Hg, CO = 41 ± 3 ml/min, means $\pm$ SEM, n = 16). Then 3 periods
each of 4 min duration of asphyxia with intermittent 4 min periods of normal respiration
were performed in 16 rats. 5, 10, 15, and 20 minutes after the last period of asphyxia, haemodynamic data were again recorded and expressed normalized to the initial measurement. For statistical analysis the data were compared (Student's t-test) with a control
group (n = 16) with identical protocol except the periods of oxygen deficiency.

Results

Ten minutes after the last period of asphyxia the basal haemodynamic parameters were
recovered to the pre-asphyxial values, respectively, comparable to the control animals
(Fig. 1). There was only a slight significant difference in left ventricular systolic pressure.
Mean aortic pressure, ejection fraction and dp/dt$_{max}$ were completely normalized. The
cardiac output was also comparable in the postasphyxial and in the normal rats. At 15
and 20 min of postasphyxial recovery there was no significant difference at all from the
controls. Therefore, the calculated oxygen demand of the hearts was also identical in both
groups, as shown in Fig. 2.

The biochemical data of the postasphyxial hearts are illustrated in Fig. 3. The phosphocreatine content and the total sum of the adeninenucleotides of the postasphyxial cardiac tissue are expressed as percentages of the normal values in our animal model. The

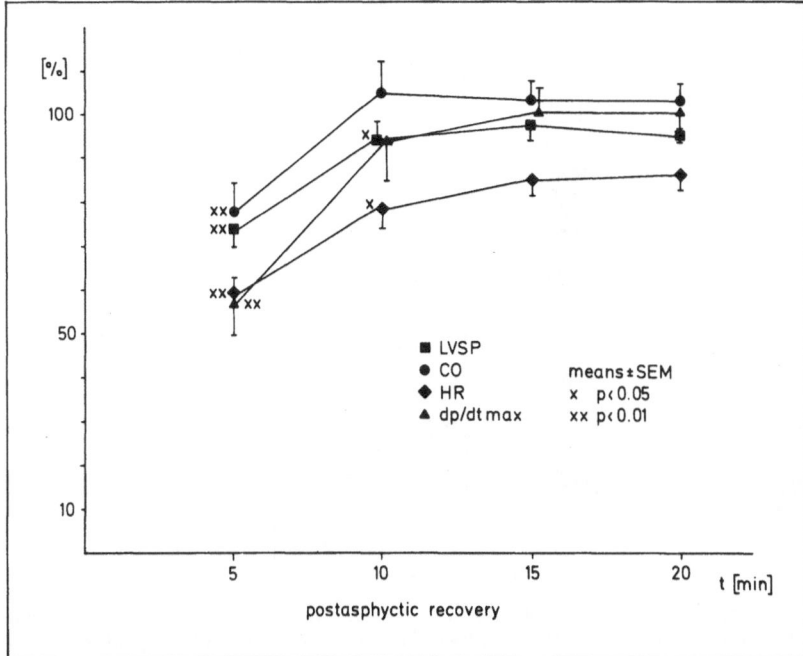

Fig. 1

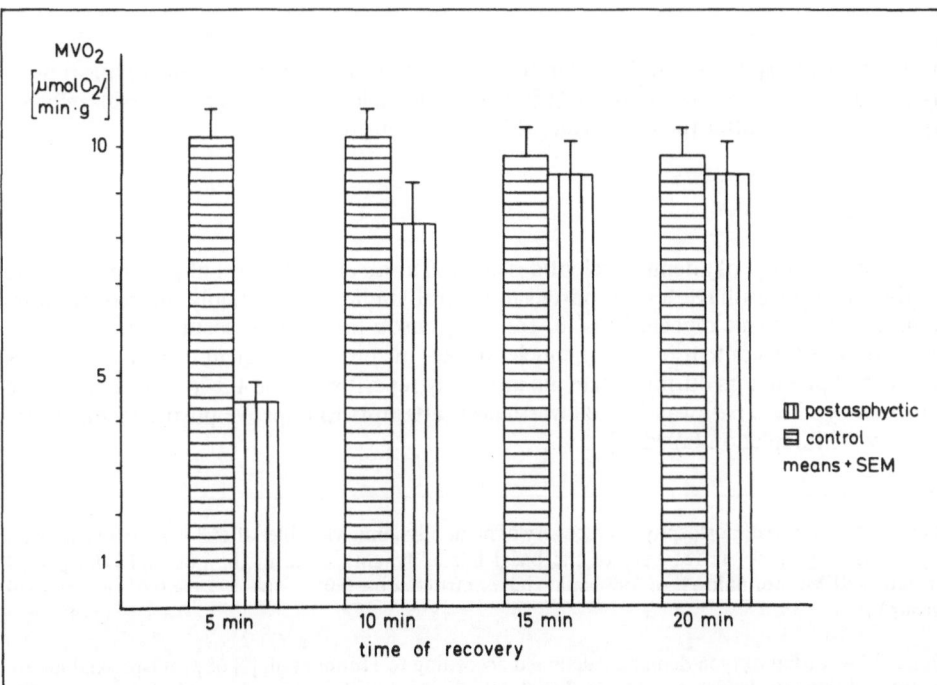

Fig. 2

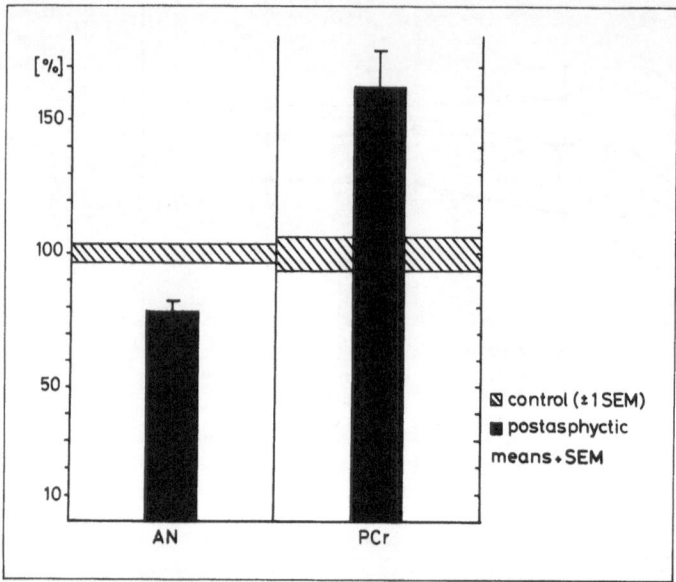

Fig. 3. Sum of adeninenucleotides and level of phosphocreatine 20 min after the last period of oxygen deficiency (n = 10) in comparison to the tissue content of control animals (n = 6). Reduction of adeninenucleotides (AN) and overshoot of the phosphocreatine level (PCr) were both significant (p < 0.05).

level of the phosphocreatine was significantly (p < 0.05) elevated showing a typical postischaemic overshoot, whereas the total adeninenucleotide content was still reduced (p < 0.05) 20 min after the last period of oxygen deficiency.

Discussion

A reduced content of adeninenucleotides, as well as an elevated level of phosphocreatine during reperfusion have been reported in several studies on reversibly injured dysfunctional postischaemic myocardium [8, 14]. The reason for the long lasting postischaemic reduction of the content of adeninenucleotides – in spite of sufficient energy supply – is the loss of precursors during ischaemia and early reperfusion and the low capacity of the de-novo synthesis pathway [17]. Nevertheless, attempts to improve postischaemic function by ATP repletion failed [6].

Fig. 1. Time course of postasphyxial recovery of haemodynamic variables after 3 × 4 min of asphyxia demonstrating complete recovery of the basal haemodynamic state in the postasphyxial period (means + SEM, normalized to pre-asphyxial control; × p < 0.05 resp. × × p < 0.01 vs. control group).

Fig. 2. Myocardial oxygen demand calculated according to Hütter et al. [7] of postasphyxial hearts compared to controls (means + SEM). The data indicate that the energy demand of the contractile apparatus is comparable to controls for at least the last ten minutes of the experiments.

The reason for the elevated phosphocreatine level, however, is not as well understood. A normal creatinekinase activity in hearts after hypoxia with supernormal phosphocreatine levels was observed by Takeo and Sakanashi [16]. The recovery of the phosphocreatine indicates at least functionally intact mitochondria [10, 11]. This assumption is supported by the finding that in irreversibly injured cardiac tissue, the phosphocreatine content during reperfusion is low. An overshoot of the phosphocreatine, however, was found in tissue with histological signs of only moderate injury [14]. It was always associated with a decreased function of the respective region of the heart. It is known that the absolute level of phosphocreatine – in the case of working mitochondria – is not directly related to myocardial function. However, Bittl et al. have recently shown [1] that the flux of the creatinekinase reaction is related to the systolic function and therefore to the myocardial oxygen demand. If the capacity of phosphocreatine production is not sufficient for the energy demand, on the other hand, the level of phosphocreatine might be indicative of the energy-turnover as observed in an experiment by Giesen and Kammermeier [4]. These authors found a markedly increased cardiac work and energy demand, but a decreased phosphocreatine level after isoproterenol administration. It might be hypothesized that the elevated postischaemic phosphocreatine is due to a reduced turnover at the level of the contractile apparatus. Therefore, a decreased postischaemic cardiac function and a reduced energy demand might cause an increased content of phosphocreatine. This should occur regardless whether a decreased turnover is the cause or an effect of postischaemic dysfunction.

A decrease in the free ADP at the sarcomeres as another hypothesis for a possible reason for both postischaemic dysfunction and phosphocreatine overshoot, was discussed by Swain et al. [15]. Calculation of the free ADP in our experiments according to Kammermeier [9] revealed an insignificantly decreased free ADP (if Pi is about normal [3]), but an unaltered free energy change of ATP hydrolysis. In contrast to other experiments, there was no decreased basal function seen in our experiments, as associated with the decreased free ADP by Swain et al. [15].

In our experiments, however, we found a phosphocreatine overshoot, in spite of the normalization of postasphyxial basal parameters of heart function and a normal oxygen demand. The calculation of the oxygen demand did not take possible differences of the efficiency between normal and postischaemic myocardium into account, but it is unlikely that postischaemic myocardium works more efficiently on a lower demand of energy for the same work compared to normal myocardium. The calculated demand therefore would rather be an under- than an overestimation. Thus, our experiments could not confirm a close or even causal relationship between the phosphocreatine overshoot and a postischaemic dysfunction. Both findings in postischaemic myocardium seem to be independent phenomena, not strictly linked by the energy turnover necessary for basal function or the myocardial oxygen demand. Compartmentation of high energy phosphates [5, 12, 13] or selective ischaemic damage of regulatory cellular functions, might be possible explanations for the phosphocreatine overshoot phenomenon which requires further detailed investigation.

References

1. Bittl JA, Ingwall JS (1985) Reaction rates of creatine kinase and ATD synthesis in the isolated rat heart. J Biol Chem 260:3512–3517
2. Ellis RJ, Gardner CR (1980) Determination of myocardial high-energy-phosphates using bioluminescence. Anal Biochem 105:354–360

3. Fischer M, Buchwald A, Winkler B, Schaper W (1985) Is inorganic phosphate (Pi) a limiting factor for rapid postischemic ATP repletion? Circulation [Suppl III] 72:358 (abstract)
4. Giesen J, Sondermann M, Juengling E, Kammermeier H (1980) Time dependent partial loss of the effects of isoproterenol on function and energy metabolism of isolated rat hearts. Basic Res Cardiol 75:515–525
5. Gudbjarnason S, Mathes P, Ravens KG (1970) Functional compartmentation of ATP and creatine phosphate in heart muscle. J Mol Cell Cardiol 1:325–339
6. Hoffmeister HM, Mauser M, Schaper W (1985) Effect of adenosine and AICAR on ATP content and regional contractile function in reperfused canine myocardium. Basic Res Cardiol 80:445–458
7. Hütter JF, Piper HM, Spieckermann PG (1985) An index for estimation of oxygen consumption in rat heart by hemodynamic parameters. Am J Physiol 249:729–734
8. Isselhard W (1968) Metabolism and function of the heart during acute asphyxial and in postasphyxial recovery. Acta Anaesth Scand 29:203–216
9. Kammermeier H, Schmidt P, Jüngling E (1982) Free energy change of ATP-hydrolysis: A causal factor of early hypoxic failure of the myocardium? J Mol Cell Cardiol 14:267–277
10. Lange R, Ingwall JS, Hale SL, Alker KJ, Kloner RA (1984) Effect of recurent ischemia on myocardial high energy phosphate content in canine hearts. Basic Res Cardiol 79:469–478
11. Reibel DK, Rovetto MJ (1978) Myocardial ATP synthesis and mechanical function following oxygen deficiency. Am J Physiol 234:620–624
12. Saks VA, Kupriyanov VV, Elizarova GV (1980) Studies of energy transport in heart cells. J Biol Chem 255:755–763
13. Saks, VV, Ventura-Clapier R, Huchua ZA, Proebrazhensky AN, Emelin IV (1984) Creatine kinase in regulation of heart function and metabolism I. Further evidence for compartmentation of adenine nucleotides in cardiac myofibrillar and sarcolemnal coupled ATPase-creatine kinase. Biochim Biophys Acta 803:254–264
14. Schaper J, Mulch J, Winkler B, Schaper W (1979) Ultrastructural, functional, and biochemical criteria for estimation of reversibility of ischemic injury: A study on the effects of global ischemia on the isolated dog heart. J Mol Cell Cardiol 11:521–541
15. Swain JL, Sabina RL, Hines JJ, Greenfield JC Jr, Holmes EW (1984) Repetitive episodes of brief ischaemia (12 min) do not produce a cumulative depletion of high energy phosphate compounds. Cardiovasc Res 18:264–269
16. Takeo S, Sakanashi M (1983) Possible mechanisms for reoxygenation-induced recovery of myocardial high-energy phosphates after hypoxia. J Mol Cell Cardiol 15:577–594
17. Zimmer HG, Trendelenburg C, Kammermeier H, Gerlach E (1973) De-novo-synthesis of myocardial adeninenucleotides in the rat. Acceleration during recovery from oxygen deficiency. Circ Res 32:635–642

Authors' address:

Dr. H.M. Hoffmeister, Medizinische Universitätsklinik, Abt. III, Otfried-Müller-Straße 2, 7400 Tübingen

Changes in cardiovascular adrenoceptor response in rats subsequent to myocardial infarction*

D. Türck and P. Dominiak

Physiologisches Institut der Universität München

Summary

Since sympatho-adrenal activity was greatest on the second day after experimental myocardial infarction in the rat, we investigated the responsiveness of cardiovascular adrenoceptors in pithed rats at this time and also determined the number of beta-adrenoceptors in the myocardium of the rats. In addition, the sympathetic outflow was measured, as a presynaptic parameter for estimating sympathetic nervous activity.

The number of beta-adrenoceptor binding sites and the frequency responses of the heart were unchanged by myocardial infarction on the second day. There was also no difference from control animals with respect to sympathetic outflow. However, the rise in diastolic blood pressure, caused by electrical stimulation of the sympathetic nervous system, was far more pronounced in animals with infarction than in controls. In contrast, rats with myocardial infarction exhibited an attenuated response of diastolic blood pressure to infused noradrenaline.

The observed effects can possibly be explained by changes of alpha-adrenoceptors, or perhaps by an enhanced degradation of catecholamines in the endothelium.

Introduction

It has been reported by several laboratories that alpha- and beta-adrenoceptor binding sites in the heart of different species were increased after *acute* myocardial infarction [1, 2, 4, 5]. In a very recent paper, the increased number of beta-adrenoceptor binding sites after acute myocardial infarction could be explained as an externalization process of beta-adrenoceptors on the plasmalemmal membrane [4].

In a previous study concerning the time course of the sympatho-adrenal activity after experimental myocardial infarction in rats (2 h, 2, 7 and 21 days after myocardial infarction), we could demonstrate that two days subsequent to myocardial infarction the adrenal medullary activity was at its highest [2].

Since the "up and down" regulation of beta- (and possibly alpha-) adrenoceptors is dependent on the local and circulating concentrations of catecholamines, and on receptor exposition to these [6], it was of interest to investigate the beta-adrenoceptor binding number in rat hearts two days after experimental myocardial infarction. The results of these binding studies were to be compared with those of cardiovascular adrenoceptor responsiveness to endogenous noradrenaline and adrenaline and to different exogenous adrenoceptor agonists, as assessed by blood pressure and heart rate effects. Furthermore, we measured the stimulated sympathetic outflow at the same time, because of its relevance to presynaptic sympathetic activity [7].

---

* Dedicated to Prof. Dr. E. Gerlach on the occasion of his 60th birthday.

Methods

Myocardial infarction: rats were anaesthetized with methohexital, 15 mg/kg intravenously, subsequently intubated and artificially respirated. After thoracotomy between the 5th and 6th rib, the pericardium was opened and the heart lifted through the incision. The left coronary artery was undersewed with a silk suture (6/0) near the base of the heart and ligated. Sham operation was identical, but without ligation.

Two days after myocardial infarction or sham operation, rats were divided into different groups for the following experiments:

1. Sympathetic outflow and internal stimulation of the cardiovascular system with catecholamines: rats were pithed with a steel rod coated with enamel except for the length of the Th1–4 segment. The indifferent electrode was placed between the skin and musculature of the right hindlimbs. Rats were artificially respirated and both carotid arteries were cannulated with PE 50 catheters, one for measuring blood pressure and the other for collecting blood samples. Both vagus nerves were transected at the neck. d-Tubocurarin (3 mg/kg) was injected intravenously to block the neuro-muscular junction. 30 min later the sympathetic nerves of (mainly) the heart were stimulated preganglionically by electrical impulses of stepwise increased frequency (from 0.1 up to 10.0 Hz). Impulse trains were generated by a HSE T stimulator (10 V, 1 ms, 30 s duration). Heart rate and blood pressure were recorded continuously by means of a Statham pressure transducer (P 23 Db) and a Gould Brush recorder, whereas blood samples (0.6 ml) for determination of the sympathetic catecholamine outflow were collected before (0 Hz) and during stimulations with 0.1, 1.0 and 10.0 Hz. Plasma noradrenaline and adrenaline were assayed by using HPLC and electrochemical detection. Desoxyepinephrine served as an internal standard.

2. Cardiovascular adrenoceptor responsiveness was determined by measuring the increases in blood pressure and heart rate (Statham P 23 Db) of pithed rats, when noradrenaline (0.1–100 nmol/kg) and fenoterol (0.06–6.25 nmol/kg) were injected i.v. in incremental doses.

3. Beta-adrenoceptor binding studies were performed in highly enriched sarcolemmal membrane fractions of the left and right ventricle and of the interventricular septum according to the method of Mukherjee et al. [5], using ^{3}H-dihydroalprenolol as ligand.

Results are presented as the mean ($\bar{X}$)±SEM. Significance was calculated by Student's t-test and U-test.

Results

As shown in Table 1, there was no significant difference in plasma noradrenaline concentrations between rats with myocardial infarction (MI) and sham operated controls (C) during preganglionic electrical stimulation. However, before stimulation (basal value) we could observe a significantly lower level of plasma noradrenaline in MI rats as compared to controls. In contrast, during stimulation with 1.0 Hz the plasma adrenaline concentrations were significantly enhanced in rats with MI.

Figure 1 presents the stimulation response curves for increases in systolic and diastolic blood pressure during electrical stimulation. The systolic blood pressure increase of MI rats was slightly diminished vs. controls at low stimulation rates, while on the other hand, the diastolic blood pressure increase of MI rats was markedly enhanced.

The response of the heart rate was similar in both groups of rats (Table 2).

Table 1. Stimulated sympathetic outflow in rats subsequent to myocardial infarction

	Hz			
	0	0.1	1.0	10.0
Noradrenaline (pg/ml)				
Sham	171 ± 32.0	204 ± 54.7	539 ± 126.4	2523 ± 595
(n)	(7)	(6)	(4)	(7)
MI	74 ± 10.6^{a}	111 ± 20.5	574 ± 180.5	2183 ± 453
(n)	(6)	(6)	(5)	(6)
Adrenaline (pg/ml)				
Sham	<20	<20	36 ± 15.7	607 ± 263
(n)	(6)	(6)	(3)	(7)
MI	<20	<20	295 ± 73.7^{a}	844 ± 206
(n)	(7)	(7)	(5)	(6)

$^{a} = p < 0.05$; Sham = sham occluded rats; MI = myocardial infarction; Hz = preganglionic electrical stimulation. Noradrenaline and adrenaline were determined from plasma.

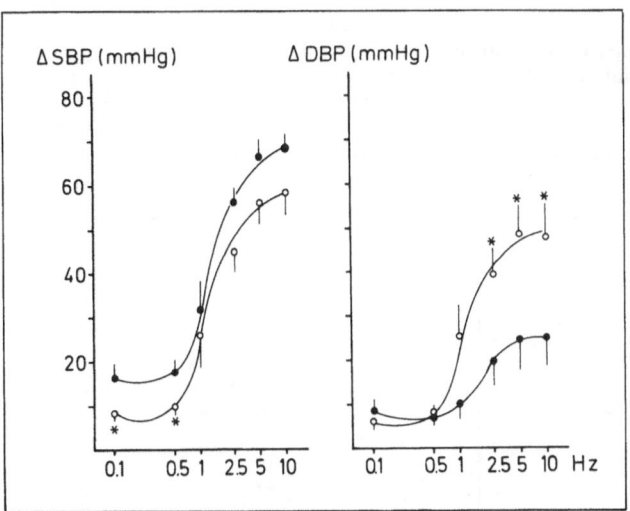

Fig. 1. Increases in systolic blood pressure (SBP) and diastolic (DBP) in pithed rats during incremental electrical stimulation (Hz), showing sham operated animals (●) and rats with myocardial infarction (○). $* = p < 0.05$. (n = 7).

Table 2. Heart rate increase in rats subsequent to myocardial infarction

	Hz					
	0.1	0.5	1.0	2.5	5.0	10.0
Sham	81±13.0	77±10.9	97±14.3	125± 6.6	120± 6.2	110± 6.6
MI	62± 6.4	61± 8.9	99±17.6	131±12.5	131±12.2	124±12.3
n=7						

	NA					
	0.1	0.3	1.0	3.3	10.0	33.3	100
Sham	3±0.4	9±1.1	22±3.2	51±4.4	82±4.8	99±5.5	109±6.6
MI	6±2.3	12±2.2	26±3.1	43±4.8	69±4.2	92±5.7	110±7.5
n=8						

	FE				
	0.06	0.21	0.63	2.1	6.25
Sham	3±0.9	9±1.8	30±7.9	58±8.1	91±6.2
MI	3±0.6	7±1.0	21±4.0	48±7.6	75±6.6
n=8					

Hz = preganglionic electrical stimulation; NA = dose response curves with noradrena-line (nmol/kg); FE = dose response curves with fenoterol (nmol/kg); Sham = sham occluded rats, MI = myocardial infarction.

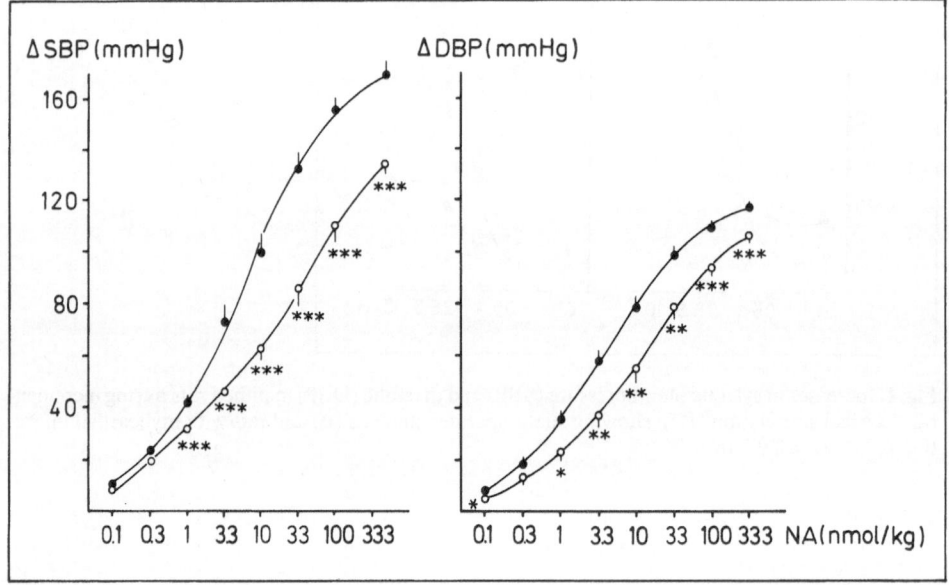

Fig. 2

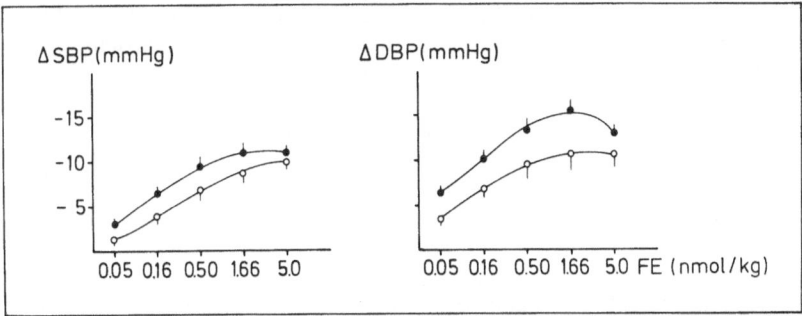

Fig. 3. Decreases in systolic blood pressure (SBP) and diastolic (DBP) in pithed rats after incremental i.v. doses of fenoterol (FE), showing sham operated animals (●) and rats with myocardial infarction (○). (n = 8).

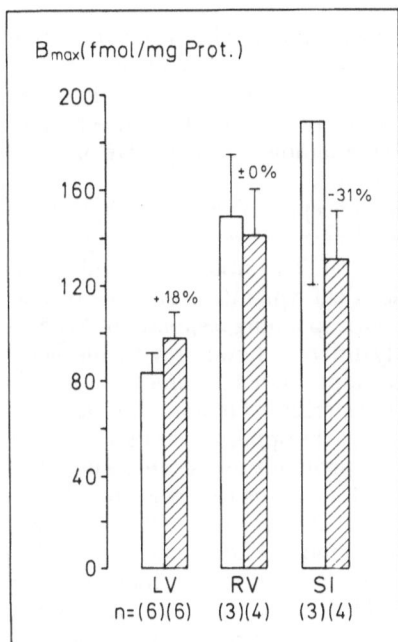

Fig. 4. Beta-adrenoceptor binding sites (B_max), calculated from the Scatchard plots, of the left ventricle (LV) and right ventricle (RV) and of the interventricular septum (SI), showing sham operated animals (white columns) and rats with myocardial infarction (hatched columns); n represents the number of binding studies (2–3 hearts per binding study).

The changes in blood pressure arising from incremental doses of noradrenaline are depicted in Fig. 2. In rats with MI a significant reduction in both systolic and diastolic blood pressure increase could observed. As already shown above for internal noradrenaline, the heart rate changes of MI rats were again similar to those of the shams (Table 2).

Fig. 2. Increases in systolic blood pressure (SBP) and diastolic (DBP) in pithed rats after incremental i.v. doses of noradrenaline (NA), showing sham operated rats (●), rats with myocardial infarction (○). * = $p < 0.05$, ** = $p < 0.01$, *** = $p < 0.001$. (n = 8).

The blood pressure increases due to fenoterol injections were only marginally different between both groups of rats (Fig. 3). The heart rate changes after fenoterol injections revealed no differences between MI rats and controls (Table 2).

In Fig. 4, the changes in beta-adrenoceptor binding sites after MI and sham operation are depicted for different parts of the rat heart. It is evident that the differences between MI rats and controls are only marginal.

Discussion

The studies on sympathetic outflow demonstrate that two days after experimental myocardial infarction (MI), no differences prevailed between rats with MI and sham operated animals (controls) with respect to stimulation-induced release of noradrenaline. The significant lower plasma noradrenaline concentrations of MI rats prior to electrical stimulation (basal values) can most likely be attributed to the lower noradrenaline content of the heart after MI, as previously determined by us [2].

In contrast to noradrenaline, the release of adrenaline due to electrical stimulation was enhanced after infarction. This finding can be explained by the already reported elevation of adrenaline content and the increase in tyrosine hydroxylase activity of the adrenal medulla two days after MI [2]. Both the increased rate of catecholamine biosynthesis and the enhanced release of adrenaline upon stimulation, are suggestive of an elevated sympatho-adrenal tone, two days subsequent to MI.

Whereas a higher number of myocardial beta-adrenoceptor binding sites have been found following acute experimental infarction in several species [2, 4, 5], the changes were only marginal on the second day post infarction in the rat. This result is in good agreement with our observations on heart rate effects of electrical sympathetic stimulation and of infusions of noradrenaline, since in each case the corresponding responses of MI animals and controls were practically identical. Markedly different, however, were the increments in diastolic blood pressure in the groups, with respect to endogenous and exogenous noradrenaline. In MI rats, diastolic blood pressure increased far more with electrical stimulation than in controls; with noradrenaline infusions the opposite was the case. Two principal mechanisms can be proposed to explain these apparently discrepant results. The stronger diastolic pressure response of the rats with MI towards electrical stimulation could arise from an increased number of vascular alpha-adrenoceptors. Changes in the number of alpha-adrenoceptor binding sites have been reported after acute myocardial infarction [1], just as for beta-adrenoceptor binding sites. The alternative, a reduction in the number of vascular beta-adrenoceptors, seems unlikely in our model, in view of the binding studies performed on the heart and the unaltered pressure response to fenoterol.

The reduced response of diastolic blood pressure to i.v. infused noradrenaline could conceivably be the result of an enhanced metabolism of noradrenaline by the vascular endothelium or the liver. The elevated sympatho-adrenal activity of rats with MI may well induce additional catecholamine degrading enzyme activity in the endothelial cells, thereby causing a faster removal of infused noradrenaline, especially during passage through the pulmonary bed (for rev. see 3).

Further detailed studies are, of course, required to support this hypothesis.

References

1. Corr PB, Shayman JA, Kramer JB, Kipnis RJ (1981) Increased alpha-adrenergic receptors in ischemic cat myocardium. J Clin Invest 67:1232–1236
2. Dominiak P, Türck D (1986) Alterations of β-adrenoceptors subsequent to myocardial infarction. Basic Res Cardiol 81 (Suppl 1): 243–251
3. Gerlach E, Nees S, Becker BF (1985) The vascular endothelium: a survey of some newly evolving biochemical and physiological features. Basic Res Cardiol 80:459–474
4. Maisel AS, Motulsky HJ, Insel PA (1985) Externalization of β-adrenergic receptors promoted by myocardial ischemia. Science 230:183–186
5. Mukherjee A, Bush LR, McCoy KE, Duke RJ, Hagler H, Buja LM, Willerson JT (1982) Relationship between β-adrenergic receptor numbers and physiological responses during experimental canine myocardial ischemia. Circ Res 50:735–741
6. Stiles GL, Caron MG, Lefkowitz RJ (1984) β-adrenergic receptors: biochemical mechanisms of physiological regulation. Physiol Rev 64:661–743
7. Yamaguchi I, Kopin IJ (1979) Plasma catecholamine and blood pressure responses to sympathetic stimulation in pithed rats. Am J Physiol 237:H305–H310

Authors' address:

Prof. Dr. med. P. Dominiak, Physiologisches Institut, Universität München, Pettenkoferstraße 12, 8000 München 2

Cardioprotection
by anti-ischaemic and cytoprotective drugs

L. Szekeres

Department of Pharmacology, University Medical School of Szeged, Hungary

Summary

Cardioprotective drugs are agents that prevent or moderate harmful consequences of impaired cardiac energetics, such as [1] sudden coronary death (SCD) due to early post-occlusion ventricular fibrillation (EPVF), [2] development of incapacitating myocardial necrosis. Cardioprotection may be due to anti-ischaemic action, correcting the imbalance between vascular supply and myocardial demand for blood, but also to cytoprotective effect, preserving cellular integrity in the presence of factors damaging structure and function of the cardiac cell membrane such as ischaemia, ionic imbalance and that of pH, etc.

Neither anti-ischaemic nor cytoprotective effect alone, or in combination, are sufficient to warrant full cardioprotection, i.e. both prevention of SCD and limitation of infarct size. Thus the beta-blocker pindolol which is anti-ischaemic in its effect reducing myocardial O_2 demand and protects from SCD and EVFP, failed to limit infarct size. Even interventions of a mainly cytoprotective type of action protecting from SCD and EPVF, such as the linoleic acid-rich diet, or lidocaine, were unable to limit infarct size. 7-oxoPGI$_2$ (anti-ischaemic and cytoprotective) failed to protect from SCD, VF and did not limit infarct size. On the other hand the nonsteroidal anti-inflammatory drugs which, like salicylates or sulfinpyrazon, reduce myocardial O_2 demand and protect from post-occlusion SCD and EPVF, effectively limiting infarct size.

A. Importance and mechanism of sudden cardiac death, anti-ischaemic and cytoprotective action

In the developed industrialized countries, nearly one third of mortality among men under the age of 65 years is due to sudden cardiac death (SCD) following acute myocardial infarction (AMI). It is mostly the result of primary ventricular fibrillation (VF), an electrical accident appearing on the basis of an electrical instability of the heart predisposing to ventricular arrhythmias. In addition, the trigger effect of a precipitating factor is also assumed to play a role in the rise of VF [10, 11].

Instability mostly develops on account of a chronic ischaemic heart disease and manifests itself by electrophysiological, biochemical and morphological inhomogeneity of myocardial fibres. The precipitating factor disrupting the normal sequence of cardiac contractions and initiating a self sustaining dysrhythmic activity could be, amongst others, an acute ischaemia superimposed on the old one, activating latent ectopic pacemaker areas. Therefore, protection of the heart includes both measures aimed at maintaining electrical stability as well as prevention of changes capable of precipitating fatal dysrhythmias.

Cardioprotective drugs can be defined as agents protecting against harmful consequences of acute myocardial ischaemia. Accordingly they should firstly prevent sudden cardiac death due to local myocardial ischaemia; secondly, if possible, limit the size of the developing infarct and thirdly, reduce the probability of recurrent infarction or infarct extension.

The concept of anti-ischaemic action will be used often in the following. Drugs of very different chemical structure and pharmacological profile can be ranged among the anti-ischaemic agents on the basis of their common action of correcting the imbalance between vascular supply of an myocardial demand for blood. This latter definition reflects more precisely the essentials of ischaemia than the myocardial O_2 supply:O_2 demand ratio because accumulating metabolites are also taken into consideration. I would like to briefly mention that most anti-ischaemic drugs are also cardioprotective but this is not a general rule, as it will be shown later.

The cytoprotective drugs act primarily by stabilizing the cell membranes. They tend to preserve the cellular integrity against injurious influence of toxic agents or metabolites – thus preventing or moderating harmful consequences, such as electrolyte shifts, cell swelling, liberation of lysosomal enzymes and of free radicals.

The mechanism of the possible ways to protect the heart is summarized in Table 1. Accordingly, the anti-ischaemic drugs are effective either by reducing myocardial O_2 demand or by increasing myocardial O_2 supply to the ischaemic area. The first can be done by reducing haemodynamic work and its main determinants, such as the heart rate, contractility, preload and afterload. Another possibility is to reduce cell metabolism directly, e.g. by diminishing Ca^{2+} influx to the cell, or by inhibition of lipolytic enzymes or by hypothermia.

Myocardial O_2 supply can be improved by enhanced flow to the ischaemic area, affecting determinants of this latter action such as an augmented collateral flow, facilitated by vasodilatation in the major vessels supplying collaterals; further by an increase in coronary perfusion pressure and in subendocardial driving pressure. Improvement of myocardial O_2 supply can also be attained by moderating or eliminating influences which may inhibit myocardial circulation, such as an increase in heart rate, in left ventricular end-diastolic pressure, in peripheral resistance, or in the tendency to platelet aggregation or to thrombus formation. In addition to these measures, increasing O_2 or substrate content of the blood or increasing their diffusion rate can be helpful to some extent.

The cytoprotective action is realized by stabilization of the cell membranes. This might prevent abnormal transmembrane electrolyte shifts; furthermore cell swelling and also liberation of lysosomes. A diet induced reorganisation of the membrane phospholipid structure may exert beneficial effects similar to those of the direct membrane stabilizing agents. The processes involved in ischaemic and reperfusion injury are closely related to the appearance of free radicals mainly because of the failure of the natural defence mechanism. Protection can be afforded either by preventing production of free radicals or by introducing free radical scavengers.

The third way to protect the heart from harmful influences is directed towards improvement of rheological factors in the blood, as well as toward prevention of thrombosis. Inhibition of platelet aggregation and of endothelial swelling and inflammation can be helpful – whereas restoration of flow in the case of established thrombi is only possible by means of active thrombolytic therapy, angioplasty or surgical revascularization. If the injury is not life-threatening and there is enough time, ischaemia-induced natural growth and enlargement of coronary anastomoses may limit infarct size.

It is obvious that all classifications shown here are more or less arbitrary and each group could be included into other categories as well. It is also conspicuous that many drugs have more than one beneficial effect, not rarely interacting with different cardioprotective mechanisms. Moreover, different drugs may share the same sites of action, thus giving the possibility of reasonable drug combinations.

Table 1. The mechanism of cardioprotective action

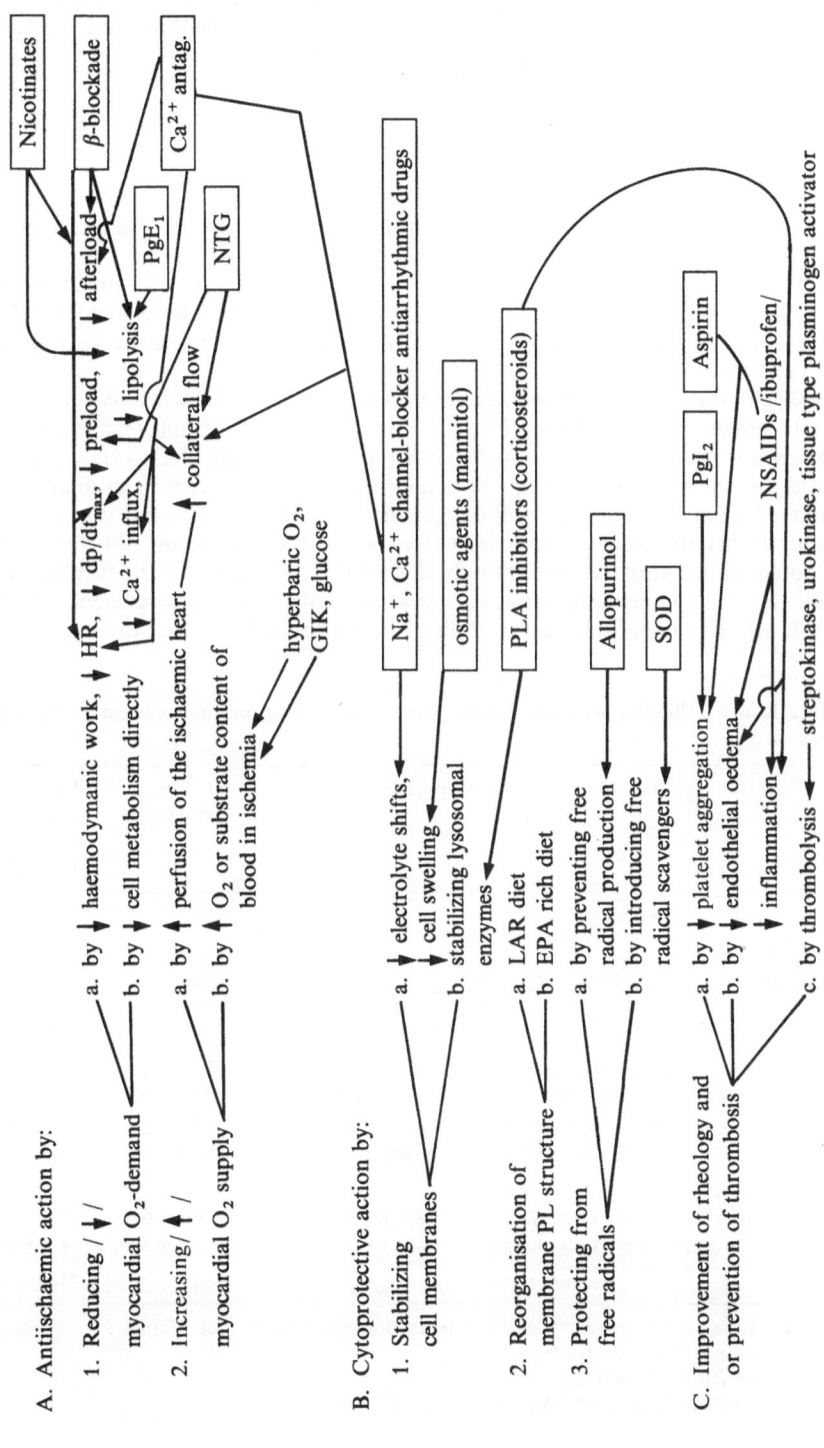

PL = phospholipids; HR = heart rate; NTG = nitroglycerin; PLA = phospholipase; SOD = superoxid dismutase; NSAID-s = non-steroidal anti-inflammatory drugs; GIK = glucose-insulin-potassium solution

In the following, we want to analyse a few special cardioprotective actions representing anti-ischaemic, cytoprotective and antiplatelet mechanisms with the aim of elucidating whether they are able to meet the main requirements of cardioprotection, namely prevention of sudden cardiac death (SCD) due to early post-occlusion ventricular fibrillation (EPVF) as well as limitation of infarct size.

B. Effects of LAR diet

Example of cardioprotection mainly by cytoprotective mechanism is the reorganisation of the membrane phospholipid structure.

We were able to show that a linoleic acid rich (LAR) diet, i.e. 12% sunflower oil in addition to normal diet, fed for 12 weeks to male rats, significantly reduced lethality in the conscious rat infarction model along with a diminished occurrence of VF and of VT [7, 16] (Table 2).

Nearly half the rats on LAR did not develop arrhythmia at all, whereas in the control group all animals were arrhythmic. This antiarrhythmic and cardioprotective action is at least partly due to a direct "membrane stabilizing" action of the diet, as shown by a significant decrease in the maximal rate of depolarisation which was present in atria taken from treated animals, bathed in well-oxygenated nutrient solution, but was even more pronounced when the heart was exposed to hypoxic solution for 5 min. Shortening of the action potential duration was present in both situations, so probably the fast Na^+ channels and the K^+ conductance are involved in this effect. We have evidence that this change in selective membrane permeability to ions could be related to corresponding

Table 2. Influence of linoleic acid rich (LAR)[a] diet on coronary occlusion induced changes in the conscious rat

Main actions of diet	Consequences of coronary occlusion		Control occlusion	Occlusion in rats fed LAR diet for 12 weeks
Reduction of sudden death	% lethality		81	19[d]
Antiarrhythmic action	% occurrence of	VF	88	28[d]
		VT	96	48[d]
		No arrhythmia	0	43[d]
Action on electro- physiological parameters	Duration of 50% repolarisation (ms)	Normal O_2 supply	22.3 ± 0.5	13.9 ± 0.2[d]
		5′ hypoxia	19.5 ± 0.4	12.3 ± 0.2[d]
	Max rate of depolarisa- tion (dV/dt max) in V/s	Normal O_2 supply	62.5 ± 2.1	53.6 ± 1.5[b]
		5′ hypoxia	55.5 ± 1.7	39.1 ± 1.0[d]
FFA in membrane phospholipids (PL)	% FFA in total heart PL docosatetraenoic acid:	$22:4, n-6$	0.6 ± 0.5	5.6 ± 0.7[b]
	Docosahexaenoic acid:	$22:6, n-3$	7.9 ± 2.4	1.6 ± 0.4[b]
	$n-6/n-3$ FFA		4.8	31.9

[a] LAR diet = 12% sunflower oil in addition to normal laboratory diet. Values are means $\pm$ SE, significant difference from control level:
[b] $p < 0.05$; [c] $p < 0.01$; [d] $p < 0.001$
VF = Ventricular Fibrillation; VT = Ventricular Tachycardia.

changes in FFA composition of the membrane phospholipids, since the diet markedly raised the ratio of n-6 FFA's (mainly arachidonic and docosatetranoic acid) to n-3 FFA's (docosahexanoic acid) from 4.8 to 31.9. It is worthy of note that the LAR diet increased survival as early as one week after feeding [13, 15]; furthermore, pretreatment with indomethacine, but not with aspirin, abolished the cardioprotective action of the LAR diet [7].

Since the above changes in FFA composition of myocardial membrane phospholipids (PL) were not associated with alteration of high energy phosphates, this type of cardioprotection can be attributed mainly to a cytoprotective mechanism. In the following we shall discuss another form of cardioprotection which is based on both an antiischaemic and cytoprotective mechanism.

Table 3. Late effects of a stable prostacyclin analogue $= 7$-oxo-PGI$_2$Na

Type of action type of intervention and that of preparation	Drug treatment	Changes in % of initial value							
		Control	Treatment maximal value	Minutes after completion of drug administration					
				90	120	150	180	210	240
A. Anti-ischaemic action		Elevation of ST segment in the epicardial electrogramm in % of initial value							
1. Angina model in the anaesthetized dog (5' frequency-loading in critically stenosed hearts)	250 µg/kg 7-oxo-PGI$_2$ infusion for 60 min	12 mV	+524[d]	−10[b]	−29[c]	−50[d]	−41[c]	−15[N.S.]	
2. Vasopressin angina in the anaesthetized rat 2.0 IU/kg i.v. vasopressin induced T wave elevation in the ECG	50 µg/kg i.p. 7-oxo-PGI$_2$	Effect on vasopressin induced T wave elevation in % of initial value −70[d]							
3. Protection of rat heart from ischaemic metabolic changes due to 1 min incubation in Ringer solution. Excision 2 h after 7-oxo-PGI$_2$ treatment	50 µg/kg i.p. 7-oxo-PGI$_2$	Effect on ATP and CP content of the heart in mmol/kg wet weight 3.72 ±0.29 ← ATP → 4.31[b] ±0.07 ← ATP 2.64 ±0.20 ← CP → 3.42[b] ±0.25 ← CP							
4. Ischaemic (non perfusable) area in the 6th min after LAD occlusion in anaesthetized rats	50 µg/kg i.p. 7-oxo-PGI$_2$	Effect on the nonperfusable area as % of total wet weight of the ventricles 34.3 ±1.3 — 23.9[d] ±4.2 —							
B. Cytoprotective action		Effect on action potential duration (APD) 90% and effective refractory period (ERP) in ms							
1. Prolongation of 90% APD and of ERP in the isolated papillary muscle of rabbits (pacing 60/min)	10^{-8} mol/l 7-oxo-PGI$_2$	171.8 ±6.8	180[N.S.] ±6.6	←APD$_{90}$→ 211[b] ±9.8	← APD$_{90}$→ 210.4[b] ±8.7	—	210.4[b] ±8.9		
		190.7 ±6.76	198[N.S.] ±5.8	←ERP→ 245.0[c] ±10.7	← ERP → 238.2[c] ±7.4	—	241.8[c] ±8.6		
2. K$^+$-loss and Na$^+$ gain of isolated rat hearts subjected to 60 min global ischaemia followed by 5 min reperfusion	50 µg/kg i.p. 7-oxo-PGI$_2$	K$^+$ and Na$^+$ content of hearts in µg/g dry weight 6850[a] — K$^+$ — 8400[b] 3100[a] — Na$^+$ — 2350[b]							

[a] Values obtained after 60 min ischaemia and 5 min reperfusion. Values are when indicated, mean $\pm$ S.E.M. Asterisks represent significant difference from control level; [b] $p < 0.05$; [c] $p < 0.01$; [d] $p < 0.001$.

C. Late effects of 7-oxoPGI$_2$

In earlier experiments, we have observed that PGI$_2$ and its stable analogue 7-oxoPGI$_2$, exert a late anti-ischaemic effect in the dog model of angina, expressed here as percentage protection from ST-segment elevation in the epicardial ECG evoked by frequency loading of the heart with critical stenosis of the left anterior descending coronary artery (LAD). This appears at a time when the characteristic effects of PGI$_2$ and its analogue, namely the vasodilatory and the platelet aggregation inhibitory effect, are already over (Table 3). In anaesthetized rats 2 h after administration of 7-oxoPGI$_2$, marked protection from ischaemic T-wave elevation was found. Similar pretreatment with 7-oxoPGI$_2$ significantly moderated ischaemic diminution of the myocardial ATP and CP content, if the excised heart were incubated in Ringer solution for 1 min. Finally, in similarly pretreated anaesthetized rats, occlusion of the LAD significantly reduced the nonperfusable area (corresponding to the area of supply of the LAD, i.e. the ischaemic area) in the 6th min after occlusion as compared with the nontreated controls [14]. Thus the late anti-ischaemic effect of 7-oxoPGI$_2$ could be clearly demonstrated in several species. But in addition we could show the presence of a marked cytoprotective action too. Among the indicators of cytoprotective action, the earliest one is the leakage of K$^+$ out of the cell and gain of intracellular Na$^+$. To show this, isolated Langendorff hearts were prepared from rats pretreated with 7-oxoPGI$_2$ 2 h before excision of the heart. Control perfusion for 20 min was followed by 60 min global ischaemia and finally, reperfusion with iso-osmotic sucrose was introduced for 5 min. Pretreatment significantly moderated ischaemic K$^+$ loss and Na$^+$ gain in tissues.

Another important indicator of a direct membrane action is our recent finding, that in isolated rabbit heart papillary preparations 2–4 h after washing out 10^{-8} mol/l 7-oxoPGI$_2$ from the perfusion fluid, the 90% repolarisation time of the action potential, as well as the effective refractory periods, were significantly prolonged as compared with the nontreated control papillary muscles.

D. Comparison of effects on SCD, arrhythmias and infarct size of different cardioprotective interventions

Some agents exert their anti-ischaemic action by reducing myocardial O$_2$ demand, as antilipolytic substances. It is well known that ischaemia stimulates lipolysis and decreases FFA oxidation at the same time. The resulting elevated serum FFA level, as well as the ischaemic hypoperfusion and the diminished wash-out of myocardial lipid-intermediates, may induce a drastic elevation (up to seven times the original value) of the myocardial FFA level [17]. Myocardial FFA accumulation exerts a negative inotropic effect on the myocardium [1] but in spite of this enhances myocardial O$_2$ uptake (MVO$_2$) and consequently increases the extension of the ischaemic and infarcted area after coronary occlusion [5].

Since antilipolytic agents inhibit TG-hydrolysis in both myocardial and adipose tissue [2] their administration will reduce FFA serum level and myocardial uptake of FFA from blood and also cAMP-dependent lipolysis from the myocardium. Thus, at least on a theoretical basis, antilipolytic drugs may exert an anti-ischaemic and perhaps indirectly an antiarrhythmic action.

Salicylates possess such an antilipolytic action but they are also non-steroidal anti-inflammatory drugs (NSAID) and most NSAID are potent antiplatelet agents as well.

Aspirin is known to inhibit PG-synthesis by blockade of the cyclo-oxygenase enzyme, and it also exerts a lipolytic effect in humans both with and without diabetes. Both aspirin and sodium salicylate inhibit isoproterenol-induced lipolysis, reduce serum FFA level and myocardial FFA uptake [18, 19]. In the dog, both substances reduced extension of the ischaemic area. On the basis of the above data, we were interested in comparing the protective effect of LAR diet (lidocaine being a typical membrane stabilizing antiarrhythmic drug) of the beta adrenoceptor blocking pindolol, as well as that of 7-oxoPGI$_2$ as well as the effects of NSAID including aspirin and Na-salicylate on survival rate, arrhythmias and infarct size in the conscious rat infarction model. This simple method for studying the acute phase of myocardial infarction in the conscious rat has been previously elaborated by us [6, 8], rendering it possible to study mortality, as well as the type, onset and duration of arrhythmias, the size of the ischaemic and infarcted area, respectively, in a relatively homogeneous population, using a sufficient number of animals for statistical analysis.

We have been prompted, among others, to study the limitation of infarct size, by the well known fact that in acute myocardial infarction the risk of sudden death increases with the size of the infarcted area. Accordingly, measures reducing the extent of local myocardial ischaemia may prevent the rise of life threatening ventricular arrhythmias leading to ventricular fibrillation and sudden death. In many cases the same intervention e.g. the blockade of the beta adrenoceptors may also reduce the ultimate size of necrosis developing in the area made ischaemic by coronary occlusion [16]. Infarct size is also a major determinant of heart failure and shock and predicts, according to Dunning [3] a lethal outcome more reliably than any arrhythmia or conduction disturbance.

In the present experiments gross infarct size determination was carried out by means of the nitrotetrazolium blue method in slices of hearts excised under pentobarbitone anaesthesia (45 mg/kg) 16 h after occlusion.

Comparison of survival rate, incidence of VF and size of the infarct demonstrate that LAR diet, pretreatment with pindolol or lidocaine, protect from sudden death due to cor-

Table 4. Effect of linoleic acid rich diet, lidocaine and pindolol, and 7-oxo-PGI$_2$ and different NSAID on survival rate, incidence of ventricular fibrillation (VF) and size of infarct area after coronary occlusion in the conscious male rat

Treatment	Dose	Survival rate %	Incidence of VF %	Infarction area as % of ventricles
Control		31 $(n=29)$	72 $(n=29)$	27.1$\pm$1.0 $(n=93)$
Linoleic acid rich diet	For 2 weeks	64 $(n=14)$[a]	36 $(n=14)$[b]	30.4$\pm$3.3 $(n=5)$[N.S.]
Lidocaine[c]	2 mg/kg i.v.	64 $(n=14)$[a]	21 $(n=14)$[b]	24.7$\pm$3.3 $(n=7)$[N.S.]
Pindolol[c]	0.015 mg/kg i.v.	63 $(n=13)$[a]	32 $(n=19)$[b]	21.8$\pm$3.6 $(n=12)$[N.S.]
Control		14 $(n=14)$	86 $(n=14)$	13.5$\pm$2.5 $(n=14)$
7-oxo-PGI$_2$ [e]	50 µg/kg i.p.	18 $(n=27)$[N.S.]	74 $(n=27)$[N.S.]	12.3$\pm$2.0 $(n=27)$[N.S.]
Control		31 $(n=29)$	72 $(n=29)$	27.1$\pm$1.0 $(n=93)$
Sulfinpyrazone[d]	10 mg/kg p.os	80 $(n=10)$[a]	20 $(n=10)$[b]	17.6$\pm$4.0 $(n=8)$[b]
	50 mg/kg p.os	65 $(n=20)$[a]	15 $(n=20)$[b]	15.5$\pm$3.7 $(n=13)$[b]
Aspirin[d]	20 mg/kg p.os	64 $(n=11)$[a]	64 $(n=11)$[N.S.]	18.8$+$4 $(n=7)$[b]
Na-salicylate[d]	20 mg/kg p.os	92 $(n=12)$[a]	31 $(n=12)$[b]	17.2$\pm$4.1 $(n=11)$[b]
Indomethacine[d]	1 mg/kg p.os	90 $(n=10)$[a]	40 $(n=10)$[b]	17.6$\pm$4.7 $(n=9)$[b]

Significantly different from control: [a] Calculated by the chi-squared method. [b] Calculated by Student's unpaired t-test. [N.S.] = non significant. In parentheses = number of experiments. [c] Given 10 min prior to occlusion. [d] Given 60 min before occlusion. [e] Given 120 min before occlusion.

onary occlusion and reduce the incidence of VF but failed to limit infarct size (Table 4). Moreover, 7-oxoPGI$_2$, shown to exert anti-ischaemic and cytoprotective effect 2 h after its administration, failed to affect survival, incidence of VF or infarct size.

On the other hand, the nonsteroidal anti-inflammatory drugs, in doses effectively increasing survival rate, have all significantly diminished the infarct size and, except for aspirin, also decrease the incidence of ventricular fibrillation [12].

Although aspirin and some other drugs possessing nonsteroidal anti-inflammatory action, such as indomethacine, sulfinpyrazon are cyclo-oxygenase inhibitors and act at least partly via inhibition of TXA$_2$ liberation and thus of platelet aggregation, the anti-ischaemic effect of sodium salicylate in not inhibiting cyclo-oxygenase [19] suggests that the antilipolytic action of these agents may also play a part in their anti-ischaemic effect. These results indicate that the effect of drugs on the infarct size and on survival after acute coronary occlusion are not closely related to each other.

In conclusion, we could show some of the many ways of protecting the ischaemic heart as well as of preventing primary VF. No single one of these measures seems to give a satisfactory, fully protective effect; however, it can be hoped that an adequate combination of these drugs could result in an optimal cardioprotection. Although the factors preventing SCD are essentially the same as those playing a role in limitation of the infarct size, further efforts are needed to solve this latter problem.

References

1. Borbola J Jr, Papp J Gy, Szekeres L (1974) Effects of octanoate on the electrical activity of Purkinje fibres. Experientia 30:262
2. Christian DR, Kilsheimer GS, Pettett G, Paradise R, Ashmore J (1969) Regulation of lipolysis in cardiac muscle: a system similar to the hormone-sensitive lipase of adipose tissue. In: Weber G (ed) Advances in enzyme regulation, vol 7. Pergamon Press, pp 71
3. Dunning AJ (1981) Pharmacotherapy at the onset of acute myocardial infarction. In: Van Zwieten PA, Hugenholtz E, Schönbaum E (eds) Progress in pharmacology, vol 4, No 2: Drug treatment of myocardial infarction. Fischer Verlag, Stuttgart New York, p 1
4. Hjalmarson A (1981) Beta-adrenoceptor drugs in the treatment of acute myocardial infarction. Progress in pharmacology, vol 4/2. Gustav Fischer Verlag, Stuttgart, p 10
5. Katz AM, Messineo FC (1981) Lipid-membrane interactions and the pathogenesis of ischemic damage in the myocardium. Circ Res 48:1
6. Leprán I, Koltai M, Szekeres L (1983) Coronary artery ligation, early arrhythmias, and determination of the ischaemic area in conscious rate. J Pharmacol Meth 9:219
7. Leprán I, Nemecz Gy, Koltai M, Szekeres L (1981) Effect of linoleic acid rich diet on the acute phase of coronary occlusion in conscious rate. Influence of indomethacin and aspirin. J Cardiovasc Pharmacol 3:847
8. Leprán I, Siegmund W, Szekeres L (1979) A method of acute coronary occlusion in anaesthetized closed chest rats. Acta Physiol Acad Sci Hung 53:190
9. Nachlas MM, Shnitka TK (1963) Macroscopic identification of early myocardial infarcts by alterations in dehydrogenase activity. Am J Pathol 42:379
10. Szekeres L (1986) Sudden death due to acute myocardial infarction. CRC Press Inc, Boca Raton, Florida
11. Szekeres L (1986) Pharmacotherapy of sudden cardiac death due to acute myocardial ischaemia. Cor Vasa 28:123
12. Szekeres L, Koltai M, Leprán I (1984) Is drug induced reduction of the infarct size essential for protection against sudden coronary death? In: Paton W, Mitchell J, Turner P (eds) IUPHAR 9th international congress of pharmacology, proceedings vol 3. Macmillan London Basingstoke, p 247

13. Szekeres L, Koltai M, Papp J Gy, Takáts I, Tósaki Á, Udvary É, Fazekas T (1986) Protection of the ischaemic heart. Prevention of primary ventricular fibrillation. In: Szekeres L, Papp J Gy (eds) Pharmacological protection of the myocardium. Pergamon Press, Oxford Akadémiai Kiadó, Budapest, p 33
14. Szekeres L, Koltai M, Pataricza J, Takáts I, Udvary É (1984) On the late antiischemic action of the stable PgI$_2$ analogue: 7-oxo-PgI$_2$-Na and its possible mode of action. Biomed Biochim Acta 43:135
15. Szekeres L, Leprán I, Boros E, Takáts I, Koltai M (1982) The effect of non-steroidal anti-inflammatory drugs and of linoleic acid-rich diet on early arrhythmias resulting from myocardial ischaemia. In: Parratt JR (ed) Early arrhythmias resulting from myocardial ischaemia. Macmillan, London, p 239
16. Szekeres L, Leprán I, Koltai M (1980) Influence of linoleic acid rich diet on the incidence of arrhythmias and death in the acute phase of myocardial infarction in conscious rats. In: Förster W (ed) Prostaglandins and thromboxanes. VEB Gustav Fischer Verlag, Jena, GDR, p 33
17. van der Vusse GJ, Roemen Th HM, Prinze FW, Coumans WA, Reneman RS (1982) Uptake and tissue content of fatty acids in dog myocardium under normoxic and ischemic conditions. Circ Res 50:538
18. Vik-Mo H, Mjos OD (1976) Myocardial metabolism and performance during sodium salicylate infusion in dogs. Scand J Clin Lab Invest 36:763
19. Vik-Mo H, Mjos OD (1977) Effect of sodium salicylate and acetylsalicylic acid on epicardial ST-segment elevation during coronary artery occlusion in dogs. Scand J Clin Lab Invest 37:287

Authors' address:

Prof. Dr. László Szekeres, Dept. Pharmacology, University Medical School Szeged, H-6701 Szeged, Dóm tér 12. Hungary

Myocardial protection by antioxidant during permanent and temporary coronary occlusion in dogs

E. Röth, B. Török, Zs. Pollák, Gy. Temes, and G. Morvay

Institute of Experimental Surgery, University Medical School, Pécs, Hungary

Summary

In an ischaemic heart model the lipid peroxidation, scavenger state and ultrastructure were studied, to determine the action of a new antioxidant of dihydroquinoline type (MTDQ-DA). In dog experiments, the left descending coronary artery (LAD) was ligated permanently (30 minutes, 1, 2 or 3 hours) or temporarily (30 minutes, 1 or 2 hours of ischaemia followed by 1 hour of recirculation). The experimental protocol involved two groups: control animals without antioxidant treatment and animals treated with antioxidant infusion during the ischaemic and reperfusion period. In both groups, the thiobarbituric acid reactive product, the malondialdehyde (MDA), reduced glutathione (GSH) and superoxide dismutase (SOD) were measured, to illustrate the injured or scavenged state of the membrane system. In nontreated animals the permanent and temporary LAD increased the MDA content, decreased GSH concentration (mainly during reperfusion) and reduced SOD activity. Treatment with MTDQ-DA diminishes the characteristic biochemical changes. According to ultrastructural investigations, irreversible alterations (Ca deposits in the mitochondria, disruption of intramitochondrial membranes, hypercontraction bands) occurred only in the control group. Antioxidant therapy is able to reduce the myocardial damages both quantitatively and qualitatively.

Introduction

In recent years, it has been appreciated that free radicals play an important role in the complex pathological events of ischaemic myocardial injury [3, 16, 21, 22, 29, 32, 24]. It was recognized that there are some possibilities for producing oxyradicals during ischaemia [20, 26, 28], moreover, after reperfusion or reoxygenation the free radicals production could be increased markedly [4, 6, 12, 14, 18, 25]. The dangerous radicals initiate a chain reaction by peroxidation of polyunsaturated fatty acids (PUFA) in the membrane system, accompanied by several functional abnormalities [6, 16, 24, 37].

Some experiments confirm that antioxidant compounds can stop this dangerous reaction and in this way the myocardium could be protected from severe ischaemic or reperfusion injuries [4, 5, 11, 13, 18, 32, 39, 40].

In our earlier experiments, it was found [29, 39] that in acute experimental myocardial infarction, a new dihydroquinoline antioxidant material (MTDQ-DA) was able to decrease the amount of breakdown products of PUFA and to preserve the endogenous scavenger compounds.

The purpose of this study was to investigate the effects of permanent and temporary coronary ligatures in dogs with and without antioxidant therapy.

Material and methods

Experimental protocol is seen in Fig. 1. Mongrel dogs were divided in two groups: Group I of nontreated control animals receiving only saline infusion during experiment; in Group II MTDQ-DA was administered by drop infusion (150 mg/kg, in saline). In both groups, permanent and transient coronary ligature were performed. The surgical procedure was as follows: under general anaesthesia ($N_2O:O_2$ in a ratio of $3:1 +$ Halothane), following left thoracotomy the pericardium was opened and the left descending coronary artery (LAD) was ligated below the first major coronary branch. Permanent ligatures lasted for 30 minutes, 1, 2 or 3 hours, and the temporary coronary ligatures were released after 30 minutes, 1 or 2 hours, then recirculation was maintained for 1 hour.

Biochemical measurements

In the tissue homogenates the following determinations were made:
a. TBA reactive materials (mainly malondialdehyde – MDA –) according to Placer [27]
b. Reduced glutathione (GSH) according to modified Sedlak method [35]
c. Superoxide dismutase (SOD) according to Misra and Fridovich [23].

The measured parameters (MDA and GSH) were expressed as percentages of normal heart tissue values, SOD activity in unit/g of wet tissue. Student's t-test was used to evaluate the differences between the control and experimental groups. The results gained were considered significant if $p < 0.05$.

Electron microscopic preparation

Tissue samples (1×1 mm block) were fixed in glutaraldehyde and osmium tetroxide, embedded in Durcupan and double stained with uranyl-acetate and lead citrate. Ultra-thin sections were evaluated by Jeol 100C electron microscope.

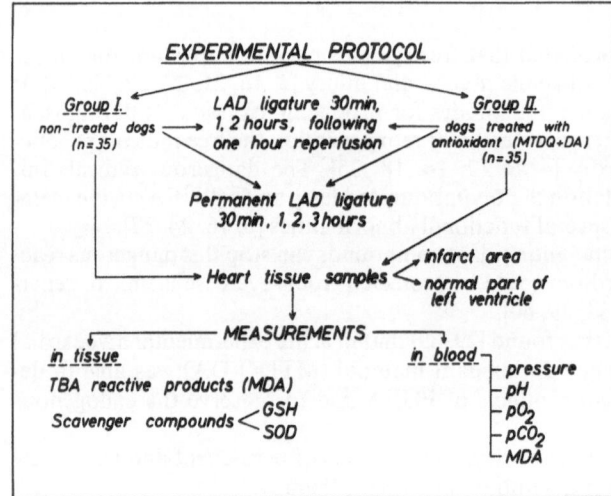

Fig. 1. Scheme of experimental groups.

Results

In Fig. 2 the alterations of MDA after permanent and temporary coronary ligature with or without antioxidant treatment are seen. It is obvious that in the infarct area this breakdown product of PUFA increased significantly after permanent ischaemia. This value was almost the same after one hour of reperfusion in the hearts of nontreated animals. In those animals treated with MTDQ-DA this elevation occurred neither during ischaemia nor in the reperfusion period.

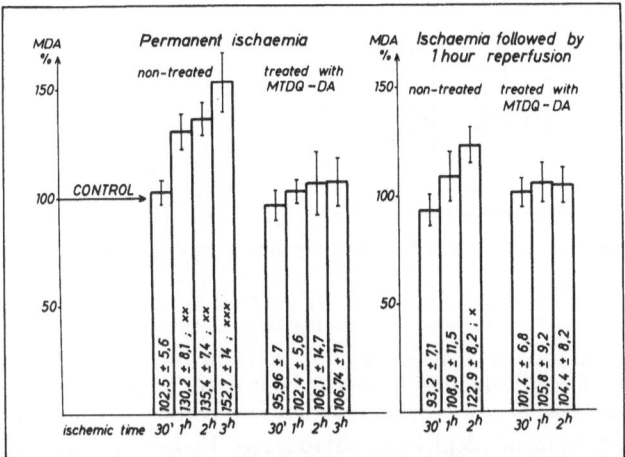

Fig. 2. Time course of MDA changes in infarct area with and without antioxidant treatment. Differences were considered significant at $p < 0.05 = x$; $p < 0.02 = xx$; $p < 0.01 = xxx$.

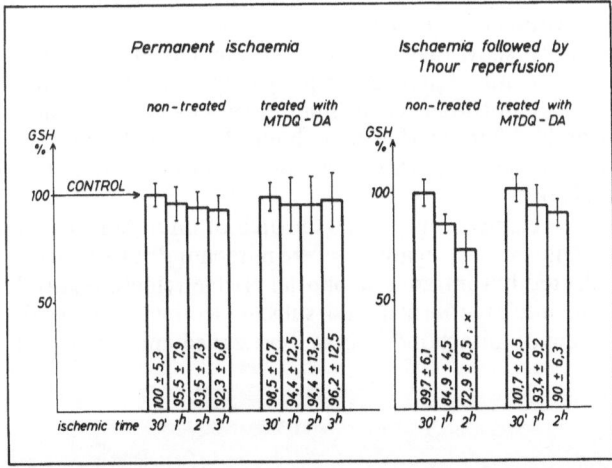

Fig. 3. Alterations of GSH in infarct area. Treatment with MTDQ-DA protects the thiol compound from ischaemic depletion.

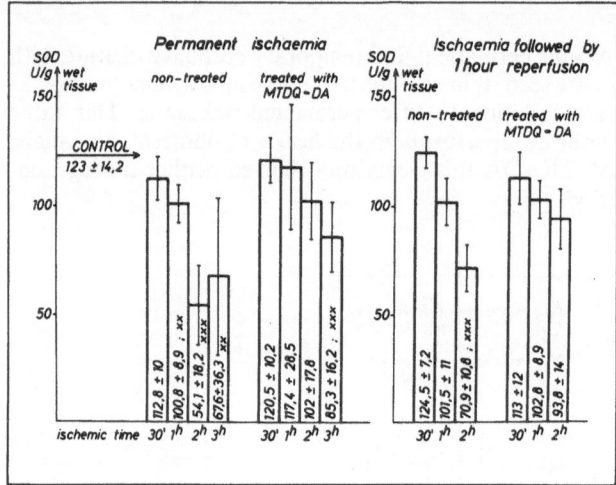

Fig. 4. MTDQ-DA treatment diminishes reduction of SOD activity during permanent and temporary coronary ligature.

Figure 3 summarizes the alterations of the reduced glutathione. In untreated animals, during permanent LAD ligature a relatively slow decrease of this thiol compound can be observed, even after 3 hour coronary ligature. In the case where reperfusion was maintained after 1 and 2 hours of ischaemia, this depletion is more severe. Following treatment with antioxidant infusion, the GSH is better preserved during permanent and/or temporary ligature; it remains about at 90% of the normal value.

The activity of SOD in hearts without treatment (Fig. 4) is decreased very sharply after 2 and 3 hour permanent ligature in the infarct area; the antioxidant treatment helps to preserve this scavenger enzyme at a higher level. Reperfusion leads only to a moderate decrease of SOD activity in untreated dogs, so it is obvious that antioxidant treatment has a beneficial effect on the preservation of this enzyme.

Electron microscope investigations show that in nontreated animals many types of tissue alterations develop in the whole infarct area. After permanent coronary ligature, mitochondria are swollen, the intramitochondrial membrane system is disorganized (Fig. 5), myofibrils show large I bands. After reperfusion without MTDQ-DA treatment, a serious form of abnormal contractions appears and mitochondria contain large amounts of amorphous calcium deposits.

The antioxidants moderate these changes and reduce the area of infarction (Fig. 6). Even after 2 or 3 hours of permanent coronary ligature, severe mitochondrial alterations cannot be seen, the matrix are cleared but intramitochondrial cristae remain relatively well organized. After reperfusion, intermyofibrillar and subsarcolemmal oedema develop, the internal membrane system of mitochondria is well preserved and calcium deposits cannot be demonstrated.

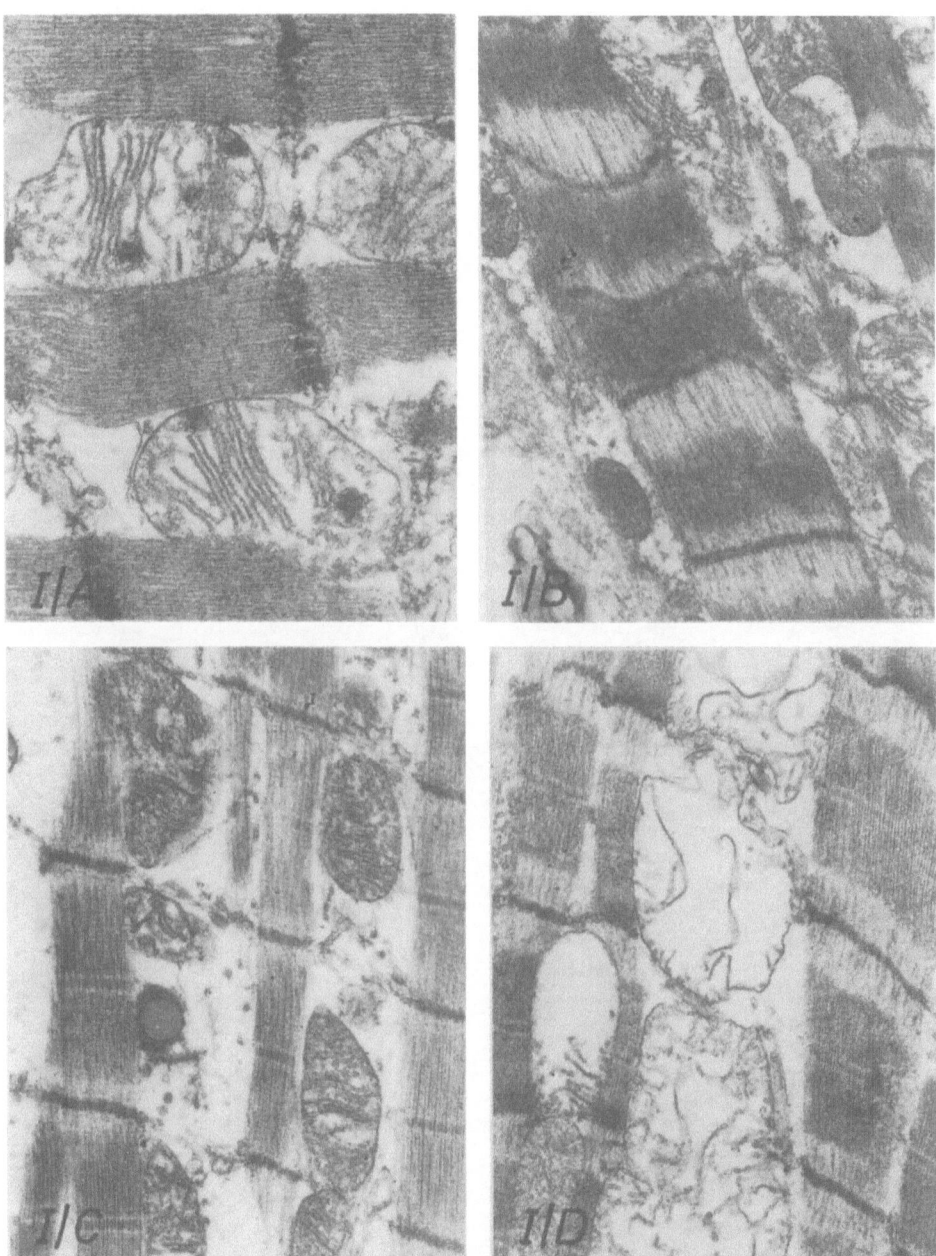

Fig. 5. Transmission electron micrographs from ischaemic areas of nontreated hearts. I A: Ca deposits in mitochondria, ruptured cristae (1 h LAD + 1 h reperfusion). I B: Hypercontracted sarcomeres with stretched zones (2 h LAD + 1 h reperfusion). I C: Intermyofibrillar oedema, swollen mitochondria (1 h LAD). I D: Large I bands, disrupted intramitochondrial cristae (2 h LAD).

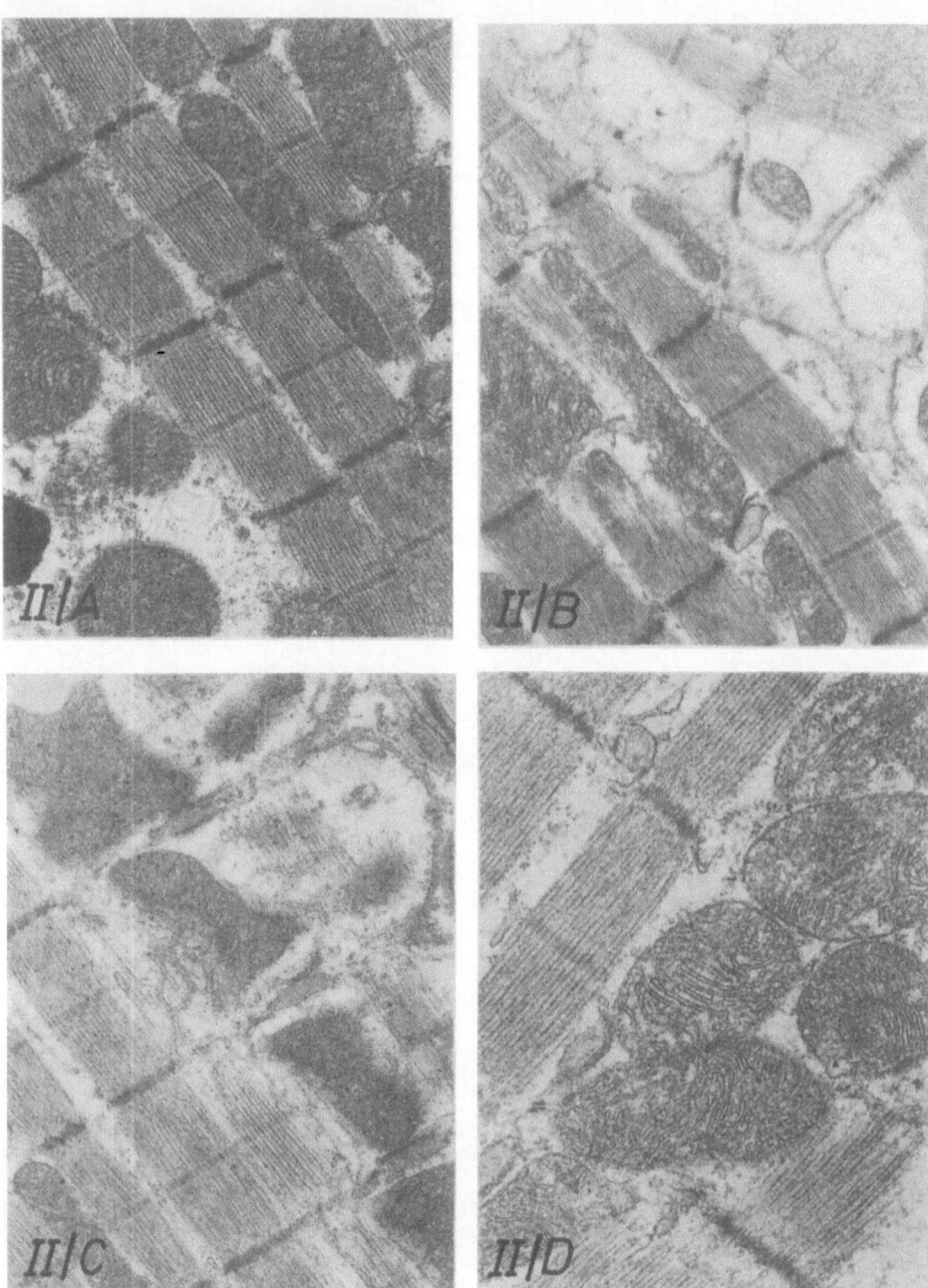

Fig. 6. Transmission electron micrographs from treated hearts: II A: Myocytes appear normal, glycogen is evident (1 h LAD + 1 h reperfusion). II B: Subsarcolemmal oedema, slightly swollen mitochondria (2 h LAD + 1 h reperfusion). II C: Mitochondria contain well organized cristae, myofibrillar structure is normal (1 h LAD). II D: In swollen mitochondria the cristae are well preserved (2 h LAD).

Discussion

Production of free radicals and peroxidation of PUFA in the membrane system is continously going on under normal conditions [8, 29]. In healthy tissue the capacity of endogenous scavengers is in equilibrium with the generation of reactive oxygen metabolites. During hypoxic conditions the large univalent reduction of O_2 results in more free radicals and, in such cases, the functional level of scavengers (SOD, catalase, GSH, GPD) are reduced due to the tissue injury [2, 9, 22]. It must also be noted that in the hyperoxic situation the available abundant molecular oxygen leads to burst of O_2 radicals, causing the exhaustion of endogenous scavengers. Finally, the peroxidation of lipid components causes cellular, mitochondrial, sarcoplasmatic, lysosomal and sarcolemmal membrane damage [3, 12, 16]. The release of lipid peroxidation (LP) products into the extracellular space aggravates this injury process (our hypothesis is summarized in Fig. 7).

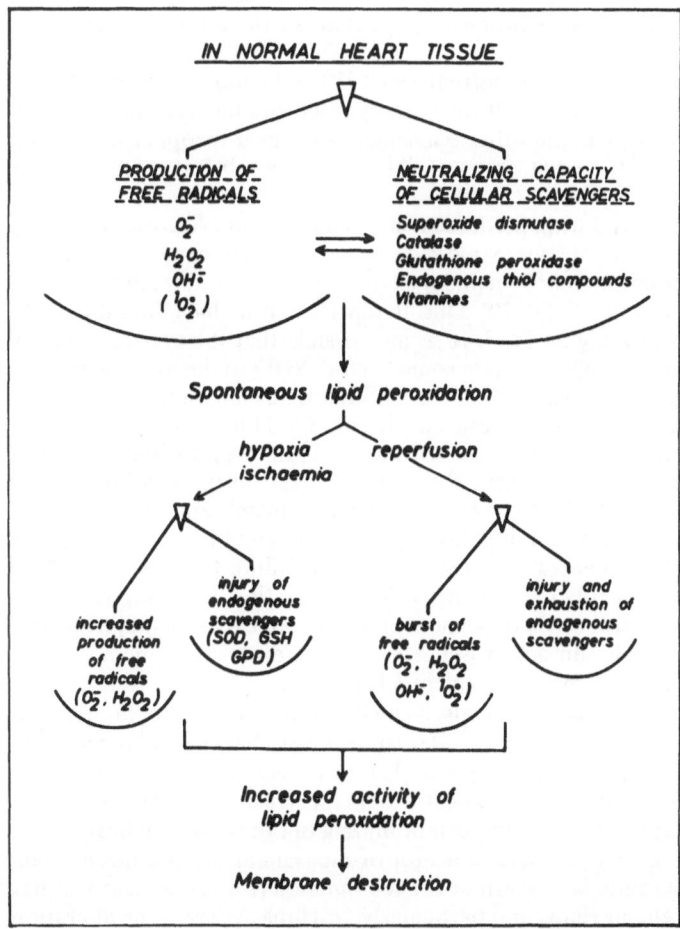

Fig. 7. Scheme of lipid peroxidation in heart tissue during normal and pathological conditions.

According to our results, it was unexpected that without antioxidant treatment the elevations of MDA during both permanent and temporary coronary ligature were almost the same. This finding is not identical with data in the literature [13, 14, 21].

We suppose that during re-establishment of coronary circulation, the end product of PUFA is washed out from the infarct area to the systemic circulation. Therefore, between 5 and 30 minutes of reperfusion an increase of MDA can be found in the systemic venous blood (according to our experiments now in progress). Gauduel [12] measured also a higher MDA release in the rat heart perfusion model. In his experiments, the re-admission of oxygen promotes a biphasic malondialdehyde release two minutes and ten minutes after oxygen administration. Rao [28] could demonstrate an increased MDA release during permanent coronary ligature into the coronary sinus. Anyway, the increase of MDA content, in spite of the different experimental conditions, is always a characteristic marker of the membrane deterioration.

Depletion of GSH during reperfusion cannot be sufficiently explained by the same mechanism. During temporary ligature, a slight GSH elevation can be measured in the systemic venous blood (data as yet unpublished). According to Curello [7] GSH release into the circulating blood does not quantitatively account for the cellular reduction of GSH. Increased content of GSSG in the ischaemic tissue after reperfusion suggests a metabolic utilization of GSH. According to current views [19, 36] endogenous glutathione may act in at least three different ways: (i) it can directly react with the free radicals forming GSSG; (ii) it is able to regenerate other scavengers such as α-tocopherol and ubiquinone, also a product of oxidized glutathione; (iii) it serves as a substrate of glutathione peroxidase scavenging the H_2O_2.

SOD is markedly decreased during permanent coronary ligature. According to present opinion, the high production of superoxide radicals decreases the capacity of this enzyme, moreover, the anaerobic metabolism and lack of ATP synthesis directly injure this essential scavenger enzyme [1, 3, 18, 20, 29]. During reperfusion we have found that the SOD activity is preserved at a higher level. It seems possible that the systemic arterial blood reaching the coronary circulation gives some "extra" SOD to the hypoxic tissues.

Ultrastructural changes are dependent on the suffered ischaemic time. They can be very variable, reversible and irreversible, respectively [10, 30]. During permanent coronary ligature mainly the mitochondrial cristae are injured, in the reperfusion state many contraction abnormalities appear in the myofibrils and Ca deposits in the mitochondria [15, 17, 31]. In the ischaemic state the basis of degradation of membrane structure is the increase of LP which alters the permeability, fluidity, structure and function of the phospholipid bilayer. Under a lack of energy, the Na^+-K^+-pump failure leads to a loss of cell volume regulation and subsequently to extensive oedema formation. Ca pump failure induces a loss of membrane control of Ca movement and sarcomere tension [38]. Hess [16] demonstrated that free radicals can act in ischaemic myocardium with uncoupling of ATP synthesis from sarcoplasmic reticulum calcium transport.

Focusing our interest on the treated animals, it can be concluded that our antioxidant compound, MTDQ-DA, protects the myocardial tissue during ischaemia and reperfusion in both biochemical and morphological respects. The low level of MDA refers to decreased LP even after 2 and 3 hours coronary ligature and reperfusion. The almost perfect preservation of GSH content is the most important finding of our results. In heart tissue the reduced glutathione is a very sensitive indicator of the cellular defense mechanism. A low value of GSH/GSSG ratio is deleterious for cell function [14, 19]. Recently, it has been suggested that intracellular thiols and particularly GSH may prevent the alteration of Ca homeostasis, reducing lipid peroxidation of the membrane [7]. We suppose that structural preservation of the treated hearts (lack of intramitochondrial Ca accumulation, no hypercontraction) are due to storage of this endogenous thiol compound.

As to the mechanism of scavenger action, we assume that MTDQ-DA reacts with deleterious oxygen metabolites and lipid peroxides, and moreover, releases oxygen in situ, as follows:

$(MTDQ = HN \langle A \rangle NH)$

$$1.1 \quad ROO \cdot + \cdot N \langle A \rangle NH \quad \rightarrow RO \cdot + \cdot ON \langle A \rangle NH$$

$$1.2 \quad ROO \cdot + \cdot ON \langle A \rangle NH \quad \rightarrow RON \langle A \rangle + O_2$$

$$1.3 \quad RON \langle A \rangle NH + ROO \cdot \rightarrow RON \langle A \rangle N \cdot + ROOH$$

$$1.4 \quad ROO \cdot + RON \langle A \rangle N \cdot \quad \rightarrow RO \cdot + RON \langle A \rangle NO \cdot$$

$$1.5 \quad ROO \cdot + RON \langle A \rangle NO \cdot \rightarrow RON \langle A \rangle NOR + O_2$$

$$1.6 \quad RO \cdot + \cdot N \langle A \rangle NH \quad \rightarrow RON \langle A \rangle NH$$

$$2.1 \quad R \cdot + RON \langle A \rangle NO \cdot \quad \rightarrow RON \langle A \rangle NOR$$

$$2.2 \quad 2 RON \langle A \rangle N \cdot \quad \rightarrow RON \langle A \rangle N - N \langle A \rangle NOR .$$

Thus MTDQ-DA can inactivate the source of free radicals and improve oxygen supply. In this context, it is understandable that during permanent and temporary coronary ligature GSH and SOD levels remain at a higher level and structural alterations of the heart tissue are less.

Another aspect of our experiments is that these hearts treated with antioxidant have an improved electrical stability (malignant arrhythmias and ventricular fibrillation decrease; ref. 39). Summing up, it is obvious that the protection of the heart from the functional, biochemical, morphological consequences of ischaemia is a complex process; nonetheless, we still do not know exactly the first step in these defense mechanisms. However, it can be established that our non-toxic scavenger material may give protection against free radical mediated injury of the heart.

References

1. Akizuki S, Yoshida S, Chambers D, Eddy I, Parmley I, Yellon D, Downey J (1984) Blockage of the O_2 radical producing enzyme xanthine oxidase, reduces infarct size in the dog. Fed Proc 43:540 (abstr.)
2. Bulkley GB (1983) The role of oxygen free radicals in human disease processes. Surgery 94:407
3. Burton KM, McCord JM, Ghai G (1984) Myocardial alterations due to free-radical generation. Am J Physiol (Heart Circ Physiol 15) 246:H776–H783
4. Burton KP (1985) Superoxide dismutase enhances recovery following myocardial ischemia. Am J Physiol (Heart Circ Physiol 17) 248:H637–H643
5. Casale AS, Bulkley GB, Bulkley BH, Flaherty JT, Gott VL, Gardner TJ (1983) Oxygen free-radicals scavengers protect the arrested, globally ischemic heart upon reperfusion. Surg Forum 34:313–316
6. Chemnitius JM, Sasaki Y, Burger W, Bing RJ (1985) The effect of ischemia and reperfusion on sarcolemmal function in perfused canine hearts. J Mol Cell Cardiol 17:1139–1150
7. Curello S, Ceconi C, Bigoli C, Ferrari R, Albertini A, Guarnieri C (1985) Changes in the cardiac glutathione status after ischemia and reperfusion. Experientia, Birkhäuser Verlag, Basel, 41:42–43

8. Del Maestro RF (1980) An approach to free radicals in medicine and biology. Acta Physiol Scand 492:153–168
9. Fridovich I (1979) Hypoxia and oxygen toxicity. Adv Neurol 26:255–259
10. Ganote CE (1983) Contraction band necrosis and irreversible myocardial injury. J Mol Cell Cardiol 15:67–73
11. Gardner TJ, Stewart JR, Casale AS, Downey JM, Chambers DE (1983) Reduction of myocardial ischemic injury with oxygen-derived free radical scavengers. Surgery 94:423–427
12. Gauduel Y, Duvelleroy MA (1984) Role of oxygen radicals in cardiac injury due to reoxygenation. J Mol Cell Cardiol 16:459–470
13. Guarnieri C, Ferrari R, Visioli O, Caldarera CM, Nayler WG (1978) Effect of α-tocopherol on hypoxic-perfused and reoxygenated rabbit heart muscle. J Mol Cell Cardiol 10:893–906
14. Guarnieri C, Flamigni F, Caldarera CM (1980) Role of oxygen in the cellular damage induced by re-oxygenation of hypoxic heart. J Mol Cell Cardiol 12:797–808
15. Hearse DJ, Humphrey SM, Nayler WG, Slade A, Border D (1975) Ultrastructural damage associated with re-oxygenation of the anoxic myocardium. J Mol Cell Cardiol 7:315–324
16. Hess ML, Manson NH, Okabe E (1982) Involvement of free radicals in the pathophysiology of ischemic heart disease. Can J Physiol Pharmacol 60:1382–1389
17. Jennings RB, Reimer KA (1981) Lethal myocardial ischemic injury. Am J Pathol 102:241–255
18. Jolly SR, Kane WJ, Bailie MB, Abrams GD, Lucchesi BR (1984) Canine myocardial reperfusion injury: Its reduction by the combined administration of superoxide dismutase and catalase. Circ Res 54:277–285
19. Kosower EM (1978) The glutathione status of cells. Academic Press, International review of cytology 54:109–160
20. McCord JM, Roy RS (1982) The pathophysiology of superoxide: role in inflammation and ischemia. Can J Physiol Pharmacol 60:1346–1352
21. McCord JM (1984) Are free radicals a major culprit? In: Hearse DI, Yellon DM (eds) Therapeutic approaches to myocardial infarct size limitation. Raven Press, New York, pp 209–218
22. Meerson FZ, Kagan VE, Kozlov YuP, Belkina LM, Arkhipenko YuV (1982) The role of lipid peroxidation in pathogenesis of ischemic damage and the antioxidant protection of the heart. Basic Res Cardiol 77:465–485
23. Misra HP, Fridovich I (1972) The role of superoxide anion in the antioxidation of epinephrine and a simple assay for superoxide dismutase. J Biol Chem 247:3170–3175
24. Myers ML, Bolli R, Lekich RF, Hartley CJ, Roberts R (1985) Enhancement of recovery of myocardial function by oxygen free-radical scavengers after reversible regional ischemia. Circulation 72:915–921
25. Otani H, Tanaka H, Inoue T, Umemoto M, Omoto K, Tanaka K, Sato T, Osako T, Masuda A, Nonoyama A, Kagawa T (1984) In vitro study on contribution of oxidative metabolism of isolated rabbit heart mitochondria to myocardial reperfusion injury. Circ Res 55:168–175
26. Parks DA, Granger DN (1983) Oxygen-derived radicals and ischemia-induced tissue injury. In: Greenwald RA, Cohen G (eds) Oxyradicals and their scavenger systems, vol II. Cellular and medical aspects. pp 135–144
27. Placer ZA, Cushman LL, Johnson BC (1966) Estimation of product of lipid peroxidation (malondialdehyde) in biochemical systems. Anal Biochem 16:359–364
28. Rao PS, Cohen MV, Mueller HS (1983) Rapid communication. Production of free radicals and lipid peroxides in early experimental myocardial ischemia. J Mol Cell Cardiol 15:713–716
29. Röth E, Török B, Zsoldos T, Matkovics B (1985) Lipid peroxidation and scavenger mechanism in experimentally induced heart infarcts. Basic Res Cardiol 80:530–536
30. Schaper J, Mulch J, Winkler B, Schaper W (1979) Ultrastructural, functional, and biochemical criteria for estimation of reversibility of ischemic injury: A study on the effects of global ischemia on the isolated dog heart. J Mol Cell Cardiol 11:521–541
31. Schaper J, Schwarz F, Kittstein H, Kreisel E, Winkler B, Hehrlein FW (1980) Ultrastructural evaluation of the effects of global ischemia and reperfusion on human myocardium. Thorac Cardiovasc Surgeon 28:337–342
32. Schlafer M, Kane PF, Wiggins VY, Kirsh MM (1982) Possible role for cytotoxic oxygen metabolites in the pathogenesis of cardiac ischemic injury. Circulation [Suppl I] 66:85–92

33. Schlafer M, Kane PF, Kirsh MM (1982) Superoxide dismutase plus catalase enhances the efficacy of hypothermic cardioplegia to protect the globally ischemic, reperfused heart. Thorac Cardiovasc Surg 83:830–839
34. Scott JA, Khaw BA, Locke E, Haber E, Homcy Ch (1985) The role of free radical-mediated processes in oxygen-related damage in cultured murine myocardial cells. Circ Res 56:72–77
35. Sedlak J, Lindsay RH (1986) Estimation of total protein-bound and non-protein sulphydryl groups in tissue with Ellman's reagent. Anal Biochem 25:192–205
36. Siesjö BK (1981) Cell damage in the brain: a speculative synthesis. J Cereb Blood Flow Metab 1:155–185
37. Stewart JR, Blackwel WH, Crute SL, Loughlin V, Greenfield IJ, Hess ML (1983) Inhibition of surgical induced ischemia/reperfusion injury by oxygen free radical scavengers. J Thorac Cardiovasc Surg 86:262–266
38. Török B, Röth E, Trombitás K (1982) Ultrastructural changes of the subendocardium in ischemic and cardioplegic states before and after reperfusion. Eur Surg Res 14:17–26
39. Török B, Röth E, Bär V, Pollák Zs (1986) Effect of antioxidant therapy in experimentally induced heart infarcts. Basic Res Cardiol 81:167–180
40. Werns SW, Shea MJ, Driscoll EM, Cohen Ch, Abrams GD, Pitt B, Lucchesi BR (1985) The independent effects of oxygen radical scavengers on canine infarct size reduction by superoxide dismutase but not catalase. Circ Res 56:895–898

Authors' address:

Dr. E. Röth, Institute of Experimental Surgery, University Medical School,
H-7643 Pécs (Hungary)

Promising reduction of ventricular fibrillation in experimentally induced heart infarction by antioxidant therapy

B. Török, E. Röth, B. Mezey, Gy. Temes, K. Tóth, and Zs. Pollák

Institute of Experimental Surgery, Institute of Anesthesiology and Intensive Therapy, University Medical School, Pécs, Hungary

Summary

Studies were undertaken using a synthetic free radical scavenger (MTDQ-DA) on regional ischaemic dog hearts; it was found that the rate of malignant ventricular arrhythmias and fibrillation after coronary ligature unexpectedly decreased. According to experiments on 22 dogs, the intravenous MTDQ-DA therapy decreases the unfavourable ECG consequences of left anterior descending branch ligature: already 5 to 10 minutes after drug administration the ST segment elevation, the QT interval lengthening and the occurrence of ventricular extrasystoles and salvos are diminishing. The so-called epicardial ST map ameliorates rapidly. MTDQ-DA as a blocking agent of free radicals is able to prevent the irritative stimuli around and in the border zone of an infarct, has a vigorous anti-arrhythmogenic effect and greatly reduces the electric heterogeneity. These unexpected results may lead to a promising therapy for the acute heart infarction.

Introduction

Sudden occlusion of a major coronary artery evokes a chain of pathological events [12, 13, 16, 27, 28]. The injury may produce more and more progressive deterioration of mechanical and electrical activity of the heart. Should the process turn into a haemodynamic "vicious circle" [3], the development of acute cardiac failure is almost unavoidable. Moreover, the increased electrical instability predisposes to the frequently fatal complication of ventricular fibrillation [2, 6, 8, 10, 19, 22, 23, 25].

Lipid peroxidation is known to be involved in many biological damage processes [5, 7, 9, 14, 17, 21, 26). As has been proposed earlier, not only does lipid peroxidation cause destruction of myocardial membranes in ischaemic state but it also gives rise to chemical products that can propagate the extension of the damage [15, 18, 20]. Since antioxidant therapy has been proved to protect the membrane structure against peroxidative deterioration [1, 14, 24], there was an obvious need to investigate this effect on the electrical susceptibility.

Experiments were undertaken to display the most characteristic ECG events during acute regional myocardial ischaemia, combined with treatment by exogenous scavenger of dihydroquinoline type (MTDQ-DA) and to analyse the effects of the applied antioxidant therapy.

Material and methods

The experiments were performed on 22 mongrel dogs, each weighing 12 to 15 kg.

Premedication and narcosis

After intramuscular injection of Droperidol (0.7 mg/kg body wt.) plus Fentanyl (0.015 mg/kg body wt.) superficial narcosis was carried out by intravenous sodium-hexobarbital. After endotracheal intubation, respiration was maintained with oxygen-nitrous oxide mixture at a ratio 1:4.

Surgical procedures and epicardial ECG mapping

Following left thoracotomy the pericardium was opened and the left anterior descending coronary branch (LAD) exposed for definitive ligature.

Epicardial ECG mapping was carried out with 24 obtuse unipolar Ag/AgCl electrodes and recorded from points of a 4×6 matrix [25]. ST elevations in mV, the sum of ST elevation (ΣST), QT time periods and disturbances of rhythmicity (ventricular extrasystoles and salvos) examined and calculated by means of an R-40 digital computer. To construct an isopotential map, the ST segment levels were also computed. Also, the conjectural markers of the "electrical instability" were measured and computed. Experimental steps were:
- recording of normal electrograms
- ligation of LAD
- recording of ischaemic effect on ECG
- rapid intravenous injection of the synthesis scavenger (MTDQ-DA = 6.6′-methylene-bis 2,2-dimethyl-4-methane sulphonic acid sodium-1,2 dihydroquinoline; 150 mg/kg) followed by a slow drop infusion of the same quantity throughout the experiments
- recording of scavenger effects on ECG.

Student's t-test was used for estimation. All statistical results (mean ± SEM) were considered significant if $p < 0.05$.

Results

Many known consequences were seen after coronary ligation in the experiments (cyanotic decoloration, contractility defect and distension of infarct areas, decrease of blood and perfusion pressure, etc.). Nevertheless, it was not the aforementioned alterations, but the electrical instability that represented the main danger, due to the severe circulatory failure. Within minutes, severe ST segment deviation, QT prolongation, negative T wave, ventricular flutter, polytopical extrasystolia and ventricular fibrillation appeared. The acutely registered pathophysiological changes were quite stable after experimental coronary occlusion in dogs.

Figure 1a shows the time course of changes of ST elevation in 2 dogs. As seen, the intravenous MTDQ-DA therapy causes decreases of ST elevations in both big and small infarct cases as early as 5 to 10 min after the beginning the drug administration.

In Fig. 1b the typical lengthening and distribution of the QT interval are shwon after acute coronary ligature and MTDQ-DA therapy. The applied drug begins to be efficacious 5 to 10 min after the intravenous administration and this effect remains durable.

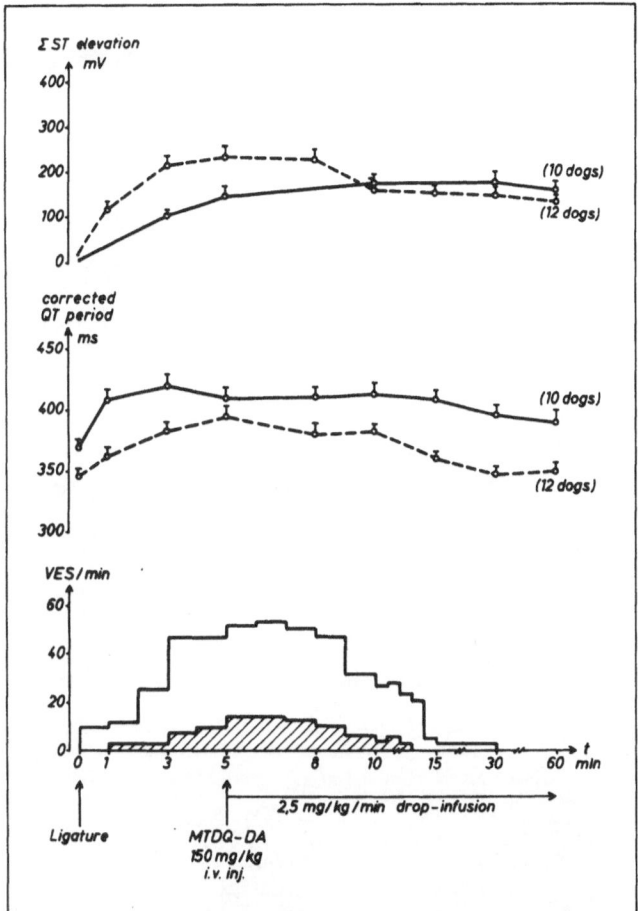

Fig. 1 a. Permanent ST elevations (mean ± SEM) in dogs undergoing coronary occlusion below the first major oblique branch. Solid line, summarized data from 10 control experiments, dashed line, 12 experiments with 4 cases of main stem occlusion. **b** Corrected QT intervals $\dfrac{\text{(QT measured)}}{\sqrt{\text{cycle time}}}$. QT prolongation is diminishing after intravenous MTDQ-DA therapy (dashed line) in comparison to the non-treated animals (solid line). **c** Distribution of ventricular extrasystoles (VES) after coronary occlusion + MTDQ-DA therapy, showing the number of ventricular extrasystoles/min (blank columns) and salvos/min (hatched columns).

Figure 1 c shows a typical distribution of ventricular extrasystoles after coronary occlusion combined with MTDQ-DA therapy. Electrical irritations arise at the marginal zones and within minutes tachycardia and cumulated polytopical extrasystolia appear. This electrical instability is permanent in the acute phase. In the figure, both the sum of ventricular extrasystoles and the salvos from one experiment after MTDQ-DA administration can be seen. The antioxidant therapy decreases or terminates the occurrence of ventricular extrasystoles 5 to 15 min after intravenous injection.

Figure 2 represents a series of epicardial mappings from one experiment. The coronary ligature causes rapid potential differences and subsequent current flow arising at the

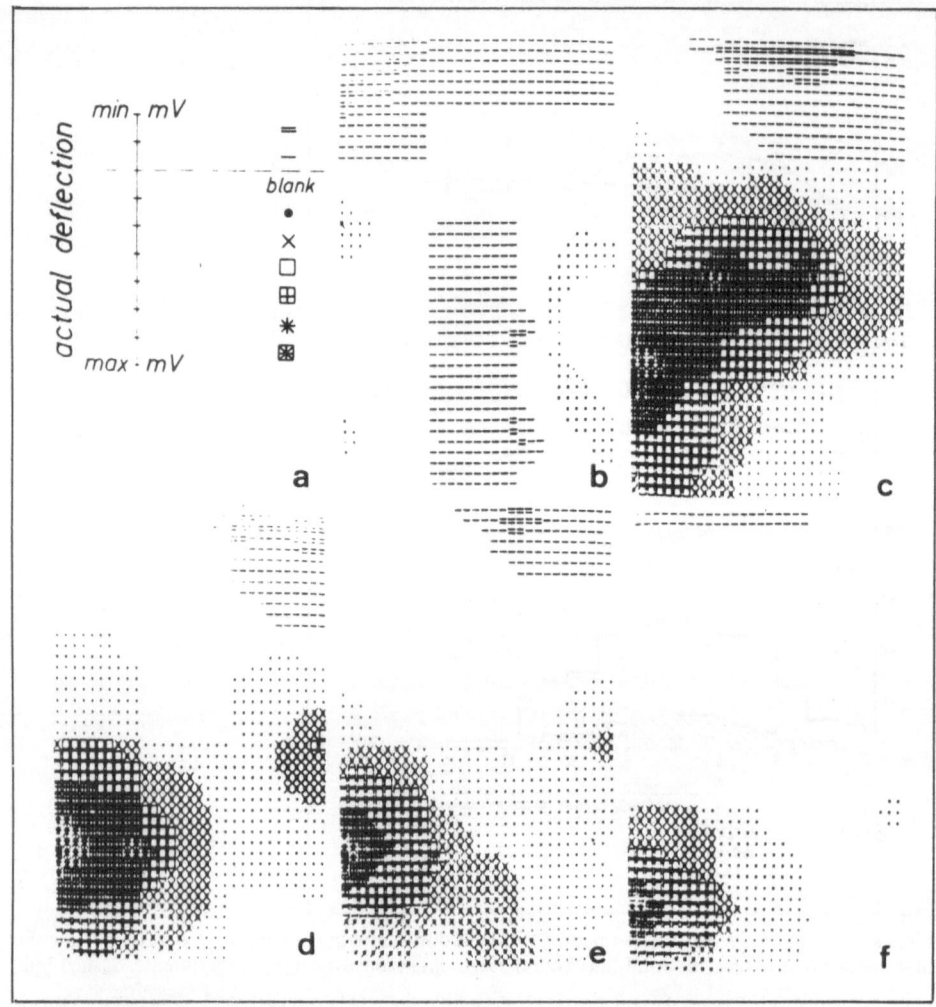

Fig. 2. Series of epicardial ST mapping from one experiment after coronary ligature. a. The symbols represent the different ST values actually measured in mVs. The whole scale of the ST shift has been automatically ranged by the computer into 2 negative and 7 positive ranges. b. Control ST map from a dog heart with open thorax and pericardium. c. ST map 10 min after coronary ligature. d, e, f. ST maps 5, 10 and 20 min, respectively, after intravenous antioxidant therapy.

boundary between the normal and ischaemic tissues. Since the different ST values, measured in mVs, are automatically scaled by the computer into 2 negative and 7 positive ranges, an exciting epicardial map is outlined by the different symbols. As seen, the MTDQ-DA therapy rapidly ameliorates the epicardial ST map. The disappearance or decrease of ST segment shift indicates the remission of the pathological current flow and so the reduction of the electrical instability.

The outlined data express the trend typical for the entire experiments.

Discussion

Acute regional ischaemia of the heart may lead to fatal malignant ventricular arrhythmias and fibrillation. Investigations using morphological methods, combined with electrophysiological studies, have proved that the infarct pattern may influence the myocardial susceptibility [2, 8]. According to numerous studies [4, 11, 29] it can be suggested that the heterogeneous ischaemia or hypoxia predisposes to malignant arrhythmias. Such derangement is especially present in the border-zone between the normal and ischaemic myocardium; namely, the border zone contains structures with different electrophysiological properties [10, 11, 19, 25]. It seems obvious that this inhomogeneity plays an important role in the increased electrical instability following heterogeneous blood, oxygen and substrate supply.

The consecutive loss of cell membrane function in the regional ischaemic myocardium might be characterized by rapid ST elevation, QT prolongation, increase of ventricular excitability, conduction disturbances and tachycardia [2, 6, 8, 10, 19, 22, 23]. Electrical instability occurs also in hearts where differently damaged areas exist side by side. The potential gradient existing at the boundary of the ischaemic and non-ischaemic segments is determined by the recording electrode with respect to this boundary. By this means, sharp electrical differences can be observed in regional ischaemic hearts which indicates a spatial dissociation in the development of ischaemia-induced disorders.

Since homogeneously (global) ischaemic hearts display a relative electric stability, it means that there are no intact areas from which current can flow into the ischaemic areas to causes malignant arrhythmias. The acute heart infarct is, however, a very dangerous state because of its heterogeneous pattern. The increase of the perimeter of infarct area (not only on the epicardial surface but also in the deep, saw-tooth-like interlocking layers) represents a raised extension of border zone between normal and ischaemic tissues. The heterogeneity also consists of more differences, such as (i) of the altered blood supply evoking ATP and pO_2 inequality, (ii) of the mechanical activity of the myocardial fibres, especially in the border zone having a disorder of contracted and relaxed fibres and (iii) of changes of the myofibrillar membrane structures and Purkinje fibres (they have extensive sarcoplasmic membrane structures) with functional aftermath, together creating the basis of the outlined electrical instability.

Since in consequence of free radical action, an increased rate of lipid peroxidation can be found in the ischaemic state (and this is an essential contribution to the membrane disintegration) it was evidently important to attempt to block this damaging process in case of acute heart infarct.

Our intention was to eliminate the dangerous free radicals, especially at the border zone, saving the cell and membrane structures and by this means stopping or inhibiting the genesis of pathological current flow, i.e. making the border zone and the infarct a latent or silent area.

According to our results, the applied artificial scavenger antioxidant compound (MTDQ-DA) has been able to scavenge the free oxygen radical action and reverse membrane permeability changes, above all at the border-zone. The rapid intravenous injection of MTDQ-DA caused an essential protective action in experimentally induced myocardial infarction. The extrasystolia has, in almost all cases, perfectly disappeared and ventricular fibrillation did not develop. The characteristic sum of ST elevations (ΣST) and the prolongation of QT intervals registered by the epicardial electrode matrix after acute coronary occlusion have rapidly and significantly decreased. The ST isopotential map showed far less ST segment shift than the heart with focal ischaemia without treatment.

It may also be assumed that MTDQ-DA blocks primarily the irritative stimuli at the marginal zone and thus considerably eliminates the acute fatal complications. The exact mechanism is not yet known. Presumably, the compound protects the maintenance of cell membrane structures, terminates the free radical action and by this means decreases the degree of "heterogeneity" around and in the infarct area and/or "numbs" the infarction.

Our additional model experiments in vitro with EPR [1] have proved that MTDQ-DA as a non-specific membrane perturbant induces changes in the lipid region of the membranes which are transferred to the protein by influencing the rotational modes of motion and may participate in the biological oxidation-reduction processes.

By this means, the irritative impulses arising from the biomembranes (muscle and Purkinje fibres) during weakened free radical action cause only diminished ECG alterations. The MTDQ-DA seems to be not only a potent scavenger but also a vigorous anti-arrhythmogenous compound.

References

1. Belágyi J, Török B MTDQ-DA as a free radical scavenger and its effects on biological membranes. Physiol Chem Phys (to be published)
2. Boineau JP, Cox JL (1973) Slow ventricular activation in acute myocardial infarction. Circulation 48:702–713
3. Braunwald E (1976) Protection of the ischemic myocardium. Circulation [Suppl 1] 53:1–2
4. Bukauskas F (1982) Electrophysiology of the normal-to-hypoxic transition zone. Circ Res 51:321–329
5. Burton KM, McCord JM, Ghai G (1984) Myocardial alterations due to free-radical generation. Am J Physiol (Heart Circ Physiol 15) 246:H776–H783
6. Doroghazi RM, Childers R (1978) Time-related changes in the Q-T interval in acute myocardial infarction: Possible relation to local hypocalcemia. Am J Cardiol 41:684–688
7. Fantone JC, Ward PA (1982) Role of oxygen derived free radicals and metabolites in leukocyte-dependent inflammatory reactions. Am J Pathol 107:395–418
8. Grosso MA, Simson MB, Kobayashy K, Wiliamson JR, Harken AH (1983) Myocardial ischemic pattern determines predisposition to ventricular arrhythmias. Surg Forum 34:239–241
9. Hess ML, Manson NH (1984) Molecular oxygen: friend and foe. The role of the oxygen free radical system in the calcium paradox, the oxygen paradox and ischemia reperfusion injury.J Mol Cell Cardiol 16:969–985
10. Holland RP, Brooks H (1977) TQ-ST segment mapping: critical review and analysis of current concepts. Am J Cardiol 40:110–119
11. Janse MJ, Cinca J, Morena H, Fioled JWT, Kléber AG, DeVries GP, Becker AE, Durrer D (1979) The "border-zone" in myocardial ischemia. Circ Res 44:576–588
12. Jennings RB, Reimer KA (1981) Lethal myocardial ischemic injury. Am J Pathol 102:241–255
13. Katz AM, Messineo FC (1981) Lipid membrane interactions and the pathogenesis of ischemic damage in the myocardium. Circ Res 48:1–16
14. Meerson FZ, Kagan VE, Kozlov YuP, Belkina LM, Arkhipenko YuV (1982) The role of lipid peroxidation in pathogenesis of ischemic damage and the antioxidant protection of the heart. Basic Res Cardiol 77:465–485
15. Mullane KM, Moncada S (1982) The salvage of ischemic myocardium by MW755C in anesthetized dogs. Prostaglandins 24:255–266
16. Neely JR, Feuvray D (1981) Metabolic products and myocardial ischemia. Am J Pathol 102:282–291
17. Rao PS, Cohen MW, Mueller HS (1983) Production of free radicals and lipid peroxides in early experimental myocardial ischemia. J Mol Cell Cardiol 15:713–716

18. Romson JL, Hook BG, Kunkel SL, Abrans GD, Schork A, Lucchesi BR (1983) Reduction of the extent of ischemic myocardial injury by neutrophyl depletion in the dog. Circulation 67:1016–1023
19. Ross J (1979) Tissue salvage in acute myocardial infarction. Am J Surg 138:392–397
20. Rowe GT, Monson NH, Caplan M, Hess ML (1983) Hydrogen peroxide and hydroxyl radical mediation of activated leukocyte depression of cardiac sarcoplasmic reticulum. Circ Res 53:584–591
21. Röth E, Török B, Zsoldos T, Matkovics B (1985) Lipid peroxidation and scavenger mechanism in experimentally induced heart infarcts. Basic Res Cardiol 80:530–536
22. Schwartz PJ, Wolf S (1978) QT interval prolongation was predictor of sudden death in patients with myocardial infarction. Circulation 57:1074–1077
23. Taylor GI, Crampton RS, Gibson RS, Stebbins PT, Waldman MTG, Galler GA (1981) Prolonged QT interval at onset of acute myocardial infarction in predicting early phase ventricular tachycardias. Am Heart J 102:16–24
24. Török B, Röth E, Bär V, Pollák Zs (1986) Effects of antioxidant therapy in experimentally induced heart infarcts. Basic Res Cardiol 81:167–180
25. Török B, Röth E, Mezey B, Szabados S, Simor T (1983) Epicardial ECG signals following global myocardial ischemia. Basic Res Cardiol 78:593–600
26. Török B, Röth E, Tigyi A, Zsoldos T, Matkovics B, Szabó L (1984) Membrane perturbations in myocardium: oxygen radicals mediate injuries in experiments. Acta Circ Hung 25:185–192
27. Török B, Röth E, Trombitás K (1982) Ultrastructural changes of the subendocardium in ischemic and cardioplegic states before and after reperfusion. Eur Surg Res 14:17–26
28. Török B, Trombitás K, Röth E (1983) Ultrastructural consequences of reperfusion of the ischemic myocardium. Acta Morph Hung 31:315–326
29. Wong SS, Bossett AL, Cameron JS, Epstein K, Kozlovskis P, Myerburg RJ (1982) Dissimilarities in the electrophysiologic abnormalities of lateral border and central zone cells after healing of myocardial infarction in cats. Circ Res 51:486–493

Authors' address:

Prof. Dr. B. Török, Institute of Experimental Surgery, University Medical School, H-7643 Pécs, Hungary

IV. Cardiac energetics in human heart: Clinical implications

Atrial and ventricular isomyosin composition in patients with different forms of cardiac hypertrophy

M. C. Schaub and H. O. Hirzel

Dept. of Pharmacology, University of Zürich and Dept. of Medicine, University Hospital, Zürich, Switzerland

Summary

In man, various forms of compensatory and idiopathic hypertrophic states can be differentiated by haemodynamic and angiographic parameters. They are morphologically indistinguishable with regard to muscle fibre diameter and non-muscle tissue content. They are, however, accompanied by contractile dysfunction of various degrees or even by hypercontractility. In hearts subjected to chronic increase in workload the peptide pattern of the slow ventricular myosin heavy chain (HC) type VM-3 does not change, while that of the fast atrial type HC does. In atria also the ventricular type of myosin light chain-2 (VLC-2) is occurring. In certain forms of hypertrophy we found the atrial type ALC-1 occurring in the ventricular tissue, in individual cases amounting to 30% of total LC-1, on average, 12% in dilated cardiomyopathy, 6% in pressure and 3% in volume overload and 2% in cases with reduced myocardial mass due to infarction. No such increase of ALC-1 was found in hypertrophic cardiomyopathy or in coronary heart disease without infarction. The isoform expression of myosin HC and LC is thus governed independently of one another in response to altered physiological or pathological conditions. A significant correlation of the ALC-1 content in ventricles could be established with the peak circumferential wall stress. This may imply the involvement of the LC-1 in the contractile properties of the myofibrils.

Introduction

Calcium regulates heart muscle contraction by binding reversibly to the troponin-C subunit of the actin filament. The mechano-chemical properties of heart myofibrils can be further modulated by phosphorylation of the troponin-I subunit and the myosin regulatory light chain-2 (see ref. 18, for review). In addition, myosin consisting of 2 heavy chains (HC) and 4 light chains (LC), exists in various isoform species with regard to HC as well as to LC, exhibiting different enzyme properties. These have been postulated to explain, at least in part, the altered contractile properties in disease states. A shift in the isoenzyme composition may result from adaptation to altered pyhsiological or pathological mechanical requirements, or else, it could be the primary cause for altered contractile functions. In heart 3 isomyosins are found and distinguished by their HC composition, VM-1 (alpha-alpha-homodimer), VM-2 (alpha-beta-heterodimer) and VM-3 (beta-beta-homodimer) [18]. VM-1 has the highest and VM-3 the lowest ATPase activity, VM-2 being intermediate. In healthy adult man the ventricle contains predominantly VM-3 and the atrium the VM-1 type [7, 13]. The atrial LC, ALC-1 and ALC-2, display a 6–8% higher apparent molecular weight in sodium dodecyl sulphate electrophoresis, than the ventricular species VLC-1 and VLC-2 [23]. In ventricles of rodents, shifts in isomyosin composition correlate well with changes in contractile and enzymic properties in their dependence on altered mechanical demands and on the hormonal status [18]. In man, however, the situation is less clear. Shifts in myosin HC and LC isoforms have been observed in atria of hearts under increased workload [5, 19, 20, 33]. In the ventricles we have described a correlation between the shift in LC-1 isoform expression and the degree of mechanical wall stress in different forms of hypertrophy [8].

Methods

Myocardial biopsy specimens came from patients admitted for cardiac evaluation, except for 4 out of 7 controls who died from traffic accidents and whose biopsy material was obtained within 8 hours post mortem. The controls, i.e. 3 patients with uncomplicated atrial septal defect and the post mortem cases, had no cardiac history or any signs of hypertrophy or valvular pathology. None of the patients were in clinically apparent heart failure. All patients underwent right and left heart catheterization, biplane left ventricular angiography and selective coronary arteriography using standard techniques [8]. Left ventricular high-fidelity pressure measurements were obtained by means of a Millar 7F-micromanometer angio-catheter, advanced into the left ventricle through a trans-septally introduced 11.5 F Brockenbrough guiding catheter. This enabled us to calculate various functional indices. Left ventricular tissue samples were taken at postmortem examination (4 cases), during catheterization (36 cases) or at surgery (12 cases).

Muscle fibre diameter and non-muscle tissue content were determined by morphometric methods [8]. The histochemical ATPase reaction was performed on fresh tissue sections in the presence of $MgCl_2$ at pH 10.5 [12]. The myosin LC complement was assessed from two-dimensional electrophoretic resolution of total tissue homogenates by quantitative densitometry after staining the patterns [8, 19]. The stoichiometric relationship of LC to intact myosin was arrived at via tropomyosin which proved to be the most reliable reference protein and which stays in a constant molar ratio to actin and myosin. The myosin HC were isolated by an electrophoretic procedure from tissue homogenates and digested subsequently by different proteinases (papain and Staphylococcus aureus V8 proteinase) under specified conditions. The resulting peptide mixtures were resolved by one-dimensional electrophoresis in sodium dodecyl sulphate, stained and assessed by densitometry again [8]. One-way analysis of variance, modified t-tests, the Bonferroni method, and multiple regression analysis were employed for statistical evaluation of correlations between different parameters [30].

Results

Patient population

According to the clinical and haemodynamic characteristics, the patients (37 men and 15 women) were grouped into 7 controls, 4 with coronary heart disease (CHD) without infarction and 5 with previous infarction (1–5 years before the investigation), 4 with hypertrophic cardiomyopathy (HCM), 5 with dilated cardiomyopathy (DCM), 11 with volume overload (predominant aortic insufficiency) and 16 with pressure overload (predominant aortic stenosis). Controls, CHD and HCM feature normal standard dynamics except for some elevation of the left ventricular end-diastolic pressure (20 mm Hg) and an increase in left ventricular ejection fraction (>68%) in HCM. A large end-diastolic volume index of 174 ml/m^2 and highly depressed ejection fraction (35%) are characteristic for DCM.

Table 1. Indices of contractile function and left ventricular load

Disease group	Mean normalized systolic ejection rate (vol/s)	Peak velocity of contractile element shortening (length/s)	Maximal rate of pressure rise dP/dt (mm Hg/s)	Circumferential wall stress (dynes $\times 10^3$/cm²)	
				Peak	End-diastolic
Hypertrophic cardio-myopathy (HCM)	2.3[a]	1.5[a]	1946[a]	322[b]	51
Dilated cardio-myopathy (DCM)	1.5	0.9	1111	465	56
Volume overload	1.8	1.1	1475	448	69
Pressure overload	1.9	1.4	2176[a]	463	48

Significance: $p < 0.05$ [a] compared to DCM and [b] compared to all other groups.

Cases with volume overload comprise aortic insufficiencies with a highly increased end-diastolic volume index of 212 ml/m² as a consequence of a large aortic regurgitant fraction (60%). Cases with predominant aortic stenosis (pressure overload) present an increased left ventricular peak pressure of 206 mm Hg together with a pressure gradient between left ventricle and aorta of 74 mm Hg and a corresponding diminished aortic valve area of 0.8 cm² (normal range, 2–3 cm²). The age range of the patients is given in Table 2.

Mechanical and morphological parameters

The data summarized in Table 1 indicate increased contractile functions together with a relatively low peak circumferential wall stress to be characteristic for HCM. In contrast, in DCM the contractile functions are impaired while the muscle fibres have to bear a high load. This latter is equally high in volume as well as pressure overload concomitant with intermediate contractile functions. The maximal rate of pressure rise is highest in pressure overload. Despite these differences in contractile function and load the tissue has to bear, the average myofibrillar diameter stays, in all disease groups, between 28 and 34 μm i.e. 190–230% of normal. The non-muscle tissue content ranges in all disease groups between 20 and 27%, thus amounting to 5–7 times higher than normal.

Myosin heavy chain isoforms

Figure 1 shows comparisons between the histochemical ATPase reaction in the atrium and ventricle in man and rat. In man, the ventricular tissue exhibits a uniformly lower ATPase reaction than the atrium, while in the rat, both tissues stain the same. Correspondingly, the peptide patterns of isolated myosin HC from atrium and from ventricle differ in man while they are identical in the rat (Fig. 2). Reproducible, identical peptide patterns obtained by different proteinases suggests the HC to be very similar, if not identical, in the primary structure, provided the digestion experiments have been done under stringent criteria [8]. In the adult rat the predominant HC species, both in ventricle and atrium, has been reported to be of the fast myosin type VM-1 (viz. alpha-alpha-homodimer) [4, 10, 24]. The peptide pattern is thus able to differentiate between different isoforms of myosin HC having different ATPase activities.

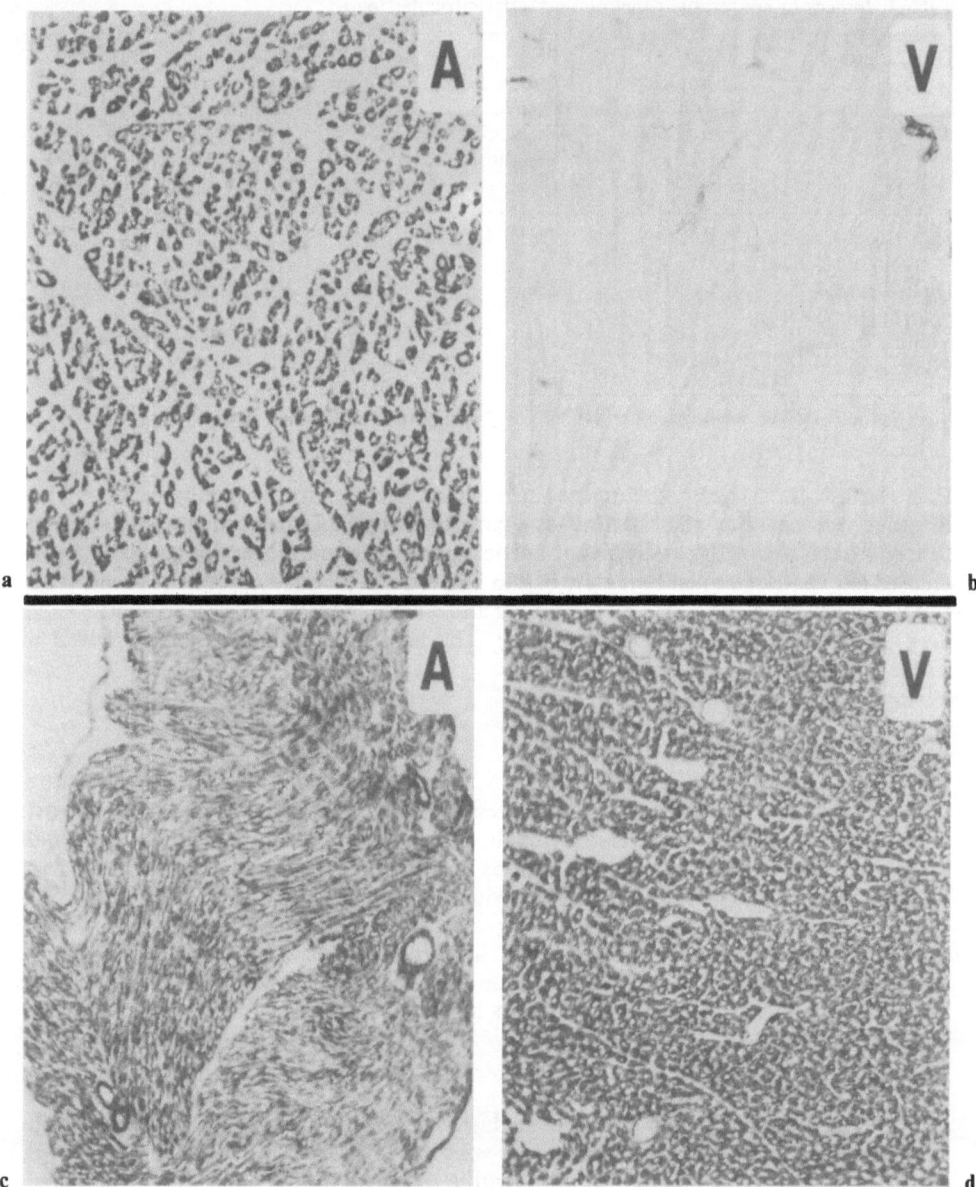

Fig. 1. Histochemical ATPase reaction in the presence of magnesium and at pH 10.5 of ventricle and atrium in man (a and b) and rat (c and d). V, ventricular tissue sections; A, atrial tissue sections. The height of each section-cut corresponds to 900 µm.

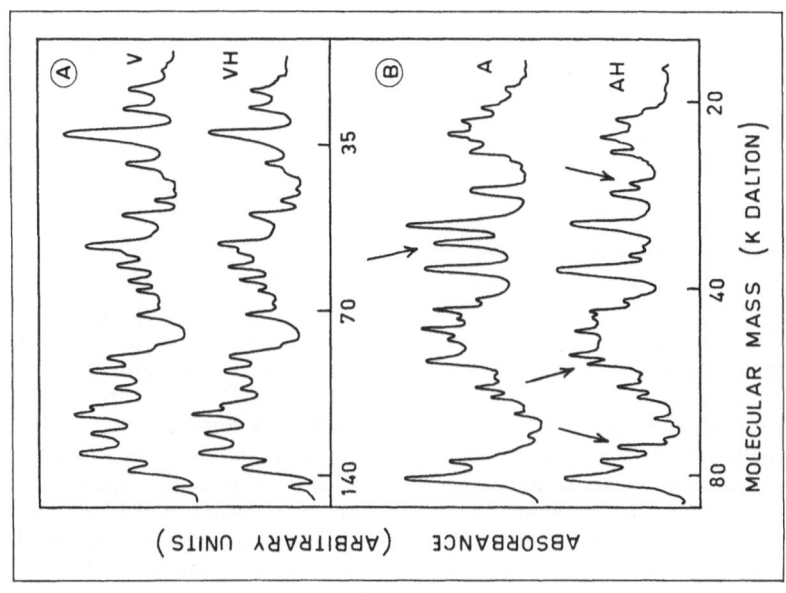

Fig. 3. Densitometric traces of electrophoretograms with peptide patterns of isolated myosin heavy chains from normal and hypertrophic human left ventricle (A) and left atrium (B) after limited digestion with papain. V and A, ventricle and atrium of healthy control person. VH and AH, ventricle and atrium of patient with pressure overload. Arrows in panel (B) indicate peptides unique to one pattern only. Coomassie staining.

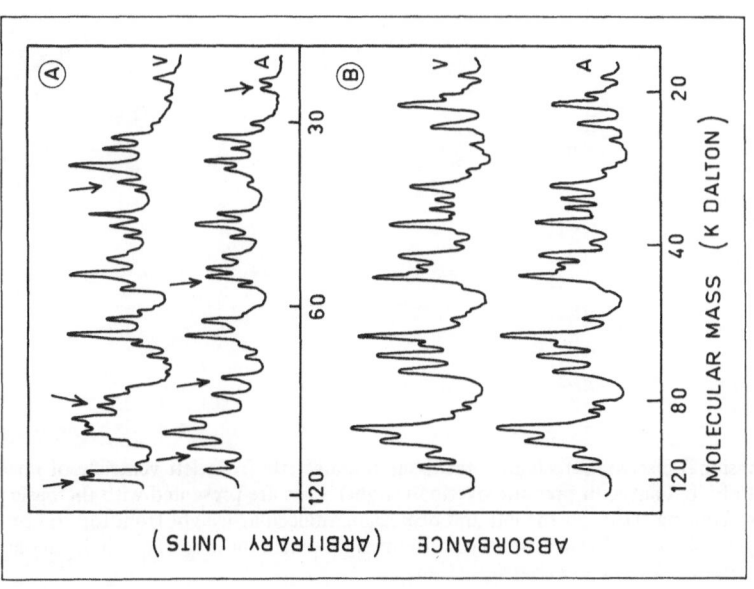

Fig. 2. Densitometric traces of electrophoretograms with peptide patterns of isolated myosin heavy chains in man (A) and rat (B) after limited digestion with papain. V, ventricle; A, atrium. Arrows in panel (A) indicate peptides unique to one pattern only. Coomassie staining.

In a number of patients with volume or pressure overload the myosin HC were isolated and digested, as well as HC from normal control ventricles. The peptide patterns were in all cases identical (Fig. 3 A). In atria of hearts working against an increased workload, however, differences occurred in the HC peptide patterns when compared to those from healthy atria (Fig. 3 B). Such qualitative differences in either new peptides occurring or missing in the pathological digestion patterns (not merely a variation in staining intensity of peptides present in both control and disease samples) indicate a shift in the expression of myosin HC isoforms in hypertrophic atria.

Myosin light chain isoforms

The most significant result in two-dimensional electrophoresis of total left ventricular tissue homogenates was the occurrence of an additional peptide above the VLC-1 in certain forms of hypertrophy (Fig. 4). In molecular weight and electrophoretic mobility it is indistinguishable from the atrial ALC-1 [8]. Densitometric evaluation of this newly occurring ALC-1 in hypertrophic left ventricles is summarized in Table 2. It was absent in controls, or only a faint spot was seen, never exceeding 0.8% of total LC-1. If, in single cases, the new ALC-1 amounted to 30% of total LC-1, it seemed to become expressed at the expense of the genuine ventricular VLC-1, since the total LC-1 content neither changed in single cases nor, on average, in the disease groups. The ALC-1 content in hypertrophic ventricles does not correlate with any haemodynamic, morphometric or mechanical parameter except for the peak circumferential wall stress [8]. Despite a considerable individual scatter, the log ALC-1 content correlated with the wall stress, irrespective of the type of hypertrophy, at a significance level better than 5%. However, comparison of the mean values of log ALC-1 content of the various hypertrophic disease groups, with the corresponding mean values for peak wall stress, yields a correlation with an r-coefficient higher than 0.96 (Fig. 5). The atrial ALC-2 type isoform was never observed in hypertrophied

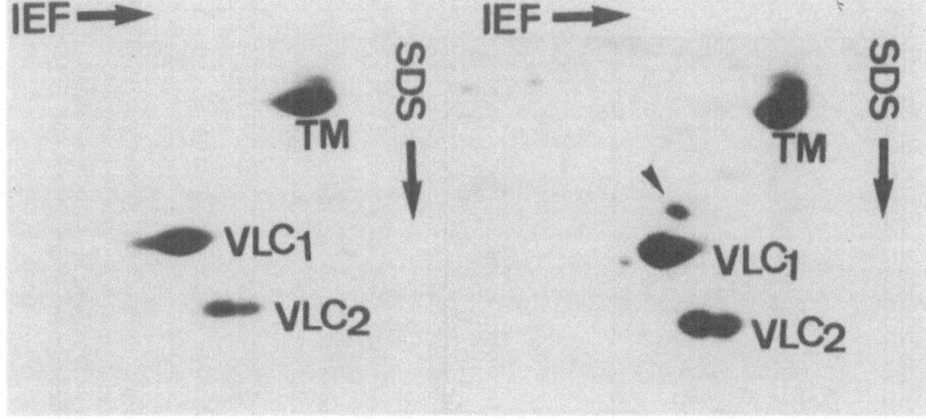

Fig. 4. Two-dimensional electrophoresis of total tissue homogenate from left ventricle of normal control (left) and of a patient with pressure overload (right). Gels are presented with the basic pH range in isoelectrofocusing (IEF) on the left and decreasing molecular weight from top to bottom in sodium dodecyl sulphate (SDS) electrophoresis. In the right panel, the arrowhead indicates atrial type myosin light chain (ALC-1). Coomassie staining.

Table 2. Total myosin light chain-1 content and fraction of ALC-1 in ventricle

Group	n	Age (years)	Total LC-1 (mol/mol myosin)	ALC-1 (% of total LC-1)	Significance with $p < 0.05$ versus
Controls	7	23–66	2.7	0.4	VO, PO, DCM
Coronary heart disease without infarction (CHD−)	4	44–68	2.6	0.6	VO, PO, DCM
Coronary heart disease with prior infarction (CHD+)	5	48–63	2.7	1.9	DCM
Hypertrophic cardio-myopathy (HCM)	4	34–48	2.5	0.3	VO, PO, DCM, CHD−
Volume overload (VO)	11	32–72	2.5	2.9	Controls, HCM, DCM, CHD−
Pressure overload (PO)	16	35–79	3.1	6.4	Controls, HCM, DCM
Dilated cardiomyopathy (DCM)	5	42–59	2.7	12.1	All other groups

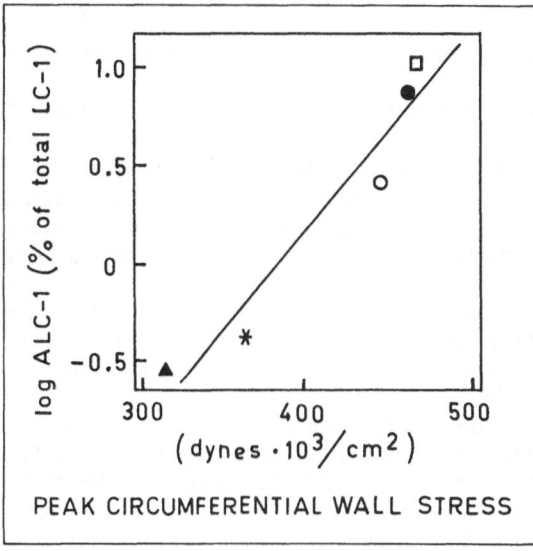

Fig. 5. Relationship between peak circumferential wall stress and log tissue content of atrial type myosin light chain (ALC-1) in ventricle, expressed as a percentage of total LC-1, showing mean values of the disease groups and controls. The peak circumferential wall stress for controls was on average 365 dynes × 10^3/cm². □ = dilated cardiomyopathy; ● = pressure overload; ○ = volume overload; ▲ hypertrophic cardiomyopathy; * = controls.

ventricles. In accordance with Cummins [5] the ventricular VLC-2 and to a lesser extent also VLC-1 was consistently observed to occur in atrial tissue in dependence on the increased workload.

Discussion

The present study indicates, in agreement with Schier and Adelstein [21], that in human hypertrophic ventricles the myosin HC isoform is not changing. The peptide technique would permit the detection of any single additional peptide originating from another HC isoform amounting to, or exceeding, 5% of total HC [8]. However, the observed peptide patterns from ventricular myosin could represent a mixture of HC isoforms. Since no quantitative changes were seen between controls and pathologically altered tissues, one has to postulate that such a putative mixture of isoforms remains unchanged. Immunological studies have demonstrated the co-existence of 1–16% of VM-1 myosin with VM-3 in human ventricles. This proportion does not, however, change in hypertrophy [7, 13]. This situation differs from that in the rat, where the ventricular VM-1 myosin changes toward the VM-3 form with lower ATPase activity under increased workload [3, 11, 14].

In contrast, in atria, the peptide technique enables us to identify a new isoform of myosin HC that becomes expressed under increased workload, or else, if the peptide pattern in controls represents a mixture of HC isoforms again, this mixture is fairly constant between individual controls (producing identical peptide patterns), and changes drastically, as indicated by the new peptides occurring in the pathological cases. The human atrium has thus preserved the ability to shift its myosin HC isoform expression in response to increased workload [19]. Immunohistochemical studies indicate that this shift involves the transition from atrial myosin with high ATPase activity (alpha-alpha-homodimer, viz. VM-1) towards the VM-3 type with low ATPase [7, 20, 33]. In small mammals (rodents) such a shift in isomyosin expression has been interpreted as an adaptational response with a myosin isoform of low ATPase activity and a concomitant lower speed of contraction [6, 22], an improved economy of force generation [1, 2] and decreased oxygen consumption [9].

The active site of myosin is comprised entirely in its HC, and the LC associated with it have no influence on its ATPase activity, as assessed in vitro [26, 28, 29], even when activated by actin [13, 21]. Nevertheless, the myofibrillar ATPase activity in hypertrophic human ventricles has been repeatedly reported to be lower than normal ([25], for review). If the contractile machine of the sarcomere were responsible for the diminished contractile function in certain forms of hypertrophy, some change in its protein composition other than the myosin HC has to be postulated then. The change in LC-1 isoform composition observed in dependence on the load the tissue has to bear, could be involved in explaining the altered contractility of intact myofibrils. The switch in the expression from VLC-1 to ALC-1, with the total LC-1 content remaining constant in hypertrophic ventricles, implies that at least in part, the VM-3 type myosin contains ALC-1.

A recent observation reports on a myosin fraction from the papillary muscles of diseased human ventricles with and without increased workload, which can be separated by prolonged runs (43 hours) in a non-denaturing electrophoretic system from the VM-3 species [25]. Its content rises from 25% to 35% of total ventricular myosin and is accompanied by a slight decrease in myofibrillar ATPase activity. No myosin of healthy controls is given for comparison. Nevertheless, if it were a newly occurring pathological HC isoform, or else, if it were always present in variable amounts, it should not escape detection in peptide patterns of cases with severe compensatory hypertrophy when compared to controls.

In skeletal muscle, the so-called alkali-LC (which correspond to the cardiac type-1 LC) have been implicated in the binding of the myosin crossbridges to actin [17, 32] and in affecting the actin-activated ATPase activity under specified conditions [31]. Furthermore, in aneural cultures of chicken embryo skeletal muscle, the alkali-LC have been shown to replace the embryonal LC type in dependence on the contractile activity without corresponding isoform transition in the myosin HC [15]. The ALC-1 isoform is also present in human fetal ventricles and disappears within the first few months after birth [5, 16]. In cases with congenital malformations associated with pressure overload, the fetal ALC-1 isoform persists, amounting to one third of total ventricular LC-1 [27] and the developmental transition to the tissue-specific VLC-1 is retarded. In developmental skeletal muscle, as well as in pathologically altered myocardium, the LC isoform composition is thus regulated, independently of the HC expression. The two situations have in common that the isoform composition of the alkali-type LC in skeletal muscle and that of the homologue LC-1 type in the heart ventricle is correlated directly with a mechanical parameter, increased contractile activity and increased wall stress, respectively.

The connection with a mechanical parameter renders the LC-1 complement of myosin a characteristic biochemical parameter for certain forms of ventricular hypertrophy which can be differentiated neither by the size of the myofibrillar diameter nor by the non-muscle tissue content. Due to this relation it probably represents a measure for the compensatory type of hypertrophy. It would then classify the state of DCM as a predominantly compensatory form of hypertrophy.

In conclusion, it seems that the atrium responds to increased workload by shifting its myosin HC and LC isoform complement toward the slow ventricular VM-3 type. The ventricle which already possesses VM-3 cannot follow the same strategy, but instead, is only able to shift the expression of its VLC-1 toward ALC-1 which is known to be associated with the state of contractility.

Acknowledgements. This work was supported by grants from the Swiss National Science Foundation (3.505.83 and 3.801.84).

References

1. Alpert NR, Mulieri LA (1981) Heart mechanics and myosin ATPase in normal and hypertrophied heart muscle. Fed Proc 41:192–198
2. Alpert NR, Mulieri LA (1982) Increased myothermal economy of isometric force generation in compensated cardiac hypertrophy induced by pulmonary artery constriction in the rabbit. Circ Res 50:491–500
3. Alpert NR, Mulieri LA, Litten RZ (1979) Functional significance of altered myosin adenosine triphosphates activity in enlarged hearts. Am J Cardiol 44:947–953
4. Clark WA, Chizzonite RA, Everett AW, Rabinowitz M, Zak R (1982) Species correlations between cardiac isomyosins. A comparison of electrophoretic and immunological properties. J Biol Chem 257:5449–5454
5. Cummins P (1982) Transitions in human atrial and ventricular myosin light chain isoenzymes in response to cardiac pressure overload induced hypertrophy. Biochem J 205:195–204
6. Ebrecht GH, Rupp R, Jacob R (1982) Alterations of mechanical parameters in chemically skinned preparations of rat myocardium as a function of isoenzyme pattern of myosin. Basic Res Cardiol 77:220–234
7. Gorza L, Mercadier JJ, Schwartz K, Thornell LE, Sartore S, Schiaffino S (1984) Myosin types in the human heart. An immunofluorescence study of normal and hypertrophied atrial and ventricular myocardium. Circ Res 54:694–702
8. Hirzel HO, Tuchschmid CR, Schneider J, Krayenbuehl HP, Schaub MC (1985) Relationship between myosin isoenzyme composition, hemodynamics and myocardial structure in various forms of human cardiac hypertrophy. Circ Res 57:729–740
9. Kissling G, Rupp H, Malloy L, Jacob R (1982) Alterations in cardiac oxygen consumption under chronic pressure overload. Significance of the isoenzyme pattern of myosin. Basic Res Cardiol 77:255–270
10. Lompre AM, Nadal-Ginard B, Mahdavi V (1984) Expression of the cardiac ventricular alpha- and beta-myosin heavy chain genes is developmentally and hormonally regulated. J Biol Chem 259:6437–6446

11. Lompre AM, Schwartz K, d'Albis A, Lacombe G, Van Thiem N, Swinghedauw B (1979) Myosin isoenzyme redistribution in chronic heart overload. Nature 282:105–107
12. Mabuchi K, Sreter FA (1980) Actomyosin ATPase. II. Typing by histochemical ATPase reaction. Muscle Nerve 3:233–239
13. Mercadier JJ, Bouveret P, Gorza L, Schiaffino S, Clark WA, Zak R, Swinghedauw B, Schwartz K (1983) Myosin isoenzymes in normal and hypertrophied human ventricular myocardium. Circ Res 53:52–62
14. Mercadier JJ, Lompre AM, Wisnewsky C, Samuel JL, Bercovici J, Swinghedauw B, Schwartz K (1981) Myosin isoenzymic changes in several models of rat cardiac hypertrophy. Circ Res 49:525–532
15. Moss P, Micou-Eastwood J, Strohman R (1986) Altered synthesis of myosin light chains is associated with contractility in cultures of differentiating chick embryo breast muscle. Dev Biol 114:311–314
16. Price KM, Littler A, Cummins P (1980) Human atrial and ventricular light chain subunits in the adult and during development. Biochem J 191:571–580
17. Prince HP, Trayer HR, Henry GD, Trayer IP, Dalgarno DC, Levine BA, Cary PD, Turner C (1981) Proton nuclear-magnetic resonance spectroscopy of myosin subfragment-1 isoenzymes. Eur J Biochem 121:213–219
18. Rupp H (1986) (ed) Regulation of heart function. Basic concepts and clinical applications. Georg Thieme Verlag, Stuttgart
19. Schaub MC, Tuchschmid CR, Srihari T, Hirzel HO (1984) Myosin isoenzymes in human hypertrophic hearts. Shift in atrial myosin heavy chains and in ventricular myosin light chains. Eur Heart J [Suppl F] 5:85–93
20. Schiaffino S, Gorza L, Saggin L, Valfre C, Sartore S (1984) Myosin changes in hypertrophied human atrial and ventricular myocardium; a correlated immunofluorescence and quantitative immunochemical study on serial sections. Eur Heart J [Suppl F] 5:95–102
21. Schier JJ, Adelstein RS (1982) Structural and enzymatic comparison of human cardiac muscle myosins isolated from infants, adults and patients with hypertrophic cardiomyopathy. J Clin Invest 69:816–825
22. Schwartz K, Lecarpentier Y, Martin JL, Lompre AM, Mercadier JJ, Swinghedauw B (1981) Myosin enzymic distribution correlates with the speed of myocardial contraction. J Mol Cell Cardiol 13:1071–1078
23. Srihari T, Tuchschmid CR, Hirzel HO, Schaub MC (1982) Electrophoretic analyses of atrial and ventricular cardiac myosins from fetal and adult rabbits. Comp Biochem Physiol 72B:353–357
24. Srihari T, Tuchschmid CR, Schaub MC (1982) Isoforms of heavy and light chains of cardiac myosins from rat and rabbit. Basic Res Cardiol 77:599–609
25. Takeda N, Rupp H, Fenchel G, Hoffmeister HE, Jacob R (1985) Relationship between the myofibrillar ATPase activity of human biopsy material and hemodynamic parameters. Jpn Heart J 26:909–922
26. Tobacman LS, Adelstein RS (1984) Enzymatic comparisons between light chain isozymes of human cardiac myosin subfragment-1. J Biol Chem 259:11226–11230
27. Tuchschmid CR, Srihari T, Hirzel HO, Schaub MC (1983) Structural variants of heavy and light chains of atrial and ventricular myosins in hypertrophied human hearts. In: Jacob R, Gülch RW, Kissling G (eds) Cardiac adaptation to heamodynamic overload, training and stress. Steinkopff Verlag, Darmstadt, pp 123–128
28. Wagner PD, Giniger E (1981) Hydrolysis of ATP and reversible binding to F-actin by myosin heavy chains free of all light chains. Nature 299:560–562
29. Wagner PD, Weeds AG (1977) Studies on the role of myosin alkali light chains. Recombination and hybridisation of light chains and heavy chains in subfragment-1 preparation. J Mol Biol 109:455–473
30. Wallenstein S, Zucker CL, Fleiss JL (1980) Some statistical methods useful in circulation research. Circ Res 47:1–9
31. Weeds AG, Taylor RS (1975) Separation of subfragment-1 isoenzymes from rabbit skeletal muscles myosin. Nature 257:54–56

32. Yamamoto K, Sekine T (1983) Interaction of alkali light chain-1 with actin: Effect of ionic strength on the cross-linking of alkali light chain-1 with actin. J Biochem (Tokyo) 94:2075–2078
33. Yazaki Y, Tsuchimochi H, Kuro-o M, Kurabayashi M, Isobe M, Ueda S, Nagai R, Takaku F (1984) Distribution of myosin isozymes in human atrial and ventricular myocardium; comparison in normal and overloaded heart. Eur Heart J [Suppl F] 5:103–110

Authors' address:

Prof. Dr. M. C. Schaub, Pharmakologisches Institut, Universität Zürich, Gloriastraße 32, CH-8006 Zürich, Switzerland

Heterogeneous regulatory changes in cell surface membrane receptors coupled to a positive inotropic response in the failing human heart*

M. R. Bristow, R. Ginsburg, E. M. Gilbert, R. E. Hershberger

Cardiology Division, University of Utah, School of Medicine, Salt Lake City, U.S.A.

Summary

The failing human ventricular myocardium undergoes heterogeneous changes at the receptor level that have some impact on the ability of the failing myocardium to respond to inotropic stimuli. In the failing human ventricular myocardium the β_1-adrenergic receptor is profoundly down-regulated, the β_2-adrenergic receptor is only slightly decreased, α_1-adrenergic receptors are unchanged and VIP receptors appear to be increased in density or affinity. These changes have implications for therapeutic strategies for heart failure and for the natural history and pathogenesis of heart muscle disease.

Introduction

Myocardial cells have evolved complex biochemical control mechanisms to regulate contraction and relaxation. One major means of modulating the force of contraction is the hormone receptor (HR)-adenylate cyclase (AC) system. This system is of interest because it plays an important role in mediating compensatory changes required of the failing heart to meet increased functional demand, and also because it is amenable to pharmacological manipulation.

The HRAC system in the human heart is outlined in Fig. 1. Into the outer hemileaflet of the cell surface membrane are inserted discrete hormone receptors, which are specific protein structures that project from the outer cell surface so that they are available for combination with hydrophilic hormones or neurotransmitters. The human myocardial cell posssesses a variety of HR coupled to AC through a stimulatory quanine nucleotide regulatory subunit (N_S), another structurally distinct polypeptide that is embedded in the membrane matrix. Current opinion is that the combination of H with R effects an affinity change in N_S such that there is an increase in GTP binding, and the combination of GTP with an N_S then results in activation of a third discrete structural entity, the catalytic subunit (C) of adenylate cyclase. Activation of C then results in increased formation of cyclic AMP from ATP. Cyclic AMP formed in this manner is available to participate in a variety of intracellular phosphorylation reactions, by activating protein kinases. One of these reactions is to phosphorylate calcium channels located in the cell membrane, which

* Supported by NIH Grant HL 13108.

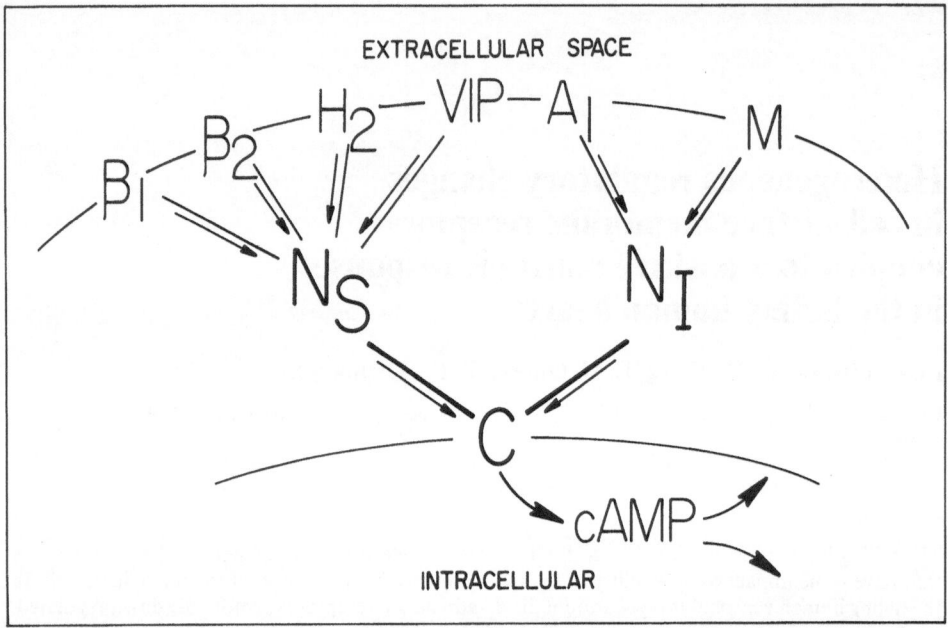

Fig. 1. Depiction of the myocardial cell surface membrane is outer and inner hemileaflets with hormone receptors protruding into the extracellular space, guanine nucleotide regulating proteins (N_S and N_I) imbedded in the membrane matrix, and the catalytic unit of adenylate cyclase (C) positioned towards the inner side of the membrane.

ultimately leads to increased calcium available for myocardial contraction and a positive inotropic response.

Balanced against these N_S coupled HR, are receptor pathways coupled to N_I, the inhibitory quanine nucleotide regulatory subunit. Just as the combination of H with R coupled to N_S facilitates an increase in C activity, the combination of H with R coupled to N_I decreases the activity of C.

As shown in Fig. 1, in human ventricular myocardium R is known to be coupled to N_S including the β_1 and β_2 adrenergic receptors [5], the H_2 histamine receptor [1] and peptide receptors including VIP [15] and glucagon [12]. Receptors coupled to N_I include the A_1 adenosine receptor and the muscarinic receptor. As heart failure develops, important changes occur in factors capable of influencing the HRAC system in myocardial cells. These changes include increased adrenergic stimulation [3], and are initially effective in maintaining cardiac output in the face of an intrinsic decrease in contractile function. Hormone receptors mediate or modulate the cellular effects of increased adrenergic activity by transmitting the cellular stimulus for an increase in force in contraction, increase in heart rate and at least part of the signal for an increase in protein synthesis.

This paper gives quantitative pharmacological data on the heart failure associated behaviour of HR coupled to AC and a positive inotropic response in human ventricular myocardium. Additionally, the behaviour of a receptor that is coupled to a positive inotropic response through non-cyclase mechanisms (α_1-adrenergic receptor), is also discussed.

Methods

1. Tissue procurement

a. Explanted human heart

Gram quantities of ventricular myocardium were removed from hearts explanted from cardiac transplant recipients or potential donors. Five to six gram aliquots were removed from the right and left ventricular free wall. Following removal of epicardial fat and sub-endocardial fibrosis, this material was homogenized and extracted to yield a crude membrane fraction that has a two to three fold enrichment of sarcolemmal markers, relative to the initial homogenate [8]. In this preparation only soluble constituents and contractile proteins are removed, which has the advantage that very little sarcolemma is lost in preparation (yield is >90%) [8]. An additional two gram aliquot was taken and prepared more gently for use in adenylate cyclase assays. A soluble fraction of this aliquot was used to measure creatine kinase activity as a cytosolic marker of viable myocardium. Finally, RV trabeculae were cut to uniform size (1–1.5 mm × 6–7 mm) and mounted in a multi-chamber muscle bath for measuring isolated tissue contractile responses to hormone agonists.

b. Studies in intact heart

We have developed methods of measuring receptors in milligram aliquots of tissue, removed by endomyocardial biopsy [9]. Using these techniques, β-adrenergic receptors can be radiolabelled using as little as 15 mg of tissue. As an adjunct of these studies, techniques for measuring the inotropic response to β-adrenergic and non-adrenergic agonists have also been developed [9].

2. Receptor radiolabelling

The total β-adrenergic receptor population was radiolabelled with both dihydroalprenolol [2] and [^{125}I]iodocyanopindolol [4, 8, 9]. β_1 vs β_2 adrenergic receptors have been identified by computer modelling of ICYP-betaxolol or ICYP 118-551 competition curves [8]. α_1 adrenergic receptors have been identified by the radioligand [^{125}I]IBE2254 [6, 16]. Radioligand bound to receptors is trapped by vacuum filtration using standard techniques [2, 4, 8, 9].

3. Adenylate cyclase activity

A modification of the original method described by Salomon et al. [14] was used [1, 4]. A "heavy" 1,085 × g, gently washed particulate fraction is suitable for measuring the response to the hormone agonists that produce an increase in adenylate cyclase activity $\geq 10\%$ above baseline. Behaviour of hormone receptors may be inferred from the characteristics of agonist stimulation (indirect measurement of hormone receptors). Adenylate cyclase stimulation was therefore used to indirectly measure the behaviour of β-adrenergic, H_2 histaminic and VIP receptors.

4. Tissue bath studies

Isolated right or left ventricular trabeculae are stable for hours when mounted in oxygenated, bicarbonate buffered physiological salt solution [2, 4, 10]. The inotropic response of a variety of agents can be assessed in this preparation, with the sensitivity of the preparation on a stimulus that produces a positive inotropic response at $\geq 10\%$ above baseline. Because of variability in individual trabeculae, three or four trabeculae per pharmacological subset were used. This necessitates the use of a multi-chamber bath, and, depending on the experiment, anywhere from an 8- to a 48-chamber bath was employed.

5. Data and statistical analysis

Receptor and enzyme data were analysed by computer using non-linear or linear weighted regression analysis as previously described [8]. Computer modelling of multiple classes of binding sites was by both the ligand program [13] and a program developed in our laboratory [8]. Dose-response curves were analysed by a modification of a repeated measures analysis of variance [8] or by analysis of covariance [2]. Differences among groups of more than two are assessed by analysis of variance and a multiple comparison test [9].

Results

1. Human ventricular myocardial receptors which markedly down-regulate in heart failure

The only example encountered thus far is the β_1 adrenergic receptor, which is decreased by 60–70% in failing human left ventricle [8]. Interestingly, human ventricular myocar-

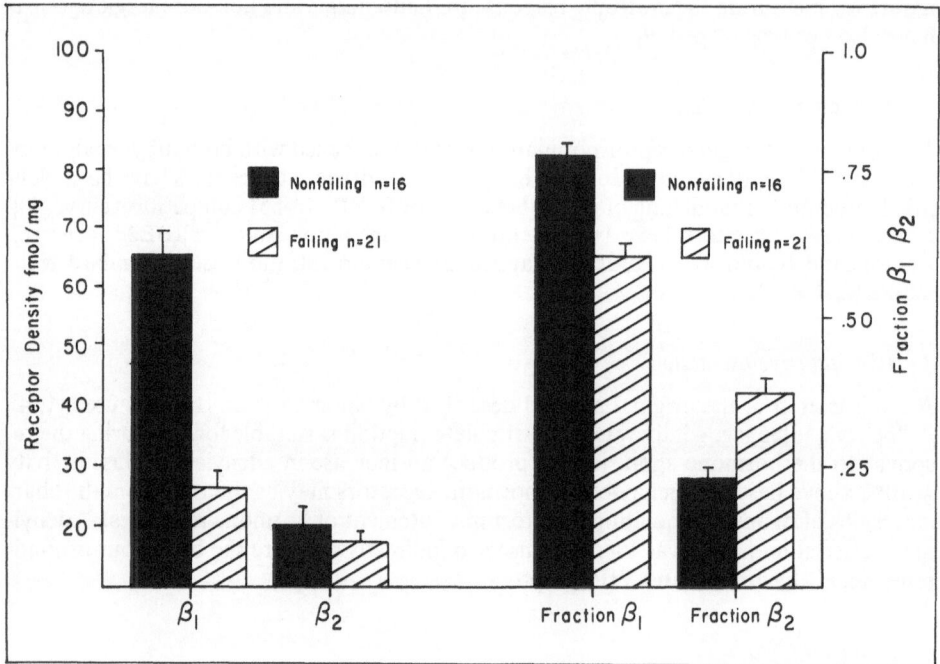

Fig. 2. β_1- and β_2-adrenergic receptor subpopulations and fractional proportion in crude membranes derived from 21 failing and 16 nonfailing left and right ventricles.

dium contains a high proportion of β_2-adrenergic receptors, some or all of which are coupled to an inotropic response [7, 8] and are therefore present on myocardial cells. In nonfailing myocardium the percentage of β_1 to β_2-adrenergic receptors is 77:23, which changes to 60:38 in the failing myocardium [8]. This can be seen in Fig. 2, where results are pooled for right and left ventricle and compared. The switch to a higher β_2 proportion in the failing heart is due to the marked reduction in β_1 receptor density, as opposed to an increase in β_2 receptor density.

2. Receptors that are minimally decreased in heart failure

An example of this is the β_2-adrenergic receptor. Radioligand binding-computer modeling techniques demonstrate a decline in β_2 receptor density of 18% (Fig. 2), which is not statistically significant. However, β_2 selective stimulation of adenylate cyclase with the specific β_2 agonist zinterol does show a statistically significant "blunting" of the maximum response, as shown in Fig. 3.

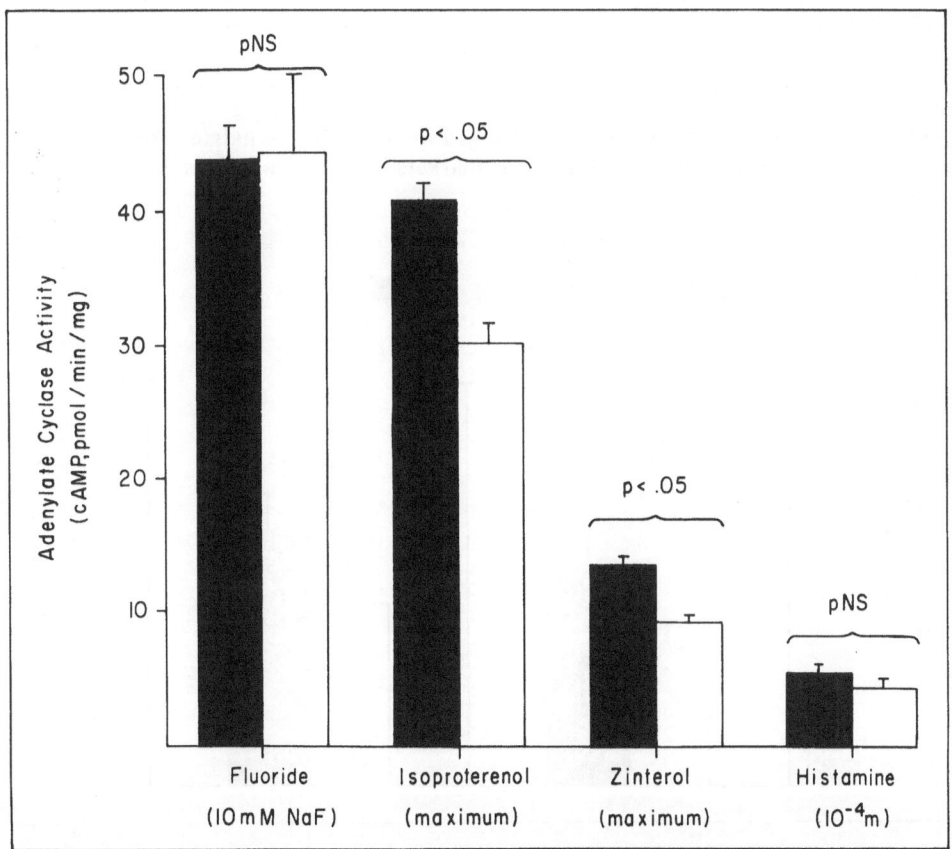

Fig. 3. Effect of various agents on adenylate cyclase response (net stimulation) in particulate fractions in nonfailing (n = 10) and failing (n = 16) human left ventricular myocardium.

3. Receptors or subunits that are unchanged in heart failure

a. α_1-adrenergic receptors

Figure 4 shows that α_1-adrenergic receptor density is unchanged in failing human ventricular myocardium. α_1 receptor density is much less than total β receptor density, and averages around 20% of the value in the non-failing myocardium.

b. Histamine H_2 receptors

In Fig. 3 are given the histamine stimulated adenylate cyclase maximum response data in nonfailing and failing human ventricular myocardium, relative to other forms of stimulation under the same assay conditions. Previous studies have shown that the H_2 receptor subtype is the histamine receptor coupled to cyclase [1]. Histamine stimulation produces a maximum response that is 10–15% of the maximum isoproterenol response when GTP is the guanine nucleotide employed in the assay conditions, and approximately 40–60% of the isoproterenol maximum when Gpp(NH)p is used as the guanine nucleotide [1]. There is no tendency for either the maximum or the ED_{50} of histamine dose-response curves to be decreased in the failing heart [1], indicating that H_2 receptors are probably unchanged by heart failure.

c. Catalytic subunit of adenylate cyclase

This discrete subunit appears to be unchanged in heart failure, judging from forskolin stimulated maximal responses [4]. Enzyme markers of sarcolemma membrane function

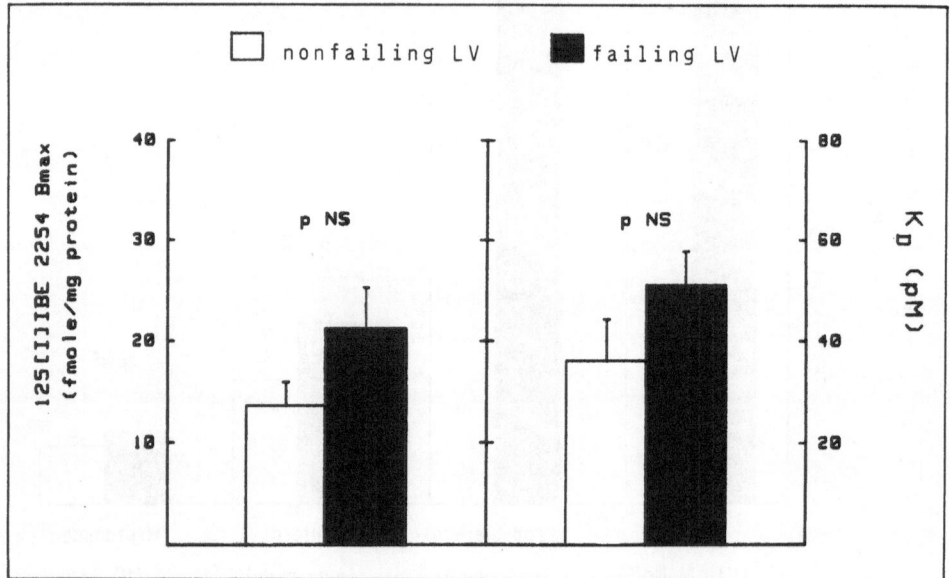

Fig. 4. α_1-adrenergic receptor density (IBE 2254 B_{max} in fmol/mg) and radioligand affinity in crude membranes derived from 8 failing and 8 nonfailing human left ventricles. The β receptor density in failing hearts was reduced by 40% (p < 0.05).

also appear to be unchanged in heart failure [8]. Creatine kinase activity as a marker of viable myocardium on cell cytoplasm is also not decreased in the failing human heart [2, 8].

4. Receptors that may be increased

We have recently reported that VIP stimulation of adenylate cyclase is enhanced in failing ventricular myocardium [11]. In this study the ED_{50} of VIP stimulated dose-response curves was 1.75-fold less in failing vs non-failing left ventricular myocardium, with 10 preparations in each group (p < 0.05). We are currently radiolabelling VIP receptors with ^{125}[I]VIP, but it is too early to conclude that an alteration in receptor density accounts for the differential sensitivity of VIP in the failing and non-failing human heart.

Discussion

The HRAC system in the failing human heart is characterized by the heterogeneous behaviour of individual components. The single most impressive finding is a profound decrease in β_1-adrenergic receptors and subsequent loss of β_1 agonist mediated responses [8]. Since the β_1-adrenergic receptor is a "norepinephrine receptor" and the failing human heart is exposed to increased quantities of norepinephrine, it is tempting to conclude that the β_1 receptors are decreased in response to chronic exposure to endogenous agonists [8]. However, other explanations are possible.

β_2-adrenergic receptors are only minimally (not statistically significantly) decreased in the failing human heart. β_2 mediated adenylate cyclase responses can be shown to be slightly blunted, either owing to greater sensitivity of this measurement compared to receptor measurements or to a decrease in receptor affinity as well as a small decrease in receptor density. Since the β_2-adrenergic receptor is an "epinephrine receptor" and epinephrine has not been shown to be statistically significantly increased in heart failure in most clinical studies, the behaviour of this receptor may simply reflect the response to exposure to hormone agonists.

Interestingly, α_1-adrenergic receptors do not appear to be decreased in density in the failing human heart. This is important because α_1 receptors have the same affinity for norepinephrine as do β_1 receptors. Therefore, simple exposure to increased tissue levels of norepinephrine would not appear to completely explain the behaviour of adrenergic receptors in heart failure. If norepinephrine is invoked as a possible explanation, additional mechanism(s) must also be involved, such as innervation of β_1 but not of α_1 receptors, or different sensitivities of β_1 and α_1 receptors to the down-regulating influence of norepinephrine.

The lack of any change in 1) histamine mediated muscle contraction or histamine mediated adenylate cyclase responses; 2) forskolin mediated adenylate cyclase responses and; 3) calcium muscle contraction responses [10] indicates that the HRAC-muscle contraction pathway is fundamentally intact in the failing human ventricular myocardium. That is, there is no general blunting of the ability of this pathway to transmit an inotropic influence other than those for which specific receptors are decreased. The finding of comparable amounts of creatine kinase activity in failing and non-failing heart indicates that the amount of viable myocardium is not reduced in heart failure. Additionally, membrane marker studies have found no difference in the amount of functioning sarcolemma in non-failing vs failing heart [8].

The finding of supersensitivity of VIP adenylate cyclase responses was unanticipated and potentially quite interesting. This hormone agonist mediates a substantial inotropic response (unpublished data) in human myocardium, and the increased sensitivity in heart failure might make this class of receptors a potentially important candidate for inotropic drug development.

In summary, components of the HRAC system in failing human ventricular myocardium demonstrate heterogeneous regulatory behaviour. The reasons for this variable behaviour probably relate to both differences in the degree of exposure to regulating influences and inherent differences in regulating mechanisms that are specific to each component.

References

1. Bristow MR, Cubicciotti R, Ginsburg R, Stinson EB, Johnson C (1982) Histamine-mediated adenylate cyclase stimulation in human myocardium. Mol Pharmacol 21:671–679
2. Bristow MR, Ginsburg R, Minobe WA, Cubicciotti RS, Sageman WS, Lurie K, Billingham ME, Harrison DC, Stinson EB (1982) Decreased catecholamine sensitivity and β-adrenergic receptor density in failing human hearts. New Engl J Med 307:205–211
3. Bristow MR (1984) The adrenergic nervous system in heart failure. N Engl J Med 311(13):850–851
4. Bristow MR, Ginsburg R, Strosberg A, Montgomery W, Minobe W (1984) Cardiovascular pharmacology and inotropic potential of forskolin in the human heart. J Clin Invest 74:212–223
5. Bristow MR, Laser JA, Ginsburg R, Minobe W (1985) β_1 and β_2 receptors are coupled to adenylate cyclase in human ventricular myocardium. Clin Res 33(2):171A
6. Bristow MR, Minobe W, Rasmussen R (1985) Differential regulation of alpha and beta-adrenergic receptors in the failing human heart. Circulation [Suppl III] 72:1315
7. Bristow MR, Ginsburg R (1985) β_2 receptors are present on myocardial cells in human ventricular myocardium. Am J Cardiol 57:3F–6F
8. Bristow MR, Ginsburg R, Umans V, Fowler M, Minobe W, Rasmussen R, Zera P, Menlove R, Shah P, Jamieson S, Stinson EB (1986) Down-regulation in heart failure. Circ Res 59:297–309
9. Fowler MB, Laser JA, Hopkins G, Minobe W, Bristow MR (1986) Assessment of the β-adrenergic receptor pathway in the intact failing human heart: Progressive receptor down-regulation and subsensitivity to agonist response. Circulation 74(6):1290–1302
10. Ginsburg R, Bristow MR, Billingham ME, Stinson EB, Schroeder JS, Harrison DC (1983) Study of the normal and failing isolated human heart: Decreased response of failing heart to isoproterenol. Am Heart J 106:535–540
11. Hershberger RE, Bristow MR (1986) Failing human left ventricular myocardium is supersensitive to vasoactive intestinal peptide. Clin Res 34(2):630A
12. Levey GS, Epstein SE (1969) Activation of adenyl cyclase by glucagon in cat and human heart. Circ Res 24:151–156
13. Munson PJ, Rodbard D (1980) LIGAND: A versatile computerized approach for characterization of ligand-binding systems. Anal Biochem 107:220–239
14. Salomon Y, Londos C, Rodbell M (1974) A highly sensitive adenylate cyclase assay. Anal Biochem 58:541–548
15. Taton G, Chatelain P, Delhaye M, Camus JC, De Neef P, Waelbroeck M, Tatemoto K, Robberecht P, Christophe J (1982) Vasoactive intestinal peptide (VIP) and peptide having N-terminal histidine and C-terminal isoleucine amide (PHI) stimulate adenylate cyclase activity in human heart membranes. Peptides 3:897–900
16. Tsujimoto G, Bristow MR, Hoffman BB (1984) Identification of alpha$_1$-adrenergic receptors in rabbit aorta with ^{125}[I]BE2254. Life Sci 34:639–646

Authors' address:

Michael R. Bristow, M.D., Ph.D., Cardiology Division, 4A-100, University of Utah Medical Center, 50 Medical Drive, Salt Lake City, UT 84132, U.S.A.

Acute and chronic changes of myocardial energetics in the mammalian and human heart *

Ch. Holubarsch, G. Hasenfuss, H. W. Heiss, T. Meinertz, H. Just

Medizinische Klinik III, Universität Freiburg, F.R.G.

Summary

In earlier studies using papillary muscles of the rat left ventricle and highly sensitive thermopiles we demonstrated that the heat liberated per gram of myocardium per unit of developed tension-time integral is decreased when the rats suffered from hypothyroidism or renal hypertension. This increase in economy of force production was shown to be associated with a decrease in myosin-ATPase activity and a change in isomyosin composition. In a recent study we showed an increase in heat per gram of mammalian myocardium per tension-time integral of 70% after application of isoproterenol.

In order to study the relationship between energy costs and developed tension-time integral in the human heart, haemodynamics and myocardial oxygen consumption were measured. The data were obtained using a Millar microtip catheter pressure transducer and the argon method. Haemodynamics and myocardial energetics were analysed in 8 patients without significant heart disease before and after application of isoproterenol and in 10 patients with dilative cardiomyopathy (NYHA II–III). During one cardiac cycle, myocardial oxygen consumption per gram of LV myocardium per beat ($M\dot{V}O_2/g \times beat$) is related to LV stress-time integral ($\int \sigma xt$). The economy of myocardial contraction (EC) was calculated by

$$EC = \frac{\int \sigma xt}{M\dot{V}O_2/g \times beat} .$$

EC was 11.3 ± 3.2 in normal and 14.3 ± 4.7 dyn $\times$ s $\times$ g/cm^2 $\times$ µcal in dilative cardiomyopathic hearts (NS). Isoproterenol decreased EC from 11.3 ± 3.2 to 5.5 ± 1.6 dyn $\times$ s $\times$ g/cm^2 $\times$ µcal in the normal hearts (p $<$ 0.01).

In the rat myocardium, changes in economy of force generation were found due to catecholamines, pressure overload and hypothyroidism. In the human heart, similar energetic changes were observed due to catecholamines. No significant differences in energy of force production were seen between normal and dilative cardiomyopathic hearts.

The effect of catecholamines in the mammalian and human myocardium is explained by changes in activation processes and in chemomechanical energy transduction at the level of the contractile proteins.

Introduction, methods and results

Myocardial energetics may be the key for understanding the fundamental mechanisms involved in cardiac hypertrophy and heart failure. Therefore, we investigated myocardial energetics in experimental animal models as well as in human hearts.

Because it is the aim of our studies to investigate possible energetic changes on the level of the myocardium, the experimental animal studies were performed on left or right

---

* The animal experiments were carried out at the Dept. of Physiology and Biophysics, University of Vermont, Burlington, U.S.A.

ventricular papillary muscles in which the energy turnover of an isometric contraction can be measured simultaneously with the force or force-time integral. This type of experiment is not yet possible in the isolated human myocardium. Therefore, myocardial oxygen consumption was measured in the human, and pressure-time, volume-time and pressure-volume data were additionally obtained. Similar to the animal experiments, it was thereby possible to analyse the economy of contraction in terms of force-time integral per myocardial oxygen consumption by normalizing all data for a unit of myocardium.

I. Acute and chronic changes of myocardial energetics in the mammalian heart

Chronic changes of myocardial energetics in the rat are shown to occur in hypothyroidism and pressure-overload hypertrophy [1, 7, 8, 9, 16]. Hypothyroidism was induced by oral application of propylthiouracil (PTU) over a period of 4 weeks. The hypothyrotic myocardium exhibits a slowed contraction phase and a small decrease in peak developed tension (Fig. 1, PTU). Pressure-overload hypertrophy resulted from renal hypertension by narrowing the left renal artery 5 weeks before experimental investigation. The hypertrophied myocardium (GOP means Goldblatt operated) shows a prolongation of the contraction phase and a small increase in peak developed tension (Fig. 1, GOP).

Simultaneously with the tension development we measured liberated initial and total activity-related heat using highly sensitive antimony-bismuth thermopiles [20, 1, 10].

Initial heat is the heat liberated during the contraction period and is therefore a measure of the amount of consumed ATP molecules.

Total activity-related heat is the sum of initial heat and recovery heat. Recovery heat represents the amount of ATP molecules which are resynthesized after the contraction period. Resting heat, which represents basal metabolism, was also measured, but is is not reported in this paper.

As a measure of isometric myocardial contraction economy, we calculated the ratio of developed tension and total activity-related heat or developed tension-time integral and total activity-related heat.

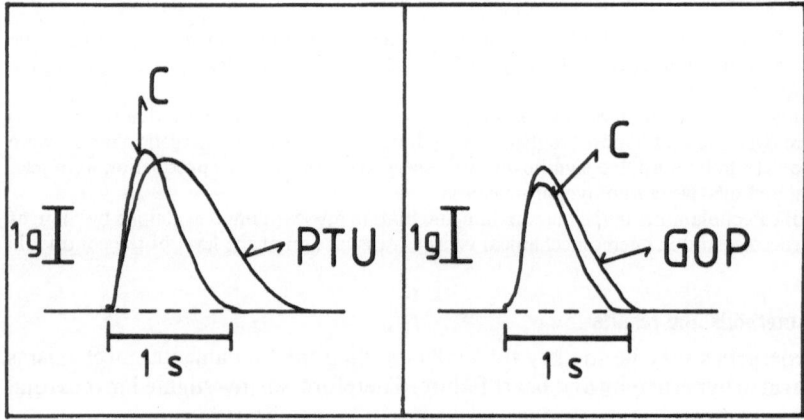

Fig. 1. Left. Representative isometric mechanograms of control (C) and hypothyrotic (PTU) rat myocardium. Note the increase in time to peak tension and relaxation time. Right. Representative isometric mechanograms of control (C) and pressure-overload (GOP=Goldblatt operated) rat myocardium. Note the increase in peak developed force.

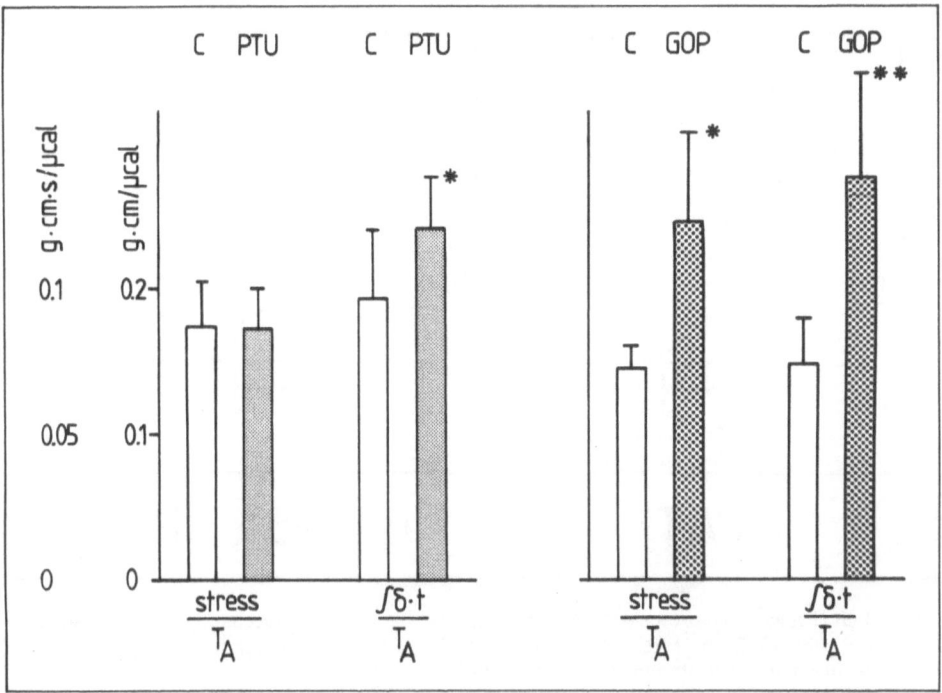

Fig. 2. Left. Economy of myocardial isometric contraction as defined by the ratio between stress and total activity-related heat (Stress/T_A), and between tension-time integral and total activity-related heat ($\int \sigma \cdot t / T_A$) in control (C) and hypothyrotic (PTU) rat myocardium. *p<0.01. Right. As in the left diagram but in control (C) and pressure-overload hypertrophied myocardium (GOP=Goldblatt operated). *p<0.05; **p<0.01.

In the Goldblatt myocardium, there is an obvious change in myocardial economy (Fig. 2, GOP): about 80% more tension or tension-time integral is generated when related to a unit of total activity-related heat. In the hypothyrotic rat myocardium (Fig. 2, PTU) we observed a similar increase in economy compared to controls only when tension-time integral is related to total activity-related heat. No difference between PTU and control is seen when stress is related to total activity-related heat. However, we believe the tension-time integral to be the better determinant of energy consumption (see III. Discussion).

In further studies, the liberated heat was separated into tension-dependent heat, i.e., the energy necessary for calcium release and re-uptake, and tension-dependent heat, i.e., the energy necessary for force generation by the cycling cross-bridges. It was found that the demonstrated increases in economy of hypothyrotic and pressure-overload hypertrophied rat myocardium are mainly due to changes in tension-dependent heat. But activation processes are also changed energetically [7, 8, 9]. The demonstrated changes of myocardial energetics are shown to be associated with alterations in the myosin isoenzyme composition. In the control myocardium, V1 myosin predominates, whereas in the PTU myocardium V3 myosin predominates. In the Goldblatt myocardium, we see about the same amounts of V1, V2, and V3 myosin [7, 9, 14, 16]. From these data, we conclude that

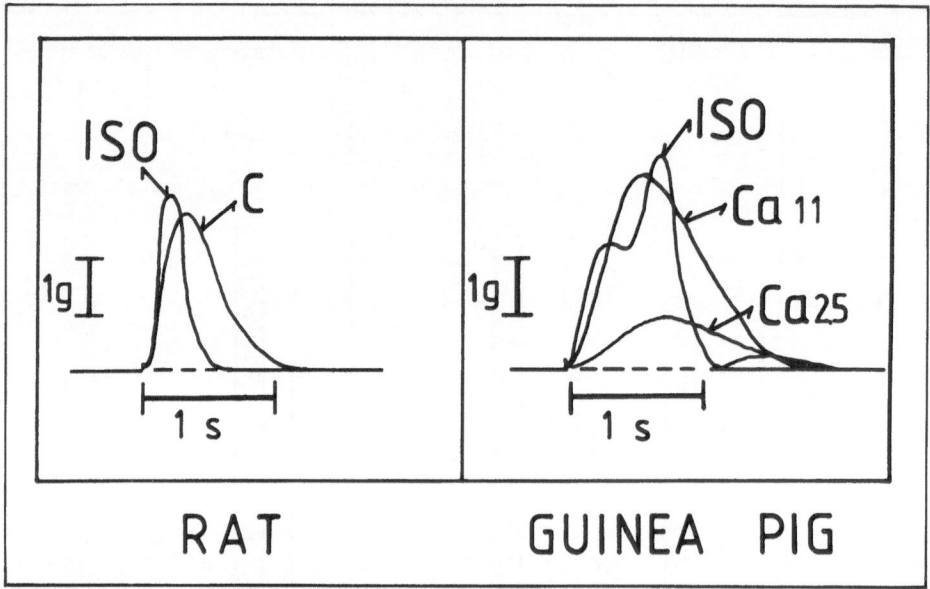

Fig. 3. Left. Effect of isoproterenol (ISO; 10^{-6} M) on the isometric mechanogram of rat myocardium (C = control). Right. Effect of isoproterenol (ISO; 10^{-6} M) and high calcium (11 mM) on the isometric mechanogram of guinea pig myocardium (Ca 2.5 = 2.5 mM calcium = control).

a change in the isomyosin pattern leads to fundamental changes of myocardial energetics. Because under conditions of hypothyroidism and pressure-overload the myocardium develops force in a more economical way than under control conditions, we interpret the underlying structural changes on the level of the cross-bridges, i.e., in the isomyosin pattern, as beneficial and therefore adaptive.

In a further study, we investigated acute changes of myocardial energetics in the normal rat and normal guinea pig myocardium. These two species were chosen because of their differences in the isomyosin pattern. In the young rat, V1 myosin predominates, in the guinea pig V3 myosin predominates.

In the rat myocardium, isoproterenol in a concentration of 10^{-6} M leads to an increase in dP/dt_{max}, and a fast relaxation (Fig. 3, rat, ISO). In the guinea pig, however, we have a quite different situation. At normal calcium concentration, peak developed tension is small and can be essentially increased by either isoproterenol (10^{-6} M) or increasing calcium concentration (from 2.5 to 11 mM). However, the contraction phases look different: with isoproterenol two contraction phases occur, an early one and a late one, presumably resulting from two different calcium releases [22].

Again, we measured total activity-related heat simultaneously with isometric tension development.

When calculating the ratio of peak developed tension and total activity-related heat in the rat myocardium, no change of isometric contraction economy is evaluated (Fig. 4, rat, ISO). However, when the tension-time integral is related to total activity-related heat we see that, per unit of tension-time integral, about double the amount of energy is consumed under isoproterenol (Fig. 4, rat, ISO).

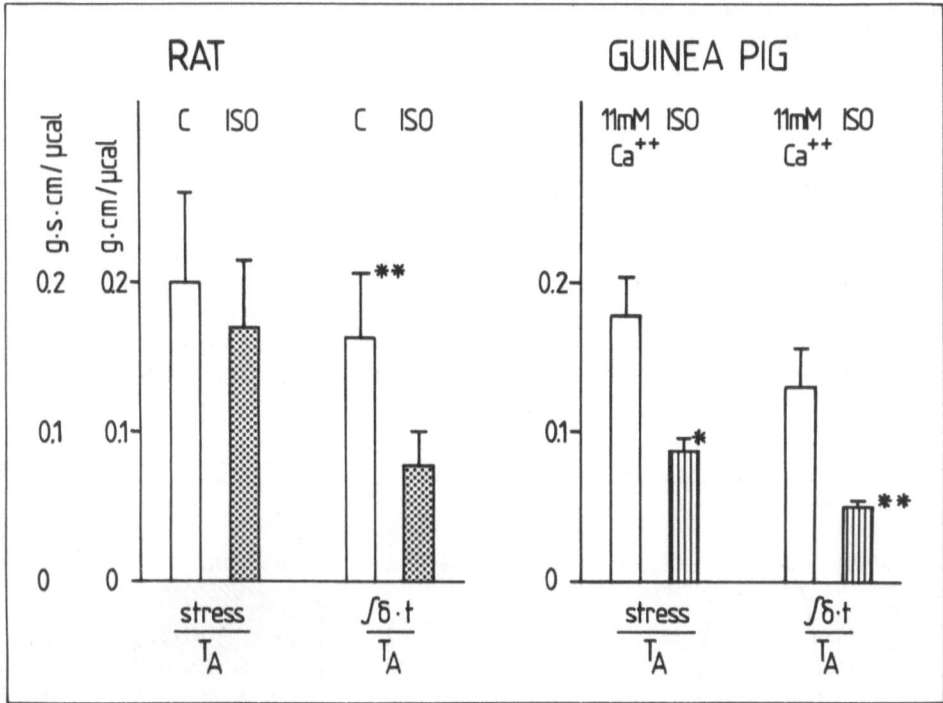

Fig. 4. Left. Economy of myocardial isometric contraction as defined by the ratio between stress and total activity-related heat (Stress/T_A), and between tension-time integral and total activity-related heat ($\sigma t/T_A$ in the rat; C = control, ISO = isoproterenol (10^{-6} M). **p<0.01. Right. As in the left diagram in the guinea pig; positive inotropism was induced by increasing the calcium concentration from 2.5 mM to 11 mM or by adding isoproterenol (ISO; 10^{-6} M) to the 2.5 mM calcium containing solution. ** <0.01; *p<0.05.

In the guinea pig myocardium, only half the tension or only half the tension-time integral is generated for a given amount of energy consumption with isoproterenol when compared to high calcium (11 mM) (Fig. 4, guinea pig). Therefore, the positive inotropic effects of isoproterenol and high calcium are energetically different, i.e., isoproterenol decreases the economy of myocardial isometric contraction.

In more detailed studies we were able to gain evidence that the demonstrated decrease of myocardial economy results primarily from changes on the level of the contractile proteins, i.e., cross-bridge kinetics [11, 12].

In conclusion, hypothyroidism and Goldblatt pressure-overload alter the rat isomyosin pattern which is associated with an increase in the economy of myocardial force generation. In the normal rat and guinea pig myocardium, isoproterenol decreases the economy of myocardial force generation. In contrast, an increase in calcium concentration can enhance peak developed tension without a change in myocardial economy.

II. Acute and chronic changes of myocardial energetics in the human heart

After our observations in mammalian papillary muscles, we were interested in asking the question of acute and chronic changes of myocardial energetics in the human heart. In order to obtain the ratio between developed tension-time integral and energy consumption, it is necessary to measure both an energy term and a mechanical term in patients.

For the analysis of the tension-time integral, we performed left ventricular angiocardiography using Millar-tip-micromanometers. From synchronously obtained pressure-volume data (the volumes were analysed every 20 ms using the area-length method) we constructed pressure-volume loops (Fig. 5 A, B, C). The area of the pressure-volume-loop represents left ventricular work, and the quotient of left ventricular work and left ventricular muscle mass yields work per gram of myocardium per beat. Furthermore, from pres-

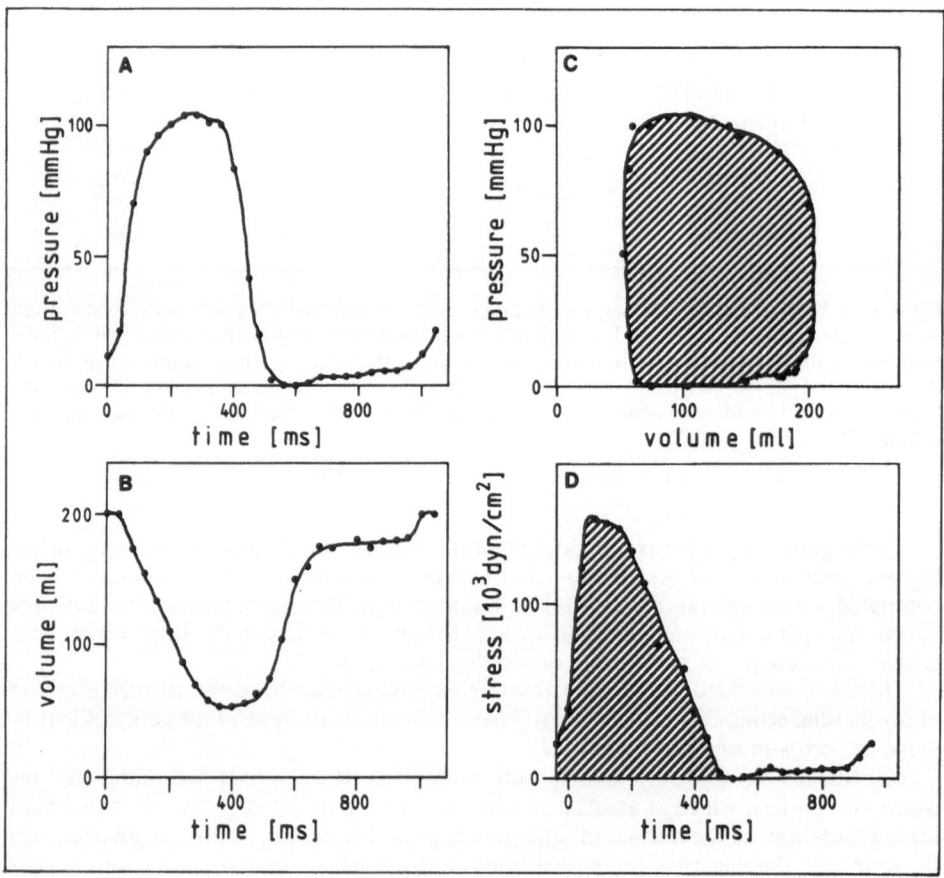

Fig. 5. A representative example of data analysis (control patient): A. Pressure versus time (Millar tip micromanometer). B. Volume versus time (simultaneous LV angiocardiography). C. Pressure-volume relationship. D. Stress-time plot. Stress was calculated from the minor and maior semiaxis, instantaneous pressure, and wall-thickness according to Mirsky [12].

sure-volume-data and wall-thickness we calculated left ventricular circumferential wall stress every 20 ms using an equation based on an ellipsoidal model, proposed by Mirsky [18]. From the stress versus time plot the systolic stress-time integral was automatically obtained [13].

As a measure of myocardial energy demand, we measured myocardial oxygen consumption using the argon method [6, 21]. A patient is connected to a argon-oxygen gas mixture, which he breathes for a period of 5 min. During this period, blood is synchronously withdrawn from the aorta – early saturation with argon – and from the coronary sinus – late saturation with argon. According to the principles of organ perfusion, originally developed by Kety and Schmidt [15], we are able to calculate myocardial blood flow in ml per gram of muscle per min from the argon concentration in the coronary venous and arterial blood samples. The oxygen concentration in the aorta and in the coronary sinus were measured so that the oxygen consumption per 100 g of muscle per min could also be calculated.

The determinants of activity-related myocardial oxygen consumption are frequency, pressure-volume work, stress-time integral and contractile state. (1) In order to be independent of frequency, all data were evaluated per beat instead of per time; (2) The stress-time integral was analysed from pressure-volume-data and wall thickness; (3) The pressure-volume work may play an important role in energy consumption. The pressure-volume integral was analysed from pressure-volume-data and divided by the muscle mass yielding myocardial work.

(4) Contractility is supposed to influence myocardial energy consumption, and indeed it is the specific aim of this study to reveal acute and chronic contractility related changes of myocardial energetics.

Economy of myocardial contraction (EC) was defined as the quotient of tension-time integral and overall oxygen consumption. Using a caloric equivalent of 5 kcal/l O_2, $M\dot{V}O_2$/beat in ml O_2/g was converted into $M\dot{V}O_2$/beat in µcal/g.

The first investigation was performed in a group of patients with dilative cardiomyopathy NYHA II–III. As a control group, patients without significant heart disease were used. These patients underwent routine heart catheterization because of atypical chest pain.

Figure 6 demonstrates a representative pressure-volume loop of the normal left ventricle with a ejection fraction of 77%, and another one of a dilative cardiomyopathic heart with an ejection fraction of only 36%. Stroke volume was decreased and end-diastolic volume was increased in dilative cardiomyopathy.

No significant change was found for myocardial blood flow, oxygen consumption per minute, and oxygen consumption per beat (no figure).

Economy of myocardial contraction did not reveal significant differences in myocardial energetics between normal and dilative cardiomyopathy (Fig. 7), although there was a tendency to even better economy in dilative cardiomyopathic hearts.

In patients without significant heart disease, isoproterenol was applied in a dosage of 1–3 µg per min so that the heart frequency was increased by more than 50%. No decisive changes occurred in the pressure-volume loops due to isoproterenol, the end-systolic volume decreased only in some patients, and therefore stroke volume increased (Fig. 6).

The myocardial blood flow was doubled, the oxygen consumption per minute was also significantly increased, whereas oxygen consumption per beat was not changed significantly. However, the stress-time integral was essentially decreased.

Therefore, the economy expressed as stress-time integral per overall oxygen consumption was decreased by about 50% (Fig. 8). When peak developed stress was used instead

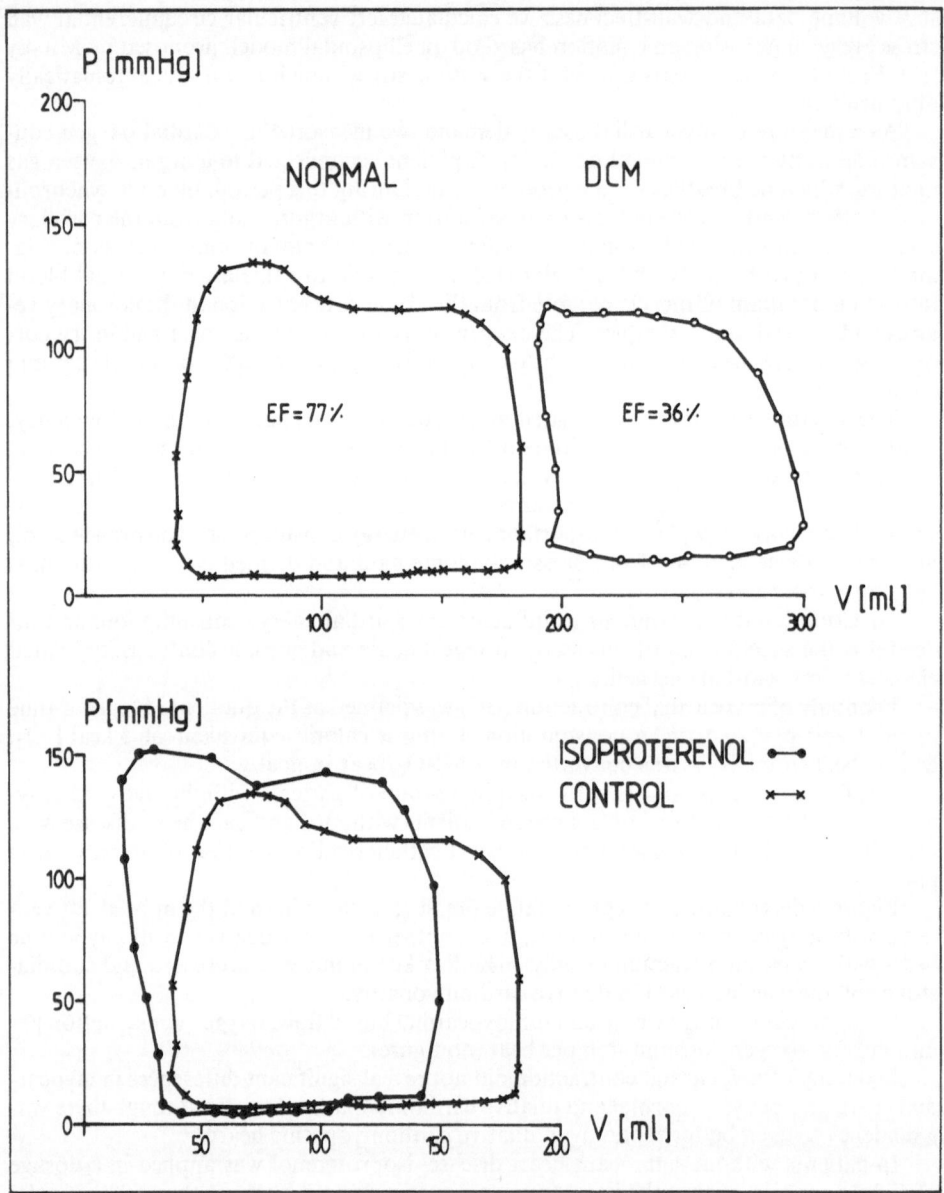

Fig. 6. Upper. Pressure-volume loop of a representative control (Normal) and dilative cardiomyopathy (DCM) of left ventricle. Lower. Pressure-volume loop of a representative control left ventricle before (Control) and after application of isoproterenol (ISO).

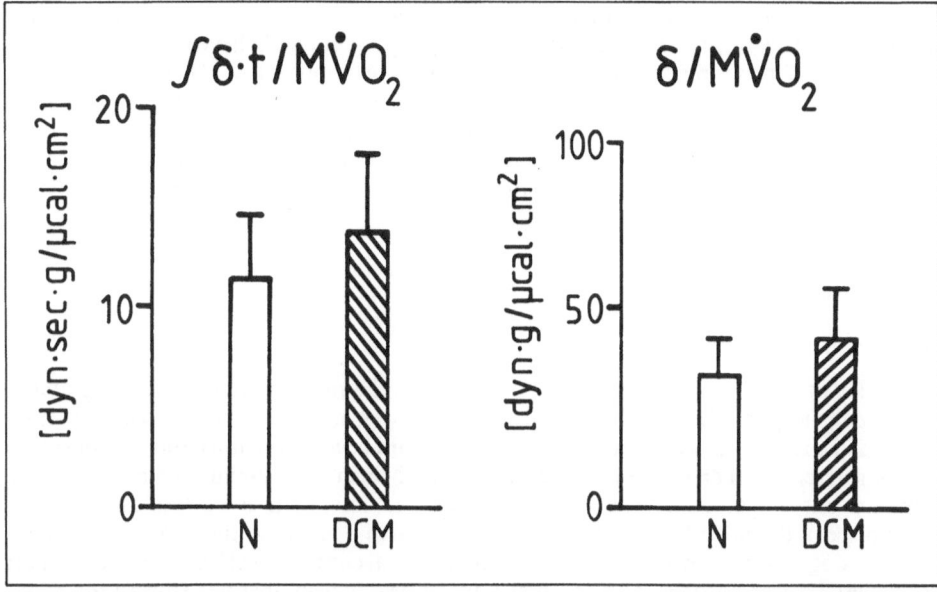

Fig. 7. Index of economy (EC $= \int \sigma \cdot t/M\dot{V}O_2$) in control patients (N) and those with dilative cardio-myopathy (DCM). No significant differences were observed. Stress versus $M\dot{V}O_2$ is also indicated.

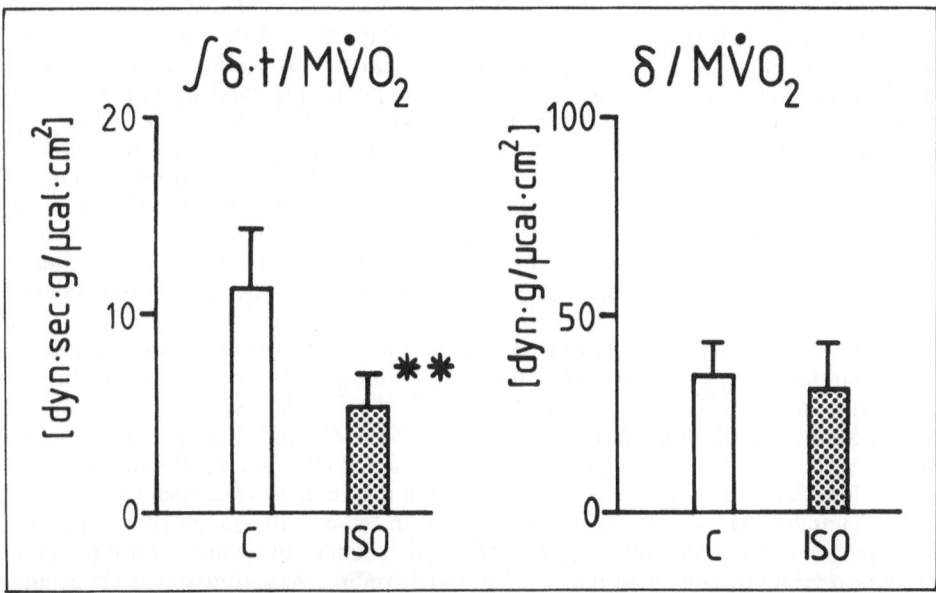

Fig. 8. Index of economy ($\int \sigma \cdot t/M\dot{V}O_2$) in patients without significant heart disease before (C) and after application of isoproterenol (ISO). **$p < 0.01$. Stress versus $M\dot{V}O_2$ is also indicated.

of stress-time integral for defining the economy of contraction, no difference in economy was found for the isoproterenol intervention (Fig. 8). Again, we believe the stress-time integral to be the better mechanical index of energy consumption (see III, Discussion).

In conclusion, we found an increased economy of myocardial stress development in patients with dilative cardiomyopathy, which was, however, not significant. Further studies are necessary to prove the significance of this change in economy.

In the normal heart, isoproterenol stimulates the myocardium, resulting in a decrease in myocardial economy of tension development.

III. Discussion

Interpretation and clinical aspects

The index of economy used in the present paper is the ratio of stress-time integral and energy consumption. If peak developed stress is chosen as the mechanical parameter, no changes in economy of myocardial contraction would become obvious under application of isoproterenol either in the experimental animal or in the human studies. There are some reasons that convinced us to use the stress-time integral as a mechanical index, instead of peak developed stress: 1. Some authors have postulated the contraction phase to be the only energy consuming process of myocardial contraction [19]. However, Alpert & Mulieri [2] clearly demonstrated that a lot of heat is also liberated during the relaxation period, indicating energy consuming cross-bridge and calcium-cycling. 2. From tetanizing experiments [8, 7, 4] we know that in the myocardium of different species there is an excellent relationship between liberated energy and stress-time integral. This indicates the time of maintenance of stress is a decisive determinant of energy demand in the myocardium. 3. As long as the contractile state is left unchanged, the correlation between heat and stress-time integral is as good as the correlation between heat and peak developed stress [5, 3]. 4. In experiments in rat, cat and guinea pig papillary muscles, it was shown that heat per peak stress is not different between species [17]. However, when preparations were experimentally tetanized and the heat output was related to stress-time integral, large species-dependent differences were observed which were attributed to different myosin isoenzyme patterns and myosin ATPases [4]. Because structures (heavy chains of myosin) are the cause of these differences, evidence for different chemomechanical transduction mechanisms between species should also become obvious in single twitches. Indeed, these differences become evident when heat is related to the stress-time integral [7, 11].

In the experimental papillary muscles studies, one is in a position to differentiate between different energy compartments: resting heat, initial heat and recovery heat under isometric conditions [1]. In the human being, only the overall oxygen consumption can be measured under physiological afterload contractions. Nevertheless, we related the tension-time integral to the overall oxygen consumption (Figs. 7 and 8). Because the pressure-volume work is supposed to be an important determinant of oxygen consumption (Table 1) and basal metabolism is also included in the overall oxygen consumption, a more refined analysis was additionally performed. The pressure-volume work per gram of myocardium was converted into mcal/g. Furthermore, the resting oxygen consumption was assumed to be constant and to comprise 15% of overall oxygen consumption of control myocardium. Both pressure-volume work per gram of myocardium and resting oxygen consumption were subtracted from the overall oxygen consumption. Then, the tension-time integral was divided by the corrected oxygen consumption value. This ratio was not significantly altered in the dilative cardiomyopathic myocardium when compared to the control myocardium, and isoproterenol decreased this ratio by a factor of about two. Therefore, the final conclusions drawn from the presented data are independent on consideration of pressure-volume work and/or resting metabolism.

How should one interpret the observed differences in economy of myocardial contraction, on the basis of the known contraction mechanisms? In the human heart studies, we are not in a position to differentiate between energetics involved in activation (calcium cycling) and those involved in force generation, by the contractile machinery (cross-bridge cycling). In the experimental animal study, however, we can separate the tension-dependent heat (cross-bridge cycling) from the tension-independent heat (activation). The changes in economy of force generation were shown to be mainly due to alterations in tension-dependent heat [7, 8, 9], although the energetics of activation processes are also changed.

The cross-bridge cycle can be described in terms of on-time (force-generating position) and off-time (non-force-generating position) [2].

If the strength of one cross-bridge is assumed to be constant, the presented results are consistent with β-receptor mediated changes in the on-time [11].

The demonstrated results in experimental animal and clinical human hearts have important implications for the clinical situation. We observed that in the experimental animals the pressure overload, i.e., high wall stress, induces a reduction in shortening speed and an increase in economy of myocardial contraction. Catecholamines are shown to drive the myocardium into the opposite direction, i.e., to increase the speed of shortening and to reduce the economy of myocardial contraction. In chronic heart failure, wall stress is increased. Therefore, catecholamines are suggested to be in-appropriate for increasing myocardial performance in chronic heart failure from an energetic point of view. Further studies are necessary to prove the effect of the well-known digitalis and other new cardiotonic drugs with respect to economy of myocardial contraction.

References

1. Alpert NR, Mulieri LA (1982) Increased myothermal economy of isometric force generation in compensated cardiac hypertrophy induced by pulmonary artery constriction in rabbit. A characterization of heat liberation in normal and hypertrophied right ventricular papillary muscles. Circ Res 50:491–500
2. Alpert NR, Mulieri LA, Litten RZ (1983) Isoenzyme contribution to economy of contraction and relaxation in normal and hypertrophied hearts. In: Jacob R (ed) Cardiac adaptation to hemodynamic overload, traning and stress. Steinkopff Verlag, Darmstadt, pp 147–157
3. Barclay JK, Gibbs CL, Loiselle DS (1979) Stress as an index of metabolic cost in papillary muscle in cat. Basic Res Cardiol 54:594–603
4. Gibbs CL, Loiselle DC (1978) The energy output of tetanized cardiac muscle: Species differences. Pflügers Arch 373:31–38
5. Gibbs CL (1978) Cardiac energetics. Physiol Rev 58:174–254
6. Heiss HW, Barmeyer J, Wink K, Hell G, Cerny FI, Keul J, Reindell H (1976) Studies of the regulation of myocardial blood flow: Training effects on blood flow and metabolism of the healthy heart at rest and during standardized heavy exercise. Basic Res Cardiol 71:658
7. Holubarsch Ch, Goulette RP, Litten RZ, Martin BJ, Mulieri LA, Alpert NR (1985) The economy of isometric force development, myosin isoenzyme pattern and myofibrillar ATPase activity in normal and hypothyroid rat myocardium. Circ Res 56:78–86
8. Holubarsch Ch, Goulette RP, Mulieri LA, Alpert NR (1983) Heat liberation in experimentally induced tetanic contractions of myocardium from normal and Goldblatt rats. In: Jacob R (ed) Cardiac adaptation of hemodynamic overload, training and stress. Steinkopff Verlag, Darmstadt, pp 158–166

9. Holubarsch Ch, Litten RZ, Mulieri LA, Alpert NR (1985) Energetic changes of myocardium as an adaptation to chronic hemodynamic overload and thyroid gland activity. Basic Res Cardiol 80:582–593
10. Holubarsch Ch, Alpert NR, Goulette RP, Mulieri LA (1982) Heat production during hypoxic contracture of rat myocardium. Circ Res 51:777–786
11. Holubarsch Ch, Hasenfuss G, Blanchard E, Alpert NR, Mulieri LA, Just H (1986) Myothermal economy of rat myocardium: Chronic adaptation versus acute inotropism. Basic Res Cardiol [Suppl 1] 81:95–102
12. Holubarsch Ch, Hasenfuss G, Just H, Blanchard E, Mulieri LA, Alpert NR (1987) Positive inotropism and energy consumption in rat and guinea pig myocardium. Influence of calcium, isoproterenol and UDCG-115. Circ Res (in preparation)
13. Holubarsch Ch, Hasenfuss G, Just H (1986) Assessment of myocardial function by calculation of two energetic parameters from pressure-volume relations and wall thickness in human ventricles. Clin Cardiol 9:292–295
14. Jacob R, Ebrecht G, Holubarsch Ch, Rupp H, Kissling GS (1983) Mechanics and energetics in cardiac hypertrophy as related to the isoenzyme pattern of myosin. In: Alpert NR (ed) Perspectives of cardiovascular research, vol 7. Myocardial hypertrophy and failure. Raven Press, New York, pp 553–569
15. Kety SS, Schmidt CF (1948) The nitrous oxide method for the quantitative determination of cerebral blood flow in man: Theory, procedure and normal values. J Clin Invest 27:476
16. Loiselle DS, Wendt IR, Hoh JFY (1982) Energetic consequences of thyroid-modulated shifts in ventricular isomyosin distribution in the rat. J Muscle Res Cell Motil 3:5–23
17. Loiselle DS, Gibbs CL (1979) Species differences in cardiac energetics. Am J Physiol 237:H90–H98
18. Mirsky I (1979) Elastic properties of the myocardium: A quantitative approach with physiological and clinical applications. In: Berne RM (ed) Handbook of Physiology, pp 497–531
19. Monroe RG (1964) Myocardial oxygen consumption during ventricular contraction and relaxation. Circ Res 14:294–300
20. Mulieri LA, Luhr G, Treffry J, Alpert NR (1977) Metal-film thermopiles for use with rabbit right ventricular papillary muscles. Am J Physiol 233:C136–C156
21. Rau G (1969) Messung der Koronardurchblutung mit der Argon-Fremdgasmethode. Arch Kreisl-Forsch 58:322
22. Rosenshtraukh LV, Bogdanor KYu, Zakharov SI (1980) The complex structure of mammalian heart contraction. J Mol and Cell Card [Suppl 1] 12:138

Authors' address:

Doz. Dr. Christian Holubarsch, Dept of Medicine, Cardiology, University of Freiburg, Hugstetter Str. 55, D-7800 Freiburg, F.R.G.

Cardiac energetics in clinical heart disease

B. E. Strauer

Department of Medicine, Division of Cardiology, University of Marburg, F.R.G.

Summary

Systolic wall stress is the main determinant of myocardial O_2 consumption in chronic clinical heart disease as well as following acute inotropic pharmacological interventions. The fundamental relationship between stress and O_2 consumption may be modified by alterations in myocardial contractility and/or left ventricular function parameters; these, however, contribute to the overall energy demand by a maximum 10–15%. The therapeutic consequences in chronic heart disease – with regard to left ventricular function and myocardial energy demand – has implications for the degree of left ventricular hypertrophy and dilatation.

Introduction

Myocardial oxygen consumption, which reflects almost totally the overall energy demand of the heart, is determined by at least 5 factors, i.e. systolic wall stress, contractility, heart rate, shortening and the oxygen demand necessary to maintain basal myocardial metabolism. The question is open as to what extent each of these five factors contributes to the overall energy demand of the heart, in various forms of clinical heart disease, especially in cardiac hypertrophy due to chronic pressure and volume overload and in clinical situations associated with different inotropic states or contractility. It is therefore the aim of this study to systematically analyze the oxygen needs of the heart in hypertrophic heart disease (pressure and volume overload) as well as in clinical heart disease with varying contractility (acute inotropic intervention, chronic changes in contractility) in order to determine the potential role of the main determinants of myocardial oxygen consumption under these pathological conditions.

Patient population and methods

Studies were performed during diagnostic cardiac catheterizations in a total of 259 patients:

Aortic stenosis	$n = 14$
Aortic incompetence	$n = 12$
Mitral incompetence	$n = 18$
Combined mitral valve lesions	$n = 32$
Hypertrophic obstructive cardiomyopathy	$n = 12$
Dilatative cardiomyopathy	$n = 10$
Hypertensive heart disease	$n = 92$
Hyperthyroidism	$n = 5$
Acute myocardial infarction	$n = 10$
Systemic collagen diseases (systemic lupus erythematosus etc.)	$n = 22$
Immune complex vasculitis	$n = 22$
Paraproteinemia	$n = 9$

Table 1. Patient population. Alterations in myocardial oxygen consumption induced by alterations in myocardial contractility. Acute, pharmacological interventions

Beta receptor blockade (5 mg Atenolol iv)	n = 19
Beta receptor stimulation (1 mcg/min Isoproterenol iv)	n = 22
Vasodilatation (20 mg Hydralazine iv)	n = 19
Contractility increase by digoxine (0.01 mg/kg iv)	n = 20
prenalterol (7 mcg/kg · min iv)	n = 12
amrinone (1.5 mg/kg iv)	n = 4

Table 2. Patient population. Alterations in myocardial oxygen consumption induced by alterations in myocardial contractility. Chronic clinical heart disease

Dilatative Cardiomyopathy	n = 10
Paraproteinemia	n = 9
Immune complex vasculitis	n = 22
Systemic collagen diseases (SLE, sclerodermia, dermato-myositis)	n = 22
Acute myocardial infarction	n = 10
Hyperthyreoidism	n = 5

The clinical measurements included right and left heart catheterisation, left ventriculography and coronary angiography (except for the hyperthyroid patients), oxymetry, thermodilution and the determination of coronary blood flow using the gas chromatographic argon technique by the analysis of argon in the coronary sinus, as well as in the arterial blood flow. The methodological details have been published previously. Acute inotropic interventions were performed in 116 patients using beta receptor blockade (atenolol), beta receptor stimulation, calcium antagonism, vasodilatation and contractility increase by digoxin, prenalterol and amrinone (Table 1). The analysis of the influence of the contractile state on myocardial oxygen consumption was performed in 78 patients (Table 2). Data are given as mean values ± SEM, significances were calculated by Student's t-test.

Results and discussion

I. *Degree of cardiac hypertrophy*

During the development of cardiac hypertrophy, ventricular wall mass increases. This may have at least three consequences, related to intraventricular volume, wall thickness, and wall mass. Firstly, in compensated pressure overload, as in arterial hypertension and in aortic stenosis, the ventricular wall becomes thickened and the mass is augmented, whereas end-diastolic volume remains normal or only increases mildly. This leads to an increase in mass-to-volume ratio, and is termed concentric hypertrophy. Here the ventricular response is appropriate to the pressure load burdened upon the ventricular wall. Secondly, in decompensated pressure overload, both ventricular mass and volume increase, whereas wall thickness may remain unchanged or be only moderately increased. This leads to ventricular dilatation with constancy or even decrease in the mass-to-volume ratio, and is termed eccentric hypertrophy. Here, ventricular response is inappropriate, since the heart dilates out of proportion to the changes in wall thickness.

Thirdly, in normally shaped pressure overload and also in hypertrophic obstructive cardiomyopathy, excess increase in wall thickness and mass may occur and may thereby narrow the intraventricular cavity. Consequently, the mass-to-volume ratio increases. The type of excess hypertrophy is inappropriate, and is termed irregular, inappropriate hypertrophy. Here, the ventricular response is also inappropriate, since the wall thickness is out of proportion to intraventricular volume changes.

These three types of hypertrophy occur in experimental and in clinical hypertrophy and are associated with different cardiac dimensions, wall stress, ventricular function, and MVO$_2$.

II. *Appropriateness of left ventricular hypertrophy*

The size of left ventricle, as represented by the end-diastolic volume, shows an inverse relationship with left ventricular (LV) function, as evidenced by the ejection fraction. With increase in end-diastolic volume, the ejection fraction decreases. However, the steepness of this relationship varies in different patient groups; the lowest decrease in function with increase in end-diastolic volume, that is aortic incompetence, is present in volume overload, whereas the greatest decrease in function is found in chronic pressure overload due

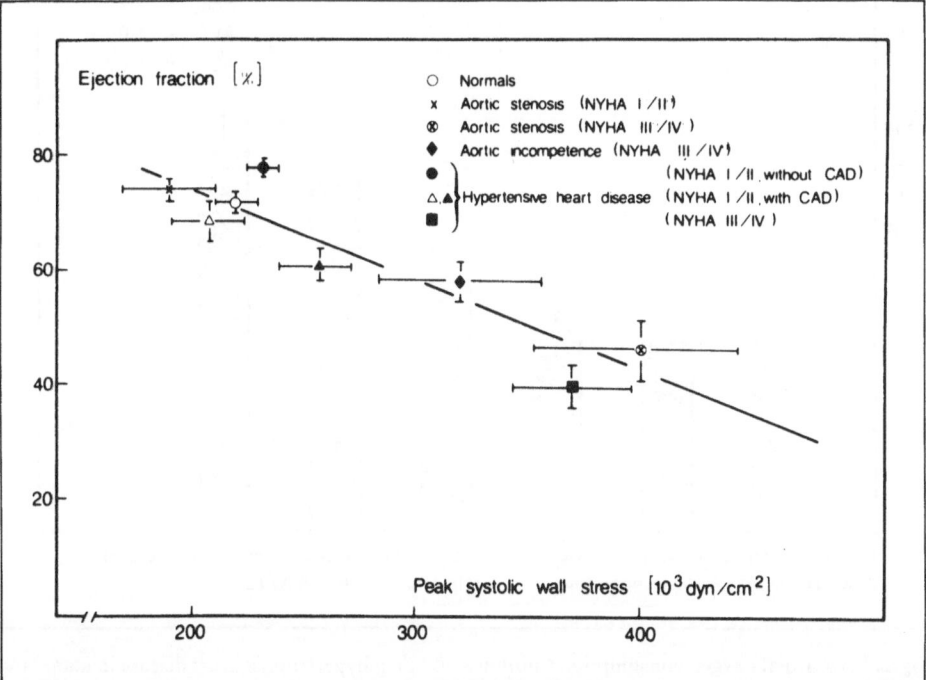

Fig. 1. Relationship between systolic wall stress (afterload) and function (ejection fraction) of the left ventricle.

to aortic stenosis and essential hypertension as well as coronary artery disease. This means that considerable decrease in LV function occurs at a small LV enlargement already in chronic pressure overload, whereas in chronic overload even a large increase in LV size may not affect normal LV function. From the diagnostic point of view, these relationships may help in evaluating LV function from LV size, provided one knows the clinical reason for LV hypertrophy or dilatation.

The different course of these characteristics is probably related to different LV contractile state, or to different LV loading conditions, or both. Except for changes in contractility, the typical alterations in LV afterloading conditions may decrease LV function with an increase in heart size. Accordingly, the relationship between wall stress and function can be determined (Fig. 1). With an increase in systolic wall stress, LV function, as evidenced by the ventricular ejection fraction, decreases. Doubling the stress leads to a reduction in the ejection fraction of approximately 50%. Since systolic wall stress results from systolic pressure and from the mass-to-volume ratio, it is comparable with the ventricular afterload that is imposed on the left ventricular wall. It is therefore reasonable to assume that left ventricular size, end-diastolic volume, and systolic wall stress are important determinants of ventricular performance and energetics.

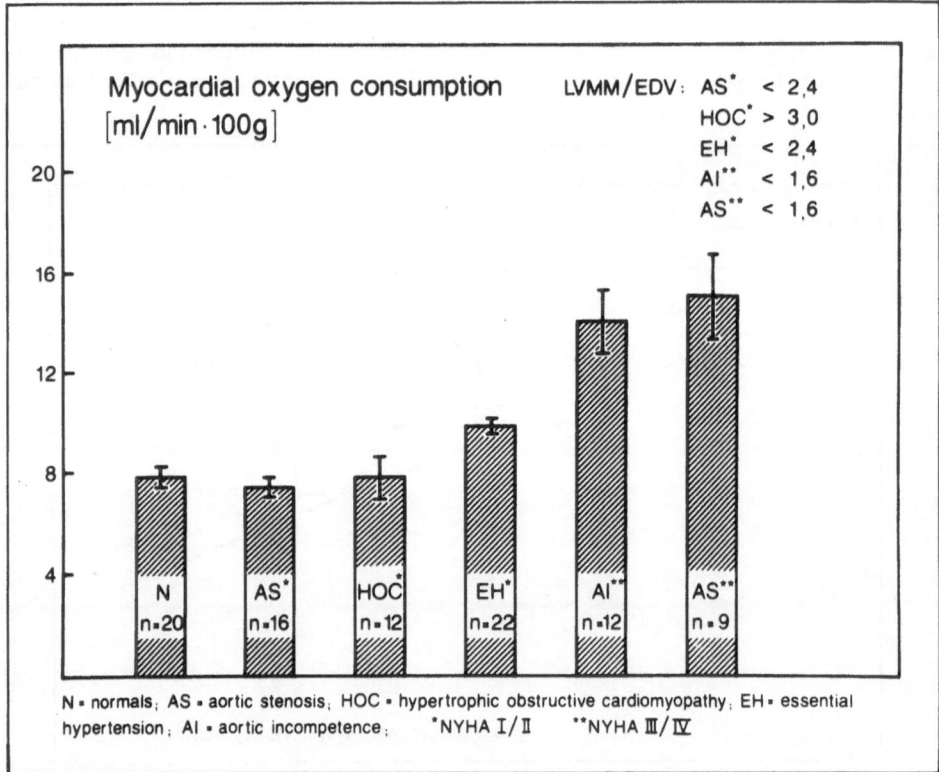

Fig. 2. Myocardial oxygen consumption (ml/min × 100 g) in hypertrophic heart disease in man. N = normals; AS = aortic stenosis (compensated); HOC = hypertrophic obstructive cardiomyopathy; EH = hypertensive heart disease (essential hypertension); AI = aortic incompetence; AS = aortic stenosis (decompensated); * NYHA I/II; ** NYHA III/IV.

III. *Myocardial oxygen consumption*

Myocardial oxygen consumption per weight unit is quite different in clinical heart disease (Fig. 2). It is lowest in normotensive patients with coronary artery disease, it is normal in concentric and clinically compensated LV hypertrophy, even in extreme pressure load, and it is increased in dilated hearts with aortic valve disease (Fig. 2). There was no correlation between the oxygen consumption and ventricular function parameters, as cardiac index, isovolumic contractility indices and ejection phase parameters. However, a significant correlation was found between both left ventricular mass and systolic wall stress and myocardial oxygen consumption (Figs. 3, 4). Patients with decompensated aortic valve disease were within the upper range of this relationship. Extrapolation to zero stress resulted in an intercept of 3.28 ml/min × 100 g, a value which corresponds quite well with the oxygen consumption of the empty beating heart.

In hypertrophic heart disease an inverse non-linear relationship exists between the mass-to-volume ratio and peak systolic wall stress (Fig. 5). The largest mass to volume ratio was found in hypertrophic cardiomyopathy, and lowest values were present for decompensated pressure and volume overload, due to aortic valve disease. Concentric LV hypertrophy, due to essential hypertension and aortic stenosis, was within this correlation, whereas normotensives were shifted to lower systolic stress at equal mass-to-volume ratio, that is, to a lower isobaric relationship.

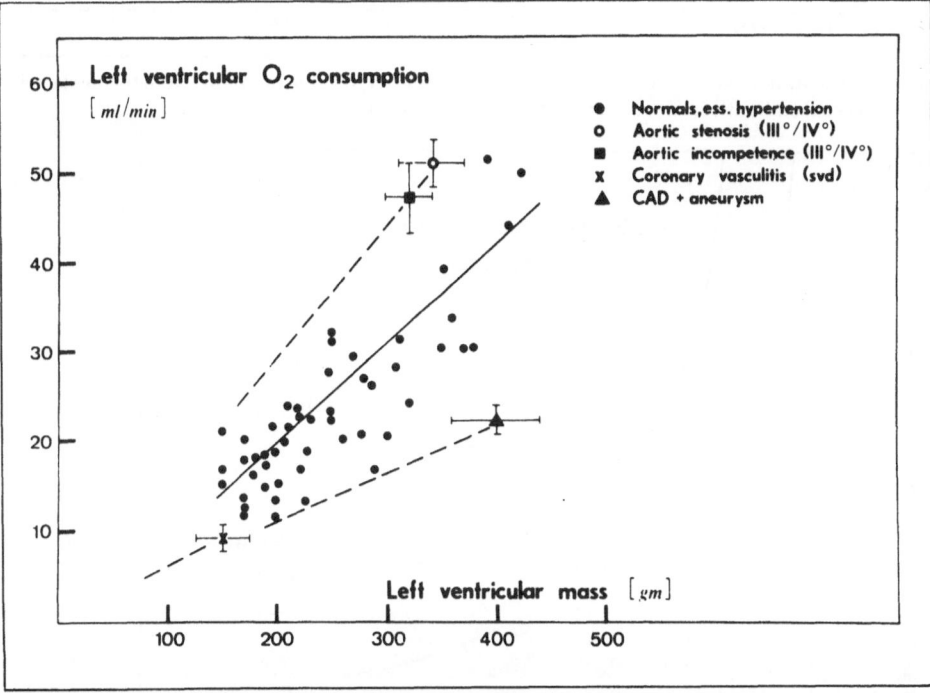

Fig. 3. Relationship between left ventricular mass (LVM) and total left ventricular oxygen consumption (LVO₂). Note the dependence of LVO₂ from LVM; also that at comparable LVM the total LVO₂ is higher in pressure and volume overload and is lower in coronary artery disease (vasculitis) in infarcted hearts (loss of contractile substance).

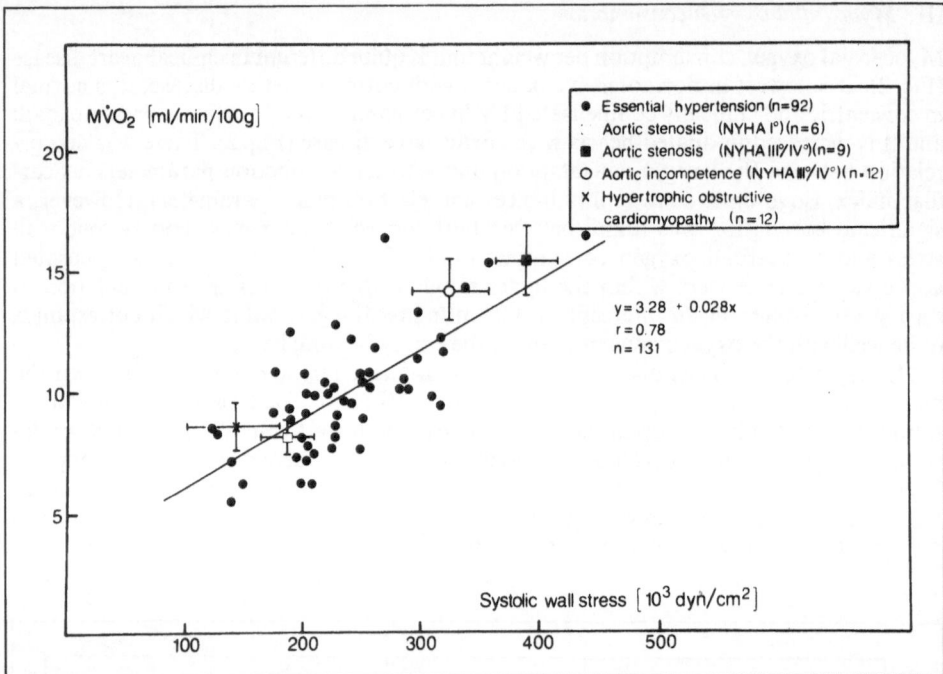

Fig. 4. Correlation of systolic wall stress and myocardial oxygen consumption. Note the linear relationship between both variables in pressure and in volume overload.

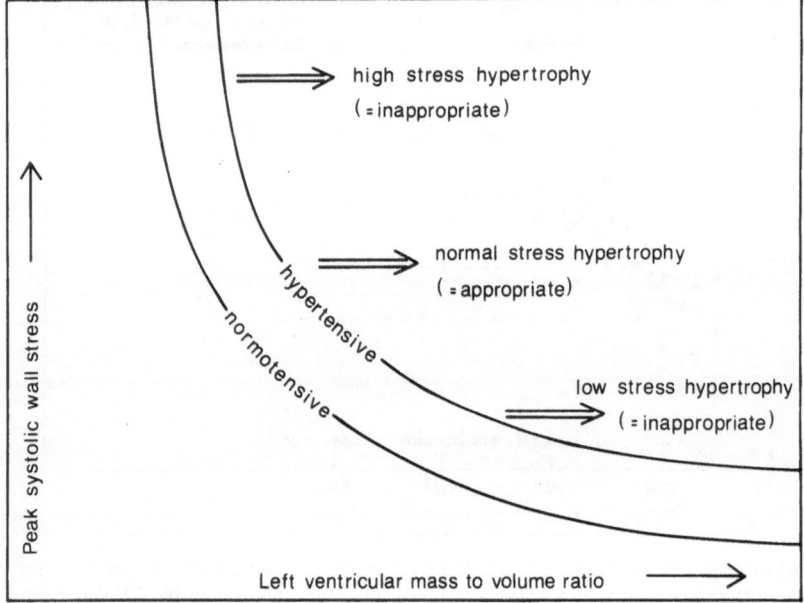

Fig. 5. Diagrammatic representation of the appropriateness of LV hypertrophy in chronic hypertrophic heart disease in man.

With regard to these characteristics, three types of LV hypertrophy (see above) may be classified:
(1) appropriate hypertrophy which keeps systolic wall stress normal even at extreme pressure load, as a result of an appropriate increase in the mass-to-volume ratio parallel to pressure load;
(2) inappropriate, i.e. low stress hypertrophy, which is associated with marked increase in LV mass, out of proportion to intraventricular volumen, and
(3) inappropriate, i.e. high stress hypertrophy, which is characterized by excess dilatation out of proportion to ventricular mass development. Thus, at least two forms of inappropriate hypertrophy may occur in chronic hypertrophic heart disease.

From the metabolic point of view, high stress hypertrophy has an increased oxygen consumption per unit mass with impairment of LV function. In contrast, low stress hypertrophy may have normal or even decreased oxygen consumption per LV unit mass and normal LV function. This helps to explain, for example, the existence of normal, decreased or increased oxygen consumption per LV unit mass in chronic pressure or volume overload. Despite large pressure load, oxygen consumption may be normal or even decreased in aortic stenosis as long as heart size and systolic wall stress are normal or decreased, respectively. On the other hand, the same pressure load may have high oxygen consumption, when LV dilatation occurs and when systolic wall stress increases.

IV. *Coronary haemodynamics in hypertrophic and coronary heart disease*

In the course of diagnostic coronary blood flow, measurements of (a) coronary flow, (b) coronary resistance and (c) coronary reserve (this is the ratio of coronary resistance at control conditions and under maximum coronary vasodilatation, as achieved by dipyridamole: 0.5 mg/kg i.v.) were analyzed in a total of 351 patients. Coronary reserve averaged 4.9 in the normals.

Coronary reserve was considerably reduced in patients with normotensive coronary artery disease, as well as in hypertensives associated with concomitant coronary artery stenosis (Table 3). This reserve reduction may reasonably be referred to the pathologically increased vascular component of coronary vascular resistance, due to narrowing of

Table 3. Coronary reserve (ratio of coronary resistance at rest and following maximum coronary dilatation by dipyridamole: 0.5 mg/kg i.v.)

	n	Coronary reserve
Normals	12	$4 \cdot 9 \pm 0 \cdot 3$
Coronary artery disease (CAD) (normotensive)	186	$2 \cdot 01 \pm 0 \cdot 12^c$
Essential hypertension (with CAD)	52	$1 \cdot 98 \pm 0 \cdot 10^c$
Essential hypertension (without CAD)	22	$3 \cdot 3 \pm 0 \cdot 22^b$
Aortic stenosis (NYHA I/II)	6	$4 \cdot 92 \pm 0 \cdot 39^a$
Aortic stenosis (NYHA III/IV)	9	$2 \cdot 56 \pm 0 \cdot 4^b$
Hypertrophic cardiomyopathy	12	$4 \cdot 0 \pm 0 \cdot 3^a$
Dilatated cardiomyopathy	6	$2 \cdot 9 \pm 0 \cdot 31^b$
Systemic collagen diseases	22	$1 \cdot 61 \pm 0 \cdot 22^c$

[a] NS; [b] $p < 0.01$; [c] $p < 0.001$.

the coronary lumen. In this regard, no significant differences in coronary reserves were found for coronary artery stenosis in the normotensive and in the hypertensive state. However, coronary reserve may be reduced even with normal coronary arteriogram due to an abnormal increase of the myocardial (extravascular) component of coronary vascular resistance.

In dilated cardiomyopathy with normal coronary arteriogram and normal myocardial oxygen consumption, the coronary reserve of the left ventricle was diminished up to 40% of normal controls, thus indicating the abnormal influence of large end-diastolic stress and of muscle fibre stretch on diastolic coronary perfusion.

Except for coronary artery disease (abnormal vascular component of coronary resistance) and dilated heart disease (abnormal myocardial component of coronary resistance), a variety of diseases exist in which coronary reserve may be reduced despite normal coronary arteriogram, due to (a) disorders of coronary microcirculation (inflammatory, non-inflammatory), (b) rheological diseases and (c) metabolically induced alterations of coronary microcirculation (Table 3). One of the most common of these diseases is hypertensive heart disease. Even in young hypertrophied hypertensives without coronary artery disease, a significant reduction in the coronary reserve was found which showed a 34% reduction from the normal controls. An increased coronary and ischaemic risk may be induced in patients with hypertensive LV hypertrophy, even without coronary artery stenosis. Thus, one of the first vascular changes, essential hypertension, seems to influence the coronary circulation.

This hypertension-related reduction in coronary reserve seems not to be hypertrophy-dependent, but to be specific for hypertensive hypertrophy since, except for decompensating aortic stenosis, normal coronary reserve is found in significant myocardial hypertrophy due to clinically compensated aortic stenosis and in hypertrophic cardiomyopathy (Table 3). Thus, the vascular lesions induced by high blood pressure in the coronary vascular bed, may considerably reduce coronary reserve at arteriolar and capillary level, whereas hypertrophy itself may be without limiting significance with regard to oxygen supply of the myocardium.

V. Cardiac reserve capacities
Relations among coronary, metabolic and systolic stress reserves

A close relationship seems to be present among the coronary, metabolic and systolic stress reserves of the left ventricle (Table 4). This obviously helps to explain the limited LV function or reserve capacity or both, in disorders associated with coronary (for example coronary artery disease), metabolic (for example hyperthyroidism) and systolic wall stress abnormalities (for example, in decompensated left ventricular pressure over-

Table 4. Systolic stress (T_{syst}), myocardial oxygen consumption (MVO_2) and coronary flow (V_{cor}). Minimum and maximum values which may be obtained in man

	Minimum	Maximum	Maximum / Minimum
T_{syst} (10^3 dynes/cm^2)	100 ±12	450 ±46	4.5 = stress reserve
MVO_2 (ml/min × 100 g)	5.2± 0.3	24 ± 2.9	4.6 = metabolic reserve
$\dot{V}_{cor}$ (ml/min × 100 g)	79 ±12	392 ±26	4.9 = coronary reserve

load). Because the human heart works almost exclusively aerobically, an increase in left ventricular function may be accomplished preferably by an increase in myocardial oxygen supply. In coronary artery disease, the coronary reserve is limited because of coronary stenosis. Accordingly, the coronary factor is the limiting factor that diminished the oxygen availability to the myocardium, thereby restricting both the metabolic and systolic reserves. In hyperthyroid heart disease an increased MVO_2 occurs due to excess increase in heart rate, cardiac output and contractility, thus narrowing both the coronary and systolic stress reserves. Here, the increased metabolism seems to play the limiting role. In decompensated LV pressure overload, as has been shown previously, both the coronary and metabolic reserves are reduced because the stress level is already elevated at rest. Thus, increased LV loading conditions at rest may skim of maximal stress capacity.

It is concluded from these results that systolic wall stress reflects the major determinant of the degree of LV hypertrophy and plays a dominant role in both left ventricular function and myocardial energy balance. The coronary, metabolic and systolic stress reserves of the human heart seem to be closely correlated with each other, and it is reasonable to assume that determination of coronary reserve gives important insight into the left ventricular metabolic and working capacity reserves.

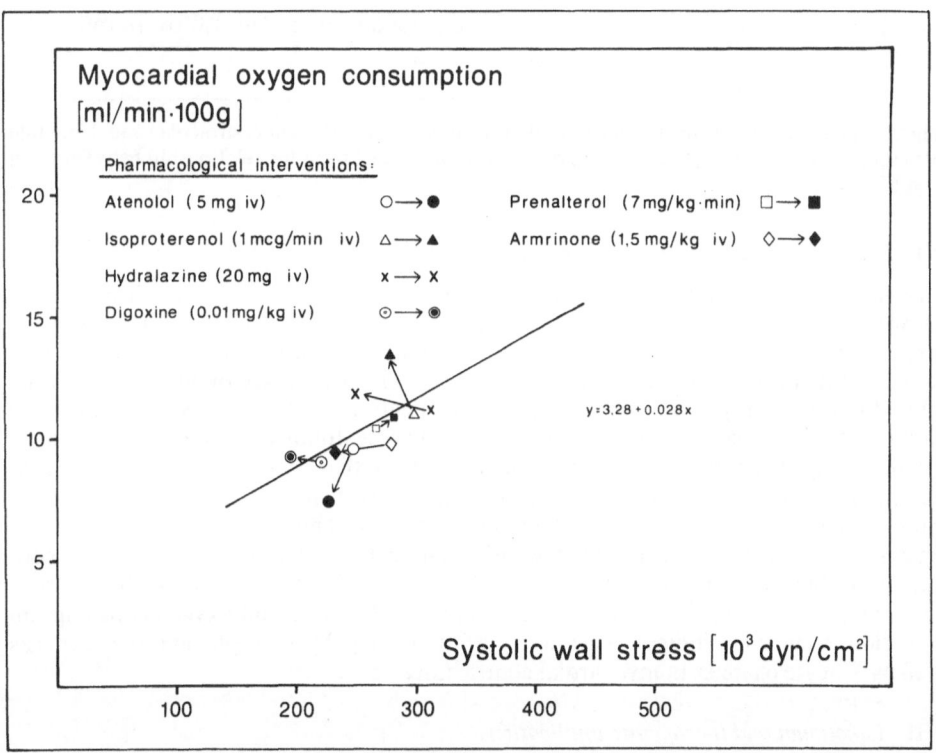

Fig. 6. Changes in MVO_2 following acute inotropic intervention. Note that there are only minor (10–15%) deviations from the fundamental relationship between stress and MVO_2.

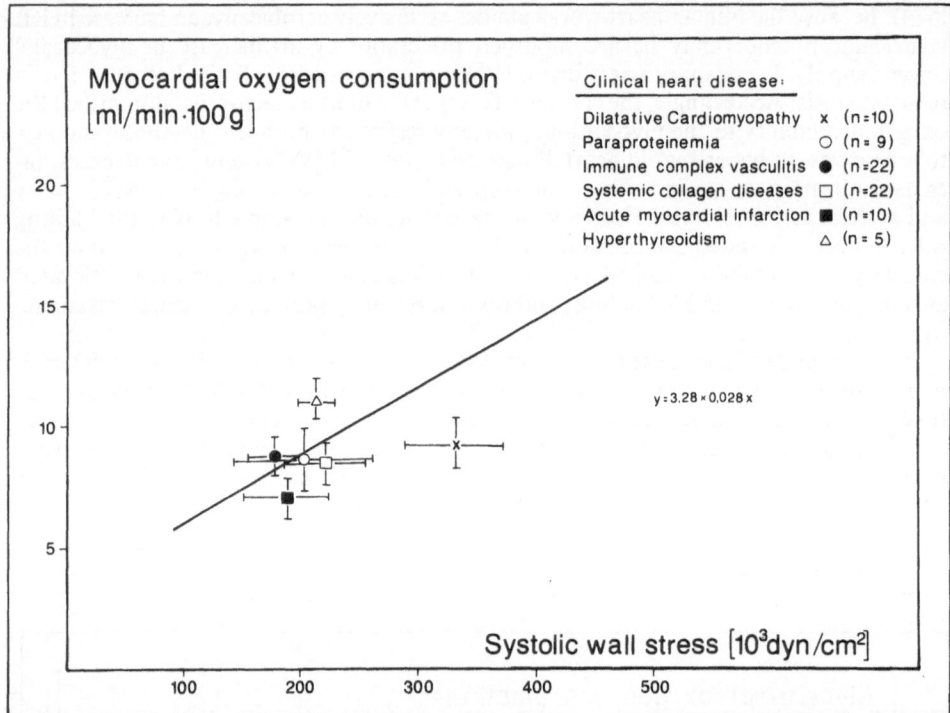

Fig. 7. Changes in MVO$_2$ in chronic clinical heart disease with different contractile state. Note that also under chronic conditions there are only minor deviations from the relationship between stress and MVO$_2$.

VI. *Contractility and myocardial oxygen consumption*

The dependence of myocardial oxygen consumption on both acute and chronic alterations in myocardial contractility, was analysed with regard to changes in myocardial oxygen consumption, considering the fundamental relationship between systolic wall stress and energy demand. In chronic heart disease, which increased or decreased contractility, there were only minor deviations from the relationship between systolic wall stress and myocardial oxygen consumption (Fig. 6). Likewise, following acute inotropic intervention (Fig. 7) the maximum alterations between both variables (stress, oxygen consumption) did not exceed 10–15%. This behaviour was even more relevant when correction for changes in heart rate was taken into account. Thus, acute as well as chronic changes in contractility may modify the basic relationship between stress and oxygen consumption, but in a quantitative sense, not by more than 10–15%. Systolic wall stress, therefore, can be considered the main determinant of myocardial oxygen consumption in various clinical heart diseases, with or without cardiac hypertrophy and in chronic as well as in acute changes in myocardial contractility.

VII. *Functional and therapeutic implications*

The inverse relationship between wall stress and function may be modified by inotropic interventions. This implies that independently of changes in systolic wall stress, there may

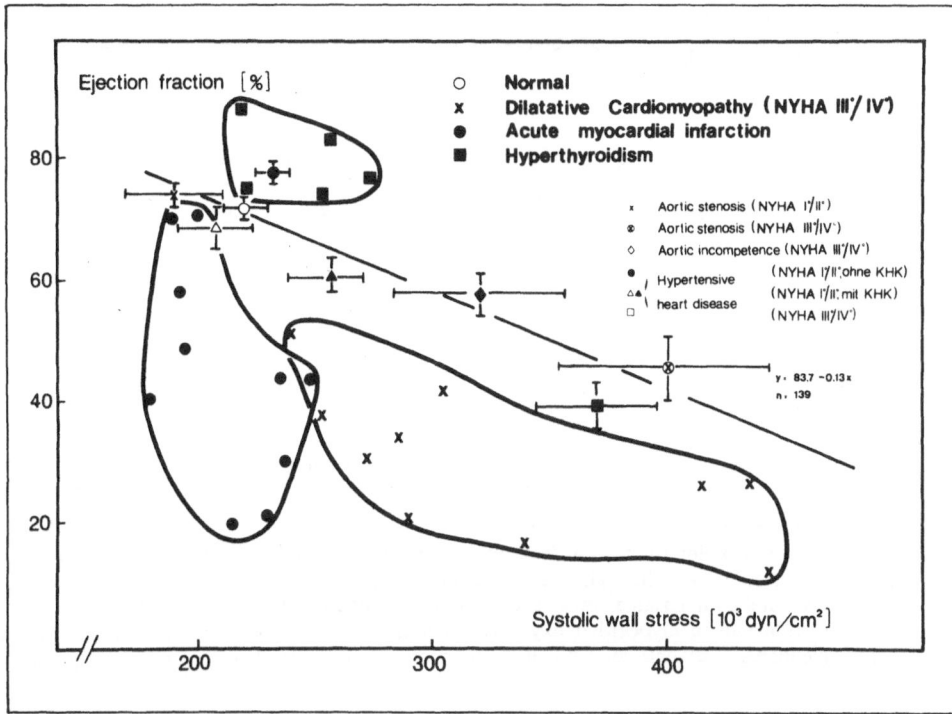

Fig. 8. Relationship between systolic wall stress and ejection fraction for chronic pressure and volume overload (regression line) as well as for acute myocardial infarction, for dilatative (congestive) cardiomyopathy, and for hyperthyroidism. Note the increase to decrease in ejection fraction and comparable systolic wall stress when the contractile state of the heart is changed.

exist both increases and decreases in left ventricular function at various contractile states. In acute myocardial infarction considerable decrease in function at normal wall stress occurs as the result of a loss in contractile substance, that is, because of myocardial necrosis. Similarly, in dilatative cardiomyopathy a lower ejection fraction at equal systolic wall stress was found when compared with chronic LV pressure and volume load. Conversely, in hyperthyroidism there is an increased LV ejection fraction (Fig. 8).

An improvement in ventricular function may be induced either by changes in load (e.g. afterload reduction), according to the load-dependent increase in ejection fraction, or by positive inotropic interventions, according to the contractile-dependent and load-independent upward shift of stress-function relationship, or by a combination of both, that is, by unloading and contractile involvements.

The therapeutic implications – as may be derived from our studies – also embrace another point of view, i.e. the clinical significance and value of a reduction of preload and/or afterload in clinical heart disease, with regard to changes in systolic wall stress, left ventricular function and myocardial oxygen consumption.

According to the unlinear relationship between the mass-to-volume ratio and systolic wall stress, preload reduction has proven to be far more effective in the dilated heart (eccentric hypertrophy) than in the compensated, concentrically hypertrophied state. The mean values of our studies demonstrate that a reduction in the end-diastolic volume (pre-

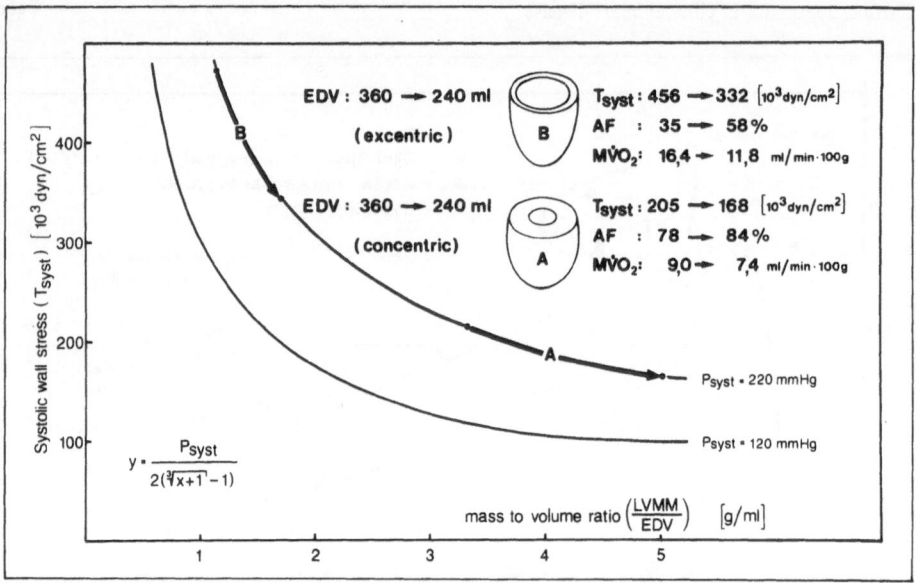

Fig. 9. Influence of preload reduction (end-diastolic volume, 360 ml–240 ml, pressure = const.) on mass-to-volume ratio, systolic wall stress, ejection fraction, and MVO₂. Note that only minor changes occur in a compensated concentrically hypertrophied heart with regard to changes in the ejection fraction and in myocardial oxygen consumption, whereas in a dilated heart there are considerable increase in ventricular function and decrease in myocardial oxygen consumption.

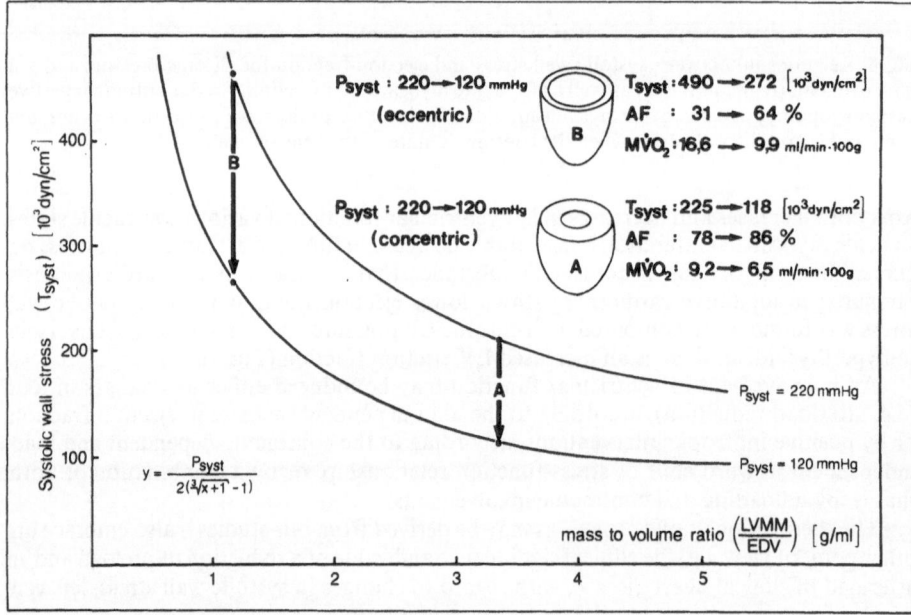

Fig. 10. Influence of afterload reduction (decrease in systolic pressure from 220 to 120 mm Hg, EDV = constant) in the compensated concentrically hypertrophied heart. Note that there are only minor changes in ventricular function and myocardial oxygen consumption, whereas in a dilated heart there are considerable increases in ejection fraction (from 31% to 64%) and marked decrease in MVO₂ (from 16.6 ml/min × 100 g to 9.9 ml/min × 100 g).

load) from 360 to 240 ml reduces systolic wall stress in the dilated heart (eccentric hypertrophy) from 456 to 332 units, improves left ventricular ejection fraction from 35 to 58 per cent and lowers myocardial oxygen consumption from 16.4 to 11.8 units (Fig. 9). In contrast, in concentric hypertrophy the same preload reduction exerts only a minor improvement on these three variables.

Similar results have been found for afterload reduction. In dilated hypertensive heart disease (eccentric hypertrophy) decrease in systolic blood pressure from 220 to 120 mm Hg results in a decrease in systolic wall stress from 490 to 272 units, in an improvement in left ventricular ejection fraction from 31 to 64 per cent and in a marked decrease in myocardial oxygen consumption from 16.6 to 9.9 units. In contrast, the same blood pressure decrease leads to only minor changes in these three variables in the concentrically hypertrophied heart (Fig. 10).

The therapeutic rationale which may be derived from the influence of changes in ventricular geometry on left ventricular function and myocardial energy demand implies the maintenance of a normal, or the normalization of an increased, systolic wall stress, which is dependent on the load conditions of the heart, that is, preload and afterload. Systolic wall stress represents the main determinant of both cardiac function and energetics and its normalization is of superior clinical importance for the diseased, dilated heart.

References

1. Alpert NR (1971) Cardiac hypertrophy. Academic Press, New York London, pp 1–23
2. Folkow B (1982) Physiological aspects of primary hypertension. Physiol Rev 62:347–504
3. Ford LE (1976) Heart size. Circ Res 39:297–303
4. Hood WP (1971) Dynamics of hypertrophy in left ventricular wall of man. In: Alpert NR (ed) Cardiac hypertrophy. Academic Press, New York London, pp 445–452
5. Kochsiek K, Heiss HW, Tauchert M, Strauer BE (1971) Koronarreserve und Sauerstoffverbrauch bei hypertrophischer obstruktiver Cardiomyopathie. Verh Dtsch Ges Inn Med 27:880–883
6. Kochsiek K, Tauchert M, Cott L, Neubaur J (1970) Die Koronarreserve bei Patienten mit Aortenvitien. Verh Dtsch Ges Inn Med 76:214–218
7. Linzbach AJ (1960) Heart failure from the point of view of quantitative anatomy. Am J Cardiol 5:370–382
8. Linzbach AJ (1981) Structural adaptation of the heart in hypertension and the physical consequences. In: Strauer BE (ed) The heart in hypertension. Springer-Verlag, Berlin Heidelberg New York, pp 243–250
9. Meerson FS (1969) Hyperfunktion, Hypertrophie und Insuffizienz des Herzens. VEB Volk und Gesundheit, Berlin
10. Strauer BE, Tauchert M, Cott L, Kochsiek K, Bretschneider HJ (1970) Simultane Bestimmung des Sauerstoffverbrauches und der Coronardurchblutung des linken Ventrikels bei Mitral- und Aortenklappenfehlern mit einem neuen hämodynamischen Parameter und der Argon-Fremdgasmethode. Verh Dtsch Ges Inn Med 76:217–220
11. Strauer BE, Brune I, Schenk H, Knoll D, Perings E (1976) Lupus cardiomyopathy: cardiac mechanics, hemodynamics, and coronary blood flow in uncomplicated systemic lungs erythematosus. Am Heart J 92:715–722
12. Strauer BE (1977a) Die quantitative Bestimmung der Koronarreserve zur Diagnostik koronarer Durchblutungsstörungen. Internist (Berlin) 18:579–587
13. Strauer BE, Scherpe A (1978) Ventricular function and coronary hemodynamics after intravenous nitroglycerin in coronary artery disease. Am Heart J 95:210–219
14. Strauer BE (1983) Das Hochdruckherz, 2nd ed. Springer, Berlin Heidelberg New York, pp 1–184

15. Strauer BE (1985) Progression und Regression der Herzhypertrophie beim arteriellen Bluthoch-
    druck. Z Kardiol [Suppl 7] 74:171–178
16. Strauer BE, Klepzig M, Motz W (1986) Reversal of left ventricular and coronary hypertrophy
    following antihypertensive treatment. In: Kaufmann W (ed) Primary hypertension. Springer,
    Berlin Heidelberg New York, pp 115–125

Author's address:

Prof. Dr. B. E. Strauer, Abteilung für Innere Medizin – Schwerpunkt Kardiologie,
Universität Marburg, Klinikum Lahnberge, D-3550 Marburg, F.R.G.

Influence of phosphodiesterase inhibition on myocardial energetics in dilative cardiomyopathy

G. Hasenfuss, Ch. Holubarsch, W. H. Heiss, T. Bonzel, T. Meinertz, H. Just

Department of Internal Medicine, III (Cardiology), Medical Clinic,
University of Freiburg, F.R.G.

Summary

The effects of inhibition of phosphodiesterase by enoximone on left ventricular haemodynamics and myocardial energetics were investigated in 10 patients with idiopathic dilative cardiomyopathy. After intravenous administration of enoximone, there was a significant reduction of left ventricular systolic pressure from 126 ± 21 to 93 ± 16 mm Hg, of left ventricular end-diastolic pressure from 16 ± 8 to 5 ± 3 mm Hg and of left ventricular end-diastolic volume from 287 ± 54 to 215 ± 69 ml. Left ventricular pressure-volume work decreased significantly from 12.1 ± 3.6 to 7.6 ± 2.8 mm Hg·l. Heart rate was 87 ± 17 before and 103 ± 18 min^{-1} after administration of enoximone (p<0.01). Left ventricular systolic stress-time integral, a major determinant of myocardial oxygen consumption, decreased by 49% from 91 ± 32 to $46 \pm 15 \, 10^3$ dyn·s/cm^2 (p<0.01). In contrast myocardial oxygen consumption per beat was reduced by only 18%, from 138 ± 28 to 113 ± 17 μIO$_2$/100 g (p<0.01). The economy of myocardial contraction as calculated by the ratio of systolic stress-time integral to myocardial oxygen consumption per beat was 675 ± 192 before and 370 ± 128 dyn·s·100 g/cm^2·μIO$_2$ after administration of enoximone. In conclusion, the phosphodiesterase inhibitor enoximone exhibits vascular and myocardial effects. The myocardial effects result in decreased economy of myocardial contraction. The possible molecular mechanisms of these energetic changes are discussed.

Introduction

Increased concentration of cyclic AMP in the myocardial cell results in increased transcellular calcium influx in response to depolarization [1] as well as increased release of calcium from the sarcoplasmic reticulum [2]. Therefore, elevation of intracellular cyclic AMP concentration is a well known pharmacological principle to augment contractile force of the myocardium [3]. From a therapeutic point of view two different mechanisms of increasing cyclic AMP concentration exist: Firstly, the stimulation of myocardial β1-receptors which results in increased cyclic AMP formation, and secondly, the inhibition of phosphodiesterase which results in decreased cyclic AMP degradation [3]. Regarding the energetic consequences of pharmacologically increased cyclic AMP, in the failing human heart, it was recently supposed that myocardial oxygen consumption may increase, resulting in accelerated progression of the underlying heart failure [4].

Accordingly, the aim of the present study was to evaluate the influence of phosphodiesterase inhibition, induced by the recently developed substance enoximone, on myocardial energetics in human dilative cardiomyopathic hearts. Enoximone was shown to be a potent inhibitor of phosphodiesterase FIII, the cyclic AMP specific cardiac phosphodiesterase [5].

Methods

Patients

The influence of phosphodiesterase inhibition on myocardial energetics was investigated in 10 patients with idiopathic dilative cardiomyopathy (NYHA II–III) undergoing routine diagnostic heart catheterization. 6 patients were male and 4 were female; the mean age was 49 ± 9 years (range of 28 to 60 years).

Cardiac catheterization

Previous medications were withheld at least 24 hours before the study. Cardiac catheterization was started with coronary angiography to exclude coronary heart disease. Next left ventricular and aortic pressures were recorded, and left ventriculography with simultaneous pressure measurements was performed by a Millar microtip catheter pressure transducer (50 frames per second). The projection was a caudal angulated 45 ° right anterior oblique view. Left ventricular volume was calculated using the method of Sandler et al. [6]. Left ventricular wall thickness was calculated as the average thickness of a 2–3 cm segment of the free wall one third of the distance from the apex. Left ventricular muscle mass was calculated according to a modification of the method of Rackley et al. [7]. Instantaneous left ventricular wall stress was calculated every 20 ms throughout one cardiac cycle using the ellipsoid model of Mirsky [8]. The systolic stress-time integral was calculated by integrating instantaneous stress values from end-diastole to end-systole. The pressure-volume loop was obtained by plotting instantaneous volume versus pressure during one cardiac cycle, and left ventricular pressure-volume work was taken as the area of the pressure-volume loop.

Measurement of myocardial oxygen consumption

Myocardial blood flow (MBF; ml/min/100 g) was calculated using the argon method [9, 10]. Myocardial oxygen consumption ($M\dot{V}O_2$; mlO_2/min/100 g) was determined as the product of MBF and arterial-coronary-sinus oxygen content difference (mlO_2/100 ml). Arterial-coronary-sinus oxygen content difference was derived from oxygen saturation measurements by oxymetry (AO unisat oximeter).

Calculation of economy of myocardial contraction

$M\dot{V}O_2$ (mlO_2/min/100 g) was divided by heart rate to obtain the oxygen consumption of one cardiac cycle ($M\dot{V}O_2$/beat; μlO_2/100 g). The economy of myocardial contraction was calculated by the ratio of the systolic stress-time integral and $M\dot{V}O_2$/beat.

Enoximone administration

During control conditions, left ventriculography with simultaneous pressure measurement was performed, and myocardial blood flow and aortic pressure were measured. Thereafter enoximone was infused at a dose of 1 mg/kg at a rate of 12.5 mg/min. Five minutes after completion of the enoximone infusion, the aortic pressure was read again. If the mean aortic pressure had fallen by more than 15% or was below 75 mm Hg, no additional enoximone was infused. Otherwise, an additional dose of 1 mg/kg at 12.5 mg/min was administered. 5 patients received 2 mg/kg, 5 patients 1 mg/kg. 15 minutes after completion of the enoximone infusion left ventriculography with simultaneous pressure measurement and measurement of myocardial blood flow were repeated.

Statistical analysis

Values are expressed as mean ± one standard deviation. Comparisons between measurements before and after administration of enoximone were made using the paired t-test. A value of $p < 0.01$ was considered statistically significant.

Results

Following the administration of enoximone, a significant decrease occurred in left ventricular systolic pressure from 126 ± 21 to 93 ± 16 mm Hg, left ventricular end-diastolic pressure from 16 ± 8 to 5 ± 3 mm Hg, left ventricular end-diastolic volume from 287 ± 54 to 215 ± 69 ml and left ventricular pressure-volume work from 12.1 ± 3.6 to 7.6 ± 2.8 mm Hg·l (Fig. 1). Heart rate was 87 ± 17 before and 103 ± 18 min^{-1} after enoximone ($p < 0.01$).

Left ventricular systolic stress-time integral decreased by 49% from 91 ± 32 to $46 \pm 15 \, 10^3$ dyn·s/cm^2 ($p < 0.01$) after administration of enoximone. In contrast, $M\dot{V}O_2$ did not significantly change (12.0 ± 2.4 versus 11.7 ± 2.5 mlO$_2$/min/100 g), and $M\dot{V}O_2$/beat

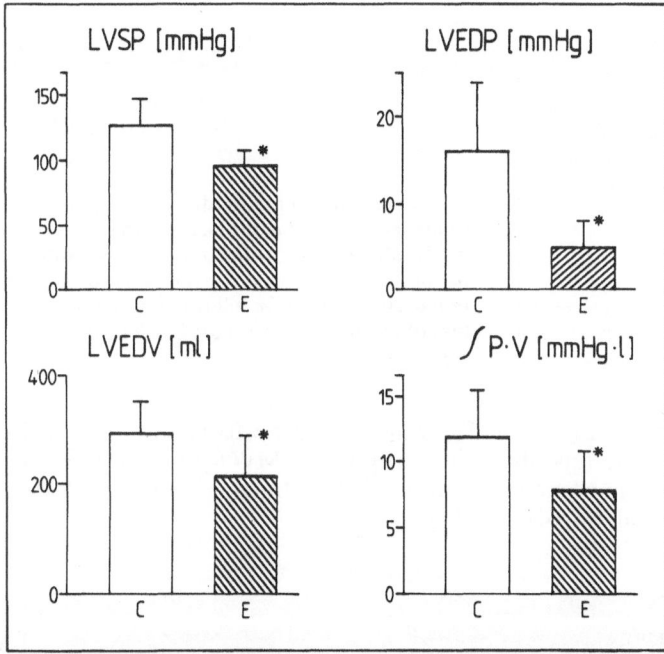

Fig. 1. Influence of enoximone on left ventricular haemodynamics. LVSP = left ventricular systolic pressure; LVEDP = left ventricular end-diastolic pressure; LVEDV = left ventricular end-diastolic volume; ∫P·V = left ventricular pressure·volume work; *$p < 0.01$; C = control; E = enoximone.

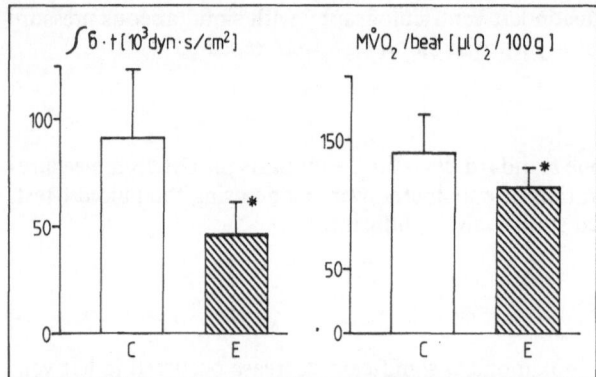

Fig. 2. Influence of enoximone on left ventricular systolic stress-time integral ($\int \sigma \cdot t$), a major determinant of $M\dot{V}O_2$, and myocardial oxygen consumption per beat. $\int \sigma \cdot t$ decreased by 49% due to enoximone; in contrast, myocardial oxygen consumption per beat ($M\dot{V}O_2$/beat) decreased only by 18%. *p < 0.01; C = control; E = enoximone.

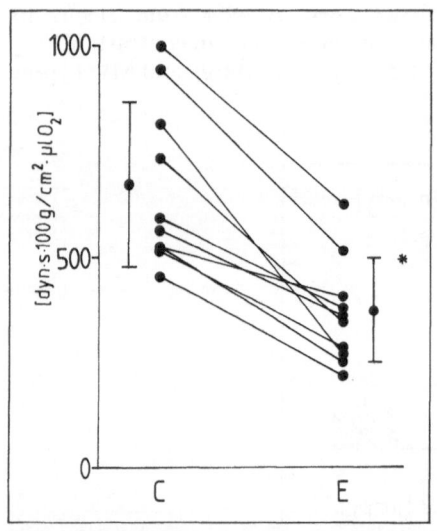

Fig. 3. Economy of myocardial contraction was evaluated by dividing the systolic stress-time integral by the myocardial oxygen consumption per beat. The economy of myocardial contraction decreased by 45% following the administration of enoximone. *p < 0.01; C = control; E = enoximone.

decreased by only 18% (from 138 ± 28 to 113 ± 17 $\mu IO_2/100$ g; p < 0.01) (Fig. 2). This discrepancy between changes in systolic stress-time integral and $M\dot{V}O_2$/beat due to enoximone resulted in a decrease in economy of myocardial contraction from 675 ± 192 to 370 ± 128 dyn · s · 100 g/cm^2 · μIO_2 (p < 0.01) (Fig. 3).

Discussion

Methodological considerations

Myocardial energy consumption is mainly determined by heart rate, systolic stress-time integral and velocity of myocardial contraction [11, 12]. Economy of myocardial contraction can be analyzed by relating myocardial oxygen consumption to its determinants. From the following considerations we evaluated the economy of myocardial contraction by the ratio of systolic stress-time integral and myocardial oxygen consumption per beat:

1. As was recently shown, changes in stress-time integral are linearly related to changes in myocardial oxygen consumption of the human heart [12]. Changes in pre- or afterload result in corresponding changes of left ventricular stress-time integral and myocardial oxygen consumption.

2. Heart rate can be taken into account by normalizing myocardial oxygen consumption to one cardiac cycle, assuming that oxygen consumption per beat is not substantially influenced by changes in heart rate itself [13, 14].

3. A given increase in contractile force can be achieved with variable rise in the velocity of myocardial contraction [15]. Furthermore, a given increase in velocity of myocardial contraction can be achieved with variable energetic costs depending on the inotropic interventions applied [16]. That is, the extra energy consumption for the increase in contractile force is variable and depends on the respective inotropic intervention. Since the intention of the use of inotropic interventions in heart failure is to increase contractile force, primarily, and not to increase velocity of myocardial contraction, the latter should not be included when changes in economy due to inotropic interventions are analysed.

Accordingly, calculating the ratio of stress-time integral to $M\dot{V}O_2$/beat enables evaluation of the energetic cost of different inotropic interventions, independent of changes in left ventricular pre- and afterload.

Clinically available phosphodiesterase inhibitors are known to possess several pharmacological actions besides their phosphodiesterase inhibiting properties (inhibition of calcium reuptake by the sarcoplasmic reticulum, adenosine-antagonistic action, sensitization of the contractile apparatus to calcium [3]). Nevertheless, since it was shown that enoximone is a potent inhibitor of phosphodiesterase FIII, the cyclic AMP specific cardiac phosphodiesterase [5], our energetic findings may predominantly represent the energetic changes due to inhibition of this enzyme in the myocardial cell.

Myocardial energetics

The purpose of inotropic interventions in the treatment of acute or chronic heart failure is to increase the reduced contractile force of the myocardium. On the intracellular level an increase in contractile force may result from the recruitment of additional cross-bridges or changes in cross-bridge kinetics [17]. Additional cross-bridges may be recruited by increasing the cytosolic calcium ions or by increasing the sensitivity of the acto-myosin system to calcium ions [18, 19, 20]. Concerning cross-bridge kinetics, contractile force may be increased by prolonging the on-time (force producing state) or by increasing cycling rate of the cross-bridges. Since the amount of ATP-consumption depends on the quantity of calcium-cycling and the frequency of cross-bridge cycling [17], the energetic costs of inotropic interventions depend on the molecular mechanism by which force is increased.

In the present study it was demonstrated that the administration of enoximone resulted in decreased economy of myocardial contraction. Left ventricular systolic stress-time integral was considerably reduced due to peripheral vasodilation following the administration of enoximone. In contrast, only a small decrease was observed in myocardial oxygen consumption per beat. This discrepancy between changes in myocardial oxygen consumption and its main determinant indicates increased energy demand of the myocardium per unit developed stress-time integral due to enoximone. Gross myocardial oxygen consumption did not increase in these patients since the overall reduction in left ventricular wall stress due to the vasodilating effects of enoximone (preload and afterload) compensated for the decreased economy at the myocardial level.

The present energetic findings in the human heart are in accordance with animal experiments, performed in the guinea pig myocardium in which increased heat liberation per unit developed stress-time integral due to enoximone has been described [21]. Since similar energetic changes have also been reported due to β1-stimulation in the rat and guinea pig myocardium [20, 21, 22], the decrease in economy of myocardial contraction due to β1-stimulation, as well as phosphodiesterase inhibition, most likely results from increased intracellular cyclic AMP. Cyclic AMP increases cytosolic calcium ions and therefore raises the ATP consumption for calcium pumping [17]. Furthermore, cyclic AMP may decrease the economy on the level of the contractile proteins by increasing the cross-bridge cycling rate at the cost of the on-time [20]. This assumption is further supported by the finding that increased cyclic AMP concentration results in increased enzymatic activity of myosin [23].

Conclusions

Inotropic drugs which mainly act by inhibiting phosphodiesterase may decrease the economy of myocardial contraction in dilative cardiomyopathic hearts. The effect of these drugs on gross myocardial oxygen consumption, depends on the drug induced changes in left ventricular working conditions which may, in part, compensate for decreased economy at the level of the myocardium.

References

1. Sperelakis N (1984) Cyclic AMP and phosphorylation in regulation of Ca^{++} influx into myocardial cells and blockade by calcium antagonistic drugs. Am Heart J 107:347–357
2. Kranias EG, Solaro J (1983) Coordination of cardiac sarcoplasmic reticulum and myofibrillar function by protein phosphorylation. Fed Proc 42:33–38
3. Scholz H (1984) Inotropic drugs and their mechanisms of action. J Am Coll Cardiol 4:389–397
4. Packer M, Medina N, Yushak M (1984) Hemodynamic and clinical limitations of long-term inotropic therapy with amrinone in patients with severe chronic heart failure. Circulation 70:1038–1047
5. Kariya T, Wille LJ, Dage RC (1982) Biochemical studies on the mechanism of cardiotonic activity of MDL 17043. J Cardiovasc Pharmacol 4:509–514
6. Sandler H, Dodge HT (1968) The use of single plane angiocardiograms for the calculation of left ventricular volume in man. Am Heart J 75:325–334
7. Rackley CE, Dodge HT, Coble YD, Hay RE (1964) A method for determining left ventricular mass in man. Circulation 29:666–671
8. Mirsky I (1979) Elastic properties of the myocardium: A quantitative approach with physiological and clinical applications. In: Berne RM (ed) Handbook of physiology. The cardiovascular system. American Physiological Society, Washington DC, pp 497–531
9. Rau G (1969) Messung der Koronardurchblutung mit der Argon-Fremdgasmethode. Arch Kreisl-Forsch 58:322–398
10. Heiss HW, Blümchen G (1982) Durchblutungsmessung am Coronargefäßsystem. In: Reindell H, Roskamm H (eds) Herzkrankheiten. Springer, Berlin Heidelberg New York, pp 413–432
11. Braunwald E (1971) Control of myocardial oxygen consumption. Physiologic and clinical considerations. Am J Cardiol 27:416–432
12. Laskey WL, Reichek N, Sutton MSJ, Untereker WJ, Hirsfeld JW (1983) Myocardial oxygen consumption in left ventricular hypertrophy and its relation to left ventricular mechanics. Am J Cardiol 52:852–858

13. Maxwell GM, Castillo CA, White DH, Crumpton CW, Rowe GG (1958) Induced tachycardia: Its effect upon the coronary hemodynamics, myocardial metabolism and cardiac efficiency of the intact dog. J Clin Invest 37:1413–1418
14. Boerth RC, Cowell JW, Pool PE, Ross J (1969) Increased myocardial oxygen consumption and contractile state associated with increased heart rate in dogs. Circ Res 24:725–732
15. Graham TP, Cowell JW, Sonnenblick EH, Ross J, Braunwald E (1968) Control of myocardial oxygen consumption: Relative influence of contractile state and tension development. J Clin Invest 47:375–385
16. Graham TP, Ross J, Cowell JW, Sonnenblick EH, Clancy RL (1967) Myocardial oxygen consumption in acute experimental cardiac depression. Circ Res 21:123–138
17. Alpert NR, Mulieri LA (1984) Hypertrophic adaptation of the heart to stress: A myothermal analysis. In: Zak R (ed) Growth of the heart in health and disease. Raven Press, New York, pp 363–379
18. Fabiato A (1983) Calcium-induced release of calcium from the cardiac sarcoplasmic reticulum. Am J Physiol 245:C1–14
19. Solaro RJ, Rüegg JC (1982) Stimulation of Ca^{++} binding and ATPase activity of dog cardiac myofibrils by AR-L 115BS, a novel cardiotonic agent. Circ Res 51:290–294
20. Holubarsch Ch, Hasenfuss G, Blanchard E, Alpert NR, Mulieri LA, Just H (1986) Myothermal economy of rat myocardium, chronic adaptation versus acute inotropism. Basic Res Cardiol 81:95–102
21. Holubarsch Ch, Hasenfuss G, Just H, Mulieri LA, Alpert NR (1986) Energetic costs of inotropic interventions in guinea pig papillary muscles. Circulation 74:326
22. Hasenfuss G, Holubarsch Ch, Just H, Blanchard E, Mulieri LA, Alpert NR (1987) Energetic aspects of inotropic interventions in rat myocardium. (This book)
23. Winegrad S, Weisberg A, Lin LE, McClellan G (1986) Adrenergic regulation of myosin adenosine triphosphate activity. Circ Res 58:83–95

Authors' address:

Gerd Hasenfuß, M.D., Universität Freiburg, Innere Medizin, III (Cardiology), Hugstetter Str. 55, D-7800 Freiburg, F.R.G.

Predicting postoperative haemodynamics in valve patients

W. A. Baxley

University of Alabama Medical Center Birmingham, Alabama, U.S.A.

Summary

Patients with combined valve and myocardial disease often have poor haemodynamic status early postoperatively. This occurs in spite of normalization of the left ventricular work load by technically uncomplicated valve replacement. Therefore an algorithm was developed for predicting postoperative left ventricular performance, based on the E_{max} concept, (end-systolic wall stress/volume relationship as a load-independent ventricular function parameter). Load changes effected by valve normalization were included in the predictive methodology, with ventricular function assumed unchanged by surgery. The algorithm was tested in 12 valve patients who had less than 10% change in heart rate and left atrial pressure pre- vs postoperatively. Preoperative data were obtained during catheterization with quantitative ventriculography. The predicted data were compared to measured data on postoperative day I. There were non-significant differences between the means of predicted and of measured postoperative left ventricular stroke volume, end-systolic volume, end-systolic stress, stroke work, and aortic pressure. The postoperative myocardial function parameter fell by 3–20% below preoperative values in 8 patients not requiring high-dose catechol support and rose by 3–36% in those requiring support. This pilot study suggests the feasibility of a predictive haemodynamic algorithm in surgical valve disease.

Introduction

One clinical problem in severe valvular heart disease is the wide range of haemodynamic status encountered early postoperatively [17, 20]. Many patients, particularly those who are relatively young, with stenotic valvular disease, good ventricular function, and clear indications for surgery, have no postoperative problems whatsoever. However, those with some indication of decreased ventricular function together with mitral or aortic regurgitation, may have poor postoperative haemodynamic status or even fatal inability to wean from cardio-pulmonary bypass, in spite of an excellent surgical procedure. Theoretically, valve surgery is done to "normalize" the workload on the left ventricle and hence improve pulmonary congestive phenomena and cardiac output capability. In the unique case of mitral stenosis, the operation will augment left ventricular filling and hence work, but in other forms of valve disease, surgery should decrease resting ventricular work. A number of previous studies bearing on this problem have been somewhat helpful but have been primarily epidemiological in format [2, 4–7, 9, 13, 19, 22]. That is, they have compared various preoperative measurements such as echo-determined chamber size or ejection fraction, with repeat measurements postoperatively, or with subjective clinical outcome.

In the present study we have taken a different approach to this problem, and have attempted to develop methodology for predicting what the haemodynamic status will be postoperatively for individual patients. The hypothesis is that early postoperative relationships between preload (expressed in this clinical setting as left atrial or left ventricular end-diastolic pressure), afterload (aortic pressure) and cardiac output can be predicted from preoperative ventricular function parameters considered, together with the expected ventricular work load changes, to be caused by correction of the valve abnormality.

Methods

1. *Experimental design*

Methodology for the study involves construction of left ventricular pressure-volume curves [14] for one cardiac cycle, together with calculation of an index of circumferential wall stress (stress per 100 g myocardium). This stress index (σ), rather than absolute stress, was utilized, since change from control values was involved, rather than inter-patient comparisons. The parameter stress-index/volume at end-systole (σ_{es}/Ves) was used to quantitate myocardial function. This functional parameter is related to the E_{max} concept but may be considered less precise, as E_{max} requires measurements made at several afterloads [8, 13, 16, 19, 21]. E_{max} has been defined as the end-systolic pressure volume relationship which has been found to be linear over a range of physiological loads. However, it encompasses a "Vo" quantity, or theoretic volume at 0 end-systolic pressure. By utilizing a single point (σ_{es}/Ves), we assume "Vo" to be 0. Recent literature has suggested this to be a valid assumption for clinical purposes [15].

The study utilized preoperative measurements in patients with valve disease. Heart catheterization was performed and left ventricular chamber volumes determined from biplane cine angiograms exposed at 30 Hz, with simultaneous pressures recorded from a high fidelity catheter tip manometer. X-ray tubes were at 90 °, with filming during quiet respiration. Premature and postpremature beats were avoided for volume calculations. Stroke work was calculated from the area within the pressure-volume curve. Forward stroke volume was calculated by the Fick method. Wall stress index was calculated utilizing the volumes for one cardiac cycle, by formulae described below. Assuming the ventricular function parameter σ_{es}/Ves unchanged, we then calculated the aortic end-systolic pressure (afterload) required postoperatively to keep the forward stroke volume unchanged. That is, any valvular regurgitant flow would be abolished by surgery. Only patients were utilized who had less than 10% change in left atrial pressure and heart rate postoperatively compared to preoperative values, to minimize the effect of preload or contractility variation. Predicted post-operative stroke work was also calculated from these variables. Finally, to test the hypothesis, predicted values were compared to measured values obtained in the cardiac surgical intensive care unit one day after operation. Postoperative cardiac output was measured by the dye-dilution technique, with left atrial and aortic pressures recorded from indwelling lines. Left ventricular end-diastolic volume and mass were assumed unchanged from the preoperative status.

2. *Mathematical concepts.*

Wall stress [11] was calculated from the formula: $\sigma = \dfrac{PB}{H}\left[1 - \dfrac{B^3}{A^2(2B+H)}\right]$ where P = intra ventricular pressure, H = wall thickness, and A and B are chamber major and minor semi-diameters, based on an ellipsoid-of-revolution ventricular model. Of major importance to the mathematical development of this methodology is that $\sigma = P$ multiplied by a complex function of the chamber radii and wall thickness:

$$\sigma = Pf(A, B, H).$$

Furthermore, chamber dimensions are a function of volume, V:

$$\sigma = Pf(V, H) \quad \text{or} \quad \frac{\sigma}{P} = f(V, H) \tag{1}$$

In 1969, Hugenholtz [12] described the complex mathematical relationship between wall thickness and chamber dimensions through the cardiac cycle, for a given fixed mass value in one individual patient: $0.2387\, M = H(2AB + A^2) + H^2(2A + B) + H^3$. ($M$ = left ventricular mass). As the ventricle contracts, wall thickness increases as described by this relationship. Wall thickness, H, can therefore be calculated at any point in the cardiac cycle utilizing the chamber dimensions at that instant and the patient's mass value (or 100 g as an arbitrary value to calculate stress index, in this study). This relationships is complex and the solution involves repetitive estimates: H = function of (A,B,M) or $H =$ (A,B,M). Since dimensions A and B are functions of volume V for an individual patient, and since $M = 100$ g for this study,

$$H = f(V) \tag{2}$$

These two formulas can now be combined into a unified and more useful relationship. Substituting Equation (2) into Eq. (1) $\frac{\sigma}{P} = f(V)$. Each of these parameters can be calculated at 30 Hz intervals and the relationship $\frac{\sigma}{P}$ vs V depicted graphically. Figure 1 shows an example of this relationship in one patient who had a ventricular chamber that was rather spherical in appearance, with the curve extrapolated to 0. Others with more elliptical shapes were shifted in the upward direction and others that changed shape during contraction were less linear.

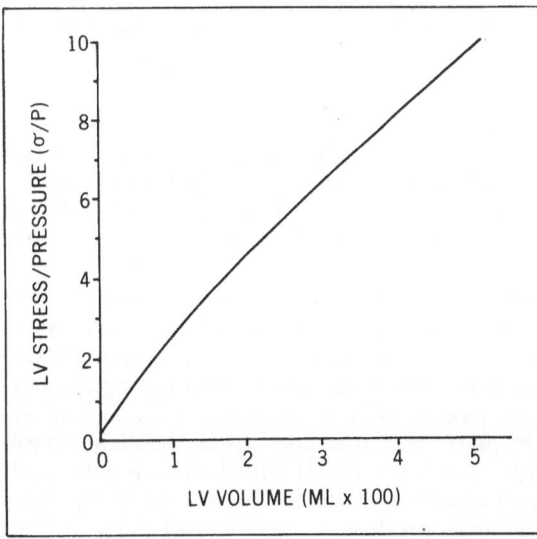

Fig. 1. Example of left ventricular stress/pressure (a dimensionless number) compared to volume for an example patient. This is not a true linear relationship, although it appears somewhat so, because of the rather spherical shape of this ventricle. The curve has been constructed according to formulae described in the text. From this curve, stress can be determined from any volume and pressure.

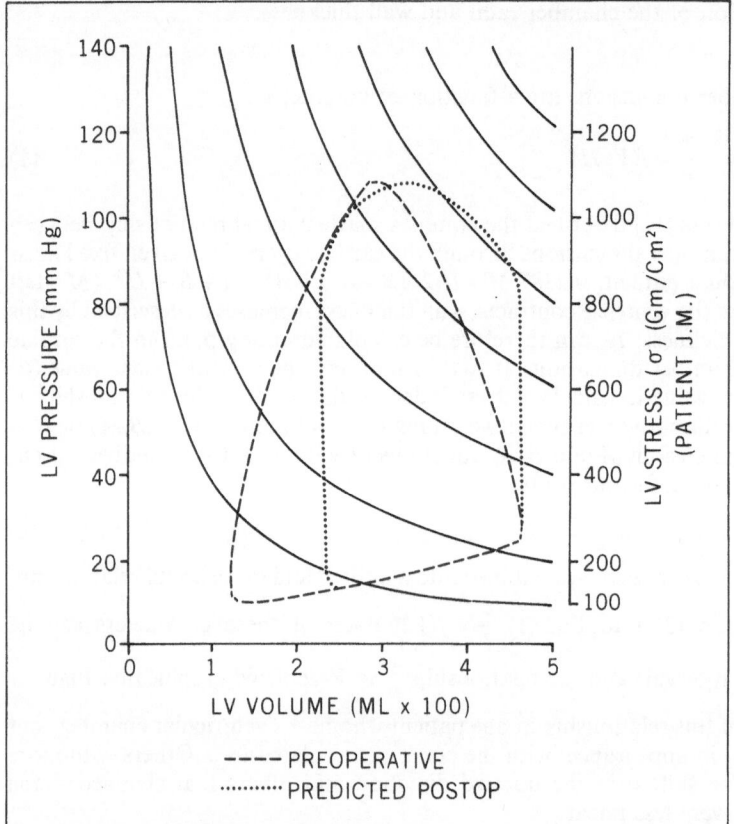

Fig. 2. This complex figure is basically a left ventricular pressure-volume curve. However, the "stress isobars" (right vertical axis) have been calculated for an example patient from the curve in Fig. 1. Also shown is the measured preoperative "loop" as well as the anticipated change after valve surgery.

One can display this same 3-dimensional relationship involving pressure, volume and stress onto a more familiar left-ventricular pressure-volume format. For each patient, a pressure value and a volume value define a stress value, and hence "stress isobars" can be so constructed. Figure 2 shows an example of these stress isobars engrafted onto pressure-volume axes.

Onto this 3-dimensional display, one can now enter the example pressure-volume curve for one cardiac cycle. Notice that in the example, of a patient with mitral insufficiency, shown in Fig. 2, volume begins to decrease early in systole, before aortic valve opening, and continues late after aortic closure. One can assume that postoperatively, this phenomenon will be abolished and the curve will have a predictable shape change. This type of three-dimensional display helps to conceptualize how peak systolic stress will be increased by surgery, a phenomenon previously described [3, 19]. Left ventricular work is quantified by the area within the respective curves [14]. Such a graphic analysis permits construction of a curve relating predicted postoperative end-systolic aortic pressure and

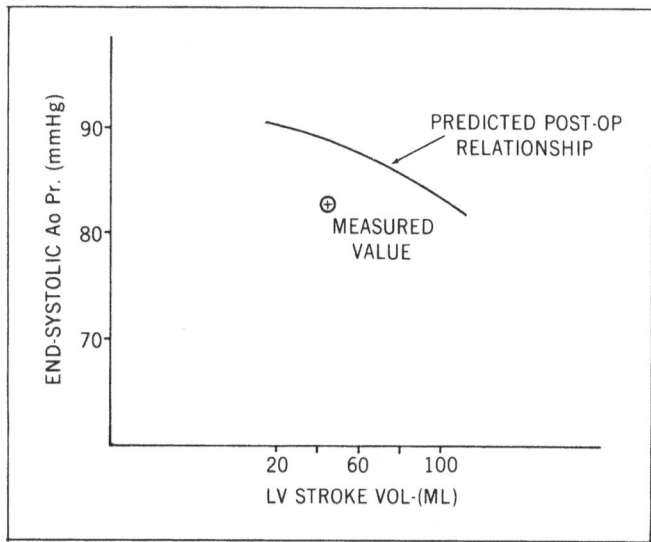

Fig. 3. Example of the predicted postoperative relationship between aortic pressure and stroke volume for one patient in the series. In constructing the curve, the functional parameter σ/V at end-systole is assumed unchanged with surgery, with the curve of Fig. 1 defining the relationships. The measured postoperative value is compared to the curve.

stroke volume, assuming pre-load remains unchanged, and it permits comparison with the actual measured postoperative value as shown in the example, Fig. 3.

Results

Twelve patients were studied, chosen on the basis of severe valve disease with class 3 or 4 heart failure preoperatively; adequate data acquisition; less than 10% change in heart rate and left atrial pressure preoperatively compared to postoperatively; and postoperative survival. There were 7 males and 5 females, altogether aged between 33 and 79. Six had mitral insufficiency, 4 mixed mitral and aortic disease, 1 aortic insufficiency and 1 mitral stenosis. Table 1 shows the wide range of haemodynamic findings in the 12 patients.

Table 1. Preoperative haemodynamics range (mean ± 1 standard deviation)

Left ventricular end-diastolic volume	118–552 ml (314 ± 145)
Cardiac index	1.16–5.60 l/min/m² (2.39 ± 1.16)
Aortic pressure (end systole)	58–103 mmHg (84 ± 16)
Regurgitant flow	0–119 ml/stroke
Ejection fraction	8–64% (41 ± 16)
Stroke work	46–260 gm-m/stroke (118 ± 56)
End systolic stress index	108–806 gm/cm² (386 ± 202)
σ/V (end systole)	1.40–3.00 (2.00 ± 0.50)

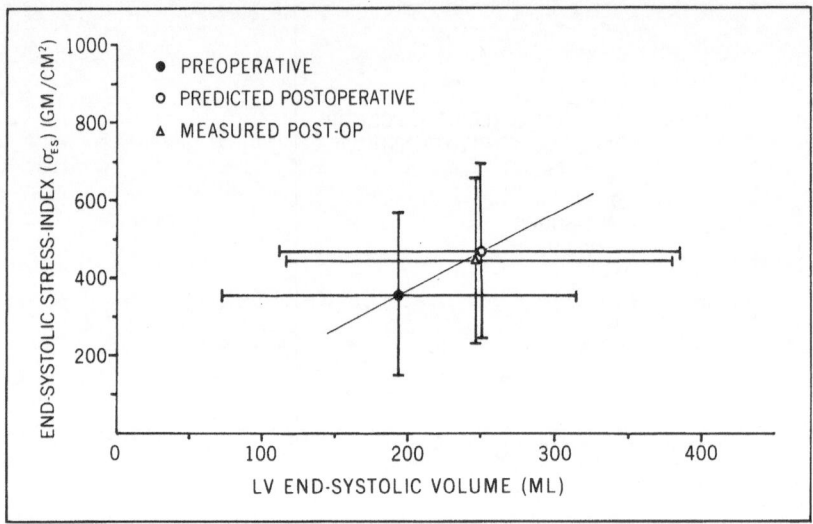

Fig. 4. Comparison of end-systolic stress to end-systolic volume for the pre-operative, predicted postoperative, and measured postoperative states. (Mean and standard deviation, n = 12 patients.) By definition, the predicted value is on the line through 0, as σ_{es}/V_{es} was assumed unchanged.

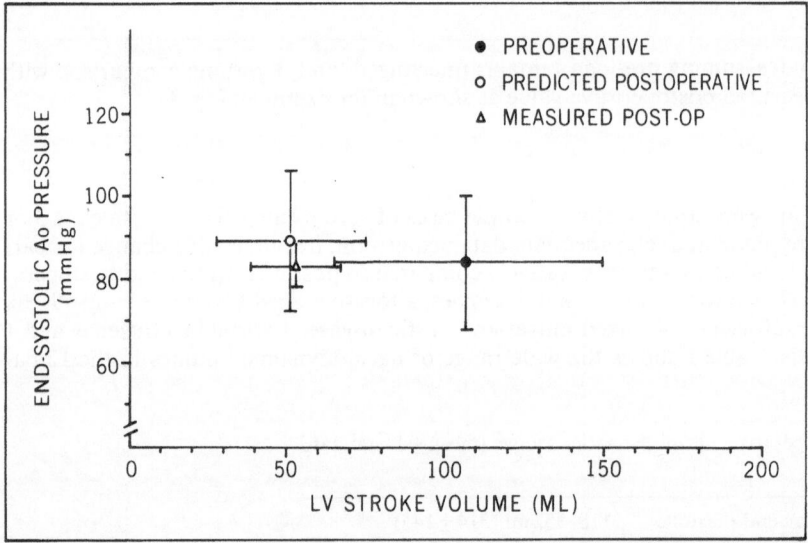

Fig. 5. Relationships between stroke volume and end-systolic aortic pressure for preoperative, predicted postoperative and measured postoperative states. (Mean and standard deviations, n = 12 patients.)

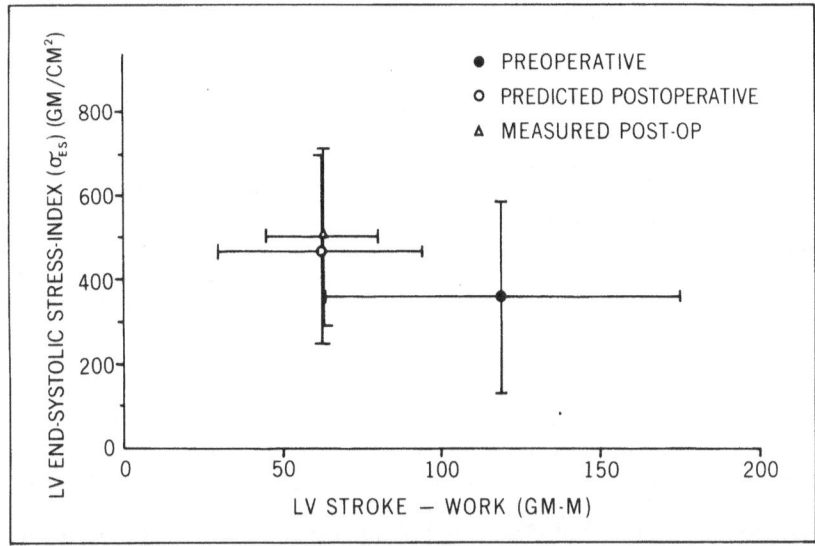

Fig. 6. Relationships between end-systolic stress and stroke work for the preoperative, predicted postoperative and measured postoperative states (mean and standard deviation, n = 12 patients). Stroke work has fallen drastically with surgery though end-systolic stress has increased, both close to predicted values.

Preoperative predictions of early postoperative status assumed no regurgitant valvular flow (cardiac output unchanged from preoperative state), and no change in left ventricular functional status (σ/V unchanged at end-systole). Figures 4–6 show the comparisons of preoperative measurements, predicted postoperative and true postoperative measurements. There was no significant difference between the means of the predicted and measured postoperative variables shown ($p < 0.05$). Postoperatively, 4 patients required no catechol support, 4 required low-dose dopamine (< 4 µg/Kg/min) and 4 high-dose dopamine (> 4 µm/Kg/min). The shift in the myocardial functional parameter σ_{es}/V_{es} with surgery for these 3 sub groups is of interest: σ_{es}/V_{es} fell by 3–20% in the zero-dose and low-dose catechol-support groups and rose by 3–36% in the high-dose support group. Figure 4 shows this relationship: since σ_{es}/V_{es} was assumed unchanged, preoperative and predicted postoperative values were by definition on the line passing through 0. Mean measured σ_{es} rose from 386 ± 202 to 449 ± 216 g/cm² with surgery and end-systolic volume from 192 ± 121 to 247 ± 132 ml. Other changes in mean measured values with surgery, as shown in Figs. 5 and 6, are a decrease in stroke volume from 107 ± 42 to 52 ± 15 ml, a decrease in stroke work from 118 ± 56 g-m to 59 ± 18, and a non-significant change in aortic end-systolic pressure from 84 ± 16 to 83 ± 6 mm Hg.

Discussion

This pilot study represents an attempt to predict the early postoperative haemodynamic status of valve patients, based on preoperative measurements of ventricular status and the assumption that valve function will be normal after surgery. Postoperative measurements were then compared to predicted values to test the predictive methodology. The results in this small patient series generally reflect success in predicting the expected early postoperative afterload requirement (end-systolic aortic pressure), left ventricular stroke volume, end systolic volume, and change in end-systolic wall stress. However, a number of sub-topics warrant discussion.

The methodology of this predictive technique is original, complex, and time consuming, but does lend itself to computer analysis. Of major importance in the method is the mathematical combination of 2 formulas or concepts: the first is that wall stress equals pressure multiplied by a function of ventricular volume and wall thickness for an individual patient's particular ventricular configuration [11]: the second is that wall thickness for a constant mass of myocardium is also a function of an individual patient's ventricular volume through the heart cycle [12]. Combination of these formulae yields a 3-dimensional relationship between stress, volume and pressure, unique for each patient (Fig. 1). Use of this curve, together with the patient's end-systolic stress/volume relationship as a load-independent functional parameter, are major components of the predictive algorithm. Finally, use of the preoperative forward cardiac output as an estimate of postoperative flow requirements completes the input data for the predictive methodology.

Left ventricular function may be expected to decrease in a transient manner from cardiopulmonary bypass and the stress of surgery. Indeed, in those patients not requiring high-dose postoperative catechol support, the functional parameter σ_{es}/V_{es} had fallen 3–20% below preoperative values. High-dose catechol support appeared to have overcome this decrease and resulted in positive shifts of $+4$ to $+37\%$ in those patients requiring this intervention postoperatively. The small mean decrease in this functional parameter for the total patient group appears to have been one cause for the slightly lower than predicted postoperative aortic pressures. That is, the left ventricles required slightly lower afterloads than predicted, to maintain the required cardiac output. The predicted stroke volume mean for the group closely approximated the measured postoperative value, suggesting that peripheral flow requirements may be relatively unchanged from the resting preoperative state. Predicted and measured mean stroke volume fell dramatically with correction of valvular disease, primarily because of those with regurgitant lesions. This resulted in a marked fall in stroke work, close to predicted values. End-systolic wall stress did increase with surgery, a phenomenon previously described [3].

Limitations of the study

1. The patient population consisted of only 12; not enough for clear statistical analysis. However, these were ill patients and data collection and analysis were both difficult and complex. This is a pilot study, however, with a larger series planned.

2. Left ventricular end-diastolic volume was assumed unchanged postoperatively. However, only patients with end-diastolic pressures changing less than 10% with surgery were chosen for analysis. Future study will address this assumption, as cardio-pulmonary bypass may alter the diastolic characteristics of the left ventricle [10].

3. σ_{es}/V_{es} was chosen as a load-independent parameter to quantity myocardial function. True E_{max}, more clearly established as such a parameter by previous studies [8, 13, 16, 19, 21], requires more complex preoperative procedures not done in these patients. However, recent literature suggests that σ_{es}/V_{es} may be adequate for practical usage [15].

Conclusions

This pilot study suggests that early postoperative haemodynamic status can be predicted from preoperative measurements. The predictive algorithm utilized is mathematically complex but lends itself to computerization techniques. A decrease in ventricular function intraoperatively is an expected complicating factor but appears quantifiable and reversible with catechol support. If larger and more completely analysed populations corroborate this methodology, the following would be expected: patient selection for valve surgery could be improved and operative risk-stratification enhanced. The need for postoperative inotropic support (or intra aortic balloon assist) could be anticipated. "Target" postoperative aortic blood pressures could be formulated to permit more rational use of afterload-reducing agents. Finally, the relationship between early and late postoperative ventricular functional status could be more clearly defined.

References

1. Bonow RO, Rosing DR, McIntosh CL, Jones M, Maron BJ, Lan KKG, Lakatos E, Bacharach SL, Green MV, Epstein SE (1983) The natural history of asymptomatic patients with aortic regurgitation and normal left ventricular function. Circulation 68:509
2. Carroll JD, Gaasch WH, Zile MR, Levine HJ (1983) Serial changes in left ventricular function after correction of chronic aortic regurgitation. Am J Cardiol 51:476
3. Eckberg DL, Gault JH, Bouchard RL, Karliner JS, Ross J (1973) Mechanics of left ventricular contraction in chronic severe mitral regurgitation. Circulation 47:1252
4. Forman R, Firth BG, Barnard MS (1980) Prognostic significance of preoperative left ventricular ejection fraction and valve lesion in patients with aortic valve replacement. Am J Cardiol 45:1120
5. Gaasch WH, Andrias CW, Levine HJ (1978) Chronic aortic regurgitation: The effect of aortic valve replacement on left ventricular volume, mass and function. Circulation 58, No. 5:825
6. Gaasch WH, Carroll JD, Levine HJ, Criscitiello MG (1983) Chronic aortic regurgitation: Prognostic value of left ventricular end-systolic dimension and end-diastolic radius/thickness ratio. J Am Coll Cardiol I (3):775
7. Gault JH, Covell JW, Braunwald E, Ross J (1970) Left ventricular performance following correction of free aortic regurgitation. Circulation 42:773
8. Grossman W, Braunwald E, Mann T, McLaurin LP, Green LH (1977) Contractile state of the left ventricle in man as evaluated from end-systolic pressure-volume relations. Circulation 56:845
9. Hammermeister KE, Fisher L, Kennedy JW, Samuels S, Dodge HT (1978) Prediction of late survival in patients with mitral valve disease from clinical, hemodynamic, and quantitative angiographic variables. Circulation 57:341
10. Hess OM, Ritter M, Schneider J, Grimm J, Turina M, Krayenbuehl HP (1984) Diastolic stiffness and myocardial structure in aortic valve disease before and after valve replacement. Circulation 69, No. 5:855
11. Hood WP, Rackley CE, Rolett EL (1968) Wall stress in the normal and hypertrophied human left ventricle. Am J Cardiol 22:550
12. Hugenholtz PG, Kaplan E, Hull E (1969) Determination of left ventricular wall thickness by angiocardiography. Am Heart J 78:513
13. Kumpuris AG, Quinones MA, Waggoner AD, Kanon DJ, Nelson JG, Miller RR (1982) Importance of preoperative hypertrophy, wall stress and end-systolic dimension as echocardiographic predictors of normalization of left ventricular dilatation after valve replacement in chronic aortic insufficiency. Am J Cardiol 49, No. 5:1091
14. McKay RG, Aroesty JM, Heller GV, Royal H, Parker A, Silverman KJ, Kolodny GM, Grossman W (1984) Left ventricular pressure-volume diagrams and end-systolic pressure-volume relations in human beings. JACC 3:301
15. McKay RG, Aroesty JM, Heller GV, Royal HD, Warren SE, Grossman W (1986) Assessment of the end systolic pressure-volume relationship in human beings with use of a time-varying elastance model. Circulation 74, No. 1:97

16. Mehmel HC, Stockins B, Ruffmann K, Olshausen K, Schuler G, Kuber W (1981) The linearity of the end-systolic pressure-volume relationship in man and its sensitivity for assessment of left ventricular function. Circulation 63:1216
17. Peterson (1983) The timing of surgical intervention in chronic mitral regurgitation. Cathet Cardiovasc Diagn 9:433
18. Phillips HR, Levine FH, Carter JE, Boucher CA, Osbakken MD, Okada RD, Akins CW, Daggett WM, Buckley MJ, Pohost GM (1981) Mitral valve replacement for isolated mitral regurgitation: analysis of clinical course and late postoperative left ventricular ejection fraction. Am J Cardiol 48:647
19. Ramanathan KB, Knowles J, Connor MJ, Tribble R, Kroetz FW, Sullivan JM, Mirvis DM (1984) Natural history of chronic mitral insufficiency: Relation of peak systolic pressure/end-systolic volume ratio to morbidity and mortality. JACC 3, No. 6:1412
20. Ross J (1981) Left ventricular function and the timing of surgical treatment in valvular heart disease. Ann Int Med 94:498
21. Sagawa K (1981) The end-systolic pressure-volume relation of the ventricle: Definition, modifications and clinical use. Circulation 63:1223
22. Zile MR, Gaasch WH, Carroll JD, Levine HJ (1984) Chronic mitral regurgitation: predictive value of preoperative echocardiographic indexes of left ventricular function and wall stress. JACC 3(2):235

Authors' address:

William A. Baxley, M.D., Cardiology 334 LHR, University of Alabama at Birmingham, Birmingham, Alabama 35294, U.S.A.

Subject index

(Numbers refer to the first page of respective articles)